NEW HORIZONS IN LOW-DIMENSIONAL ELECTRON SYSTEMS

Physics and Chemistry of Materials with Low-Dimensional Structures

VOLUME 13

The titles published in this series are listed at the end of this volume.

NEW HORIZONS IN LOW-DIMENSIONAL ELECTRON SYSTEMS

A Festschrift in Honour of Professor H. Kamimura

Edited by

H. AOKI and M. TSUKADA
Department of Physics, University of Tokyo, Japan

M. SCHLÜTER
AT&T Bell Laboratories, Murray Hill, U.S.A.

F. LÉVY
Institut de Physique Appliquée, École Polytechnique Fédérale de Lausanne, Switzerland

KLUWER ACADEMIC PUBLISHERS
DORDRECHT / BOSTON / LONDON

Library of Congress Cataloging-in-Publication Data

New horizons in low-dimensional electron systems: a Festschrift in honour of Professor H. Kamimura / edited by H. Aoki ... [et al.].
p. cm. (Physics and chemistry of materials with low-dimensional structures; v. 13)
Includes index.
1. Electronic structure. 2. High temperature superconductors. 3. Superconductivity. 4. Clathrate compounds. 5. Kamimura, Hiroshi, 1930– . I. I. Kamimura, Hiroshi, 1930– . II. Aoki, Hideo, 1950– III. Series.
QC176.8.E4N48 1991
530.4'1--dc 20
91–16710

ISBN 0–7923–1302–X

Published by Kluwer Academic Publishers,
P.O. Box 17, 3300 AA Dordrecht, The Netherlands.

Kluwer Academic Publishers incorporates
the publishing programmes of
D. Reidel, Martinus Nijhoff, Dr W. Junk and MTP Press.

Sold and distributed in the U.S.A. and Canada
by Kluwer Academic Publishers,
101 Philip Drive, Norwell, MA 02061, U.S.A.

In all other countries, sold and distributed
by Kluwer Academic Publishers Group,
P.O. Box 322, 3300 AH Dordrecht, The Netherlands.

Printed on acid-free paper

Printed in the Netherlands

PROFESSOR H. KAMIMURA

TABLE OF CONTENTS

Part II: Two-Dimensional Systems and Superlattices

Part III: High T_c Compounds

PREFACE

In *Bird of Passage* by Rudolf Peierls, we find a paragraph in which he describes his Cambridge days in the 1930s:

> On these [relativistic field theory] problems my main contacts were Dirac, and the younger theoreticians. These included in particular Nevill (now Sir Nevill) Mott, perhaps the friendliest among many kind and friendly people we met then.

Professor Kamimura became associated with Sir Rudolf Peierls in the 1950s, when he translated, with his colleagues, Peierls's 1955 textbook, *Quantum Theory of Solids*, into Japanese. This edition, to which Sir Rudolf himself contributed a preface, benefitted early generations of Japanese solid state physicists. Later in 1974/5, during a sabbatical year spent at the Cavendish Laboratory, Professor Kamimura met and began a long association with Sir Nevill Mott. In particular, they developed ideas for disordered systems. One of the outcomes is a paper coauthored by them on ESR-induced variable-range hopping in doped semiconductors. A series of works on disordered systems, together with those on two-dimensional systems, have served as building blocks for *Physics of Interacting Electrons in Disordered Systems*, in the International Series of Monographs on Physics, coauthored by Aoki and published in 1989 by the Oxford University Press.

Soon after Professor Kamimura obtained a D.Sc. in 1959 for the work on the ligand field theory under the supervision of Masao Kotani, his strong connections in the international physical community began when he worked at the Bell Telephone Laboratories in 1961/64. A more recent interaction developed into a major US–Japan collaboration project on graphite, with Mildred Dresselhaus representing the US groups. From this he pioneered a series of theoretical efforts on the graphite intercalation compounds. Another equally fruitful international collaboration germinated in the Cavendish Laboratory on the sulphur nitride polymer, a quasi-one dimensional superconductor.

These descriptions give a glimpse of the surprisingly wide spectrum of interests evinced by Professor Kamimura. On the occasion of his retirement from the Physics Department in the University of Tokyo at the age of sixty, we have edited, as a tribute to his accomplishments, a collection of articles on current research by the physicists who have interacted closely with him on one facet or another. In doing so we have had the privilege of asking leading scientists from the international community to contribute.

Even within the physics of low-dimensional systems, to which this volume is devoted, the richness in subjects and approaches is evident. The topics covered here range from intercalation compounds, semiconductor surfaces and interfaces, high T_c compounds, quasi-one dimensional compounds to

H. Aoki et al. (eds.), New Horizons in Low-Dimensional Electron Systems, xvii–xviii.

clusters and defects, thus spanning from 2D down to 0D, from superconductors to insulators, and from physics to quantum chemistry.

We hope that the facets of the physics surveyed in this volume will prove to be evocative when one explores new horizons in the condensed matter physics in the 1990s.

March, 1991 The Editors

PART I

Intercalation Compounds and Layered Materials

LOCALIZATION PHENOMENA AND CARRIER-CARRIER INTERACTION IN FLUORINE-GRAPHITE INTERCALATION COMPOUNDS

S.L. DI VITTORIO and M.S. DRESSELHAUS

Center for Materials Science and Engineering, Massachusetts Institute of Technology, Cambridge, Massachusetts 02139, U.S.A.

and

G. DRESSELHAUS

Francis Bitter National Magnet Laboratory, Massachusetts Institute of Technology, Cambridge, Massachusetts 02139, U.S.A.

Abstract. We have used fluorine-intercalated GICs to vary the amount of disorder in a two-dimensional material. The amount of disorder increases with increasing fluorine concentration. For dilute fluorine concentrations, C_xF fibers are in a weak disorder regime. A transition to a strong localization regime is observed at higher fluorine concentrations. In both regimes, we have measured the temperature-dependent resistivity and the magnetoresistance. The experimental magnetoresistance is explained using theoretical models corresponding to the regimes of weak and strong disorder.

1. Introduction

Hiroshi Kamimura has left an indelible mark upon the literature of low-dimensional systems. His impact on advances in the field of graphite intercalation compounds (GICs) has been especially important, and dates back to the mid 1970's when physicists began to develop an interest in this field. His contributions span band structure calculations of the alkali metal GICs [1–3] and the clarification of their Fermi surfaces, and explanations of anomalous effects of the anisotropic magnetism [4] and superconductivity [5] in GICs. Concurrently his work on transport in disordered systems has been fundamental and far-reaching, especially his treatment of transport phenomena at high magnetic fields, culminating in a beautiful book [6], *The Physics of Interacting Electrons in Disordered Systems*, which will influence the next generation of workers in the field.

In this article, we present recent results on high magnetic field transport in a 2D system based on an acceptor GIC where the amount of disorder is varied. The results are strongly dependent for their interpretation on the theoretical contributions of Professor Kamimura. It is hoped that the unanswered questions in our work will stimulate him into further investigations in the future.

H. Aoki et al. (eds.), New Horizons in Low-Dimensional Electron Systems, 3–23.

2. Disorder in Intercalation Compounds

2.1. General considerations

In recent years, new theories [7,8] have emerged that attempt to explain the behavior of weakly disordered solids, i.e., solids where the density of defects is below the threshold for localization of the wavefunctions. These theories, which describe modifications brought about by quantum many-body effects and carrier-carrier interaction in the electron gas, use perturbation techniques to predict some corrections to the low-temperature conductivity and magnetoresistance with respect to the behavior of a perfectly ordered metal. The magnitude of the predicted effects increases as the dimensionality decreases. Since, in one-dimensional systems, all wavefunctions are localized, however small the disorder, two-dimensional systems have come to play a crucial role in the study of these new theories of disordered metals. Two-dimensional systems are also important in that 2D is a critical dimension for localization phenomena, since it has been predicted [9] that all states are marginally localized in the non-interacting two-dimensional electron gas. Indeed, Abrahams *et al.* [9] predicted that, in a strictly two-dimensional system, all states are localized, however small the disorder, but that $(2+\varepsilon)$-dimensional systems can be metallic, however small the value of ε.

As we increase the density of defects, we eventually cross over into the insulating regime. Although the theory of the Anderson insulator is now well established [10], a host of interesting problems remain in the field of strongly localized materials. One of them is the role of carrier-carrier interaction in the Anderson insulator, and another is the effect of a magnetic field on the transport properties of a disordered and/or interacting system.

Graphite Intercalation Compounds (GICs) provide a system well-suited to study disorder effects in solids. Indeed, pristine graphite is a highly anisotropic material, whose anisotropy can be further increased by intercalation with acceptors, leading to two-dimensional behavior [11]. Electronic transport in GICs takes place through the graphene planes which, after intercalation with an acceptor-GIC, are probably one of the best approximations of a real material to ideal two-dimensionality. In the case of fluorine-GICs, the density of defects can also be easily tailored and varied by orders of magnitude, making it possible to create a well ordered metal, a weakly disordered metal or a highly disordered material on the insulating side of the metal-insulator transition. Whereas weak localization effects have been reported in the transport properties of GICs for other intercalates [12–14], fluorine is the only intercalate species that has permitted the study of so wide a range of disorder phenomena [15,16].

2.2. Special features of fluorine fiber GICs

Because of the unique nature of fluorine as an intercalate species, fluorine-intercalated GICs have very special properties among all GICs. Whereas most intercalates exhibit ionic bonding to the graphene layers, fluorine in graphite has the possibility of mixed ionic and covalent character [17]. For dilute intercalate concentrations, fluorine acts predominantly as an ionic acceptor, releasing holes into the graphene layers, as is the case for other acceptor GICs. However, for higher fluorine concentrations beyond stage I, the C-F bond shifts to a more covalent character [18], which depresses the carrier concentration and displaces some of the carbon atoms slightly away from the highly planar structure of pristine graphite. As a result, instead of the usual increase in electrical conductivity σ by an order of magnitude or more, with increasing intercalate concentration, the peak in conductivity σ_{max} obtained for C_xF compounds is only a factor of 2-3. Furthermore, whereas in typical GICs, σ drops from its maximum value σ_{max} by about 30% at stage I, σ drops by several orders of magnitude for fluorine concentrations above that corresponding to σ_{max} in fluorine GICs (see Figure 1).

In addition, fluorine uptake continues beyond the concentration of the stage I compound ($C_{4.5}F$), unlike the situation for other GICs, where stage I denotes the maximum intercalate uptake [11].

Another peculiarity of fluorine in graphite is the introduction of a large amount of defects into the graphene layers upon intercalation [15], whereas for most other GICs intercalation introduces only marginal disorder. The density of defects in F-GICs increases with increasing intercalate concentration. These defects are observed using high resolution transmission electron microscopy in terms of three characteristic features: a waviness of the graphene planes in the lattice fringe images, some defective intercalated regions and some unintercalated regions [19]. Although the origin of this disorder has not been fully elucidated at the time of writing, it is believed to be related to the nature of fluorine as an intercalate species in graphite, since the bonds between the very reactive fluorine atoms and the carbon atoms partially destroy the planarity of the graphene layers. The non-planar nature of the covalent $(CF)_n$ compound [20] is consistent with the observation of waviness of the graphene layers for high intercalate concentrations. Furthermore, the nature of the observed disorder in the graphene planes is very different from the disorder usually observed in pristine graphite or what is created by most intercalate species, namely point defects and dislocations.

Some special mention needs to be given to the use of carbon fibers for transport measurements to study localization effects. Weak localization effects were initially studied in acceptor-intercalated GIC fibers, because the fiber geometry is highly favorable for high resolution experiments, originally used to observe very small changes in $\rho(T)$ at low temperatures [12–14].

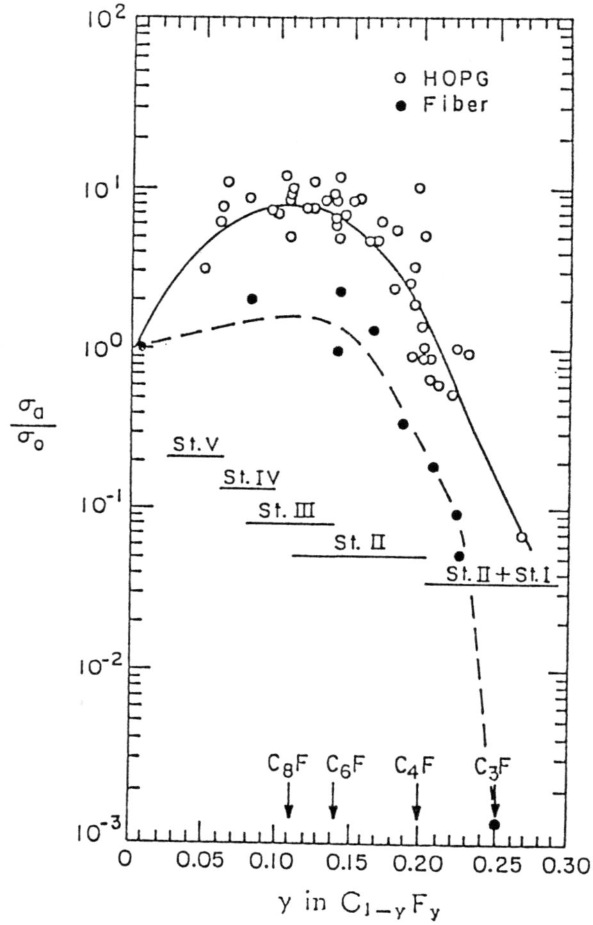

Fig. 1. The dependence of the *a*-axis conductivity of fluorine-intercalated graphite on the fluorine concentration, as measured by y in $C_{1-y}F_y$. The conductivity data are shown for both HOPG (from [18]) and fiber host materials, and the data are normalized to the conductivity of the pristine material.

Furthermore, intercalation to a given acceptor concentration was found to cause more disorder in fiber hosts than in HOPG (highly oriented pyrolytic graphite, the most common host material for GICs) [11]. To enhance the amount of disorder, early studies employed less ordered fiber host materials [21,22] for intercalation of the acceptor species [12–14]. However, in the case of the C_xF fibers, the vapor-grown fibers are chosen as the host material for intercalation because they have the highest available structural ordering in a fiber, so that the intercalation-induced defects can be studied more systematically [15]. Indeed, no other intercalated species yields as wide a disorder range as fluorine. Thus C_xF fibers have been employed in systematic studies of localization phenomena in GICs over the widest range of disorder.

3. Background on Kamimura Theories

A first understanding of the behavior of very disordered materials began with a seminal paper from Anderson [10], who showed that, in the presence of very large disorder, all wavefunctions are localized. As the disorder in conducting materials is increased, the resulting localization of the carriers induces a transition from a metallic to insulating behavior for a non-interacting electronic system. Disorder, however, is not the only effect that leads to a metal-insulator transition. Carrier-carrier interaction effects can also induce a metal-insulator transition in the absence of disorder, as exemplified by the Coulombic interaction-driven Mott transition in a periodic array of hydrogen atoms. These two phenomena of disorder and carrier-carrier interaction are present to varying degrees in all real physical systems and are basic to an understanding of the properties of disordered solids.

Kamimura has made numerous contributions to the field of disordered solids [6]. Since we cannot, for lack of space, refer to his many contributions in this paper, nor even to a small fraction of them, we will focus on two theories relevant to the experimental results presented in this paper: the effect of carrier-carrier interaction in the Anderson localization regime, and the effect of a magnetic field on the transport properties of the Anderson insulator.

In order to get a better understanding of the respective roles that disorder and correlations play in the metal-insulator transition, Kamimura [23] studied carrier-carrier interaction effects in an Anderson-localized material, using a Hamiltonian of the form

$$H = \sum_{i\sigma} \varepsilon_i c^\dagger_{i\sigma} c_{i\sigma} + \sum_{ij\sigma} t_{ij} c^\dagger_{i\sigma} c_{j\sigma} + \frac{1}{2} \sum_{ijkl\sigma\sigma'} \langle ij|\hat{U}|kl\rangle \, c^\dagger_{i\sigma} c^\dagger_{j\sigma'} c_{l\sigma'} c_{k\sigma} \quad (1)$$

The random character of the system is described by the randomness of the site energy ε_i and the transfer energy t_{ij}. The Coulombic interaction $\hat{U}$ can be divided into intrastate correlations and interstate correlations. The effect of the intrastate correlation energy [6], which varies as $U_{\mathrm{intra}} \propto 1/\xi^3$, where ξ is the localization length, over most of the disorder spectrum, is to increase the energy of the doubly occupied state with respect to that of two singly occupied states with the same ξ [24,25]. Therefore, intrastate correlations lead to the appearance of singly-occupied states even at zero temperature, with a density increasing with the parameter U/W where U is the average intrastate correlation energy (i.e., some average of U_{intra} over ξ), and W is the bandwidth. Kamimura [6] showed that another effect of the intrastate interaction is to create a peak in the density of second-electron states below the Fermi level, i.e., of states occupied by two electrons with opposite spins.

The interstate interaction leads to two quite different effects. Firstly the usual exchange interaction H_{DX} in the electron gas, corresponding to the terms

$$H_{DX} = \sum_{\alpha\beta\sigma\sigma'} U_{\alpha\beta\beta\alpha} c^{\dagger}_{\alpha\sigma} c^{\dagger}_{\beta\sigma'} c_{\alpha\sigma'} c_{\beta\sigma} \tag{2}$$

tends to align the spins of the carriers, and thus favors ferromagnetic behavior. A dynamic exchange interaction H_{YX} [25,26], particular to localized states is also present and corresponds to the terms

$$H_{YX} = \sum_{\alpha\beta\sigma\sigma'} U_{\alpha\alpha\alpha\beta} c^{\dagger}_{\alpha\sigma} c^{\dagger}_{\alpha\sigma'} c_{\beta\sigma'} c_{\alpha\sigma} \tag{3}$$

in the Hamiltonian. Physically, the energy of a system of two carriers localized at sites α and β is lowered by a virtual transfer of one of the carriers to the site of the other. This effect, however, requires two electrons with opposite spins, thus favoring an antiferromagnetic coupling of the spins. The final ordering behavior of the insulator depends on the relative strength of the two exchange interactions H_{DX} and H_{YX}.

Another important contribution by Kamimura to the theory of disordered systems was a theory for the effect of magnetic fields on the transport properties of the Anderson insulator in the variable-range hopping regime in the presence of intrasite correlations. In this regime, transport occurs through four different hopping processes [27]: hopping from a singly (doubly) occupied state to an empty (singly occupied) state. The total hopping rate is the sum of the individual hopping rates for each of the four processes. When a magnetic field is applied, it tends to spin-polarize the system, thus reducing hopping from a doubly-occupied state to an empty state and from a singly occupied state to another singly occupied state, since the Pauli principle requires opposite spins on the same site. The reduction of some of these hopping processes results in a positive magnetoresistance [28]. When the above-mentioned processes are completely eliminated, the magnetoresistance saturates. The effect of interstate interactions is to reduce this magnetic-field-induced positive magnetoresistance because of the tendency towards magnetic ordering, independent of whether the interstate interactions are ferromagnetic or antiferromagnetic. Ferromagnetic coupling eliminates the processes described above, because the system becomes spin-polarized, even in the absence of a magnetic field. If the coupling is antiferromagnetic, the spins become very hard to polarize, thus reducing the effect of an applied field. Even though the magnetoresistance is reduced by the ordering processes, the theory of Kamimura still predicts a positive magnetoresistance over the entire available magnetic field range.

TABLE I
The parameters characteristic of disorder for the C_xF fibers studied in this paper, namely the low temperature resistivity $\rho_{4.2K}$, sheet resistivity $R_\square$, and $k_F\ell$. For the strongly localized $C_{2.9}F$ fiber, the localization length is given in place of the parameter $k_F\ell$.

x in C_xF	$\rho_{4.2K}(\mu\Omega cm)$	$R_\square(k\Omega/\square)$	$k_F\ell$
5.8	16-30	0.17-0.32	40
4.5	70-140	1.2-2.6	5.5-7.0
4.1	280-540	5-10	1.7-2.0
3.6	700-1500	12-25	0.7-1.0
2.9	$5\times10^4 - 10^5$	500-1000	$\xi \approx 1000$Å

4. Resistivity

The presence of intercalation-induced disorder in acceptor-graphite intercalation compounds leads to some corrections to the low temperature transport properties of the material, and especially so in fluorine-GICs where the effect is strongest. The presence of disorder shows up most clearly in the resistivity and magnetoresistance properties. In this section, we thus use the temperature-dependent resistivity $\rho(T)$ and the dependence of the low-temperature resistivity $\rho_{4.2K}$ on fluorine concentration $1/x$ in C_xF as sensitive tools to characterize the amount of disorder in the material. The parameters used in this work to characterize the disorder are $k_F\ell$ and ξ for the metallic and insulating regimes, respectively, where k_F, ℓ and ξ are the Fermi wavevector, the carrier mean free path, and the localization length, respectively. As the system gets close to the metal-insulator transition, we expect $k_F\ell \rightarrow\approx 1$ and $\xi \rightarrow \infty$ on the metallic and insulating sides, respectively [15,16].

The parameters characteristic of the transport properties of the fibers, namely the resistivity ρ, the sheet resistivity $R_\square$ (i.e., the resistivity ρ divided by the repeat distance I_c, as obtained from X-ray measurements), and disorder parameter $k_F\ell$ (or ξ in the limit of strong disorder) are summarized in Table I.

The parameter $k_F\ell$ was obtained from the equation $k_F\ell = \pi\hbar I_c/\rho e^2$, based on the Blinowski-Rigaux model [29]. The localization length ξ was obtained from the relationship between T_0 and ξ in a 2D material, namely

$$T_0 = \frac{\beta}{k_B N(E_F)\xi^2} \tag{4}$$

where β is a constant on the order of unity.

The sheet resistivity is on the order of $500k\Omega/\square$ for the most concentrated samples that were measured ($C_{2.9}F$ and $C_{3.0}F$), thus well above the approx-

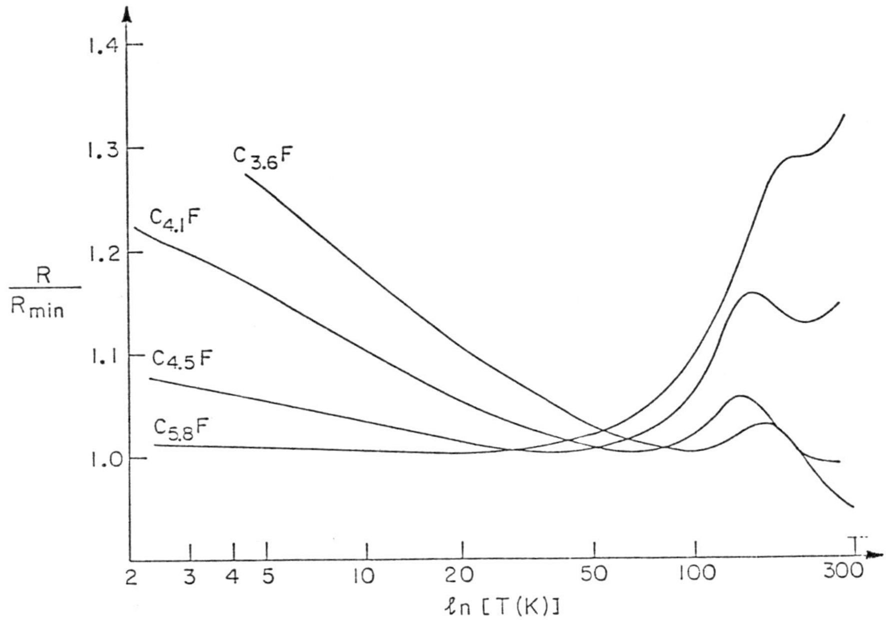

Fig. 2. The temperature-dependent resistance, normalized to the minimum in resistance R_{min}, for the fluorine-intercalated fibers C_xF in the weak disorder regime ($x \geq 3.6$). The logarithmic correction at low temperatures is characteristic of weak disorder in the fibers. As expected, the magnitude of the low-temperature correction increases with increasing fluorine concentration.

imate $25k\Omega/\square$ sheet resistivity predicted for the metal-insulator transition in a non-interacting electronic system [30].

We expect, and experimentally observe, as discussed below, a metal-insulator transition as we increase the fluorine concentration in the C_xF fibers. This transition from a metal to an insulator is not purely the effect of the increasing disorder. Because of the decreasing carrier density due to the shift to a more covalent solid, screening is poor in the concentrated C_xF fibers, and the metal-insulator transition occurs because of both disorder and carrier-carrier interaction effects.

The effects of disorder and carrier-carrier interaction can also be observed in the temperature-dependent resistivity of C_xF fibers in the metallic regime, as shown in Figure 2.

The most dilute samples exhibit metallic behavior at high temperature, i.e., a resistivity increasing with increasing temperature, and saturation at low temperatures. The bumps observed in $\rho(T)$ for the C_xF fibers for $x \geq 3.6$ in the 130K-170K temperature range (see Figure 2) are believed to be caused by a structural order-disorder transition in the fluorine layers. This effect is

TABLE II
Two parameters characteristic of the amount of disorder in the weakly localized fibers, namely the relative increase in ρ per decade of temperature $[\Delta R/R]_{\rm decade}$, and the temperature $T_{\rm log}$ corresponding to the onset of the logarithmic low-temperature correction. We note that there is no $R_{\rm min}$ nor $T_{\rm log}$ for the $C_{2.9}F$ fiber.

x in C_xF	$[\frac{\Delta R}{R}]_{\rm decade}(\%)$	$T_{\rm log}(K)$
5.8	0.4-1.2	20
4.5	6-8	40
4.1	10-15	60
3.6	15-30	90
2.9	–	–

in a different temperature range from the one where localization effects are important, and we thus will not discuss this bump in $\rho(T)$ any further.

As the disorder increases, the fibers exhibit a logarithmic correction to the resistivity at low temperatures, and the magnitude of the correction increases with increasing disorder. This logarithmic correction can be quantitatively measured in terms of the relative increase in resistivity per decade of temperature $[\Delta R/R]_{\rm decade}$, and the temperature $T_{\rm log}$ corresponding to the onset of the logarithmic correction. These two parameters are gathered in Table II for the various concentrations of C_xF fibers investigated in this paper. The logarithmic correction, characteristic of a two-dimensional material, is predicted by the theories of weak localization [7] and carrier-carrier interaction [8]. These theories describe the corrections to the resistivity at low temperatures that come about because of quantum interference effects and poor screening in disordered metals. The theories predict a correction to the resistivity

$$\Delta\rho = -R_\square^2 \frac{e^2}{2\pi^2\hbar}\left[p\ln(T) + \left(1 - \frac{3}{4}\tilde{F}_\sigma\right)\ln(T)\right] \tag{5}$$

where the first and second terms represent the weak localization and carrier-carrier interaction corrections, respectively. In Equation 5, $R_\square$ is the sheet resistivity, p gives the power law of the temperature dependence of the rate of the dominant phase-coherence-breaking mechanism $1/\tau_\phi \propto T^p$, and $\tilde{F}_\sigma$ is a parameter describing the screening in the material. Both corrections to the resistivity in Equation 5 are logarithmic, but have different magnetic field dependences, namely anisotropic negative magnetoresistance for weak localization and isotropic positive magnetoresistance for the hole-hole interaction. Since the weak localization and hole-hole interaction effects make dominant contributions to the magnetoresistance in the weak and strong

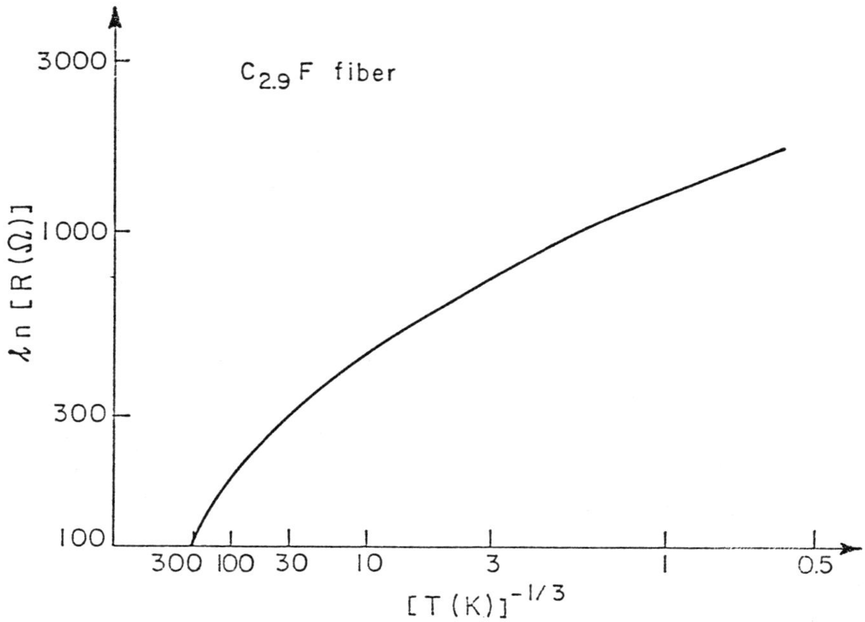

Fig. 3. The temperature-dependent resistivity for a $C_{2.9}F$ fiber, plotted as ln ρ vs $T^{-1/3}$. The experimental Mott law that is obeyed at low temperatures indicates a transition from a metallic regime to a strong localization regime.

magnetic field ranges, respectively, we are able to separate the two contributions, as discussed in the next section.

As we increase the fluorine concentration to the level of $C_{2.9}F$ and $C_{3.0}F$ fibers, the bump in the 140-170K range disappears and, as shown in Figure 3, the low-temperature resistivity departs from a logarithmic behavior. Instead, the low temperature resistivity obeys a 2D Mott law,

$$\ln(\rho) \propto \left(\frac{T_0}{T}\right)^{1/3} \tag{6}$$

signaling the metal-insulator transition. The quantity T_0, further discussed below, is a constant. Although some deviations from Mott's law are observed in the larger 0.5K to 300K temperature range, probably because of the proximity of the $C_{2.9}F$ fiber to the metal-insulator transition, or the possible onset of other transport mechanisms at higher temperatures, all experimental features point to a transition to an insulating material: the sheet resistivity is large and well in excess of the threshold for a transition to an insulating phase for the $C_{2.9}F$ fiber; the temperature-dependent resistivity obeys Mott's law at low T; the dependence of ρ on electric field E, though ohmic for low electric fields, exhibits an activationless hopping regime at high fields, i.e., $V(E) \propto E^{-1/3}$, where V is the measured voltage.

As discussed below, the magnetoresistance is also characteristic of a strong localization regime.

5. Magnetoresistance

In this section, we investigate the weak field and strong field magnetoresistance in both regimes of weak and strong disorder, with special emphasis on the magnetoresistance effects in the variable-range hopping regime. As mentioned before, two mechanisms have been predicted that contribute to the resistivity in the weak field regime, and yield the same ($\ln T$) dependence at low temperatures. Magnetoresistance measurements make it possible to analyze the two contributions separately. This analysis yields an especially large contribution of the carrier-carrier interaction ($\approx 70\%$), compared to the very minor contribution ($\approx 7\%$) in the 2D electron gas produced at the interface of a MOSFET [31]. Furthermore, many different mechanisms have been predicted for the magnetoresistance in the strong localization regime. These mechanisms depend on the magnitude of the localization length and of the carrier-carrier interaction. Thus, magnetoresistance measurements are useful for characterizing the strongly localized material electronically, as well as getting a better fundamental understanding of the changes induced by a magnetic field in the strong localization regime.

The transverse magnetoresistance for a weakly disordered $C_{3.6}F$ fiber at various temperatures in the range 2K< T <20K and in the magnetic field range 0< H <0.75 tesla is shown in Figure 4.

The points represent the experimental data, whereas the dotted curves represent a theoretical fit to the theory of weak localization using the formula [15]

$$\frac{\Delta R}{R_0} = -\lambda \frac{e^2}{2\pi^2 \hbar} \frac{2}{\pi} \int_0^{\pi/2} \left[\Psi \left(\frac{1}{2} + \frac{H_i}{H \cos \Theta} \right) - \ln \left(\frac{H_i}{H \cos \Theta} \right) \right] \mathrm{d}\Theta \qquad (7)$$

where Ψ is the digamma function, $0 < \lambda < 1$ is an inhomogeneity parameter, and $H_i = \hbar/(4eD\tau_i)$ the inelastic scattering equivalent magnetic field. In the analytical expression for H_i, D is the 2D hole diffusion coefficient and τ_i the inelastic scattering time. The integral accounts for the onion-ring geometry of the fibers. The temperature dependence of H_i is shown in Figure 5.

The magnitude of $H_i(T)$ increases with increasing temperature, but does not extrapolate to zero at zero temperature. The origin of this finite value for H_i at low temperatures will be discussed later in light of the experimental magnetoresistance in the variable-range hopping regime.

Since weak localization is an interference effect, it is characterized by an anisotropic magnetoresistance. The weak localization contribution to the magnetoresistance vanishes for an in-plane applied magnetic field, i.e., for

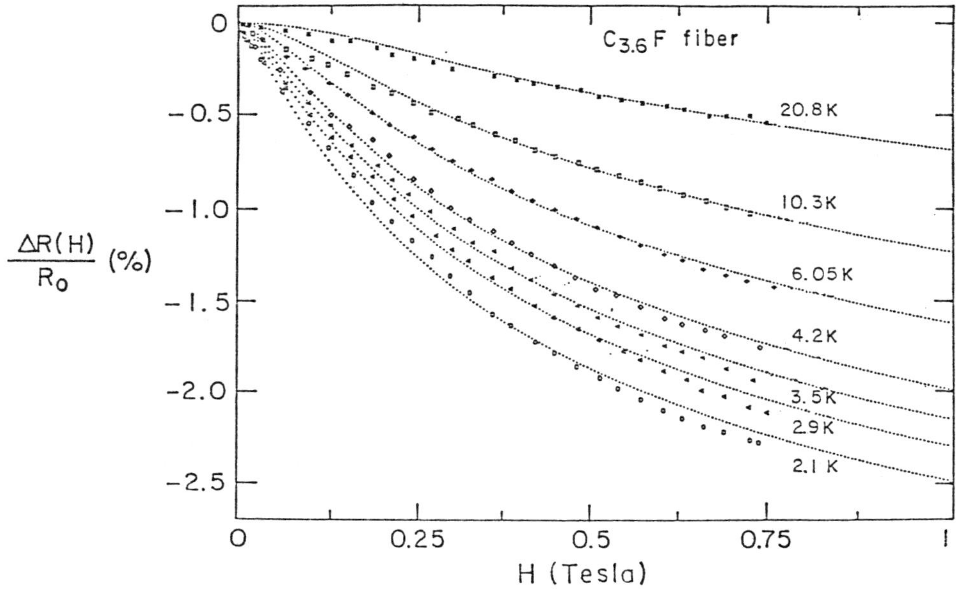

Fig. 4. The transverse magnetoresistance for a $C_{3.6}F$ fiber in the weak localization regime for various temperatures and in the field range $0< H <0.75$ tesla.

the longitudinal magnetoresistance of the fiber. Therefore, the longitudinal magnetoresistance of the fiber yields the effect of an applied magnetic field on the hole-hole interaction, as shown in Figure 6.

Since the graphene planes in the C_xF fiber are not perfectly parallel to each other, but have an angular spread of a few degrees, a small negative magnetoresistance is measured in the longitudinal magnetoresistance. Therefore, the data shown in Figure 6 were in fact extracted from transverse magnetoresistance plots to yield the hole-hole interaction contribution to the magnetoresistance [32]. The weak localization contribution was fitted in the weak magnetic field range, extrapolated to higher fields, and subtracted from the experimental transverse magnetoresistance. The continuous curves in Figure 6 represent the theoretical fit to the formula [30]

$$\frac{\Delta R}{R_0} = \lambda \frac{e^2}{2\pi^2\hbar} \frac{\tilde{F}_\sigma}{2} G\left(\frac{\mu_B g H}{k_B T}\right) \tag{8}$$

where λ is the same parameter as in Equation 7, g is the Landé g-factor and G is a positive and increasing function that behaves as H^2 for small argument and as $(\ln H)$ for large argument. From the analysis of the magnetoresistance data at both weak and strong magnetic fields, we show that the hole-hole interaction effect is dominant (about 70% of the total logarithmic temperature-dependent effect in $\Delta\rho$ given by Equation 5) over the

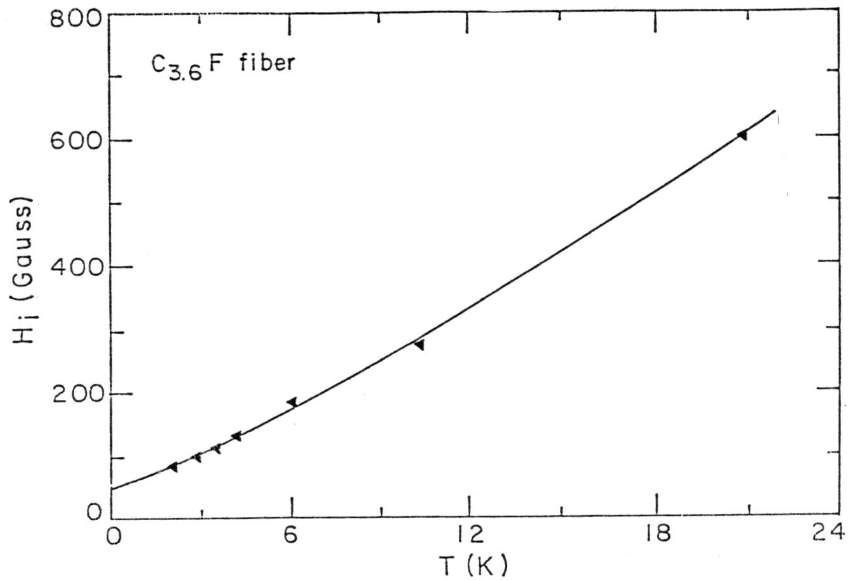

Fig. 5. The temperature dependence of the inelastic scattering equivalent magnetic field H_i between 2K and 20K, for the $C_{3.6}F$ fiber for which the transverse magnetoresistance is presented in Figure 4. The continuous curve is a guide to the eye.

weak localization contribution to the logarithmic increase in resistivity for the C_xF fibers in the weakly localized regime. This shows experimentally that the effects of both disorder and of carrier-carrier interaction are important for the C_xF fibers in the weakly localized regime and as we approach the metal-to-insulator transition.

We now turn to the experimental behavior of the magnetoresistance in the more concentrated $C_{2.9}F$ and $C_{3.0}F$ fibers, whose resistivity exhibits a variable-hopping regime at low temperatures [33]. Both longitudinal and transverse magnetoresistance curves are shown in Figure 7 for a $C_{3.0}F$ fiber at the two temperatures of 1.2K and 4.2K. The transverse magnetoresistance exhibits a negative magnetoresistance for weak fields, with a qualitative behavior very similar to that measured in weakly localized fibers. This is not surprising since all measurable parameters point to the fact that the system lies very close to the metal-insulator transition, and thus the localization length ξ and the inelastic scattering length L_i have similar magnitudes [34].

In contrast, the high field longitudinal magnetoresistance exhibits a very different behavior from that measured experimentally in the weak localiza-

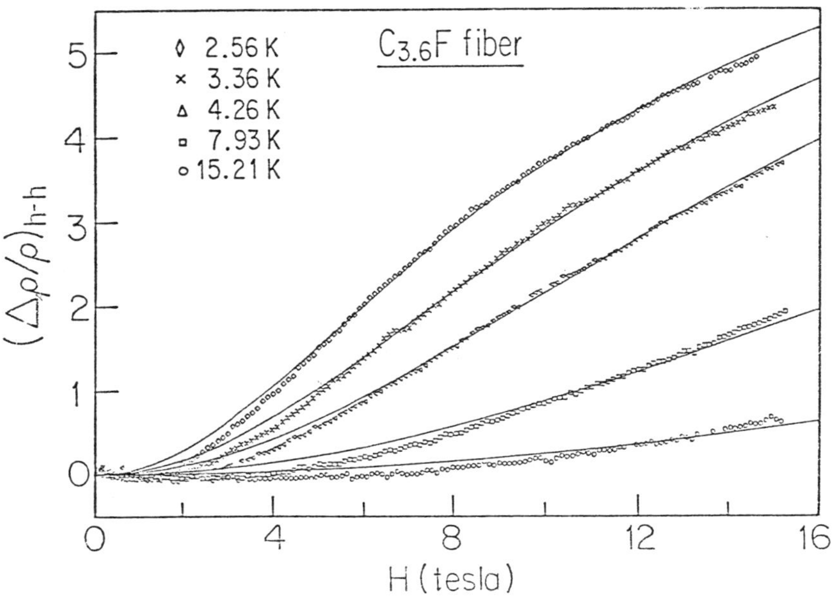

Fig. 6. The magnetic field dependence of the hole-hole interaction contribution to the magnetoresistance in a $C_{3.6}F$ fluorine-intercalated graphite fiber, for magnetic fields $0< H <15$ tesla and for several temperatures in the range $2.5< T <15$K. The weak localization contribution has been fitted to the transverse magnetoresistance in the weak field regime, then extrapolated to high fields, and subtracted from the experimental transverse magnetoresistance to yield the hole-hole interaction shown in the figure.

tion regime [33]. The longitudinal magnetoresistance saturates at high fields for T =1.2K, though the experimentally available field is not intense enough to exhibit the saturation effects in the magnetoresistance at 4.2K.

We attribute this experimental longitudinal magnetoresistance in the $C_{2.9}F$ and $C_{3.0}F$ fibers to the magnetic-field-induced suppression of some hopping processes, as predicted by the theory of Kamimura *et al.* described in the theoretical section. For a 2D system, Kamimura predicts [6] that the temperature-dependent conductivity in zero field is given by

$$\rho \propto \exp\left(\frac{T_0}{T}\right)^{1/3} \tag{9}$$

with

$$T_0 = \frac{3.0\ \eta_{2D}}{k_B N(E_F)} \frac{\xi_1^2 + \xi_2^2}{(\xi_1^2 + \xi_2^2)^2 + 4\xi_2^4} \tag{10}$$

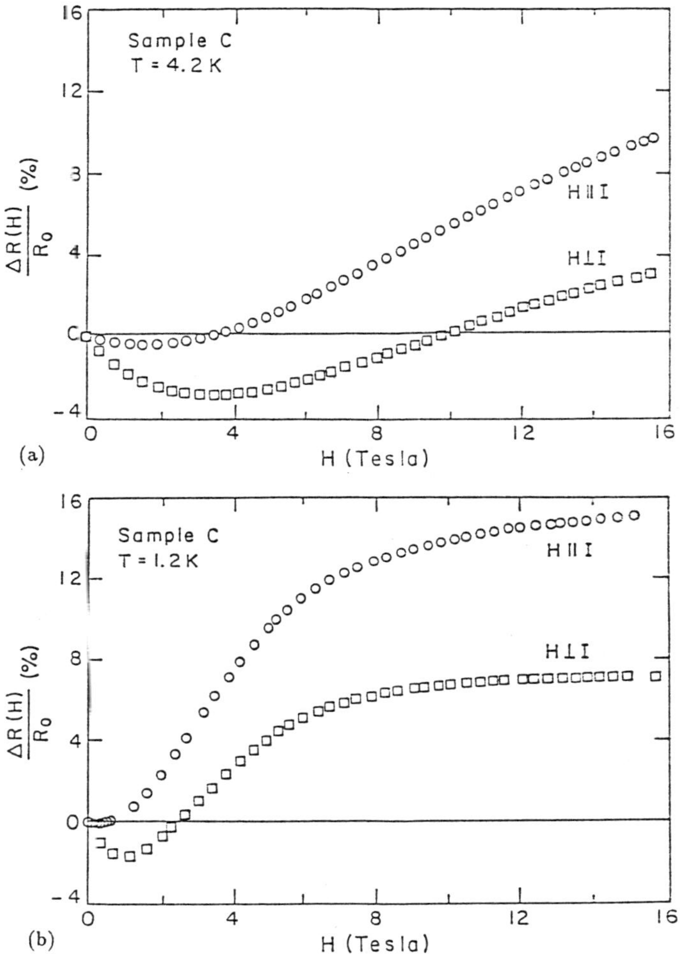

Fig. 7. The experimental transverse and longitudinal magnetoresistance at (a) 4.2K and (b) 1.2K for a strongly localized $C_{2.9}F$ fluorine-intercalated fiber in the magnetic field range $0 < H < 15$ tesla.

where $N(E_F)$ is the density of states at the Fermi level, η_{2D} is a constant related to percolation theory in 2D, and ξ_1 and ξ_2 are the localization lengths of the singly and doubly occupied localized states. The reason why singly and doubly localized states have different localization lengths lies in the additional contribution of the intrastate Coulombic energy in the case of doubly occupied states. This additional energy reduces the localization length for doubly-occupied states because of the energy dependence of the localization length $\xi(E) \propto (E_c - E)^\nu$ where E_c is the mobility edge and ν a constant parameter on the order of 1. In particular, Kamimura showed that Mott's law is still valid in the presence of intrastate correlations, an effect neglected in the original derivation of Mott's law. Experimentally, we find a typical value T_0=5K in the $C_{2.9}F$ and $C_{3.0}F$ fibers.

However, before applying Kamimura's theory to our system, we need to check that the fibers are indeed in a range of temperature T and intrastate correlation energy U where the theory applies. Since the localization length of the fibers is so large, we use the formula [6]

$$U \approx \frac{e^2}{\varepsilon \xi} \tag{11}$$

for the intrastate correlation energy, where ε is the dielectric constant. Using ε=1, which yields an overestimate of the intrastate interaction energy, we obtain an estimate $U \approx$ 10 meV. The theory of Kamimura is valid below a temperature T^* which, in 2D, is

$$k_B T^* = \left(\frac{U}{2}\right)^{3/2} (k_B T_0)^{-1/2} . \tag{12}$$

Numerically, we find a critical temperature of T^*=100K, well above the temperatures at which we carried out our experiments. We are thus in a regime where the theory of Kamimura on the effect of a magnetic field on the hopping processes should apply.

At high magnetic fields, Kamimura predicted a saturation of the magnetoresistance, and the value of the saturation magnetoresistance R_{sat}/R_0 in 2D is expressed as

$$\frac{\Delta R_{sat}}{R_0} = \exp\left\{\left(\frac{A\ \eta_{2D}}{N(E_F)k_B T}\right)^{1/3}\left[\left(\frac{\xi_1^2+\xi_2^2}{\xi_1^4+\xi_2^4}\right)^{1/3} - \left(\frac{\xi_1^2+3\xi_2^2}{(\xi_1^2+\xi_2^2)^2+4\xi_2^4}\right)^{1/3}\right]\right\} - 1 \tag{13}$$

where A is a numerical constant on the order of 1. Kamimura further derived the analytical dependence of the magnetoresistance below the saturation field, but only for the case of a three-dimensional material. Below saturation, he predicts a magnetoresistance

$$\frac{\Delta R(H)}{R(0)} = \frac{\exp(x(H)) - \exp(x(0))}{\exp(x(0))} \tag{14}$$

where the function $x(H)$ is given by the equation

$$\tfrac{23}{252}(K_1^2+K_2^2)x(H)^9+\tfrac{23}{126}K_2(K_1+2K_2)y(H)^9+$$
$$+\tfrac{23}{28}K_2(3K_1-5K_2)hy(H)^8+\tfrac{2}{35}K_2(239K_1+329K_2)h^2y(H)^7+$$
$$+\tfrac{1}{15}K_2(597K_1+677K_2)h^3y(H)^6+\tfrac{304}{5}K_2(K_1+K_2)h^4y(H)^5+$$
$$+48K_2(K_1+K_2)h^5y(H)^4+\tfrac{224}{15}K_2(K_1+K_2)h^6y(H)^3$$
$$=\tfrac{\eta_{3D}}{k_BT}\left[\tfrac{3}{10}(K_1+K_2)x(H)^5+\tfrac{3}{5}K_2y(H)^5+3K_2hy(H)^4+4K_2h^2y(H)^3\right]. \quad (15)$$

In Equation 15, η_{3D} is a constant related to 3D percolation theory, and the following definitions are used

$$h=\frac{\mu_B H}{k_BT}$$
$$y(H)=x(H)-2h$$
$$K_l=\frac{1}{6}\pi N(E_F)\xi_l^3 \quad \ell=1,2\,. \quad (16)$$

Although the exact analytical dependence given in Equation 14, 15 and 16 was derived for a three-dimensional material, all qualitative features of the magnetoresistance behavior predicted by Kamimura *et al.* are the same in two dimensions and three dimensions. The experimental longitudinal magnetoresistance for a $C_{2.9}F$ fiber is shown in Figure 8, in the field range $0< H <15$ tesla and at the three low temperatures of 1.48 K, 1.0 K and 0.79 K. The magnetoresistance behavior can be divided into three regimes: weak field ($\mu_BH/k_BT \ll 1$), intermediate field ($\mu_BH/k_BT \approx 1$), and large field ($\mu_BH/k_BT \gg 1$), depending on h, the ratio of the magnetic energy to the thermal energy. At weak fields, the Zeeman splitting between spin-up and spin-down states is small, so that we expect both spin-up and spin-down states to contribute to hopping, resulting in an H^2 analytical dependence for the magnetoresistance. On the other hand, when the magnetic energy is about equal to or larger than the thermal energy, the magnetoresistance is determined by the unbalance between the two spin-up and spin-down states, and we expect the magnetoresistance to be approximately linear in H. At very high fields, as previously mentioned, the magnetoresistance saturates. These three behaviors of the magnetoresistance are experimentally observed in Figure 8, as the magnetic field is increased.

Kamimura *et al.* further predicted that the field at which the magnetoresistance saturates H_{sat} should decrease with decreasing temperature. This effect is also observed experimentally, as seen in Figure 8. In two dimensions, he predicted the following temperature dependence for H_{sat}

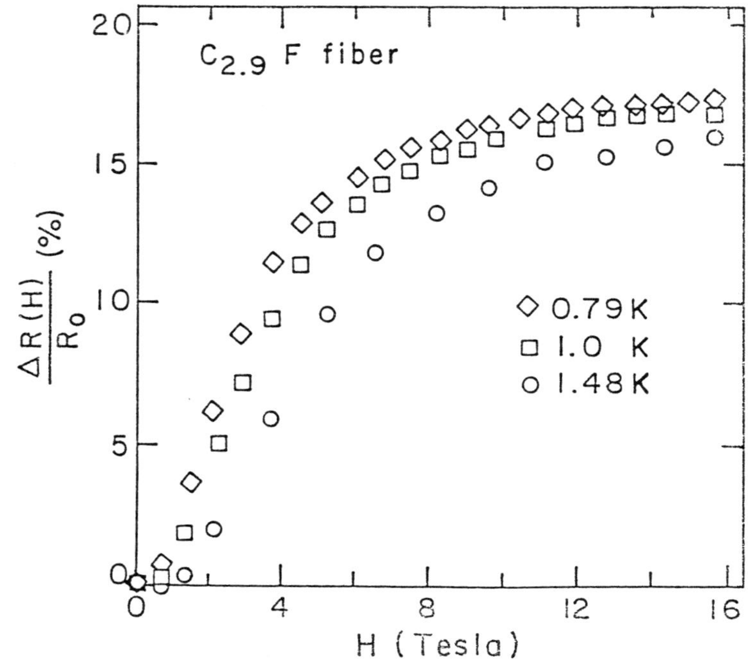

Fig. 8. The experimental longitudinal magnetoresistance for a $C_{2.9}F$ fiber in the strong localization regime in the magnetic field range $0< H <15$ tesla, and at the three temperatures of 1.48 K, 1.0 K and 0.79 K. The experimental magnetoresistance is explained by the theory of Kamimura *et al.* [27]

$$\mu_B H_{\mathrm{sat}} = \left(\frac{B\eta_{2D}}{N(E_F)} \frac{\xi_1^2 + \xi_2^2}{\xi_1^4 + \xi_2^4} \right)^{1/3} (k_B T)^{2/3} \tag{17}$$

where B is a constant on the order of 1. We can make a rough estimate for H_{sat} by using $\xi_1 \approx \xi_2$, since we are so close to the metal-insulator transition in the $C_{2.9}F$ fibers that ξ is large and the intrastate repulsion energy is small. We then obtain from Equation 17 a value of $H_{\mathrm{sat}} \approx 6$ tesla at 1 K, in good agreement with the experimental data.

The theory of Kamimura *et al.* also predicts a $T^{-1/3}$ temperature dependence for the logarithm of the saturation magnetoresistance $\ln(R_{\mathrm{sat}}/R_0)$ for a two-dimensional system, as seen in Equation 13. Although the experimental saturation magnetoresistance increases with decreasing temperature as expected, it deviates from the expected $T^{-1/3}$ temperature dependence. Incidentally, a related effect was also observed in another 2D system [35], namely the transition metal dichalcogenide 1T–TaS_2. Although the predicted saturation of the magnetoresistance with magnetic field is observed, the variation of the magnitude of the saturation magnetoresistance R_{sat}/R_0 with temperature exhibits a trend towards saturation at low temperatures (below 1K) that is not predicted by Kamimura. We want to emphasize that this saturation phenomenon as a function of temperature at low T is different from the magnetic field-dependent saturation of the resistance at high magnetic fields.

This saturation effect in temperature is strikingly similar to the one observed in the weak localization regime, and which, as mentioned above, is experimentally measured as a finite value for the inelastic scattering-equivalent magnetic field H_i at zero temperature. The origin of this saturation effect is not completely clear as of now, and is still being investigated experimentally. As mentioned in the theoretical section, interstate interactions favor spin ordering at low temperatures, which can be either ferromagnetic or antiferromagnetic, and reduce the effect of a magnetic field on the resistivity. An antiferromagnetic ordering at low temperatures is excluded, though, since we would not expect a saturation of the magnetoresistance in this case. Another possible cause for the saturation effect is the presence of a spin-relaxation effect, such as the presence of magnetic impurities or impurities with a large spin-orbit coupling. Such impurities have not been detected by chemical analysis.

6. Conclusions

We have used the special properties of fluorine-intercalated graphite fibers, unique among all GICs, namely the possibility to vary the disorder over a very wide range in order to study the resistivity and magnetotransport properties of weakly-disordered metals and insulators close to the metal-insulator transition. With a further increase in the fluorine concentration, and a corresponding increase in the density of defects in the graphene layers, we expect to change the intrastate correlation energy by a large amount. We thus expect the fibers to exhibit quantitatively and qualitatively different magnetotransport properties as they move away from the metal-insulator transition and into a regime characterized by smaller localization lengths.

Acknowledgements

The authors would like to express their gratitude to Professor J-P. Issi and Dr. Luc Piraux for introducing them to this problem in the weak localization regime and for many enlightening discussions. Special thanks are due to Professor T. Nakajima of Kyoto University and Professor M. Endo of Shinshu University for the preparation of the C_xF fiber samples and for valuable discussions. The high magnetic field experiments have been carried out at the Francis Bitter National Magnet Lab, supported by the National Science Foundation. The preparation of this manuscript was supported by the National Science Foundation Grant 88–19896–DMR.

References

1. T. Ohno, K. Nakao and H. Kamimura: 1979, *J. Phys. Soc. Jpn.* **47**, 1125.
2. H. Kamimura, K. Nakao, T. Ohno and T. Inoshita: 1980, *Physica* **B99**, 401.
3. H. Kamimura: 1986, *Ann. de Phys.* **11**, Suppl. 2, 39.
4. H. Akera and H. Kamimura: 1983, *Solid State Commun.* **48**, 467.
5. A. Shimizu and H. Kamimura: 1983, *Synthetic Metals* **5**, 301.
6. H. Kamimura and H. Aoki: 1989, *The Physics of Interacting Electrons in Disordered Systems*, Clarendon Press, Oxford.
7. P.W. Anderson, E. Abrahams and T.V. Ramakrishnan: 1979, *Phys. Rev. Lett.* **43**, 718.
8. B.L. Altshuler, A.G. Aronov and P.A. Lee: 1980, *Phys. Rev. Lett.* **44**, 1288.
9. E. Abrahams, P.W. Anderson, D.C. Licciardello and T.V. Ramakrishnan: 1979, *Phys. Rev. Lett.* **42**, 673.
10. P.W. Anderson: 1958, *Phys. Rev.* **109**, 1492.
11. M.S. Dresselhaus and G. Dresselhaus: 1981, *Adv. in Phys.* **30**, 139.
12. L. Piraux, J.-P. Issi, J.-P. Michenaud, E. McRae, and J.F. Marêché: 1985, *Solid State Commun.* **56**, 576; L. Piraux, V. Bayot, J.-P. Michenaud, J.-P. Issi, E. McRae, and J.F. Marêché: 1986, *Solid State Commun.* **59**, 711.
13. L. Piraux, V. Bayot, X. Gonze, J.-P. Michenaud and J.-P. Issi: 1987, *Phys. Rev.* **B36**, 9045.
14. L. Piraux: 1990, *J. Mat. Res.* **5**, 1285.
15. L. Piraux, V. Bayot, J-P. Issi, M.S. Dresselhaus, M. Endo and T. Nakajima: 1990, *Phys. Rev.* **B41**, 4961.
16. S.L. di Vittorio, V. Bayot, L. Piraux, J-P. Issi, M.S. Dresselhaus, M. Endo and T. Nakajima in *Defects in Materials*, MRS Symposia Proceedings, Boston, volume 209, edited by P.D. Bristowe, J.E. Epperson, J.E. Griffith, and Z. Lilienthal-Weber, p. 329.
17. T. Mallouk and N. Bartlett: 1985, *J. Chem. Soc. Chem. Commun.* **12**, 103.
18. I. Ohana, I. Palchan, Y. Yacoby, D. Davidov and H. Selig: 1988, *Phys. Rev.* **B38**, 12627; D. Vaknin, I. Palchran, D. Davidov, H. Selig, and D. Moses: 1986, *Synth. Met.* **16**, 349.
19. A. Rao, A. Fung, M.S. Dresselhaus, M. Endo, T. Nakajima and G. Dresselhaus (to be published).
20. N. Watanabe, T. Nakajima and H. Touhara: 1988, *Graphite Fluorides*, Elsevier, Amsterdam, Studies in Inorganic Chemistry, Vol. 8.
21. V. Bayot, L. Piraux, J-P. Michenaud, J-P. Issi, M. Lelaurain, A. Moore: 1990, *Phys. Rev.* **B41**, 11770.
22. L. Piraux, V. Bayot, J-P. Michenaud and J-P. Issi: 1988, *Physica Scripta* **37**, 942.
23. H. Aoki and H. Kamimura: 1976, *J. Phys. Soc. Jpn.* **40**, 6.
24. E. Yamaguchi, H. Aoki and H. Kamimura: 1979, *J. Phys. C: Sol. St. Phys.* **12**, 4801.
25. E. Yamaguchi, H. Aoki and H. Kamimura: 1980, *J. Non-cryst. Sol.* **35&36**, 47.
26. T. Takemori and H. Kamimura: 1982, *Sol. St. Commun* **41**, 885.
27. A. Kurobe and H. Kamimura: 1982, *J. Phys. Soc. Jpn.* **51**, 1904.
28. H. Kamimura, A. Kurobe and T. Takemori: 1983, *Physica* **117&118***B+C*, 652.
29. J. Blinowski and C. Rigaux: 1980, *J. Physique* **41**, 667.
30. P.A. Lee and T.V. Ramakrishnan: 1985, *Phys. Rep.* **57**, 187.

31. R.A. Davies, M.J. Uren and M. Pepper: 1981, *J. Phys. C: Sol. State Phys.* **14**, L531.
32. S.L. di Vittorio, M.S. Dresselhaus, M. Endo and T. Nakajima: 1991, *Phys. Rev. B* **43**, 1313.
33. S.L. di Vittorio, M.S. Dresselhaus, M. Endo and T. Nakajima: 1991, *Phys. Rev. B.* **43**, 12304.
34. Z. Ovadyahu: 1986, *Phys. Rev.* **B33**, 6552.
35. H. Kamimura: 1982, *Progr. of Theor. Phys., Suppl.* **72**, 206.

STRUCTURAL PHASE TRANSITIONS AND KINETIC PROCESSES IN GRAPHITE INTERCALATION COMPOUNDS

HIROYOSHI SUEMATSU
Department of Physics, University of Tokyo, Hongo 7-3-1, Bunkyoku, Tokyo 113, Japan.

Abstract. We describe the kinetic processes of phase transitions, which are characteristic of graphite intercalation compounds, from the viewpoint of microscopic structures. The novel stage transition is discussed in relation to the creation and annihilation of stage disorders at a phase boundary. We present the scaling features of nucleation and growth process in the first-order phase transition of the one-dimensional stacking orderings, and the kinetics of the domain growth in the two-dimensional (in-plane) order-disorder transition of second-order.

1. Introduction

Graphite intercalation compounds (GICs) are known to exhibit many types of structural orderings of the intercalant atoms. These orderings are classified into three hierarchical categories, i.e., the staging, in-plane two-dimensional, and stacking orderings, because they arise from interactions of different types and magnitudes and, as a consequence, the orderings occur successively at separate temperatures.

The staging structure is a one-dimensional ordering of intercalant layers resulting from the interlayer Coulombic interaction, and is generally stable up to high temperatures. The staging transition is a first-order transition since it is accompanied by a significant change of intercalant density. An exception of a second order transition, which is predicted only in theory, is a particular case of stage 2, in which the stage 1 dilute (gas) phase is transformed into the stage 2 dense (liquid) phase, like the liquefaction of a gas at the critical point [1,2].

Intercalant layers incorporated in the stage structure are in a disordered or liquid state at high temperatures, and fall into an ordered state at a low temperature. They show a large variety of two-dimensional superlattices depending on the relative strength of the intercalant-substrate interaction as compared to the intercalant-intercalant one, namely a typical system of competing interactions. A stacking ordering, by which the three-dimensional ordering is completed, takes place simultaneously with the in-plane ordering in some cases, but in other cases it occurs at a lower temperature.

Most of the order-order and order-disorder phase transitions relevant to these structures are first-order, with some exceptions. Lee, Aoki and

H. Aoki et al. (eds.), New Horizons in Low-Dimensional Electron Systems, 25–44.

Kamimura [3] predicted the possibility of the second-order transition in some of the 2×2 and 2×1 commensurate structures of stage 1 alkali metal (AM)GICs with the use of the lattice gas model in the Landau–Lifshitz approximation. Bak and Dommany [4] have discussed the case of $\sqrt{3} \times \sqrt{3}$ superlattice. These predicted transitions have not been realized in actual stage 1 crystals, and instead an order-disorder transition very close to second-order is observed in the 2×2 structure in stage 1 K and Rb GICs [5,6]. However, in the case of incommensurate structures, the order-disorder or melting transitions are found to be second-order, as in low stage (2 and 3) AMGICs [7]. In high stage $SbCl_5$ – GICs, moreover, the order-disorder transition of the $\sqrt{7} \times \sqrt{7}$ commensurate structure is found to be second-order [8].

For these phase transitions in GICs the kinetic processes are very important from the viewpoints of a particular atomic rearrangement in geometrically restricted structures and of the general problem of nucleation and domain growth in a low dimensional system. As a typical example the microscopic structure of stage ordering should depend crucially on the kinetics of atomic reconstructions. The most famous microscopic structure is the model proposed by Daumas and Herold (DH) [9], in which intercalants are inserted as islands (or domains) in every gallery of graphite and the staging transition is attained by a collective diffusion of these islands (Figure 1).

This classical model has been intensively investigated, both experimentally and theoretically, and is successful in explaining the macroscopic characteristics. In this paper we will describe the microscopic structures in the nonequilibrium process of stage transition, and the stage defects or disorders in the DH model which are inherent in the transient process of stage transition.

Because of the low dimensional nature of phase transitions, GICs are widely appreciated as model substance for studies of the kinetics of phase transitions as well as of critical behavior. The kinetics and pattern formation in first-order transitions depend crucially on the dimensionality of the system, the nature of conservation of the order parameter, and the degeneracy of the ground state p; the universality class is determined by the nature of the system. The conservation of the order parameter depends on whether its integral value over the system is conserved or not before and after the transition: For example, a two-phase separation process in a binary alloy belongs to the conserved order parameter (COP) system, while spin systems are a typical non-conserved order parameter (NCOP) system.

The kinetic process in the first-order transition generally consists of two successive processes, namely, the nucleation and growth (NG) process and the domain growth (DG) process. The nucleation of the stable phase and the growth of these nuclei in the metastable phase matrix, NG, takes place at the early stage of the transition; this process is driven by the free energy gain. After the system is completely transformed into the stable phase, the

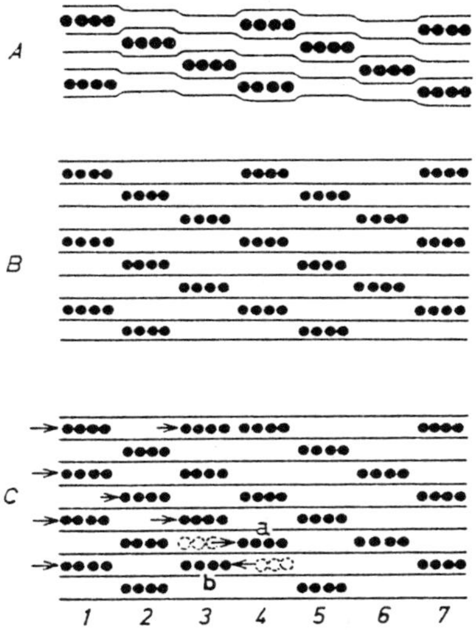

Fig. 1. The Daumas–Herold (DH) model for the stage structure and the kinetic processes of a stage transition. (A) The DH domain model for stage 3 (a realistic presentation). Solid lines and circles represent graphite layers and intercalant atoms, respectively. (B) A simple schematic presentation of the DH domains (A). (C) A kinetic process of the stage transition from stage 3 to 2. An arrow indicates the direction which a domain pointed came from. Intercalant atoms are fed at the left edge of the sample (which is in contact to the reservoir), the case corresponding to the nonconservative process. The domains, a and b, show a local reconstruction of domains, corresponding to the conservative process. The columns 1 through 3 are the transformed stage 2 region, while the columns 5 through 7 are the untransformed stage 3. The column 4 is the phase boundary, in which stage disorders such a stage mixing of stage 2 and 3 appear.

DG process begins and the stable phase domains grow in size by merging with other domains. This process is driven by the gain of interface energy between domains.

We will present recent studies of the kinetic processes of nucleation and growth in the one-dimensional stacking transition, which in our case is classified as belonging to NCOP system, and show that the scaling law in the NG process is obtained for volume fraction, grain size and structure factor. Furthermore, we will discuss an interesting example of the kinetics in the DG process in the two-dimensional (in-plane) order-disorder transition.

2. Stage Transitions

2.1. Macroscopic kinetics of stage boundary

The non-equilibrium kinetics of staging transition have so far been studied in many theoretical and experimental works. These investigations can be

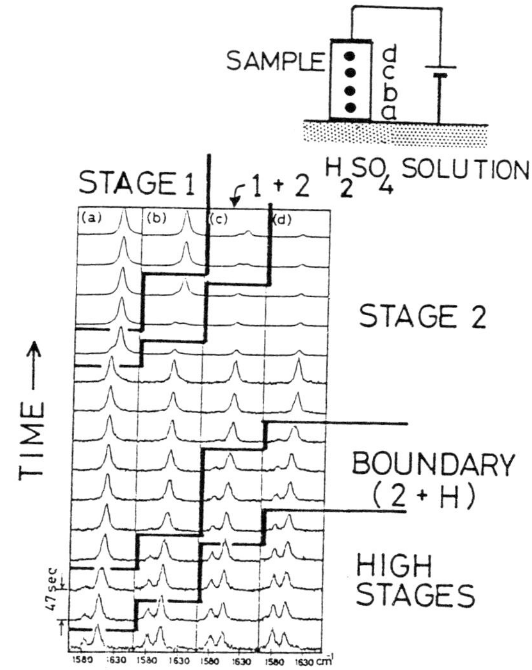

Fig. 2. The time- and space-resolved Raman spectra in the electrochemical reaction of H_2SO_4 GICs. Each column corresponds to a probed position (a to d) on a sample (inset); the spatial resolution is 0.33 mm. The spectra for the four positions are measured simultaneously at each 47 seconds. Three phases of stage 1, 2 and higher stages can be identified by the peak position of spectra, and their regions are indicated separated by thick lines. Note that phase boundaries climb up into the sample with time evolution. (Figure drawn on the basis of Nishitani's paper [10].)

classified into two categories, namely, the macroscopic kinetics of the stage boundary and the microscopic transient structures during the stage transitions. Here, we use the term "the macroscopic stage boundary" for the phase boundary between the well-ordered staging areas, but not for the domain boundary as in the DH islands. The macroscopic motion of the stage boundary is ruled by the intercalant flow in a sample, so that essential parameters in the study of the kinetics are the feed rate of intercalants across the sample edges (which is relevant to the driving potential to promote the intercalation reaction) and the diffusion rate inside the sample. Many experimental works have shown that these items are well understood in terms of the diffusion-limited kinetics and the mechanism of the collective motion of DH islands. A recent beautiful example is the observation of the temporal and spatial dependences of phase boundaries by Nishitani *et al.* using Raman scattering (Figure 2) [10].

After the onset of intercalation at the reservoir end of a sample, some well-staged regions proceed into the inside with association of phase boundaries between them. Nishitani *et al.* revealed that the observed movements of boundaries can be well described by the conventional diffusion equation. In the case of a large diffusion rate compared with the feed rate at the sample surface only two stage phases coexist in a sample [12], while more than two phases appear in the opposite case [10].

2.2. Microscopic stage disorder

At the growing front of such a phase boundary, however, we can expect the microscopic mixing of different-stage domains, or stage disorder, as shown in Figure 1 (C). Meisenheimer and Zabel have pointed out the evidence of stage disorder in their high-resolution x-ray diffraction experiments [11]. In their μ- or T-scan experiments they observed that the staging disorder is inherent in stage structures, even at thermal equilibrium, and is most significant at μ, where a stage transition takes place. These features of stage disorder in equilibrium can be well understood in the domain model for finite size stage domains and finite temperature [14].

Suematsu *et al.* [12] and Suda [13] performed non-equilibrium experiments in μ-scan from stage n_i to n_f, where one or more intermediate stages are included between n_i and n_f.

Figure 3 shows an example of the diffraction spectra for the transition from stage 6 to 4 in potassium GICs. The experiment was performed with a rapid change of vapor pressure from the equilibrium pressure of stage n_i to that of stage n_f. The diffraction spectra along the c-axis (the scattering vector perpendicular to the graphite layers) show two successive stage transitions from stage 6 to an intermediate stage 5 and from stage 5 to 4.

Figure 4 (a) and (b) show the time dependences of the volume fraction $F(n)$ of a stage n phase and the rate of stage disorder g. The parameter g is defined as follows in the Kakinoki–Komura formulation for stacking faults [15]:

$$g = \sum_{n} \sum_{n' \neq n} f(n)\ Pnn' ,$$

where $f(n)$ is the fraction of a stage n package consisting of an intercalant layer and n graphite layers, and Pnn' is the probability that a stage n' package follows after a stage n package. g is zero for the two-phase separation as well as the single phase state ($Pnn' = 0$ for $n \neq n'$), and unity for the complete mixing state of stages.

In the figure we observe a remarkable feature of stage mixing during two successive transitions; g is most significant when $f(n)$ (or $f(n')$) ~ 0.5, that is, the midpoint of the stage transition. Another noticeable feature is that the intermediate stage 5 appears as a stationary state with the single stage

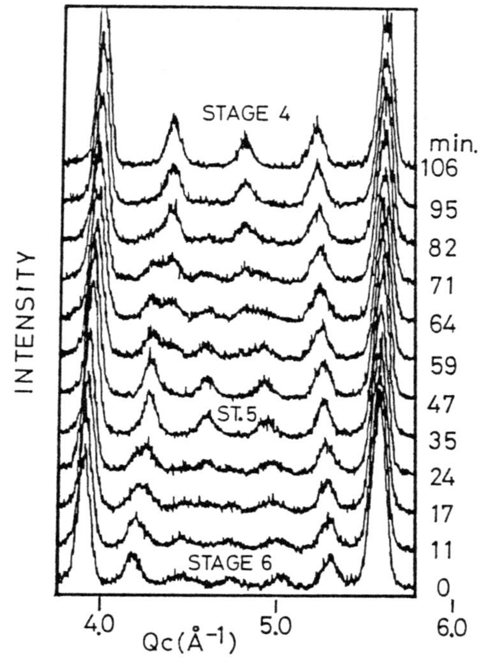

Fig. 3. The time dependence of the (001) diffraction spectra in the stage transition from stage 4 to 6 in K GICs. The spectra were taken after quenching potassium vapor pressure from the equilibrium pressure of stage 4 to that of stage 6, while the temperature of the sample T_G is held constant at 400.1°C. Starting from pure stage 6 at $t = 0$ (the bottom spectrum) the sample undergoes two successive stage transitions from stage 6 to 5 and from stage 5 to 4. Note that the intermediate, pure stage 5 exists as a stationary state ($62 \leq t \leq 95$ min). In the other time intervals two stages coexist with some stage disorder.

structure. This means that stage disorders created in the stage transition are completely dissolved after the transition. If we consider that a part of the sample was probed by an x-ray beam, these facts indicate that a single-stage region moves across the probed point from the end of a sample to the inside, which is followed by the development of the next stage region. Furthermore, the stage disorder appears mainly at the phase boundary while a region bounded by phase boundaries is almost free of stage disorder. This picture of stage transition is consistent with the result of the macroscopic studies of the phase boundary movement [10].

An interesting problem in relation to the correlation function of stage structures is a unique $f(n)$-dependence of Pnn, as shown in Figure 5.

Pnn corresponds to the nearest neighbor correlation of a stage n package. The extreme case of a perfect stage (or the perfect two-phase separation) corresponds to $Pnn = 1$, while the other case of the complete disorder or no

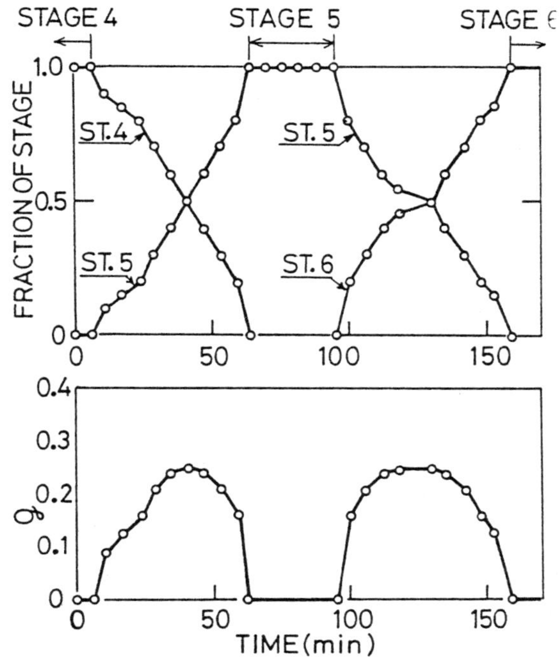

Fig. 4. (a) The time dependence of the volume fraction of stage n, $f(n)$ and (b) of the parameter of stage disorder, g, in the stage transition from stage 4 to 6 in K GICs. The line shape analysis was performed on the basis of the two-stage model using the Kakinoki–Komura formulation. Note that g has a maximum value at the midpoint of the stage transition and is zero in the pure stage 5 region.

correlation state does to $Pnn = f(n)$ (line B). The upper region, above the line B, corresponds to the case of attractive interaction between the same stage packages, while the lower region corresponds to the repulsive case. The experimental points can be represented by $Pnn = (1/2) \cdot (1 + f(n))$. Thus the nearest neighbor correlation of a stage package is found to be more than ~ 0.5 even in the early period of the transition, and is dependent on $f(n)$, namely, there is a much stronger correlation for $f(n)$ closer to 1. The entire correlation function is given by the sum of distant pair correlations beyond nearest neighbors, although it is difficult to determine in practice because of insufficient experimental resolution and the complicated calculation.

2.3. Model simulation of stage transitions

Some theoretical works have discussed the kinetic process in stage transitions from the viewpoint of microscopic structures. Kirczenow calculated the three-dimensional distribution of intercalant islands as a function of time using the Monte Carlo technique [16].

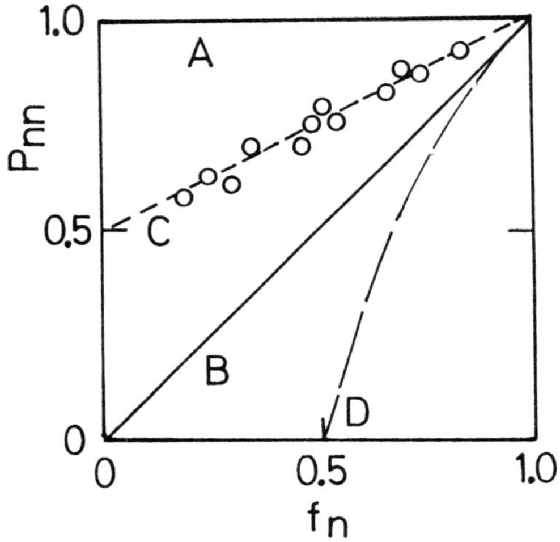

Fig. 5. The nearest neighbor correlation of a stage n package, Pnn, vs. the fraction of stage $f(n)$. The experimental data were taken for the transition from stage 4 to 6 in K GICs. The lines A and B correspond to the perfect stage (or the two-phase (stage) separation) and the complete disorder of stage, respectively, while C represents $(1/2) \cdot [1 + f(n)]$, a fitting function for experimental data. The line D is the lower limit of Pnn for the two-stage mixture model.

As shown in Figure 6 (A) and (B), the model simulation visualized the pattern formation of intercalant islands in real space, and revealed a remarkable stage disorder at the initial process of intercalation, and the in-plane reconstruction of island patterns which leads to the DH domain structure as the final ordering. The succeeding stage transition begins in the smallest domains and at domain boundaries, the feature being observed experimentally. The calculated $f(n)$ is consistent with the experimental values of Meisenheimer *et al.* On the basis of these qualitative and quantitative comparisons the model calculation can be considered to give a good picture of stage transition processes. Using another approach based on a nonlinear Langevin equation, Miyazaki *et al.* calculated the time evolution of the structure factor in the out-of-plane direction, and obtained two different results depending on the final chemical potential μ, that is, no significant stage disorder for a certain range of μ, and an intermediate stage-disordered state for other values [17].

The Monte Carlo simulation by Kirczenow has explicitly shown the simultaneous growth of in-plane domain size, as well as the development of the stage ordering, as shown in Figure 6 (B). Huster *et al.* observed in high stage Cs GICs that the development of the in-plane domain size associates

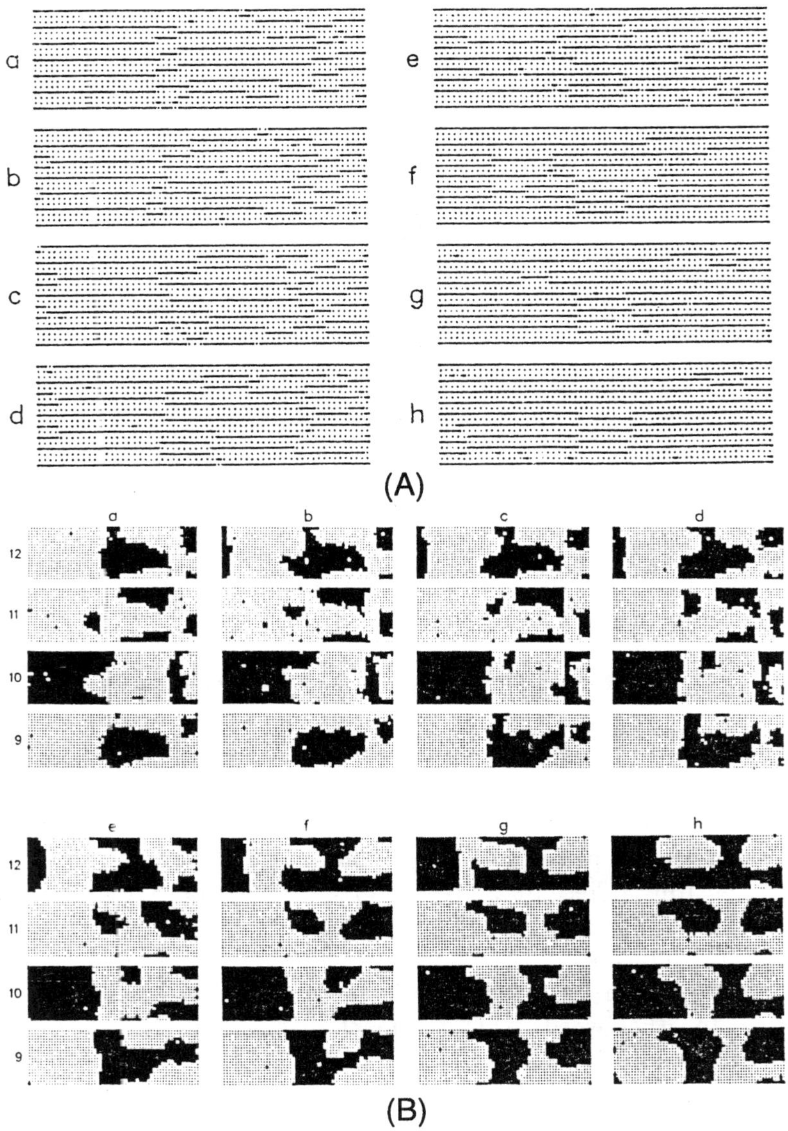

Fig. 6. The Monte Carlo simulation of the stage transition from stage 3 to 2 (after Kirczknow [16]). (A) The time evolution of an out-of-plane cross section. Intercalants are shown by solid lines. Frames, a to f, are respectively the sections after 0, 400, 800, 1200, 2000, and 4020 million Monte Carlo Steps (MMS) when the chemical potential μ was raised to the stage 3-2 boundary. Frame g is the section after 2042 MMS for a further increase of μ, and h is the result at 1055 MMS when μ is raised again near the center of the stage 2 stability region. (B) The in-plane cross sections corresponding to the time evolution of a to h in (A). Four successive intercalant layers are presented. Note the domains grow in the in-plane and also out-of-plane directions.

with a stage transition, although the situation is rather complicated in this case because the order-disorder transition of the in-plane lattice is involved [18]. The coherent length of these domains increases in the three dimensional directions, so that the microscopic domain patterns during the stage transition are three-dimensional in nature.

3. Stacking Transition: A First-Order Transition in the One-Dimensional System

3.1. Universality class of system

Some stage 1 alkali metal graphite intercalation compounds (AMGIC) have stacking transitions between different sequence states (an order-order transition) at a temperature below the melting (order-disorder) transition of intercalant layers. The stacking transition is one-dimensional (1D) in principle and is found to be first-order [6], so that the compounds are very attractive for the study of kinetics.

In the case of stage 1 Rb and K GICs the order-order transition occurs between the four-layer stacking ($\alpha\beta\gamma\delta$) phase and the two-layer stacking ($\alpha\beta$) one (Figure 7).

These phases have the same in-plane structure of the (2×2) superlattice commensurate with graphite, and the stackings α, β, γ, and δ indicated here are referred to the four non-equivalent sublattices of the superlattice (Figure 7, bottom). The four-layer stacking phase is the ground state to minimize the repulsive interaction energy between intercalant atoms, while the two-layer stacking phase becomes stable at a high temperature because of its excess entropy. These phases can be transformed from one to the other by a simple operation: The sequence $\alpha\beta\alpha\beta\ldots$ is obtained from $\alpha\beta\gamma\delta\ldots$ by the lateral displacement $1/2(\mathbf{a}_1+\mathbf{a}_2)$ of the package $\gamma\delta$, which results in the package shifting to the $\alpha\beta$ position, where $\mathbf{a}_1$ and $\mathbf{a}_2$ are in the in-plane unit vectors. In other words, this is a martensitic transition accompanying no atomic diffusion.

Thus, two sequences, $\alpha\beta\gamma\delta\ldots$ and $\alpha\beta\alpha\beta\ldots$, correspond to the two states in the one-dimensional Ising model, because we can regard the package $\alpha\beta$ (= A) as a spin-up state and $\gamma\delta$ (= B) as a spind-down state analogous to antiferro- and ferromagnets. The transition belongs to the one-dimensional system with a non-conserved order parameter of degeneracy $p = 2$. The order parameter is the number of the pairs A-A for $\alpha\beta$ phase and the pairs A-B for $\alpha\beta\gamma\delta$ phase.

We have described above the ideal model structures expected for stage 1 GICs, but whether it is actually realized or not is another thing. Metoki *et al.* [23,24] have examined the dimensionality of the system in detail as well as degeneracy of the ground state. They have observed the facts that the

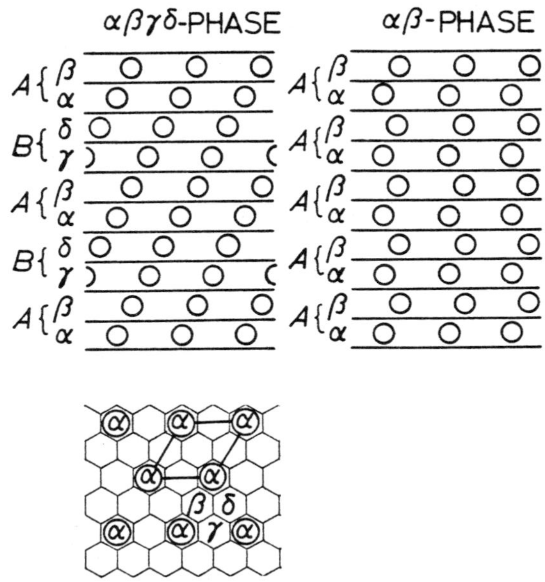

Fig. 7. The crystal structure of stage 1 Rb GIC. Stacking sequence of $\alpha\,\beta\,\gamma\,\delta$ phase and $\alpha\,\beta$ phase (above). The in-plane 2×2 superstructure and four sublattices α, β, γ, and δ (bottom). The two-layer package $\gamma\,\delta$ (= B) can be transformed to $\alpha\,\beta$ (= A) stacking by a lateral displacement of $1/2\,a$, a being the unit vector of the superlattice.

coherence length ξ_a ($\geq$ 1000 Å) in the in-plane direction is much larger than ξ_c ($\geq$ 60 Å) in the out-of-plane direction, and shows no change even in the transition, and concluded that our system can be considered as a good 1D system.

The degeneracy of the ground state is important for the domain structure in the DG process. In the case of $\alpha\,\beta\,\gamma\,\delta$ phase there are 24 stacking configurations having the same ground state energy, such as $\alpha\,\beta\,\gamma\,\delta$, $\alpha\,\gamma\,\beta\,\delta$, ... etc. However, from the analysis of the stacking faults found in the actual crystal [24] we can show that a domain of the sequence $\alpha\,\beta\,\gamma\,\delta\,\alpha\,\beta\,\gamma\,\delta\,\alpha$, for example, has only two types of two-layer stacking faults, $\alpha\,\beta$ or $\gamma\,\delta$, out of many configurations, that is,

$$\underset{\hphantom{\alpha\beta\gamma\delta}\uparrow\hphantom{\alpha\beta\gamma\delta}}{\alpha\,\beta\,\gamma\,\delta\,\overline{\alpha\,\beta}\,\alpha\,\beta\,\gamma\,\delta} \quad \text{and} \quad \underset{\hphantom{\alpha\beta\gamma\delta}\uparrow\hphantom{\alpha\beta\gamma\delta}}{\alpha\,\beta\,\gamma\,\delta\,\overline{\gamma\,\delta}\,\alpha\,\beta\,\gamma\,\delta}\,,$$

where a bar represents the stacking fault. This means that only two types of domains, $\alpha\,\beta\,\gamma\,\delta$ and $\gamma\,\delta\,\alpha\,\beta$ appear in a single crystallite in the present case; an arrow indicates a domain boundary between the two domains. Therefore, we conclude that the ground state degeneracy is $p = 2$.

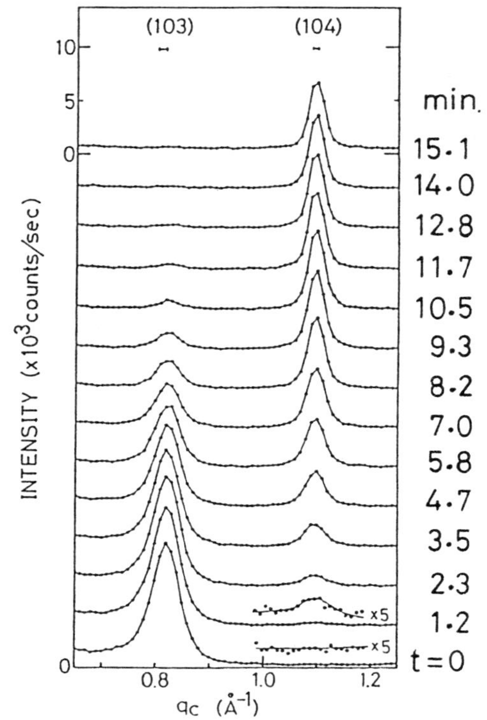

Fig. 8. The time dependence of the (101) diffraction spectra for $\Delta P = -1.6$ Pa and $T_G = 378.0$ °C. The (103) and (104) peaks respectively correspond to the Bragg diffractions of $\alpha\,\beta\,\gamma\,\delta$ and $\alpha\,\beta$ phases. Note that $\alpha\,\beta$ phase grows in intensity at the expense of $\alpha\,\beta\,\gamma\,\delta$ phase, and the peaks are sharp throughout the transition.

3.2. Scaling of volume fraction in nucleation and growth process

We will now discuss experimental results of the kinetics in the stacking transition, and describe the remarkable features of nucleation and growth process and the scaling of the order parameter.

Figure 8 shows the time dependence of the (101) x-ray diffraction spectra in the transition from the $\alpha\,\beta\,\gamma\,\delta$ to the $\alpha\,\beta$ phase. The kinetic spectra were obtained by rapidly changing the vapor pressure of rubidium P from and initial pressure P_i stabilizing the $\alpha\,\beta\,\gamma\,\delta$ phase to a final pressure P_f stabilizing the $\alpha\,\beta$ phase across the critical pressure P_c. The kinetic rate is dependent on the metastability, which is defined by

$$\Delta P = P_f - P_c\ .$$

In the figure the (103) and (104) peaks correspond to the Bragg reflections $\alpha\,\beta\,\gamma\,\delta$ and $\alpha\,\beta$ phases, respectively. They were chosen because they are less affected by the sample mosaic effect compared with the (100) and (101) diffractions. At $t = 0$ we observe only the (103) peak of $\alpha\,\beta\,\gamma\,\delta$ phase, which was stable at $t \leq 0$. As time proceeds the (104) peak of the $\alpha\,\beta$ phase, which is stable at $t > 0$, comes out and increases in intensity while the (103) peak decreases in contrast. Eventually the $\alpha\,\beta$ phase peak grows up to the maximum, and the $\alpha\,\beta\,\gamma\,\delta$ phase disappears completely.

The observed spectra present remarkable characteristics of the NG process in the first-order transition. The (104) spectrum has a sharp peak profile characteristic of the Bragg reflection with no diffuse scattering throughout the transition; this is the case even in the earliest stage of $t = 1.2$ min immediately after quenching. The evidence of a sharp peak profile indicates that nuclei of $\alpha\,\beta$ phase germinate in the matrix of a $\alpha\,\beta\,\gamma\,\delta$ metastable phase with a finite crystal size enough to give the Bragg diffraction. The absence of diffuse scattering during transition indicates that no intermediate, disordered state exists in a grain of the growing phase nor at a phase boundary. Namely, the width of phase boundary can be negligibly small compared with the grain size, and the order parameter changes discontinuously at the boundary. These features — that is, a finite crystal size of nuclei at the nucleation stage, the existence of only the stable and metastable phases and the sharp phase boundary — indicate that the process is the NG process, but not the DG one.

The volume fraction of the growing phase, $X(t)$, has a characteristic time dependence, shown in Figure 9. $X(t)$ is obtained experimentally from $I(t)/I_0$, the integrated intensity of a diffraction peak relative to the saturation value (I_0) at the completion of transition, and the rate of kinetics depends on the metastability ΔP. However, we note that the functional form of $X(t)$ for different ΔP can be scaled by a unified function when using the normalized time $\tau = t/t_0$, as shown in Figure 9, where t_0 is the time during which $I(t)$ grows to a half the saturation intensity $I_0/2$.

According to the phenomenological theory the volume fraction of transformed region $X(t)$ in NG is expressed by the equation [19–21]

$$X(t) = 1 - \exp[-\int_0^t \Gamma\, V(s)\, \mathrm{d}s]\ , \tag{1}$$

where Γ is the nucleation rate per unit volume and $V(t)$ is the volume of a grain. Assuming that Γ is constant and a grain grows with an isotropic, constant velocity v, the volume fraction for the d-dimensional system is derived as

$$X(\tau) = 1 - \exp[-\tau^{d+1}]\ , \tag{2}$$

where time t is replaced by the normalized time τ. Equation (2) represents the scaling law for the volume fraction in NG.

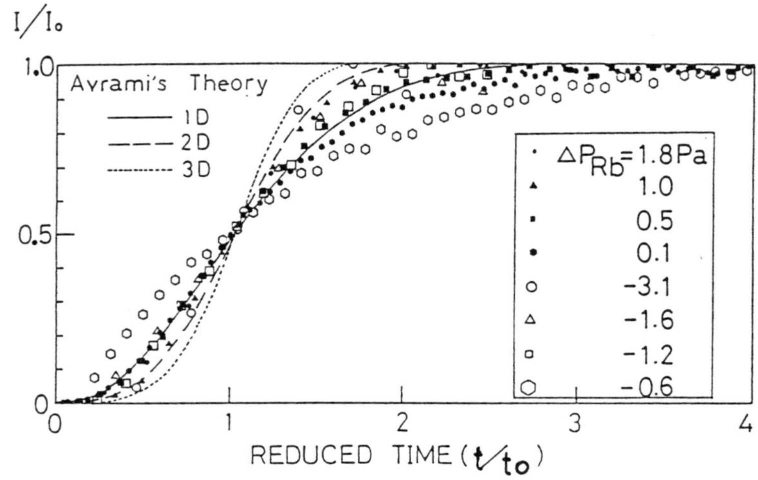

Fig. 9. The time dependence of the volume fraction of the growing phase. The data are plotted as a function of the normalized time, t/t_0. The theoretical scaling functions proposed by Avrami are shown for 1D (solid line), 2D (dashed line) and 3D (dotted line). The 1D calculated function can be fitted to a good approximation to the experimental data, except for the smallest ΔP case, which may belong to another category of system.

In the figure are shown the theoretical functions according to Equation 2. The calculated curve for $d = 1$ is in good agreement with experimental values. These facts indicate that the scaling function corresponds to the 1D NG process. This is the first observation of scaling for NG process in the 1D NCOP system. The possibility that a DG process may be responsible for the observed change of $I(t)$ can be eliminated in the present case because we observe no appreciable growth of domain size after $I(t)$ attains the saturation value, i.e., the completion of the phase transition.

At a small ΔP (< 0.6 Pa), $X(t)$ shows a much slower kinetics than the calculated function. In such a case the inhomogeneity in a sample is important because the nucleation is allowed only at limited parts of a sample where P_c is relatively low, but not in the other parts. The situation is likely the case of the cell model, and $X(\tau)$ seems to be understood in this model.

3.3. Scaling in grain size and structure factor

The grain size of the growing phase in the NG process generally increases with time, as assumed in the phenomenological theory. Although the microscopic observation of the size is not so easy in the NCOP system because of

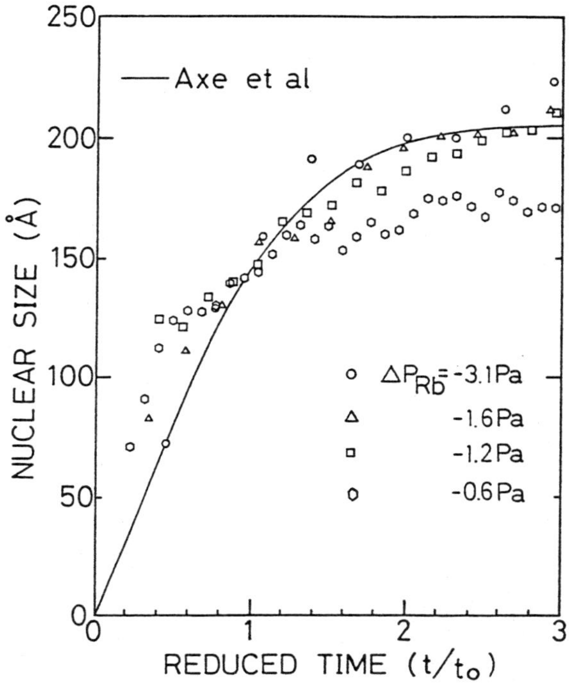

Fig. 10. The time dependence of grain size $R(\tau)$ of the growing $\alpha\,\beta$. The experimental points appear to distribute around a unified function. The solid line presents the theoretical result of Axe *et al.* for the NG process in the 1D NCOP system with the degeneracy $p = \infty$. (After Metoki [24].)

insufficient experimental resolutions in time and space, we have successfully observed the whole process in the transition.

Figure 10 shows the t-dependence of the grain size of the growing phase $(\alpha\,\beta)$, $R(t)$: $R(t)$ is experimentally determined from the linewidth of a diffraction peak along the c^*-axis, Δq_c. The observed value for different ΔP ranges from a critical size (60 Å) to a final size (300 Å), and the growing velocity of a grain depends strongly on ΔP. However, Metoki [24] has pointed out that using the normalized time $\tau = t/t_0$, $R(\tau)$ is approximately represented by a scaling function irrespective of ΔP, except for a very small ΔP. The experimental scaling function seems to be nearly logarithmic at a large τ [22]. This is in strong contrast with the consequence of the constant growing velocity $R \sim t$, the phenomenological assumption based on the constant rate of the free energy gain in 1D NG.

A possible interpretation of the slow kinetics of $R(t)$ is the impingement effect of grains. In the late stage of NG a grain impinges upon neighboring grains, and is subjected to a slowing down of its growth velocity. Therefore $R(t)$ shows a crossover from a linear t dependence to a slower dependence,

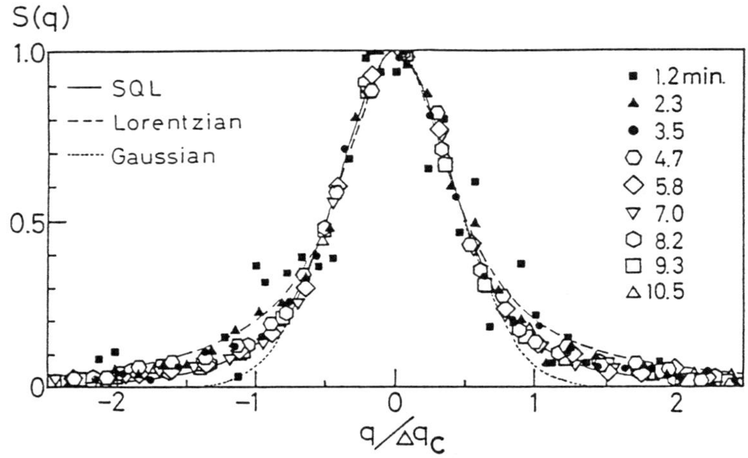

Fig. 11. The scaling function of the structure factor $S(q)$. The (104) diffraction spectra at different times are plotted as a function of the normalized wavenumber $q/\Delta q$, Δq being the linewidth FWHM. The experimental data give a best fit to the squared Lorentzian (SQL) rather than a Lorentzian or Gaussian.

and approaches a limiting value determined by the number of nuclei at $t = \infty$. The effect is dependent on the dimension of the system and the degeneracy of the ground state. The impingement is effective only when two encountering grains are in different degenerated states, so that large number of degeneracy results in more significant effect. Axe and Yamada [25] calculated $R(\tau)$ in NG for the 1D NCOP system with degeneracy $p = \infty$, instead of $p = 2$ in our case; no theoretical calculation for $p = 2$ has been made so far. They obtained the following relation by using the extended scaling assumption,

$$R(\tau) \sim X(\tau)^{1/(d+1)} . \tag{3}$$

The calculated curve for $d = 1$ is in fairly good agreement with the experimental results, as shown in Figure 10. This apparent agreement, however, may be a little strange because p assumed in the theory differs from that of our case. A reasonable interpretation is that the sample is composed of many mosaic structures, and they may prevent grains from merging with each other, which effectively acts as a large degeneracy.

Figure 11 shows the diffraction spectra of the growing phase during the transition as a function of the normalized wavenumber $\eta = (q - q_0) / q_c$. Even though the linewidth Δq_c decreases with time evolution, the structure factor $S(q)$ is found to be scaled by a unified function. In the figure we can see that the experimental $S(q)$ fits well to the squared Lorentzian $S(q) \sim$

$1/(q^2 + \Delta q_c^2)^2$, rather than the conventional function of a Lorentzian or Gaussian. This is the first experimental observation of the scaling function $S(q)$ for NG process in the 1D NCOP system.

As for $S(q)$ in the NG process we have no theoretical work exactly corresponding to our case of degeneracy $p = 2$. Instead $S(q)$ for $p = \infty$ was obtained by Axe *et al.* [25]. They take account of the correlation in a grain only, with no correlation between grains. The calculated $S(q)$ is very close to the squared Lorentzian. If we consider that the mosaic structure in our sample results in incoherent scattering, we may understand this qualitative agreement. On the other hand, for the DG process in the 1D NCOP system, Kawasaki and Nagai [26] obtained theoretically a nearly squared Lorentzian for $S(q)$ and a logarithmic t-dependence of domain size $R(t)$. It is very interesting that the ND process in the 1D system has the same kinetics in $R(t)$ and $S(q)$ as those for the DR process, though the processes have different mechanism.

In concluding this section we may mention that the scaling law exists for the volume fraction, the grain size and the structure factor in the nucleation and growth process of the first-order stacking transition, the one-dimensional NCOP system.

4. In-Plane Order-Disorder Transition. Two-Dimensional Ordering Processes

Graphite intercalation compounds are also found to be unique for studies of kinetics in the domain growth process. Intercalant layers in GICs have many 2D superlattices commensurate with the graphite substrate, and they take part in an order-disorder or melting transition, some of which are second-order transition. The interesting features are the two-dimensional nature of the system and a high configurational degeneracy of the superlattice with respect to the substrate, i.e., the structure has many equivalent ground states. Such a highly degenerate system has slow kinetics in general, because the higher degenerate system has the smaller probability of selecting domains of the same sublattice out of many other degenerate ones: the situation is very convenient for time-resolved experimental study.

Hernandez *et al.* [27] studied the kinetics of in-plane ordering in stage 4 $SbCl_5$ GIC, which has a $\sqrt{7}\times\sqrt{7}$ commensurate superlattice with degeneracy $p = 7$ (Figure 12).

They observed the time dependence of the (10) diffraction peak of the superlattice after quenching the sample from the disorder state to ordered state below T_m, the melting temperature. The superlattice peak grows rapidly in the early period of the transition, which is fitted by a power law of t, and thereafter it increases very slowly according to a $\log(t)$ dependence. The line width of the peak, which is very narrow even at short times after quenching, decreases in the course of transition (Figure 13). This implies that the

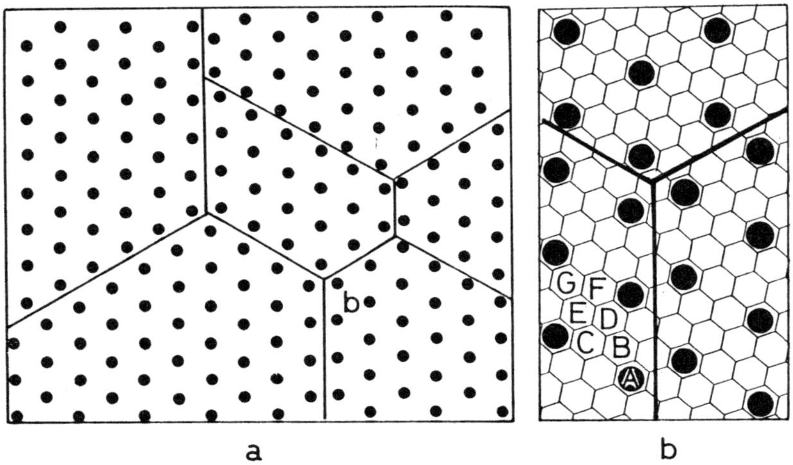

Fig. 12. (a) The domain boundaries for the $\sqrt{7} \times \sqrt{7}$ two-dimensional superlattice on graphite. Only the intercalant lattice is presented. (b) The details at the boundary region b in Figure (a). The seven sublattices A to G on graphite lattice are shown. Different types of domains, A, D and G, compete with each other at the domain boundaries, and in the domain growth process a domain is merged with another into a larger one.

superlattice domains appear immediately after quenching and grow in size. The most significant point in the experiment is the fact that the domain size shows an algebraic law, $R(t) \sim a \cdot t^n$, at the early period of transition. The observed exponent n is 0.43 ± 0.05, and consistent with the theoretical value 1/2 calculated for the NCOP system [28,29]. In the experiment they observed an anomalous decrease of domain size following after the saturation, and discussed it in relation with the thermal domain-wall roughening.

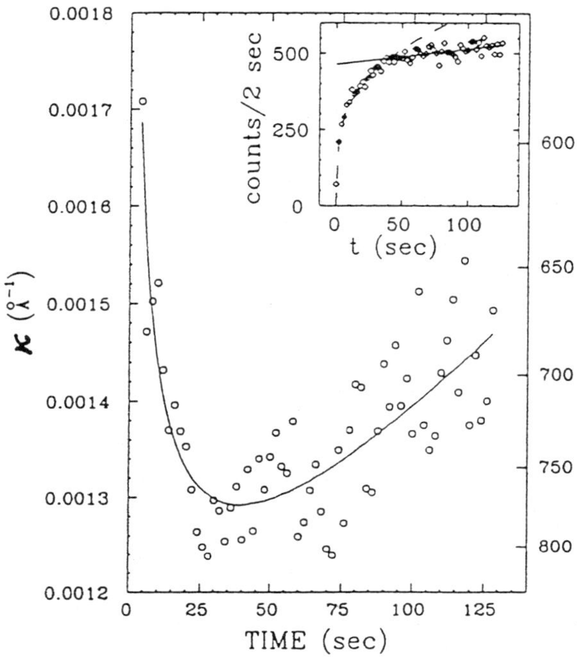

Fig. 13. The time dependence of line width κ and domain size $L = \kappa^{-1}$ observed in the melting transition of stage 4 $SbCl_5$ GIC. The time is measured after quenching from the disordered state to ordered state. The solid line represents a fitting function of $\kappa = at^{-n} + bt + c$; $n = 0.43$. Inset: the time dependence of the intensity. The initial rise is fitted by a power law and the following slow kinetics by a logarithmic dependence. (After Hernandez [27].)

Acknowledgements

The author gratefully acknowledge Dr. Naoto Metoki for stimulating and fruitful discussion. The present work has been partially supported by a Grant-in-Aid for Scientific Research from the Ministry of Education, Science and Culture.

References

1. A. Safran: 1980, *Phys. Rev. Lett.* **44**, 937.
2. S.E. Millman and G. Kirczenow: 1983, *Phys. Rev.* **B28**, 3482.
3. C.R. Lee, H. Aoki, and H. Kamimura: 1980, *J. Phys. Soc. Jpn.* **49**, 870.
4. P. Bak and E. Dommany: 1979, *Phys. Rev.* **B20**, 2818.
5. H. Minemoto and H. Suematsu: 1985, *Synth. Met.* **12**, 33.
6. N. Metoki and H. Suematsu: 1988, *Phys. Rev.* **B38**, 5310.

7. M. Mori, S.C. Moss, Y.M. Jan, and H. Zabel: 1982, *Phys. Rev.* **B25**, 1287.
8. H. Homma and R. Clarke: 1985, *Phys. Rev.* **B31**, 5865.
9. N. Daumas and A. Hérold: 1969, *C.R. Acad. Sci., Ser. C* **268**, 273.
10. R. Nishitani, Y. Sasaki, and Y. Nishina: 1988, *Phys. Rev.* **B37**, 3141.
11. M.E. Meisenheimer and H. Zabel: 1985, *Phys. Rev. Lett.* **54**, 2521.
12. H. Suematsu, K. Suda and N. Metoki: 1988, *Synth. Metals* **23**, 7.
13. K. Suda: 1988, *ME Thesis*, Tsukuba University.
14. G. Kirczenow: 1985, *Phys. Rev.* **B31**, 5376.
15. J. Kakinoki and Y. Komura: 1952, *J. Phys. Soc. Jpn.* **7**, 30; 1962, *Acta Cryst.* **15**, 292: 1965, *ibid.* **19**, 137.
16. G. Kirczenow: 1985, *Phys. Rev. Lett.* **55**, 2810; 1988, *Can. J. Phys.* **66**, 39.
17. H. Miyazaki, T. Watanabe, and C. Horie: 1986, *Phys. Rev.* **B34**, 5736.
18. M.E. Huster P.A. Heiney, V.B. Cajipe, and J.E. Fischer: 1987, *Phys. Rev.* **B35**, 3311.
19. J.D. Gunton, M.S. Miguel, and P.S. Sahni: 1983, *C. Domb and J. Lebowitz*, Phase Transitions and Critical Phenomena, Academic Press, New York.
20. A.N. Kolmogorov: 1937, *Bull. Akad. Sci. USSR, Phys. Ser.* **3**, 355.
21. M. Avrami: 1939, *J. Chem. Phys.* **7**, 1103; 1940, *ibid.* **8**, 212; 1941, *ibid.* **9**, 177.
22. N. Metoki, H. Suematsu, Y. Murakami, Y. Ohishi and Y. Fujii: 1990, *Phys. Rev. Lett.* **64**, 657.
23. N. Metoki, H. Suematsu, Y. Murakami, Y. Ohishi and Y. Fujii: 1989, *Synth. Metals* **34**, 323.
24. N. Metoki: 1990, *Dr. Thesis*, University of Tokyo.
25. J.D. Axe and Y. Yamada: 1986, *Phys. Rev.* **34**, 1599.
26. K. Kawasaki and T. Nagai: 1986, *J. Phys.* **C19**, L551.
27. P. Hernandeze, F. Lamelas, R. Clarke, P. Dimon, E.B. Sirota, and S.K. Sinha: 1987, *Phys. Rev. Lett.* **59**, 1220; H. Homma and R. Clarke: 1984, *Phys. Rev. Lett.* **52**, 629.
28. K. Kawasaki, M.C. Yalabik and J.D. Gunton: 1978, *Phys. Rev.* **A17**, 455.
29. K. Nakashima, T. Nagai, and K. Kawasaki: 1989, *J. Stat. Phys.* **57**, 758.

ELECTRONIC STRUCTURES OF HYDROGEN-POTASSIUM-GRAPHITE TERNARY INTERCALATION COMPOUNDS

KENJI NAKAO
Institute of Materials Science, University of Tsukuba, Tsukuba 305, Japan
and
SEIJI MIZUNO
Department of Engineering Science, Hokkaido University, Sapporo 060, Japan

Abstract. We present the results of the band-structure calculation for the hydrogen-potassium-graphite ternary intercalation compounds. The self-consistent numerical-basis-set linear combination of atomic orbitals method has been used within the local density functional formalism. We find that hydrogen acts as an acceptor, while potassium acts as a donor. The charge transfer to the hydrogen atom is imperfect and therefore the hydrogen $1s$ state forms a partially occupied metallic band. This hydrogen $1s$-like band coexists with the graphite π^*-like bands at the Fermi level. These results indicate that these compounds have a feature of the two-dimensional metallic hydrogen.

1. Introduction

Ternary graphite intercalation compounds (T GICs) are layered compounds in which two different species are intercalated into the galleries between layers of host graphite. The hydrogen-potassium-graphite intercalation compounds $C_{4n}KH_x$ (n is the stage number) are the most typical examples of these T GICs. In $C_{4n}KH_x$, two intercalating species (potassium and hydrogen) form separate layers (K–H–K trilayers) within the same galleries.

The simplest picture for an understanding of the electronic band structures of these layered compounds is a rigid band model. The rigid band is obtained by superposing the hydrogen and potassium bands (intercalant bands) on the graphite band folded into the smaller Brillouin zone of GIC. The relative position of the intercalant bands and the graphite band it determined by the amount of the charge transfer between the intercalant layers and the graphite layers.

A considerable number of attempts to determine the amount of the charge transfer in the stage-1 compound C_4KH_x and the stage-2 C_8KH_x have been performed experimentally [1–12]. The experiments suggest that the potassium acts as a donor and the hydrogen acts as an acceptor. However, there has been a controversial discussion. The key problem is whether the charge transfer to the hydrogen $1s$ (H $1s$) state is perfect or not. If the charge transfer is perfect, then the hydrogen band lies below the Fermi level and there exist only the two-dimensional (2D) conduction bands originating from the

H. Aoki et al. (eds.), New Horizons in Low-Dimensional Electron Systems, 45–53.

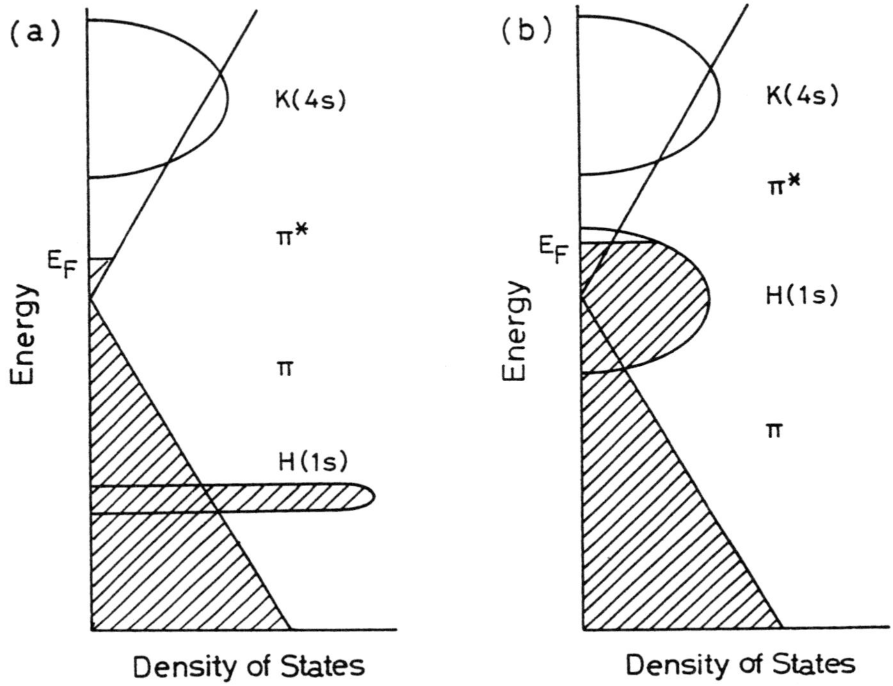

Fig. 1. Two models of the density of states for $C_{4n}KH_x$. The potassium 4s band, hydrogen 1s band, graphite upper and lower π bands are shown. In the model (a), the potassium 4s band lies above the Fermi level and therefore the charge transfer from this potassium band is perfect. On the other hand, the hydrogen 1s band lies below the Fermi level and the charge transfer to the hydrogen band is perfect. In the model (b), on the other hand, the hydrogen 1s band intersects the Fermi level and forms a metallic band.

graphite π^* state. (The schematic picture of the density of states in this case is shown in Figure 1(a).) However, if the charge transfer is imperfect, then the H 1s state intersects the Fermi level and forms a conduction band in addition to the π^* conduction bands. (The schematic picture is shown in Figure 1(b).) Several experiments, such as low-temperature specific heat [1], Shubnikov–de Haas effect [2], and optical reflectivity [3], suggest the presence of a low-lying H 1s state. On the other hand, recent experimental results of proton NMR [4–7], electrical conductivity [8], and thermoelectric power [8] are interpreted as showing the metallic character of the H 1s state.

In this article, we present the calculated band structures of C_4KH_x and C_8KH_x and show that the H 1s state forms a metallic band in both C_4KH_x and C_8KH_x.

2. Structure

C_4KH_x (C_8KH_x) has a stage-1 (stage-2) structure where K–H–K sandwiched layers are formed in the intercalants. That is, the stacking order along the c axis is –C–K–H–K–C– (–C–C–K–H–K–C–C–). The c axis repeat distance of C_4KH_x (C_8KH_x) is reported to be 8.53 Å(11.88 Å) [9]. The neutron scattering experiments on C_4KH_x [10] indicate a stacking sequence $A\phi B\chi A\phi$, where A, B denote the graphite layers and ϕ, χ denote the K–H–K trilayers. For simplicity, in this study we use the simpler stacking sequence $A\phi A\phi A\phi$ for C_4KH_x. This difference is not important because the interaction between layers is expected to be very weak. (In fact, our calculated results show that the energy band near the Fermi level has very small c-axis dispersion (Section 4).) For C_8KH_x we use the stacking sequence $AA\phi AA\phi AA\phi$ with the same reasoning.

As for the in-plane atomic arrangement, the structure of a carbon layer is a hexagonal net, as in graphite. In a potassium layer, potassium atoms form a (2×2) $R(0°)$ superlattice. However, the structure of a hydrogen layer has not been established at the time of writing. Some possible models have been proposed [1,11,12]. In this study, we use two simple models for C_4KH_x: (a) (2×2) $R(0°)$; (b) $(2\times\sqrt{3})$ $R(0°, 30°)$ in-plane superlattice model. Both models correspond to a hydrogen content $x = 0.5$.

In the stage-2 compound C_8KH_x, the unit cell in the $(2\times\sqrt{3})$ $R(0°, 30°)$ model is too large for the band structure calculation to be carried out. Therefore, we use only the simple model with the smallest unit cell (2×2) $R(0°)$ in-plane superlattice model for C_8KH_x. This model also corresponds to a hydrogen content $x = 0.5$.

Figure 2(a) shows the (2×2) $R(0°)$ in-plane superlattice model of hydrogen structure. The Bravais lattice of this model structure is base-centered orthorhombic and its space group is $D_2^6(C222)$. The primitive translation vectors $\mathbf{a} = (a/2, -\sqrt{3}a/2, 0)$, $\mathbf{b} = (a/2, \sqrt{3}a/2, 0)$, $\mathbf{c} = (0, 0, c)$, where $a = 4.92$ Å and c is the c-axis repeat distance. The x and y axes are shown in Figure 2(a) and the z axis is perpendicular to the layer plane. A unit cell is denoted by broken lines in this figure. The Brillouin zone (BZ) is a hexagonal prism, as in graphite.

Figure 2(b) shows the $(2\times\sqrt{3})$ $R(0°, 30°)$ in-plane superlattice model of hydrogen structure. The Bravais lattice of this model structure is different from that of the model (a). It is primitive orthorhombic and its space group is $D_2^1(p222)$. The primitive translation vectors are $\mathbf{a} = (a, 0, 0)$, $\mathbf{b} = (0, \sqrt{3}a, 0)$, $\mathbf{c} = (0, 0, c)$. The BZ is a rectangular parallelepiped.

Our band structure calculation is carried out for $C_4KH_{0.5}$ with the assumed structure models (a) and (b), and for $C_8KH_{0.5}$ with the assumed structure model (a).

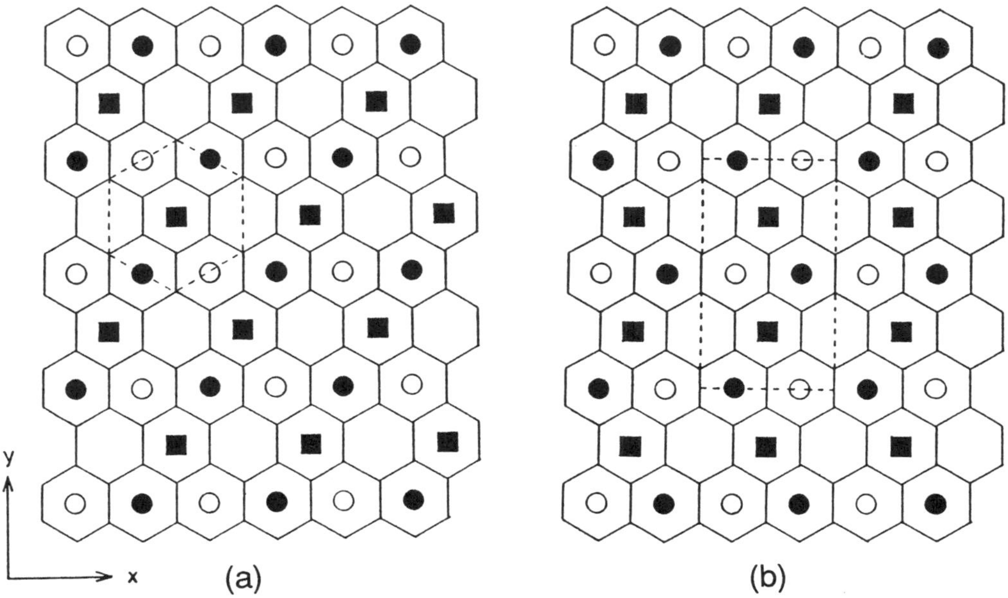

Fig. 2. Top view of the structural models: (a) (2×2) $R(0°)$; (b) $(2 \times \sqrt{3})$ $R(0°, 30°)$ in-plane superlattice model of hydrogen structure. The hexagonal net represents the carbon layer. Hydrogen atoms are denoted by squares (■). Potassium atoms are denoted by closed circles (●) and open circles (○). The stacking order along the c-axis is –C–K (●)–H(■)–K(○)–C–. Both models correspond to a hydrogen content $x = 0.5$.

3. Method of Band Structure Calculation

The numerical method used in this study is the self-consistent numerical-basis-set linear combination of atomic orbitals (LCAO) method within the Local Density Functional (LDF) formalism [13]. This method was developed by Zunger and Freeman [14] and later applied to the band-structure calculation of the binary GIC C_8K in a previous work [15]. The Hedin and Lundqvist parametrization [16] for the exchange-correlation potential was used. As a numerical basis set, we take the $1s$, $2s$ and $2p$ orbitals of carbon, and $1s$, $2s$, $2p$, $3s$, $3p$ and $4s$ orbitals of potassium and $1s$ orbital of hydrogen. The details of this method are presented in a previous work [15].

4. Results and Discussion

The calculated band structure of $C_4KH_{0.5}$ with the structure model (a) is shown in Figure 3(a). The overall band structure is similar to that obtained by superposing the hydrogen and potassium bands on graphite bands folded

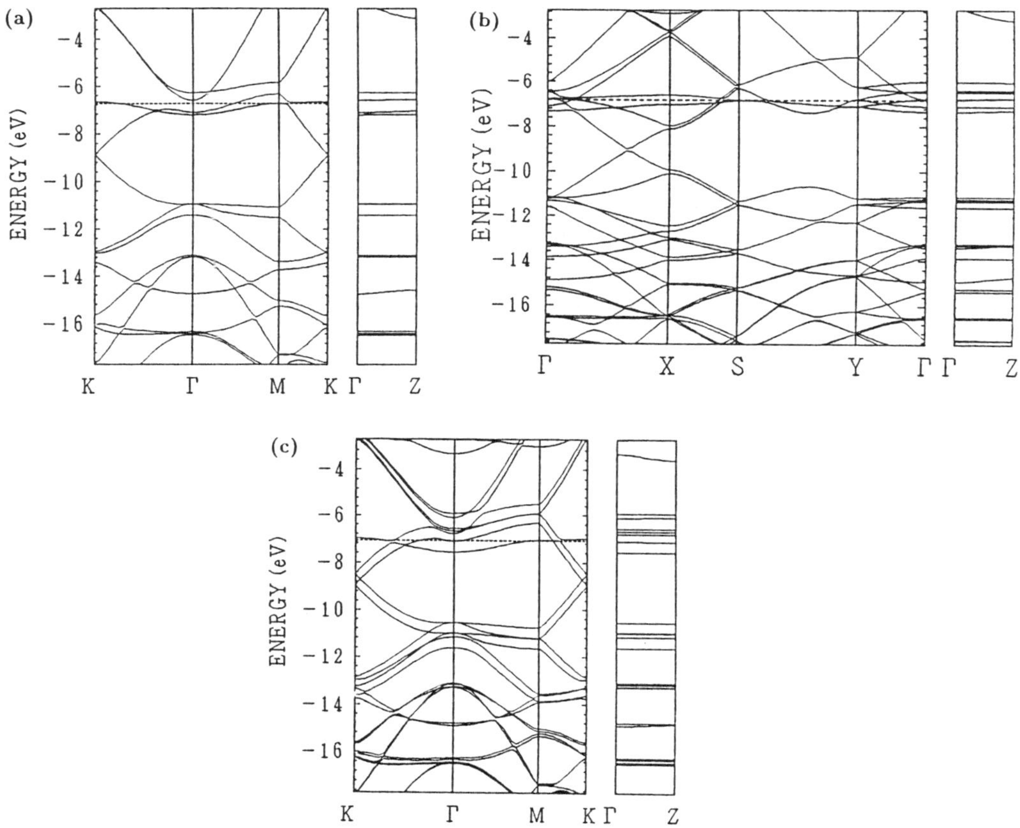

Fig. 3. Self-consistent band structures along several high-symmetry directions in the Brillouin zone: (a) $C_4KH_{0.5}$ [(2×2) $R(0°)$ model]; (b) $C_4KH_{0.5}$ [(2×$\sqrt{3}$) $R(0°, 30°)$ model]; (c) $C_8KH_{0.5}$ [(2×2) $R(0°)$ model]. Here, the dashed line represents the Fermi energy.

into the smaller BZ and $C_4KH_{0.5}$. (The band folding is described in detail in previous works [17,18].) The most remarkable feature in this band structure is the existence of two kinds of partially occupied conduction bands. These are directly shown to be a hydrogen 1s-like band and a graphite π^*-like band, from the amplitude of the wave function. That is, the hydrogen 1s-like band coexists with the graphite π^*-like band at the Fermi level in $C_4KH_{0.5}$ with the (2×2) $R(0°)$ triangular lattice model.

The H 1s-like band has very small c-axis dispersion, because of large separation (8.53 Å) between the nearest hydrogen layers. The in-plane band width of this 2D band is about 0.6 eV. Thus, the hydrogen in $C_4KH_{0.5}$ has a weakly metallic nature. Consequently, the 2D hydrogen 1s-like Fermi surface around the center of the BZ coexists with the 2D graphite π^*-like Fermi surface.

The calculated band structure of $C_4KH_{0.5}$ with the structure model (b) is shown in Figure 3(b). This band structure seems to be quite different from that of the model (a). This is due to different shape and size of the BZ. In fact, this band structure can be well identified with the hydrogen, potassium, π, π^*, and σ bands as in the model (a), using the rigid band model.

In the model (b), the unit cell contains two hydrogen atoms. Therefore, there are bonding and antibonding H 1*s*-like bands. The bonding H 1*s*-like band lies below the Fermi level and the antibonding H 1*s*-like band intersects the Fermi level. (These two bands are degenerate on the S–Y line.) Therefore, it is found that the H 1*s*- like band coexists with the π^*-like bands at the Fermi level in the model (b), as in the model (a).

The bonding and antibonding H 1*s*-like bands have very small *c*-axis dispersion because of large separation between the nearest hydrogen layers. The in-plane band width of the bonding and antibonding H 1*s* bands are about 0.5 eV and 0.2 eV, respectively. The sum of these two values is close to the band width of the H-1*s* band in $C_4KH_{0.5}$ with the model (a). Thus, the hydrogen in this crystal has also a weakly metallic nature.

Consequently, it is found that the band structures of $C_4KH_{0.5}$ with the $(2\times\sqrt{3})$ $R(0°, 30°)$ rectangular lattice model has the essentially same feature as that with the (2×2) $R(0°)$ hexagonal lattice model, though it seems to be different at sight. That is, the electronic state of $C_4KH_{0.5}$ is almost independent on the hydrogen structure.

The calculated band structure of the stage-2 compound $C_8KH_{0.5}$ with the structure model (a) is shown in Figure 3(c). This band structure is very similar to the result of the stage-1 compound $C_4KH_{0.5}$ with the (2×2) $R(0°)$ hexagonal lattice model. The only difference is the number of folded graphite bands. In $C_8KH_{0.5}$, a unit cell in our model structure contains two graphite layers. Therefore, there are two degenerate folded 2D graphite bands, if there is no interaction between the graphite layers and the potassium and hydrogen potentials are vanishingly small at the graphite layers. Due to the interaction between the graphite layers, two folded bands split off. The energy separations of the folded $\pi(\pi^*)$ bands are large (about 0.4 eV), because of a large overlap between the $\pi(\pi^*)$ orbitals of the adjacent graphite layers. On the contrary, the separations of the folded σ bands are small because of small overlap of σ orbitals.

Further, it is found that the hydrogen 1*s*-like band coexists with the graphite π^*-like bands at the Fermi level, as in the stage-1 $C_4KH_{0.5}$. This H 1*s*-like band has very small *c*-axis dispersion, because of large separation (11.88 Å) between the nearest hydrogen layers. The width of this 2D band is about 0.6 eV and therefore the hydrogen in $C_8KH_{0.5}$ has a weakly metallic

TABLE I
Total population and amount of charge redistribution.

Atom	Total population	Amount of charge redistribution
	(a) $C_4KH_{0.5}$ [(2 × 2) $R(0°)$ model]	
Potassium	18.54	+ 0.46
Hydrogen	1.29	– 0.29
	(b) $C_4KH_{0.5}$ [(2 × $\sqrt{3}$) $R(0°, 30°)$ model]	
Potassium	18.59	+ 0.41
Hydrogen	1.26	– 0.26
	(c) $C_8KH_{0.5}$ [(2 × 2) $R(0°)$ model]	
Potassium	18.54	+ 0.46
Hydrogen	1.37	– 0.37

nature. Thus, the band structure of the stage-2 compound $C_8KH_{0.5}$ has the essentially same feature as that of the stage-1 compound $C_4KH_{0.5}$, that is, the electronic state of $C_{4n}KH_x$ is almost independent of the stage number.

In the three band structures (Figure 3(a), (b), (c)), the π^* bands near the Fermi level split off around the Γ point and one of them goes down below the Fermi level. This separation cannot be understood in terms of the rigid band model. We explained this energy separation, using a simple tight binding model, in a previous work [17,18].

The total population of each atom in the crystal is obtained from the self-consistent calculation. These values for the potassium and hydrogen atoms in $C_4KH_{0.5}$ with the (2×2) $R(0°)$ and $(2\times\sqrt{3})$ $R(0°, 30°)$ structure, and $C_8KH_{0.5}$ with the (2×2) $R(0°)$ structure, are shown in Table I(a), (b), and (c), respectively. Further, we can define the amount of the charge redistribution of atoms as the difference between the above self-consistent population and the population of isolated (neutral) atoms [15,18]. These values are also shown in Table I. Here, a plus sign represents the charge transfer from the atom and a minus sign represents the transfer to the atom. These results indicate that the hydrogen atom acts an an acceptor, while the potassium atom acts as a donor in all cases. Though the K 4s-like bands lie above the Fermi level, the amount of the charge redistribution of potassium atom is not unity. The reason for this is that the K 4s components are mixed slightly into the wave functions of many graphite like and hydrogen-like states below the Fermi level.

Next, we will discuss several experiments based on our results. Enoki *et al.*

[8] measured electrical conductivity and thermoelectric power of the stage-1 and stage-2 compounds. They showed that a large electron and a small hole Fermi surfaces coexist and that potassium acts as a donor to graphite, while hydrogen plays the role of an acceptor. This result is consistent with our results.

Nomura *et al.* [4] carried out ^{1}H NMR experiment for $C_8KH_{2/3}$. Saito *et al.* [6] carried out ESR, ^{1}H NMR and ^{23}C NMR for $C_8KH_{0.61}$. These two groups observed metallic nature of hydrogen through the Korringa relation, T_1T = const, for the spin lattice relaxation T_1. These experiments suggested that the H 1s state forms a metallic band intersecting the Fermi level. This is consistent with our results. From the ESR experiment [6], the density of states at the Fermi level of carbon site in $C_8KH_{0.61}$ is obtained as 0.09 states/eV per carbon atom. Further, that of hydrogen site is obtained as 0.01 states/eV per proton from the ^{1}H NMR [6]. Therefore, it is expected that the total density of states of $C_8KH_{0.61}$ is larger than 1.45 states/eV per unit cell. Our calculated density of states of $C_8KH_{0.5}$ is 3.0 states/eV per unit cell. Very recently, Miyajima *et al.* [7] carried out proton NMR study on the stage-1 compound C_4KH_x. They observed metallic nature of hydrogen through the Korringa relation. This is also consistent with our results.

Enoki *et al.* [1] investigated the low-temperature specific heat of $C_8KH_{0.6485}$. In their measurements, the electronic specific heat coefficient γ is obtained as 0.46 mJ/K^2 mol; this value corresponds to the density of states at the Fermi level $N(E_F)$ = 3.2 states/eV per unit cell. Enoki *et al.* argue that the charge transfer to the hydrogen band is perfect, using the above value. However, our calculated result is 3.0 states/eV per unit cell for $C_8KH_{0.5}$, and this is in good agreement with the experimental result.

5. Summary

In this article, we calculated the band structures of the stage-1 compounds C_4KH_x and stage-2 compound C_8KH_x. Our band structure calculation for C_4KH_x was carried out for two assumed in-plane superlattice models of hydrogen structure: (2×2) $R(0°)$ model ($x = 0.5$); $(2 \times \sqrt{3})$ $R(0°, 30°)$ model ($x = 0.5$). The calculation for C_8KH_x was carried out for the assumed structure, (2×2) $R(0°)$ model ($x = 0.5$).

We compared the band structure of $C_4KH_{0.5}$ with the (2×2) $R(0°)$ structure with that of $C_4KH_{0.5}$ with the $(2 \times \sqrt{3})$ $R(0°, 30°)$ structure, and found that the electronic structure does not depend on the structure of the hydrogen layer. Further, we compared the band structure of $C_4KH_{0.5}$ with the (2×2) $R(0°)$ structure with that of $C_8KH_{0.5}$ with the (2×2) $R(0°)$ structure, and found that the electronic structure is also independent on the stage number.

In all cases, we estimated the amount of the charge transfer among the individual atoms and found that the hydrogen atom acts as an acceptor, while the potassium atom acts as a donor. The charge transfer to the hydrogen atom is imperfect and the H 1*s* state forms a weakly metallic band. That is, it is found that the H 1*s*-like bands coexists with the graphite π^*-like bands at the Fermi level in $C_{4n}KH_x$. These results indicate that $C_{4n}KH_x$ has a feature of the metallic hydrogen.

Acknowledgements

It is a great pleasure to dedicate the present article to Professor Hiroshi Kamimura in honor of his sixtieth birthday. We would like to thank Professor H. Suematsu, Dr. H. Hiramoto, and Dr. S. Miyajima for valuable discussions.

References

1. T. Enoki, M. Sano, and H. Inokuchi: 1985, *Phys. Rev. B* **32**, 2497.
2. T. Enoki, N.C. Yeh, S.T. Chen, and M.S. Dresselhaus: 1986, *Phys. Rev. B* **33**, 1292.
3. G.L. Doll, M.H. Yang, and P.C. Eklund: 1987, *Phys. Rev. B* **35**, 9790.
4. K. Nomura, T. Saito, K. Kume, and H. Suematsu: 1987, *Solid State Commun.* **63**, 1059.
5. S. Miyajima, T. Chiba, T. Enoki, H. Inokuchi, and M. Sano: 1988, *Phys. Rev. B* **37**, 3246.
6. T. Saito, K. Nomura, K. Mizoguchi, K. Kume, and H. Suematsu: 1989, *J. Phys. Soc. Jpn.* **58**, 269.
7. S. Miyajima, M. Kabasawa, T. Chiba, T. Enoki, and H. Inokuchi: 1990, *Phys. Rev. Lett.* **64**, 319.
8. T. Enoki, K. Imaeda, H. Inokuchi, and M. Sano: 1987, *Phys. Rev. B* **35**, 9399.
9. D. Guerard, N.E. Elalem, and C. Takoudjou: 1985, *Synth. Metals* **12**, 195.
10. W.A. Kamitakahara, G.L. Doll, and P.C. Eklund (unpublished).
11. D. Guerard, C. Takoudjou, and F. Rousseaux: 1983, *Synth. Metals* **7**, 43.
12. T. Trewern, R.K. Thomas, G. Naylor, and W. White: 1983, *J. Chem. Soc. Faraday Trans., Ser. 1* **78**, 2369.
13. P. Hohenberg and W. Kohn: 1964, *Phys. Rev. B* **136**, 864; W. Kohn and L.J. Sham: 1965, *Phys. Rev. B* **140**, 1133.
14. A. Zunger and A.J. Freeman: 1977, *Phys. Rev. B* **15**, 4716.
15. S. Mizuno, H. Hiramoto, and K. Nakao: 1987, *J. Phys. Soc. Jpn.* **56**, 4466.
16. L. Hedin and B.I. Lundqvist: 1971, *J. Phys. C* **4**, 2064.
17. S. Mizuno and K. Nakao: 1989, *Phys. Rev. B* **40**, 5771; S. Mizuno and K. Nakao: 1990, *Phys. Rev. B* **41**, 4938.
18. S. Mizuno: 1989, *Dr. Thesis*, University of Tsukuba.

ELECTRONIC BAND STRUCTURES AND MAGNETIC PROPERTIES OF Mn-INTERCALATED COMPOUNDS OF 2H-TYPE TRANSITION METAL DICHALCOGENIDES

NAOSHI SUZUKI, SHIGEKI TOMISHIMA[1] and KAZUKO MOTIZUKI
Department of Material Physics, Faculty of Engineering Science, Osaka University, Toyonaka 560, Japan.

Abstract. We present the results of LAPW band calculations carried out for the non-magnetic and ferromagnetic states of the transition metal intercalation compound $Mn_{1/4}TaS_2$. From an analysis of the partial density of states and the charge density map it is clear that the Mn $3d$ states hybridize strongly with the Ta $5d_{z^2}$ state which extends along the c-axis, but the hybridization with the S $3p$ states is rather small. It is suggested that the large spin-wave dispersion along the c-axis observed by inelastic neutron scattering measurements may be ascribed to the strong interaction between Mn atoms along the c-axis which can be caused via Ta atoms. The calculated total magnetic moment per Mn ($MnTa_4S_8$) is 4.35 μ_B, which is in good agreement with the observed value 3.9 ~ 4.2 μ_B.

1. Introduction

The layer compounds of transition metal dichalcogenides, 2H-TX_2 (T=Nb, Ta; X=S, Se), are interesting systems, both theoretically and experimentally. First they show rich variety of CDW transitions accompanying structural transformations [1], and secondly they become superconductors at low temperature. They are also important from the practical point of view because they can be host materials for intercalation compounds. In fact, many atoms and molecules can be intercalated into the so-called van der Waals gap layers between X–T–X sandwiches which consist of basic structures of $2H$–TX_2, and many efforts have been made to synthesize materials with new functions by intercalation. In particular, the intercalation of organic molecules has been studied extensively in order to explore the possibility of two-dimensional superconductors and also to obtain high T_c superconductors.

Intercalation of $3d$ transition metals has also been studied extensively and various interesting physical properties have been observed [2–7]. In the intercalation compounds M_xTX_2 with the concentration $x = 1/4$ or $1/3$ the transition metal atoms, M, form an ordered array and, depending on the intercalant species M and the concentration x, various types of magnetic ordering have been found at low temperatures. Among the many M_xTX_2 compounds, the magnetic properties of $Mn_{1/4}TaS_2$ have been investigated in detail. This intercalation compound shows a ferromagnetic transition at

[1] Present address: LSI Research & Development Laboratory, Mitsubishi Electric Corporation, 4–1 Mizuhara, Itami, Hyogo 664, Japan.

H. Aoki et al. (eds.), New Horizons in Low-Dimensional Electron Systems, 55–69.

$T_c \sim 80$ K and the saturation moment is estimated experimentally to be $3.9 \sim 4.2\ \mu_B$ per Mn atom [5,8], which is 20% smaller than that expected from a free Mn^{2+} ion. The effective moment estimated from the susceptibility data at high temperatures [8] is 4.8 μ_B per Mn, which is larger than the saturation moment. The measurement of magnetization density distribution by Parkin *et al.* [9] shows that the magnetization is principally localized on the Mn sites, but there is a non-negligible spin polarization at the Ta sites. Parkin *et al.* [9] concluded that the $3d$ electrons of Mn atoms are localized and that the reduction of the Mn moment and the conduction band spin polarization are caused by the $s - d$ mixing, which gives rise to the RKKY interaction.

Recently inelastic neutron scattering measurements have been performed on $Mn_{1/4}TaS_2$ [10]. According to the results, the spin-wave dispersion along the c-axis is much larger than that in the c-plane. In order that the observed spin-wave dispersion may be reproduced within the usual spin-wave theory of the localized spin system, one must assume that the exchange interaction along the c-axis is much larger than that in the c-plane. On the other hand, according to the theoretical estimation of the RKKY interaction on the basis of the realistic electronic band structure of TaS_2, the interplane RKKY interaction is smaller by one order of magnitude than the in-plane RKKY interaction [11]. These facts suggest that the Mn moment cannot be regarded as fully localized and that the Mn $3d$ electrons are more itinerant than was expected previously. Furthermore, with increasing temperature from zero Kelvin the magnetization decreases more rapidly than expected for the usual localized spin systems [8]. This fact may also indicate the itinerant character of the Mn $3d$ electrons. Therefore, knowledge about the electronic band structures of $Mn_{1/4}TaS_2$ is desired.

As for the electronic band structures of intercalation compounds of $2H$–TX_2, only an ASW band calculation has been carried out for the ferromagnetic state of $Fe_{1/3}TaS_2$ and $Mn_{1/3}TaS_2$ by Dijkstra *et al.* [12]. Their results show that the electronic structure of these intercalates can be understood only in first approximation in terms of the rigid band model. The purpose of the present paper is to calculate the electronic band structures of the non-magnetic and the ferromagnetic states of $Mn_{1/4}TaS_2$ in order to provide a basis for understanding its physical properties, particularly its magnetic properties, including spin-wave excitation.

Intercalation of transition metals into $1T$-type TiS_2 has been also studied extensively. Systematic band calculations carried out for M_xTiS_2 (M = Ti, V, Cr, Mn, Fe, Co, Ni) by Motizuki's group [13–19] have clarified the change of the electronic band structure of the host TiS_2 due to the intercalation as well as the bonding nature between the guest atoms and the host. Detailed results of experimental measurements have been reviewed by Inoue *et al.* [20].

2. Electronic Band Structures of Non-Magnetic $Mn_{1/4}TaS_2$

In the mother crystal 2H–TaS_2 (space group D^4_{6h}), the Ta ions form triangular lattices, each Ta ion being surrounded by six S ions coordinated prismatically. The unit cell contains two S–Ta–S sandwiches as shown in Figure 1(a) and thus two Ta and four S ions are contained in the primitive unit cell. The first Brillouin zone (BZ) is hexagonal prismatic, as shown in Figure 1(b). The lattice constants of TaS_2 are $a_0 = 3.315$ Å and $c_0 = 12.099$ Å [21]. The distances of S layers within one sandwich and between neighboring sandwiches are the same and 3.025 Å.

When Mn atoms are intercalated the basic structure of the mother crystal is unchanged and Mn ions occupy octahedral sites between the S–Ta–S sandwiches. In $Mn_{1/4}TaS_2$, the Mn ions form an ordered $2a_0 \times 2a_0$ superstructure in the c-plane. Figure 2 shows the crystal structure of $Mn_{1/4}TaS_2$ (space group D^4_{6h}). The unit cell contains twenty six ions in total (two Mn, eight Ta and sixteen S ions). There are two types of Ta and S ions (two Ta1 and six Ta2 atoms; four S1 and twelve S2 atoms); Ta2 atoms have no direct Mn neighbors, whereas each Ta1 is linearly coordinated by two Mn atoms along the c-axis; S1 atoms have no Mn neighbors while S2 has one Mn neighbor. The lattice constants of $Mn_{1/4}TaS_2$ are $a = 6.64$ Å and $c = 12.54$ Å [9]. The distances of S layers within one sandwich and between neighboring sandwiches are 3.02 Å and 3.25 Å, respectively [9].

We have performed band calculations for TaS_2 and $Mn_{1/4}TaS_2$ by using a self-consistent LAPW method of Takeda and Kübler [22]. The muffin-tin (MT) approximation has been used for the crystal potential and the exchange and correlation energies are treated in the local density approximation by Gunnarson and Lundqvist [23]. The starting charge density has been constructed by superposing the self-consistent charge densities of the neutral atoms:

$$\text{Ta:} \quad (5d)^3\,(6s)^2\,,$$

$$\text{S:} \quad (3s)^2\,(4p)^4\,,$$

$$\text{Mn:} \quad (3d)^5\,(4s)^2\,.$$

Inside the MT spheres, the angular momentum expansion is truncated at $\ell_{\max} = 7$ for the wave functions. The LAPW functions with the wave vector $|\mathbf{k}+\mathbf{G}| \le K_{\max} = 3.0 \times (2\pi/a_0)$ $[4.7 \times (2\pi/a)]$ have been used for TaS_2 [$Mn_{1/4}TaS_2$] in the expansion of the eigenfunctions. Here $\mathbf{k}$ is a wave vector in the Brillouin zone and $\mathbf{G}$ is a reciprocal lattice vector. The result shows that the number of basis functions used in the calculation is about 540 [1130] for TaS_2 [$Mn_{1/4}TaS_2$]. The radii of the MT spheres used for the calculations are 1.024 Å for Ta and 1.415 Å for S in the case of TaS_2, and 1.171 Å for Ta,

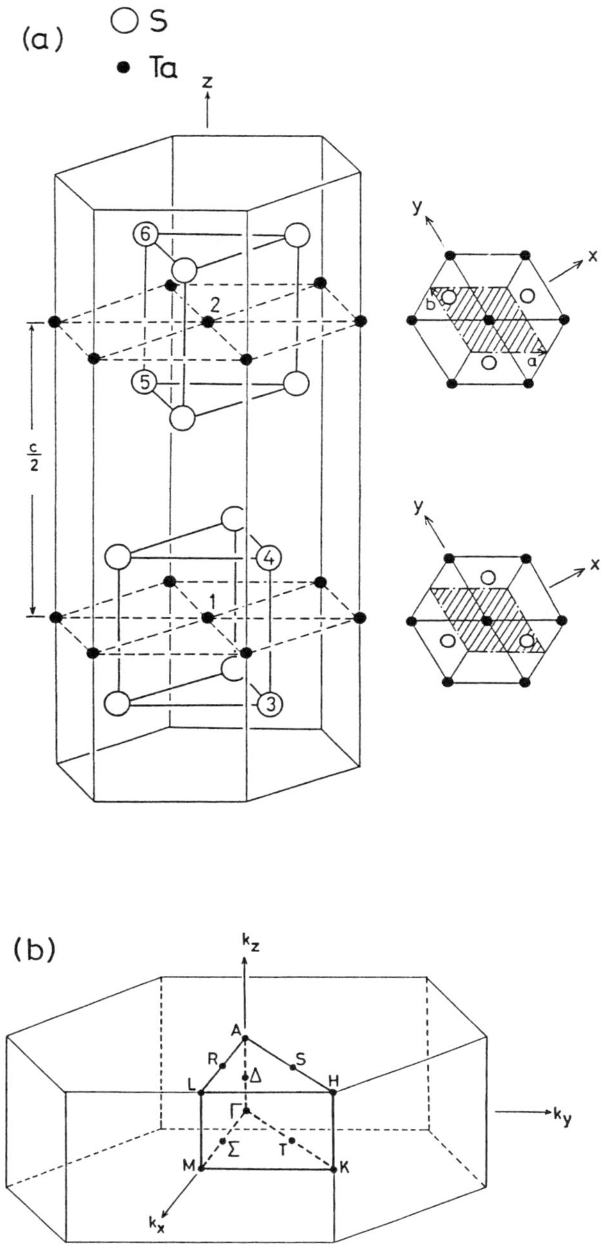

Fig. 1. (a) The crystal structure of TaS_2 and (b) the first Brillouin zone (BZ). The unit cell contains two Ta ions (1,2) and four S ions ($3 \sim 6$).

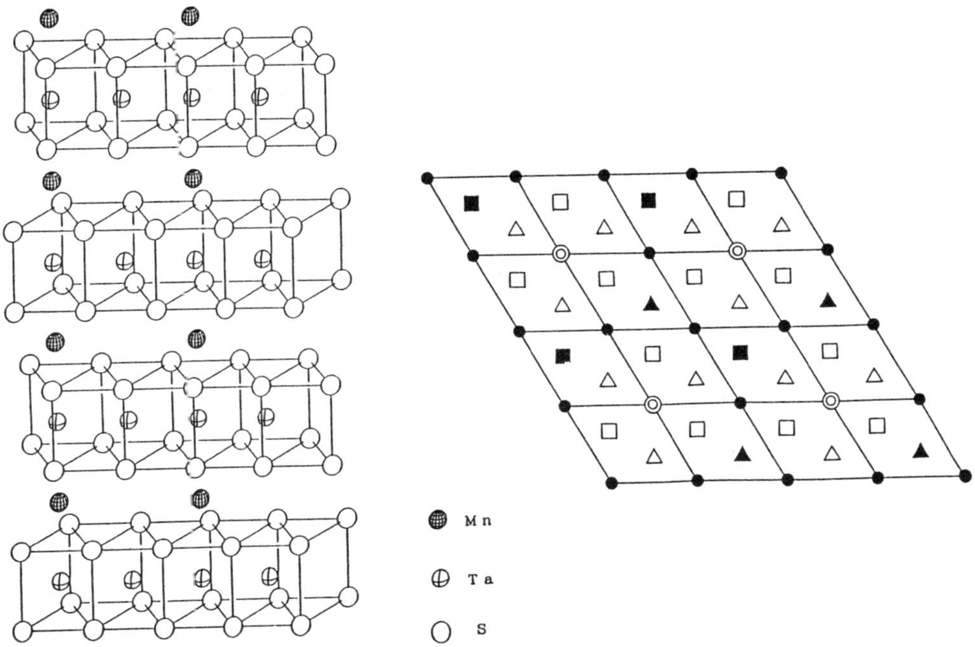

Fig. 2. The crystal structure of $Mn_{1/4}TaS_2$. The left-hand figure shows the relationship of the Mn atoms to the TaS_2 layers. The right-hand figure represents the (001) projection of the $Mn_{1/4}TaS_2$ unit cell: Mn = ◎ (z = 0, 1/2); Ta1 = ◎ (z = 1/4, 3/4); Ta2 = ● (z = 1/4, 3/4); S1 = ▲ (z = 1/8, 3/8) and ■ (z = 5/8, 7/8); S2 = △ (z = 1/8, 3/8) and □ (z = 5/8, 7/8). Ta2 atoms have no direct Mn neighbors, whereas each Ta1 is linearly coordinated by two Mn atoms along the c-axis. S1 atoms have no Mn neighbors while S2 has one Mn neighbor. The first BZ is the same as that of TaS_2.

1.274 Å for S and 1.244 Å for Mn in the case of $Mn_{1/4}TaS_2$. The potential and the charge density of the crystal have been determined self-consitently by using four **k** points, Γ, A, M and L.

We first calculated the electronic band structure of the host compound $2H$–TaS_2. Figures 3(a) and 3(b) show, respectively, the dispersion curves and the density of states obtained by our calculation. The lowest four bands consist of S $3s$ states and above a large gap there are twelve S $3p$ bands which hybridize with the Ta $5d$ states (p-d bonding bands). The next two bands, which slightly overlap in energy with the p-d bonding bands, consist primarily of the Ta $5d_{z^2}$ states and above a small gap there are eight Ta $5d$ bands which hybridize with the S $3p$ states (p-d anti-bonding bands). The Fermi level, E_F, lies in the Ta $5d_{z^2}$ bands and the density of states at E_F is $D(E_F)$ = 50.3 states / (Ryd · formula), which is in agreement with the

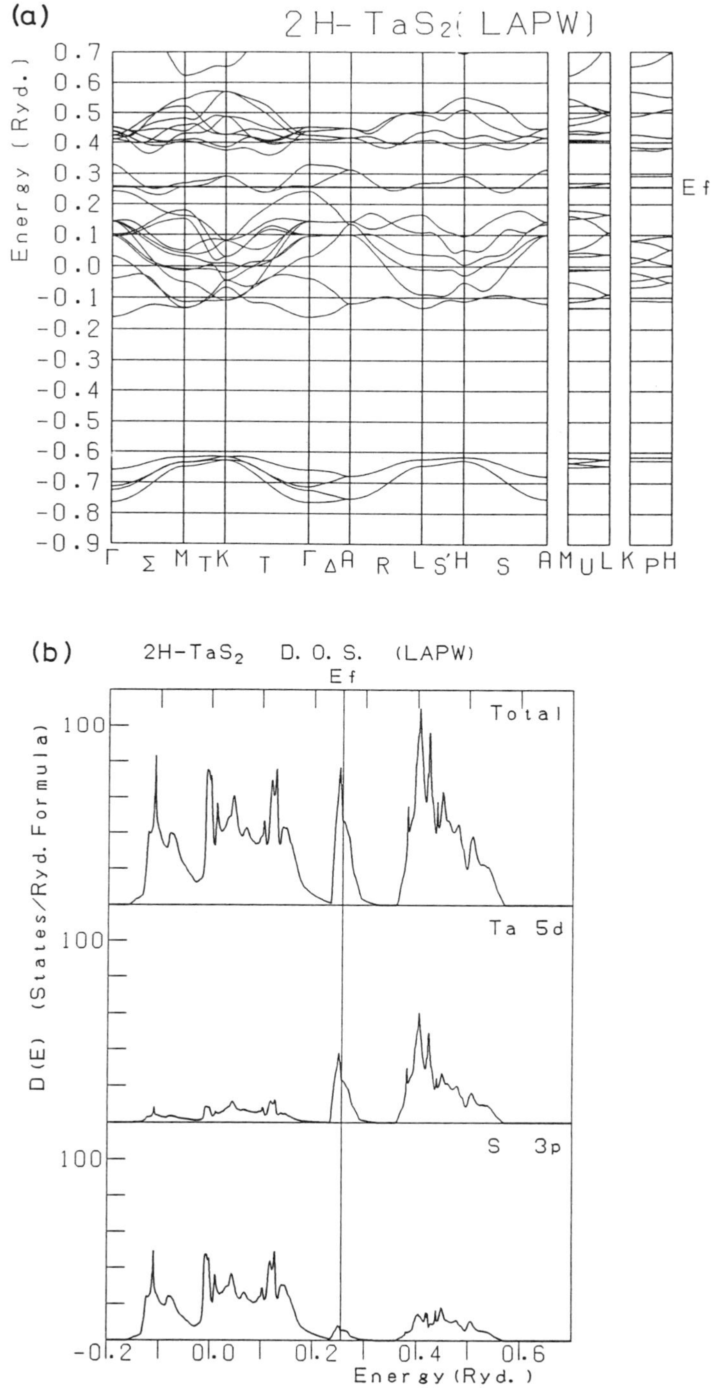

Fig. 3. (a) The dispersion curves and (b) the density of states of $2H$–TaS_2.

experimental value 49.0 states / (Ryd · formula) estimated from the specific heat measurements [24]. The width of the Ta $5d_{z^2}$ bands is 0.8 Ryd $\simeq$ 1.1 eV. It is noted here that the Ta $5d_{z^2}$ bands do not have a pure d_{z^2} character, but they hybridize with the d_{xy} and $d_{x^2-y^2}$ states, particularly away from the Γ point in the BZ.

The band structures of $2H$-TaS_2 obtained by us are similar to the results of the LMTO calculations by Guo and Liang [25] and of the ASW calculations by Dijkstra [26], but differ from those obtained by Mattheiss [27], by Kasowski [28] and by Wexler and Wooley [29]. The main difference is that in the former band structures the Ta $5d_{z^2}$ bands overlap slightly with the p-d bonding bands whereas in the latter there is a complete gap between the d_{z^2} bands and the p-d bonding bands. Experimentally, the photo-emission measurements [30] indicate an overlap of 0.1 eV between the p-d bonding and the d_{z^2} bands.

In order to see more clearly the character of the $5d_{z^2}$ bands we have also calculated the charge distribution in real space. Figure 4 shows the charge density map constructed by using the eigenfunctions of the $5d_{z^2}$, bands. Figure 4(a) represents the charge distribution in the c-plane of Ta atoms while Figure 4(b) depicts the charge distribution in the Ta atom plane parallel to the c-axis. The charge distribution around the Ta atoms extends mainly along the c-direction, but to some extent it also spreads in the c-plane, which results from the hybridization with the $d_{x^2-y^2}$ or d_{xy} states.

Now we turn to the electronic band structures of $Mn_{1/4}TaS_2$. Figure 5 shows the density of states obtained for $Mn_{1/4}TaS_2$. It can clearly be seen that the Mn $3d$ bands hybridize strongly with the Ta $5d_{z^2}$ bands, but the hybridization between the Mn $3d$ and the S $3p$ states is very small. This situation is quite different from the transition metal intercalation compounds of $1T$-TiS_2, in which the $3d$ states of the intercalated transition metal atoms hybridize well with the S $3p$ states as well as with the Ti $3d$ states [13–19]. The width of the Mn $3d$ bands is about 0.14 Ryd $\simeq$ 2 eV. The width of the $5d_{z^2}$ bands becomes wider, from 0.08 Ryd to 0.14 Ryd, upon Mn intercalation, which means that the effective transfer between Ta $5d_{z^2}$ orbitals is increased through hybridization with the Mn $3d$ states. The Fermi level lies in the Mn $3d$-Ta $5d_{z^2}$ hybridization bands and the value of $D(E_F)$ is 47.7 states / (Ryd · formula) which is a little bit smaller than that of the host TaS_2. Comparing Figure 5 with Figure 3(b) it can clearly be seen that the electronic structure near the Fermi level of host $2H$-TaS_2 is substantially modified and hence the rigid band model is not applicable to $Mn_{1/4}TaS_2$.

We have also calculated the charge distribution in $Mn_{1/4}TaS_2$ in order to clarify the bonding nature between the guest Mn atoms and the host. We have constructed the charge density map by using eigenstates in the energy range of 0.1 Ryd below the Fermi level. Figure 6(a) represents the charge distribution in the Mn atom layer and Figure 6(b) shows the charge

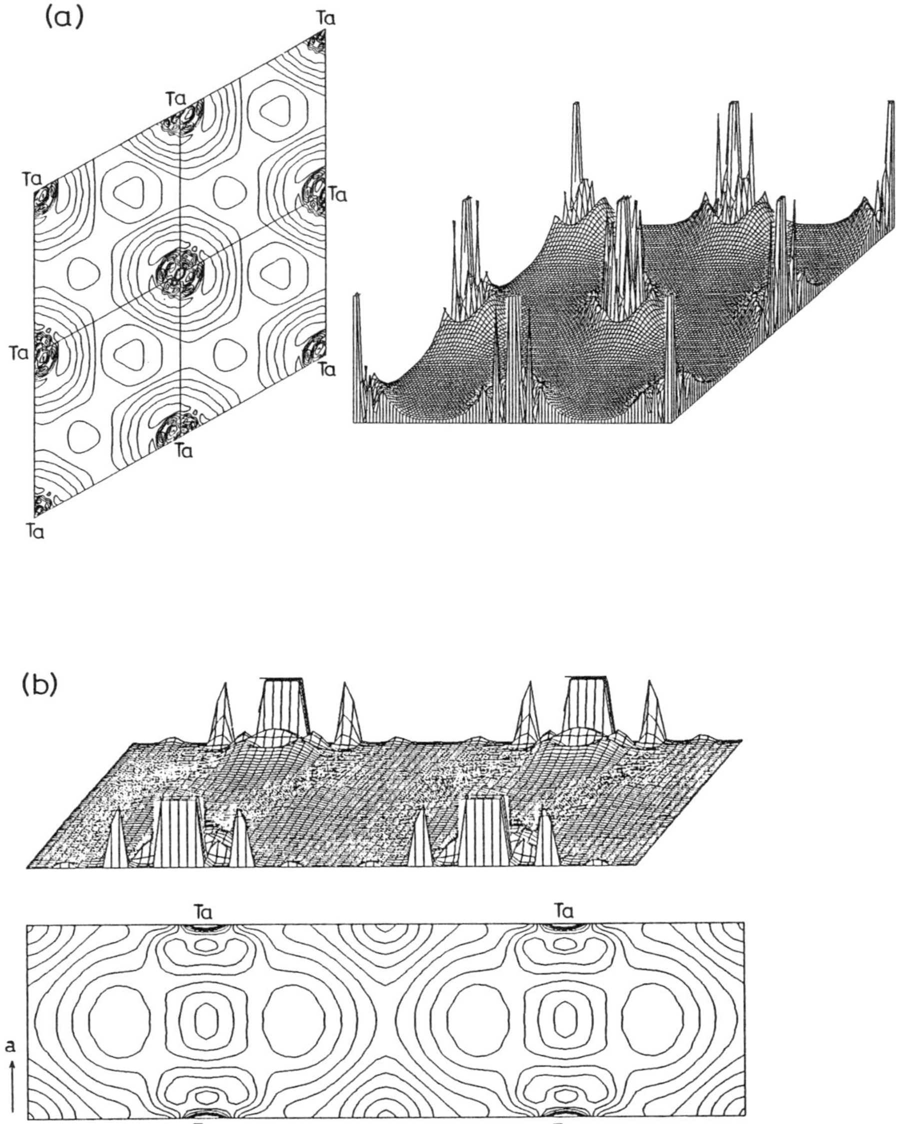

Fig. 4. The contour map and stereograph of the charge density of $2H$-TaS_2 constructed by using the eigenstates of the Ta $5d_{z^2}$ bands. (a) The charge distribution in the c-plane of Ta atoms, and (b) the charge distribution in the Ta atom plane parallel to the c-axis.

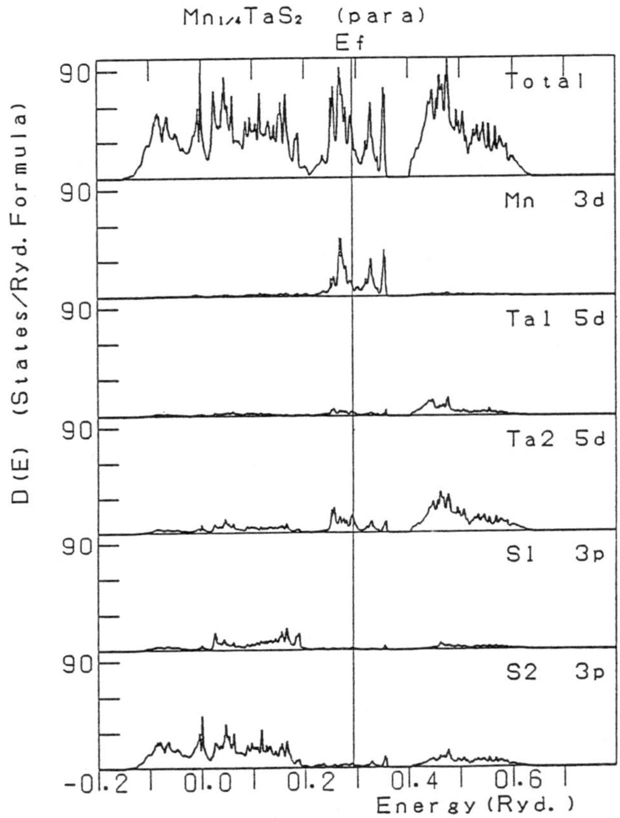

Fig. 5. The density of states of the non-magnetic bands of $Mn_{1/4}TaS_2$. The partial density of states is calculated in each muffin-tin sphere. The Ta $5d$ band of width 0.08 Ryd in the energy region near the Fermi level, E_F, is the so-called Ta $5d_{z^2}$ band. Note that the number of Ta2 atoms in the unit cell is three times that of Ta1 atoms.

distribution in the plane which contains both the Ta and Mn atoms and is parallel to the c-axis. From these two figures it is clear that the charge distribution around the Mn atoms extends towards the Ta atoms in the c-direction, but does not spread in the c-plane. This fact indicates that the intercalated Mn atoms destroy the two dimensional character of host $2H$-TaS_2 through hybridization between the Mn $3d$ and Ta $5d_{z^2}$ states. It is further noted that this strong hybridization between the Mn $3d$ and the Ta $5d_{z^2}$ states may cause a strong magnetic interaction between interplane Mn atoms, which can give rise to a large spin-wave dispersion along the c-direction.

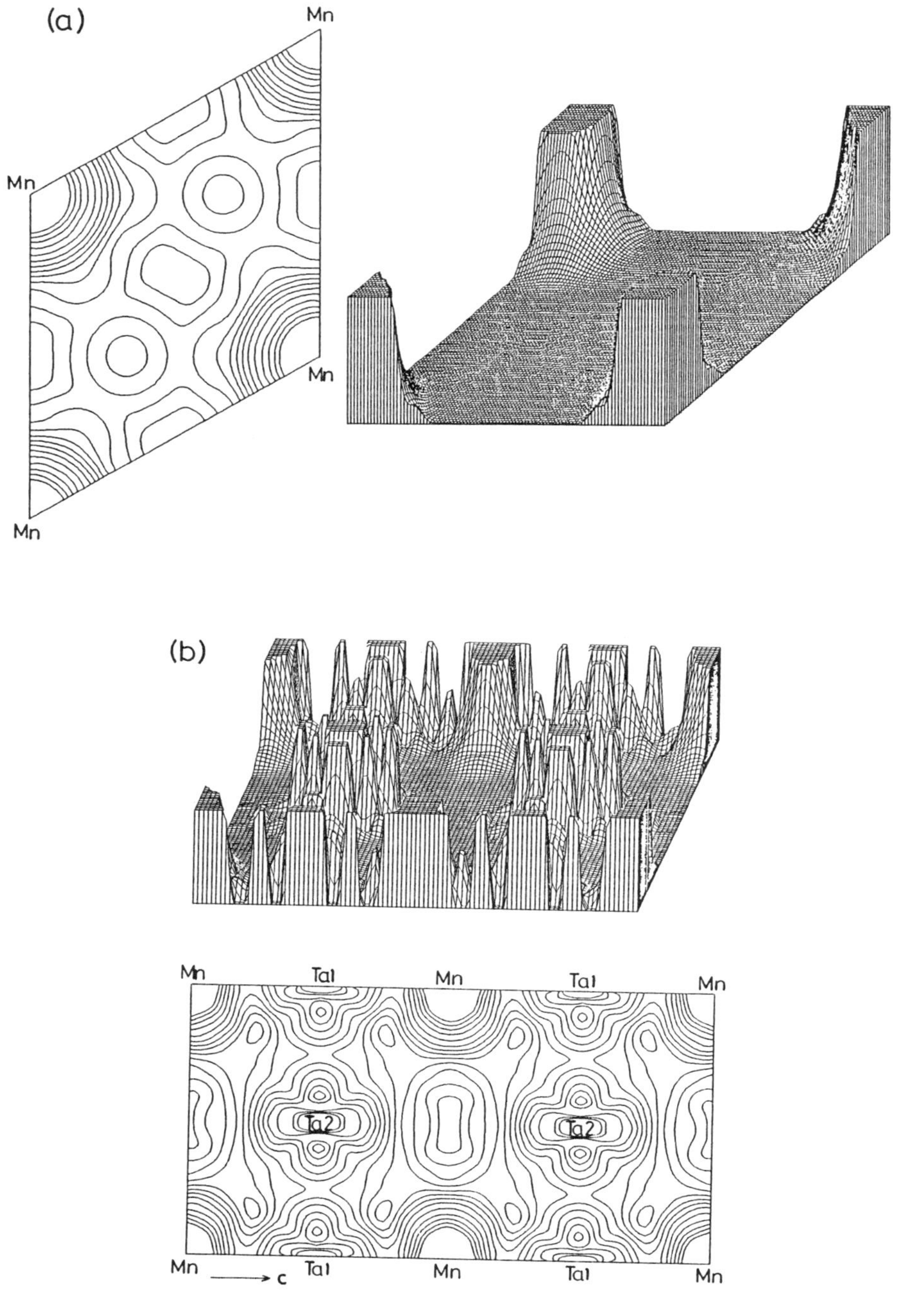

Fig. 6. The contour map and stereograph of the charge density of $Mn_{1/4}TaS_2$ constructed from eigenstates in the energy range of 0.1 Ryd below the Fermi level. (a) The charge distribution in the Mn atom layer, and (b) the charge distribution in the plane of Ta and Mn atoms which is parallel to the c-direction.

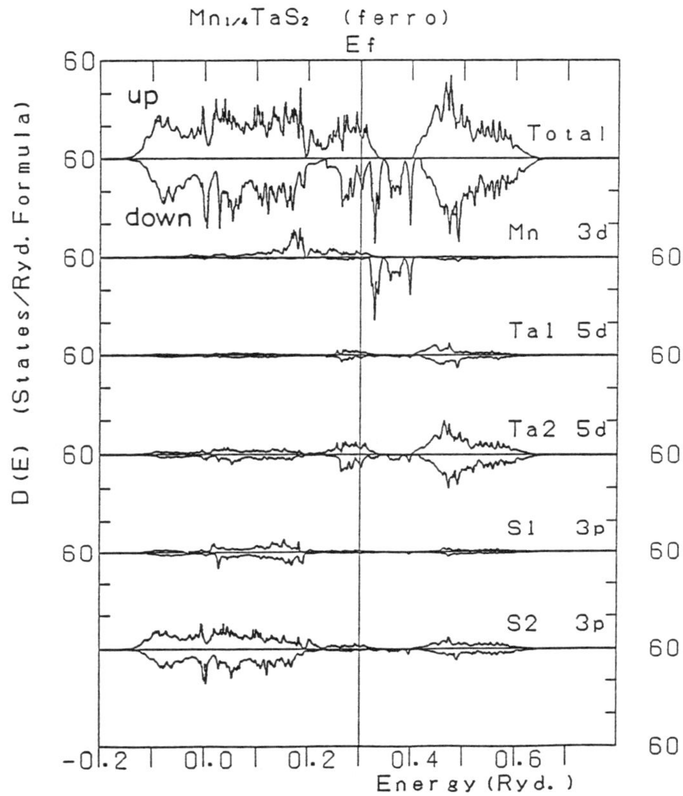

Fig. 7. The density of states of the ferromagnetic bands of $Mn_{1/4}TaS_2$.

3. Electronic Band Structures of Ferromagnetic $Mn_{1/4}TaS_2$

We have calculated also the spin polarized bands in the ferromagnetic state of $Mn_{1/4}TaS_2$ by using the LAPW method. The starting spin polarization was assumed as follows:

$$\text{Mn:} \quad (3d\uparrow)^3\,(3d\downarrow)^2 \,,$$

$$\text{Ta:} \quad (5d\uparrow)^2\,(5d\downarrow)^1 \,.$$

For ℓ_{max}, K_{max}, lattice constants and MT radii we have used the same values as for the non-magnetic states. The potentials and the charge density of the crystal have been determined self-consistently by using two **k** points, Γ and A. Figure 7 illustrates the total and partial densities of states for both the spin-up and spin-down bands. Large spin splitting occurs only for the Mn $3d$ bands. Through hybridization effects, however, magnetic moments parallel to that of Mn atoms are also induced on Ta and S atoms. The magnitude of the spin moment within each MT sphere is as follows (in unit of μ_B):

Mn: 3.61,

Ta1: 0.039, Ta2: 0.054,

S1: 0.012, S1: 0.016.

The total magnetic moment per Mn ($MnTa_4S_8$) is 4.35 μ_B, which is in good agreement with the observation 3.9 $\sim$ 4.2 μ_B [5,8]. This magnitude of the total moment is a little bit larger than that of $Mn_{1/3}TaS_2$ (4.18 μ_B) obtained by Dijkstra *et al.* [12]. According to the polarized neutron measurements [9] the magnitude of the induced moments on Ta atoms is a few percent of that of Mn atoms, but the direction of the Ta2 atom moment is opposite to that of the Mn atom. The moments of S atoms have been reported to be much smaller than those of Ta atoms [9]. The value of the total density of states at E_F has been found to be 27.32 states / (Ryd $\cdot$ formula), which is much smaller than that of the non-magnetic state. This large reduction of $D(E_F)$ arises primarily from the spin splitting of the Mn $3d$ bands.

In order to see spatial distribution of the spin moments we have constructed the contour map of the spin density, $\sigma(\mathbf{r}) \equiv \rho_\uparrow(\mathbf{r}) - \rho_\downarrow(\mathbf{r})$, by using the eigenstates below the Fermi level. The results are shown in Figure 8. In the Mn plane, as seen from Figure 8(a), a large positive spin density is principally confined within the Mn MT spheres (note that the MT radius of Mn atom is about $a/5$). Just outside the Mn MT spheres there is a region of negative spin polarization, and in the greater part of the interstitial region the spin density is weakly positive. In the Ta plane (see Figure 8(b)) the spin density is positive around the Ta atoms, but in a wider part of the interstitial region the spin density is polarized negatively.

4. Summary and Discussion

In summary, analysis of the partial density of states and the charge density map obtained for $Mn_{1/4}TaS_2$ has clarified that the Mn $3d$ states hybridize strongly with the Ta $5d_{z^2}$ states which form the conduction bands of the host compound TaS_2, and hence the rigid band model is not applicable to $Mn_{1/4}TaS_2$. Furthermore, the width of the Mn $3d$ bands is about 2 eV, which is not so narrow, and hence the Mn $3d$ electrons cannot be regarded as fully localized and should be treated rather as itinerant electrons. The magnitude of the spin moment obtained from the spin polarized band calculation for the ferromagnetic states is in good agreement with the observed saturation moment. Therefore, we may say that the ground state properties of $Mn_{1/4}TaS_2$ can be understood well on the basis of the band picture. At

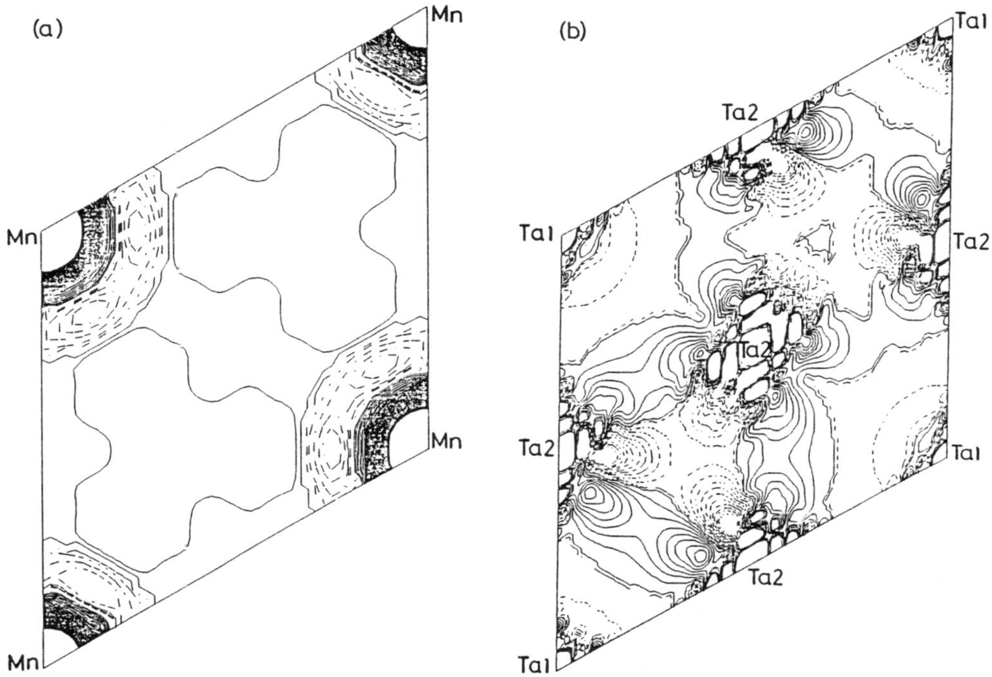

Fig. 8. The contour map of the spin density, $\sigma(\mathbf{r}) \equiv \rho_\uparrow(\mathbf{r}) - \rho_\downarrow(\mathbf{r})$, of ferromagnetic $Mn_{1/4}TaS_2$. (a) The spin density in the c-plane of Mn atoms, and (b) the spin density in the c-plane of Ta atoms. The full (dotted) curves represent the positively (negatively) spin-polarized region.

finite temperatures, however, it is naturally expected that the correlation effects, which are not fully taken into account in the band calculations with use of the local density approximation, will play an important role in magnetic properties. Application of such techniques as Moriya's spin fluctuation theory [31] to $Mn_{1/4}TaS_2$ with the use of the band structure obtained in the present paper will be an important and interesting problem.

It is also suggested that the hybridization between the Mn $3d$ and the Ta $5d_{z^2}$ states may cause a strong interaction between inter plane Mn atoms, which may give rise to a large spin wave dispersion along the c-axis, as observed by inelastic neutron scattering measurements [10]. Calculation of spin-wave excitations on the basis of the obtained ferromagnetic band structures will be one of the most interesting problems left to be solved in the future.

Acknowledgements

On of the authors (N.S.) would like to dedicate this paper with sincere gratitude to Professor Hiroshi Kamimura on the occasion of his retirement from University of Tokyo.

We express our sincere thanks to Professor A. Yanase and Dr. H. Harima of University of Osaka Prefecture for valuable comments on band calculations.

This work is supported by a Grant-in-Aid from the Ministry of Education, Science, and Culture of Japan. Financial support from the Kurata Foundation is also acknowledged greatly.

References

1. See for example, *Structural Phase Transitions in Layered Transition Metal Compounds*, K. Motizuki (ed.), D. Reidel., Dordrecht, 1986.
2. G.V. Subba Rao and M.W. Shafer, Intercalation in layered transition metal dichalcogenides, in *Intercalated Layered Materials*, by F.A. Lévy (ed.), D. Reidel, Dordrecht, 1979, p. 99.
3. A.R. Beal, The first row transition metal intercalation complexes of some metallic group VA transition metal dichalcogenides, in *Intercalated Layered Materials*, F.A. Lévy (ed.), D. Reidel, Dordrecht, 1979, p. 251.
4. R.H. Friend and A.D. Yoffe: 1987, *Adv. Phys.* **36**, 1.
5. S.S.P. Parkin and R.H. Friend: 1980, *Phil. Mag. B* **41**, 65.
6. S.S.P. Parkin and R.H. Friend: 1980, *Phil. Mag. B* **41**, 95.
7. S.S.P. Parkin and A.R. Beal: 1980, *Phil. Mag. B* **42**, 627.
8. Y. Onuki, K. Ina, T. Hirai and T. Komatubara: 1986, *J. Phys. Soc. Jpn.* **55**, 347.
9. S.S.P. Parkin, E.A. Marseglia and P.J. Brown: 1983, *J. Phys. C* **16**, 2749.
10. L.D. Cussen, E.A. Marseglia, D. McK. Paul and B.D. Rainford: 1989, *Physica B* **156** & **157**, 712.
11. N. Suzuki, Y. Yamazaki, T. Teshima and K. Motizuki: 1989, *Physica B* **156** & **157**, 286.
12. J. Dijkstra, P.J. Zijleman, C.F. van Bruggen, C. Haas and R.A. de Groot: 1989, *J. Phys. Condens. Matter* **1**, 6363.
13. T. Yamasaki, N. Suzuki and K. Motizuki: 1987, *J. Phys. C* **20**, 395.
14. N. Suzuki, T. Yamasaki and K. Motizuki: 1987, *J. Mag. Mag. Mater.* **70**, 64.
15. N. Suzuki, T. Yamasaki and K. Motizuki: 1988, *J. Phys. C* **21**, 6133.
16. N. Suzuki, T. Yamasaki and K. Motizuki: 1988, *J. Phys. (Paris)* **49** (C8), 201.
17. N. Suzuki, T. Yamasaki and K. Motizuki: 1989, *J. Phys. Soc. Jpn.* **58**, 3280.
18. N. Suzuki, T. Teshima and K. Motizuki, *Proc. 4th APPC* (Asia Pacific Physics Conference), by A.H. Ahn (ed.), World Scientific, Singapore, 1991, p. 448.
19. K. Motizuki and N. Suzuki, Intercalation compounds of transition metal dichalcogenides, in *Physics of New Materials – A Status Report*, by F.E. Fujita (ed.), Springer, Berlin, Heidelberg, to be published.
20. M. Inoue, H.P. Hughes and A.D. Yoffe: 1989, *Adv. Phys.* **38**, 565.
21. F. Jellinek: 1962, *J. Less-Common Met.* **4**, 9.
22. T. Takeda and J. Kübler: 1979, *J. Phys. F* **9**, 661.
23. O. Gunnarson and B.I. Lundqvist: 1976, *Phys. Rev. B* **13**, 4274.

24. F.J. DiSalvo, R. Schwall and T.H. Geballe: 1971, *Phys. Rev. Lett.* **27**, 310.
25. G.Y. Guo and W.Y. Liang: 1987, *J. Phys. C* **20**, 4315.
26. H. Dijkstra: 1988, *Ph.D. Thesis*, University of Groningen.
27. L.F. Mattheiss: 1973, *Phys. Rev. B* **8**, 3719.
28. R.V. Kasowski: 1973, *Phys. Rev. Lett.* **30**, 1175.
29. G. Wexler and A.M. Woolley: 1976, *J. Phys. C* **9**, 1185.
30. P.M. Williams and F.R. Shepherd: 1973, *J. Phys. C* **6**, L36.
31. T. Moriya: 1985, *Spin Fluctuations in Itinerant Electron Magnetism*, Springer, Berlin, Heidelberg.

SPIN-DEPENDENT LIGHT SCATTERING IN TWO-DIMENSIONAL MAGNETICALLY ORDERED SYSTEMS

M. BALKANSKI

Laboratoire de Physique des Solides, Université Pierre et Marie Curie, 4, Place Jussieu, Tour 13, 75252 Paris Cedex 05, France

and

Department of Physics, University of California, Irvine, Irvine, CA 92717, U.S.A.

Abstract. Experimental results of spin-dependent light scattering in 2D magnetically ordered systems are confronted with the theory of Suzuki and Kamimura.

The temperature dependence of the Raman integrated intensity of phonon bands which have become Raman active after magnetic ordering has occurred below the Néel temperature show the behaviour predicted by Suzuki and Kamimura's theory on spin-dependent phonon Raman scattering. The macroscopic aspect of this theory predicts a temperature dependence of the integrated Raman intensity which is proportional to the nearest neighbor spin correlation function $< S_0 \cdot S_1 >$. This prediction is confirmed experimentally.

Insertion of lithium into 2D magnetically ordered intercalation compounds like $FePS_3$ lowers the Néel temperature. This effect demands further theoretical attention.

1. Introduction

Recent interest in layered materials for their use as insertion cathodes in solid state batteries has motivated extensive work on the fundamental properties of the transition metal phosphorus trichalcogenides: MPS_3 with M = Fe, Ni or Zn and X = S or Se. For Fe and Ni compounds at temperatures below the Néel temperature T_N, a 2D magnetic ordering occurs in the layers and the phonon Raman scattering is strongly affected by the spin ordering. The theory of spin-dependent phonon Raman scattering has been developed by Hiroshi Kamimura and Naoshi Suzuki [1]. Our work has been strongly inspired by this theory and we shall see that the main features of the experimental results are accounted for by it. Some of our most recent results on the variation of T_N as a function of the concentration of lithium intercalated in the layered compound would need an extension of this theory in order to account for the effect of intercalation on the spin-spin coupling.

In the iron compounds, the successive layers in the c direction have a spin orientation into an antiferromagnetic order, forming a magnetic superstructure which is double with respect to the crystalline cell along the c axis. The low temperature Raman scattering spectra of $FePS_3$ [2] show new peaks attributed to inelastic scattering from zone boundary phonons [3,4]. The theory of Suzuki and Kamimura [1] implies that the microscopic spin-dependent electron–phonon interaction is due to the modulation of the

H. Aoki et al. (eds.), New Horizons in Low-Dimensional Electron Systems, 71–84.

magnetic exchange interaction by selected phonon modes. The experimental observation [2–4] is that spin order induced light scattering intensity is as high if not higher than that of the Raman active normal modes which we believe might be related to the intervention of the magnetostriction effect in the coupling of the radiation field with the phonon field.

The effect of magnetic ordering on the temperature dependence of the phonon Raman intensities was first observed [5] in the ferromagnetic semiconducting spinels $CdCr_2S_4$ and $CdCr_2Se_4$. It has been observed that certain lines of particular symmetry, due to Raman active phonon modes, exhibit an abrupt decrease in integrated intensity with increasing temperature up to the Curie temperature T_c. Resonance effects on electronic levels which split and shift at T_c have been observed [6] in $CdCr_2S_4$ and $CdCr_2Se_4$.

In another class of magnetic semiconductors, the europium chalcogenides, which crystallize in the NaCl structure, spin disorder in the paramagnetic phase is responsible for the appearance of a broad band in the Raman spectrum [7] related to the one phonon density of states. In the ordered magnetic phase of EuSe and EuTe, new strong narrow peaks appear in the Raman spectra with an intensity which depends on the temperature and applied magnetic field. These peaks are attributed [8] to zone boundary phonons activated by the folding of the Brillouin zone due to the magnetic superstructure.

Brillouin zone folding effects have also been observed [9] in the antiferromagnetic phase of VI_2 and attributed to the phonon modulation of the antiferromagnetic exchange interaction between the V^{2+} spins.

The effects of magnetic ordering on the phonon Raman scattering are different from those related to the direct excitation of spin systems such as the inelastic light scattering by spin waves [10]. Magnon Raman scattering has also been observed in the MPX_3 systems and will be reported elsewhere.

Here we will discuss recent light scattering experiments performed on MPS_3 with M = Ni or Fe and also the concentration dependence of T_N in lithium intercalated $FePS_3$.

When the Néel transition temperature is approached from above, entering from the paramagnetic into the antiferromagnetic phase, three effects are observed on the $FePS_3$ Raman spectrum. The first is the appearance of lines which are not observed in the paramagnetic phase. The second is a considerable enhancement of the integrated intensity and narrowing of certain lines which were weak and broad in the paramagnetic phase. The third is a relative enhancement and narrowing of lines present in the paramagnetic phase spectrum. In the antiferromagnetic phase of $FePS_3$ the magnetic unit cell along the c axis is double with respect to the chemical cell. The Raman intensity enhancement is related to the Brillouin zone folding.

In $NiPS_3$ the effects of magnetic ordering are quite different [11]. In the magnetically ordered phase a broad band appears in the Raman spectrum and the bands appearing on the top of the broad-band continuum show interference deformations [12]. Some Raman lines are slightly enhanced and narrowed.

$NiPS_3$ has been successfully used as cathodes in lithium solid state batteries based on the cell [13]

$$Li \mid B_2O_3 - 0.45\ Li_2O - 0.47\ Li_2SO_4 \mid NiPS_3$$

The energy density of this battery is 326 Wh/kg compared to a theoretical energy density of 1300 Wh/kg for an electron transfer of $3e^-/NiPS_3$ molecular unit.

The effect of lithium intercalation on the magnetic properties of $FePS_3$ has also been investigated [14]. Upon lithium intercalation the Néel temperature of $FePS_3$ decreased from 118 K for pure $FePS_3$ to 56 K for $Li_{0.77}FePS_3$.

In the second section of this paper we shall summarize the theoretical results of Suzuki and Kamimura of interest to the results presented in Section 3. Examining the results on different MPX_3 compounds we shall see the difference in behaviour according to structure and specific magnetic ordering. In Section 4 we will examine the effect of lithium intercalation on the spin coupling between spin-ordered layers and point to the lack of basic theoretical understanding of the effect of intercalation on the interlayer spin coupling and its relation to the variation of T_N as a function of the intercalant ion concentration.

2. Theory of Spin-Dependent Raman Scattering

Suzuki and Kamimura [1] have derived a general formulation of the phonon Raman scattering in magnetic crystals. In the transition metal compounds for which this theory would apply, the valence band is considered to be full while the conduction band is empty and the d electrons are well localized around the transition metal ion, which is also the case of MPX_3 compounds.

2.1. Macroscopic aspect of the theory

The phonon Raman scattering cross section is expressed in terms of the polarizability tensor. The total polarizability tensor $\tilde{\alpha}$ of a magnetic compound is written following the general form established by Moriya [15]. The polarizability tensor is then expended in powers of the ionic displacements $\{\delta\mathbf{R}\}$ from the equilibrium position.

The differential Raman scattering cross-section, namely, the ratio of energy flow of the scattered photon with the frequency $\omega + d\omega$ through a solid angle Ω along the direction $\mathbf{R}$ to that of the incident photon with frequency ω_0 is calculated from the well-known formula [16]

$$\frac{d\sigma}{d\Omega\, d\omega} = \left(\frac{\omega^4}{c^4}\right) \frac{1}{2\pi} \int_{-\infty}^{+\infty} < \mathbf{e}_0 \cdot \tilde{\alpha}_p(0)\, (1 - \mathbf{r} \cdot \mathbf{r})\, \tilde{\alpha}_p(t) \cdot \mathbf{e}_0 > \; e^{i(\omega-\omega_0)t}\, dt \tag{2.1}$$

where $\mathbf{e}_0$ is the unit vector parallel to the electric field vector $\mathbf{E}_0$ of the incident photon, $\mathbf{r} = \frac{\mathbf{R}}{R}$ and $< Q >$ means the statistical average.

Considering a particular Raman scattering configuration and retaining only Stokes components the leading terms in the integrated phonon Raman intensity $S(T)$ as a function of temperature is calculated as follows:

$$S(T) = (\mathbf{n}_{op} + 1)\, \{[R + M < \mathbf{S}_0 \cdot \mathbf{S}_1 > /S^2\, |^2 + |K|^2 \;\; < S_z >^2\} \tag{2.2}$$

where $< \mathbf{S}_0 \cdot \mathbf{S}_1 > /S^2$ is the reduced nearest neighbor spin correlation function. R is the spin independent phonon Raman scattering term, $M < \mathbf{S}_0 \cdot \mathbf{S}_1/S^2 >$ is the spin-dependent term and $|K|^2 < S_z >^2$ is the spin-dependent term associated with single ionic spins. This last term is shown to be considerably smaller than R and M and can therefore be omitted from the discussion of the case of MPX_3.

The reduced integrated intensity $I(T)$ can then be put in the form

$$I(T) = |R + M \; < \mathbf{S}_0 \cdot \mathbf{S}_1/S^2 > |^2\ . \tag{2.3}$$

This formula clearly shows that the temperature variation of the reduced integrated intensity $I(T)$ exhibits different kinds of patterns by the interference effect between the spin-independent term R and the spin-dependent term $M < \mathbf{S}_0 \cdot \mathbf{S}_1/S^2 >$.

Comparison with the experiment shows that for $FePS_3$ the integrated scattering intensity $I(T)$ follows the temperature variation in the spin correlation function. Figure 6 shows a plot of the temperature variations of the nearest neighbor spin correlation function obtained from the intensity temperature dependence of the phonon at 249 cm^{-1} in the light scattering spectrum of $FePS_3$ shown in Figure 2 compared to the square of the reduced magnetization as obtained by Kurosawa *et al.* [17].

2.2. Microscopic mechanisms for the spin-dependent phonon Raman scattering

The explicit expressions for the Raman tensor R, M and K are derived on the basis of two mechanisms which are supposed to be determinant for the appearance of the spin-dependent polarizability. The first one, type I, comes directly from the variation of d electron transfer energy with the relative displacement of the ions, and the other, type II, arises from that of non-diagonal exchange energy between magnetic ions.

The type I mechanism supposes that an incident photon $\hbar\omega_0$ is absorbed and a d electron in a transition metal ion i is virtually excited to the conduction band by the electric dipole transition. Then a d electron in the neighboring transition metal ion j transfers to the ith site via the intervening non-magnetic ions. Since the transfer energy t_{ij} depends on the position of the intervening non-magnetic ions as well as magnetic ions, it can be expanded in powers of the ionic displacements ($\delta\mathbf{R}$) from their equilibrium position. Then the linear terms in the expansion give the d-electron-phonon interaction. This is the intermediate state interaction. Finally the virtually excited electron in the conduction band recombines with the jth ion, returning the electron system to its ground state and a scattered photon is emitted. In this process electrons are exchanged between the ith and jth ions. Hence the polarizability tensor for the Raman scattering process can be expressed in the form of spin Hamiltonians and with it the contribution of the type I mechanism to the expressions of the spin-independent tensor R and the spin-dependent tensor M are calculated [1].

The type II mechanism is due to the variation of the non-diagonal exchange energy with the relative displacement of ions. Its contribution to the spin-dependent Raman tensor M_{II} is caculated using the methods given by Loudon [18] for two magnon Raman scattering in antiferromagnetics.

Estimates of the order of magnitude of the Raman tensors made by Suzuki and Kamimura show that $K << M_{\text{II}} < M_{\text{I}} < R$ and therefore it is plausible to neglect K and M_{II} in comparison with M_{I} and R. For any practical purposes, we should then retain the first mechanism and compare it with the experimental results.

It is clear from this discussion that, for materials in which the charge-transfer transitions from the d level to the conduction band correspond to the absorption band edge, the variation in de d electron transfer with lattice vibrations predominantly acts as the intermediate-state interaction of the Raman process. Study of the spin-dependent Raman scattering can supply information not only on the lattice dynamics but also on the electronic structure of magnetic crystals, because the intermediate-state interaction in the Raman process critically depends on the behaviour of the excited electronic states.

3. Comparison with Experiments

Raman spectra of a set of MPX_3 crystals recorded at different temperatures were published previously [3]. We shall recall here the most significant spectra in order to illustrate the theory of Suzuki and Kamimura.

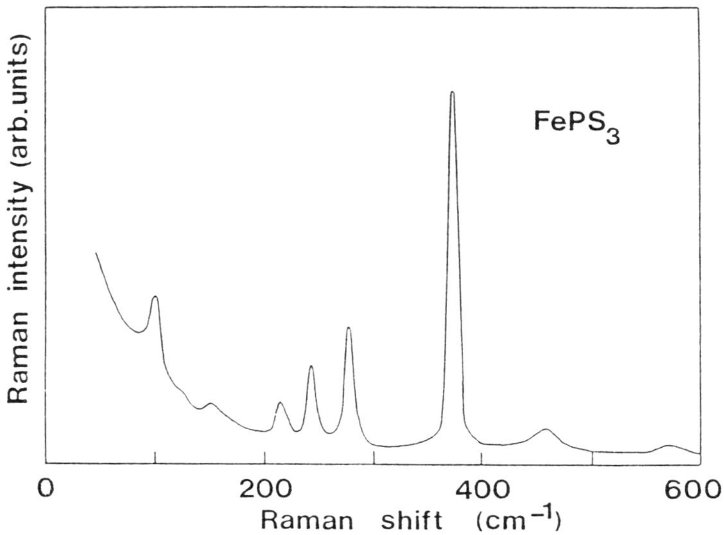

Fig. 1. Raman spectrum of $FePS_3$ recorded at room temperature, excited by the 488.0 nm line of an Ar^+ laser (Ref. [3]).

3.1. Ion phosphorus trisulfide $FePS_3$

Figure 1 shows the Raman spectrum of $FePS_3$ at room temperature. The identification of the Raman peaks is quite easy if one considers a simple pseudo-cell containing one P_2S_6 anion and two iron cations [19]. The P_2S_6 molecule belongs to the D_{3d} symmetry group and six Raman-active modes are expected. The three polarized modes at 243, 376 and 466 cm^{-1} are assigned to A_{1g}-type modes of the P_2S_6 ion. The three polarized modes at 215, 275 and 573 m^{-1} are assigned to E_g-type modes of the P_2S_6 ion. The modes in the low-frequency region below 150 cm^{-1} are related to the cation motion.

The evolution of the Raman spectra as a function of the temperature below the Néel temperature (118 K) is shown in Figure 2. At 104 K the changes in the Raman spectrum are already definite. Although relatively broad, the new bands are all present. Three new bands appear for $T \leq T_N$ at 88, 95 and 109 cm^{-1}. The enhancement and the narrowing of these bands are very important below the Néel temperature. The intensity of the 88 cm^{-1} peak increases abruptly below T_N without any frequency variation; the intensity of the peak at 249 cm^{-1} is significantly enhanced on passing through T_N, while the intensity of the peak at 380 cm^{-1} changes only slightly. Around T_N, the 88 cm^{-1} bandwidth undergoes significant broadening, whereas the 245 cm^{-1} bandwidth seems to exhibit a singularity [3].

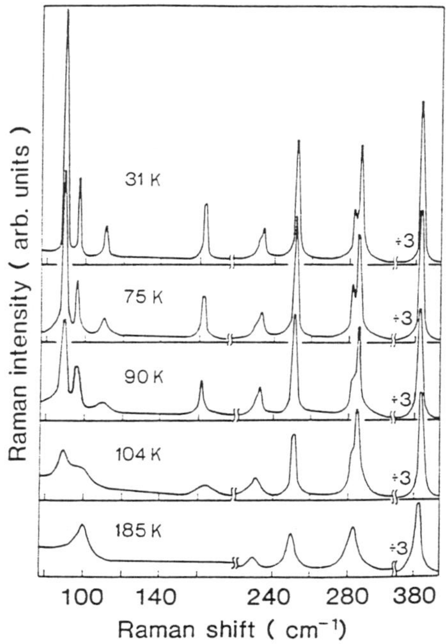

Fig. 2. Raman spectra of $FePS_3$ recorded at different temperatures using 488.0 nm laser excitation (Ref. [3]).

The main features of the temperature variation of the light scattering intensity may be summarized as follows:

(i) the appearance of peaks which are not observed at room temperature, e.g., peaks at 88, 95 and 109 cm^{-1};

(ii) the enhancement of the peak at 161 cm^{-1}, which can hardly be detected at room temperature;

(iii) the enhancement of peaks which have already been observed at room temperature (this is almost so for all the remaining peaks).

The drastic difference between the spectra at room temperature shown in Figure 1 and the spectra taken at liquid He temperature shown in Figure 3 can be accounted for by the difference between the crystallographic unit cell and the magnetic unit cell shown in Figure 4 which is a superstructure of the crystallographic unit cell. For $FePS_3$ the magnetic unit cell along the c axis is double with respect to the crystallographic unit cell. Elastic (Bragg) magnetic scattering induces the folding of the Brillouin zone. The phonon branches are folded along the Δ direction. As a consequence modes from the A point of the Brillouin zone are folded into the Γ point and can therefore be

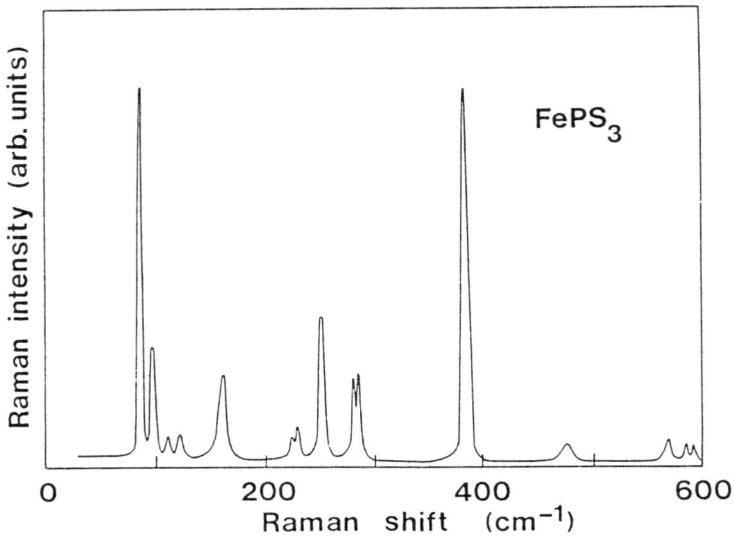

Fig. 3. Raman spectrum of $FePS_3$ at liquid-helium temperature, excited by 388.0 nm line of an Ar^+ laser (Ref. [3]).

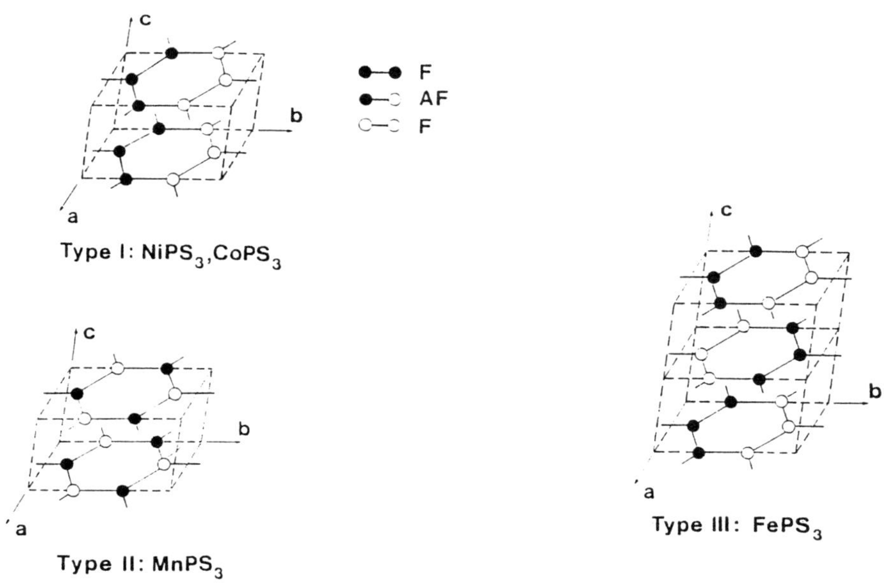

Fig. 4. Magnetic structure presented by the transition metal thiophosphates [13] (•, up atomic spins; o, down atomic spins) (Brec 1986); (a) type I, $NiPS_3$, $CoPS_3$; (b) type II, $MnPS_3$; (c) type III, $FePS_3$.

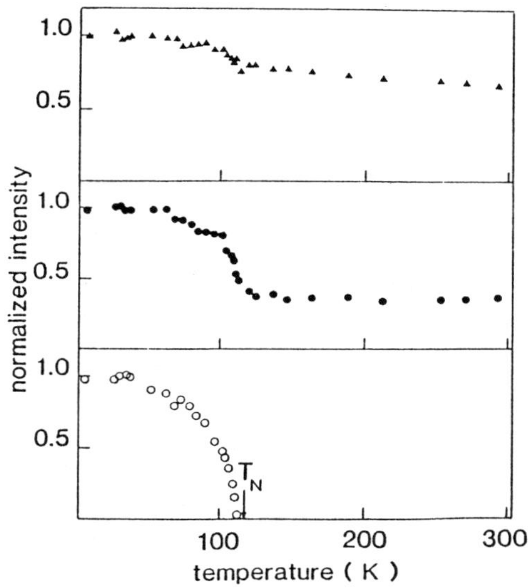

Fig. 5. Temperature dependence of the reduced integrated intensities ($I(T)/I(0)$) of the peaks at 88 cm^{-1} (○), 249 cm^{-1} (●) and 380 cm^{-1} (△) in the spectrum of $FePS_3$, excited by the 488.0 nm line. $I(0)$ is the value extrapolated at $T = 0$ K (Ref. [3]).

observed in Raman spectra. The three new peaks, at 88, 95 and 109 cm^{-1}, not observed in the paramagnetic phase but appearing in the magnetically ordered phase below the Néel temperature, correspond to folded acoustic branches. The peak at 161 cm^{-1}, infrared active at Γ, can hardly be seen at room temperature, but appears strongly enhanced in the Raman spectrum below T_N.

The spin dependent Raman scattering is demonstrated through the temperature variation of the integrated Raman intensity as shown in Figure 5. Then second term in Equation (2.3) is proportional to the nearest-neighbor spin correlation function. In Figure 6 we compare the reduced integrated intensities $I(T)/I(0)$ of the peak at 249 cm^{-1} and the square of the reduced magnetic moment. This is a clear demonstration of the predictions of the theory of Suzuki and Kamimura.

3.2. Nickel phosphorus trisulfide $NiPS_3$

Figure 7 shows the spectra of $NiPS_3$ recorded at room temperature and liquid-nitrogen temperature. Here again we observe a drastic difference between the two spectra, but in this case the characteristics are quite different. Peak enhancements are observed [11] below T_N, but the most significant fea-

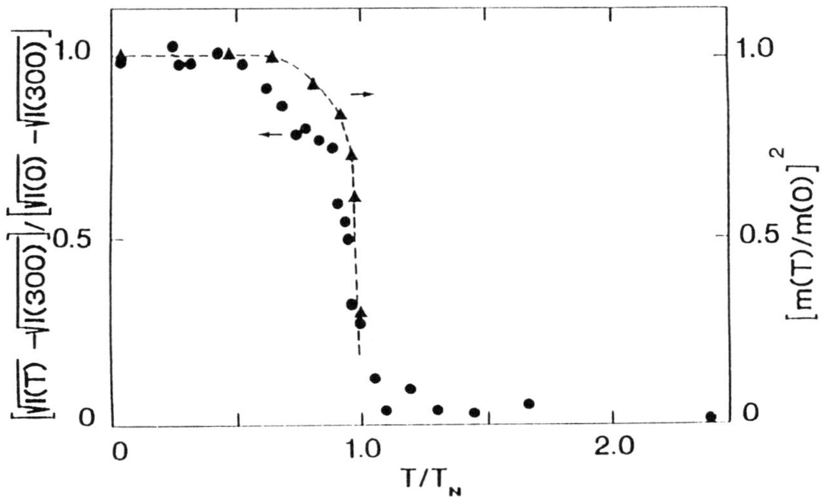

Fig. 6. Reduced nearest-neighbor spin correlation function obtained from the temperature dependence of the intensity of the phonon at 249 cm^{-1} (•) in the spectrum of $FePS_3$ (see Figure 5). The square of the reduced magnetisation [17] $m(T)/m(0)$ (Δ) is reported for comparison. The broken line is shown as a visual aid.

ture is the appearance of a very broad band centered around 500 cm^{-1} on which are superimposed phonon Raman peaks whose lineshapes are now strongly deformed and asymmetric.

4. Effect of Lithium Intercalation in the Antiferromagnetic $FePS_3$

Light scattering spectra [14] of lithium intercalated $FePS_3$ at different temperatures are shown in Figure 8. After intercalation the Raman intensity of the original Raman-active phonons decreases drastically and the background increases. The background also increases as the frequency shift increases. The electronic band structure might be modified upon intercalation as it has been shown [20] in the case of other intercalation compounds with the appearance of luminescent centers. The broad background would then be due to luminescence.

We can consider three possible causes for the decrease of the Néel temperature after intercalation with Li. First the intercalation of Li in the van der Waals gaps decreases the interlayer exchange interaction. Kurosawa *et*

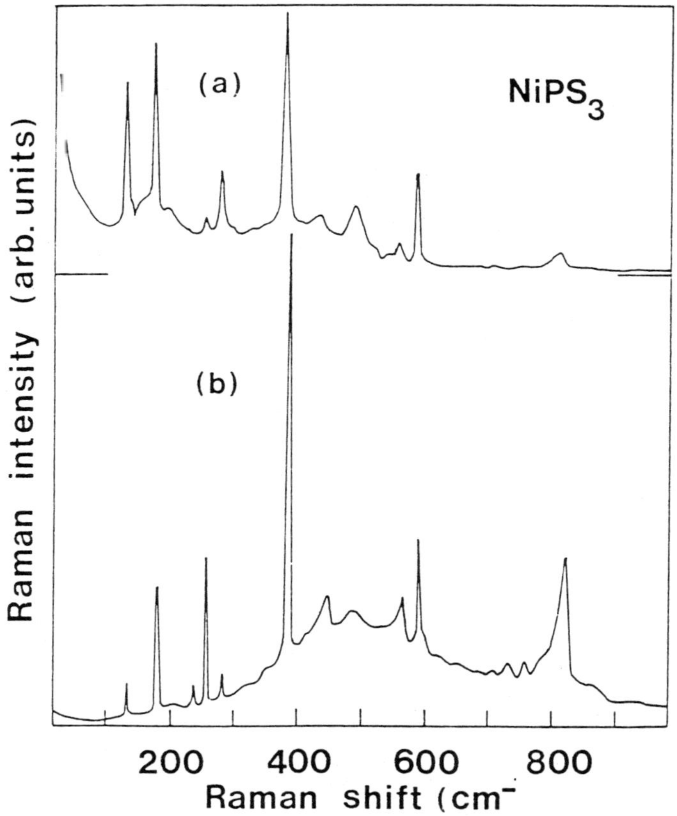

Fig. 7. Raman spectra of $NiPS_3$ [4] recorded (a) at room temperature and (b) at liquid-nitrogen temperature with 514.5 nm line of an Ar^+ laser as the exciting source.

al. [17] reported that the interlayer exchange interaction cannot be neglected in $FePS_3$, in particular when compared with the third nearest-neighbor intralayer exchange interaction, even though it is very weak. However the only decrease of interlayer exchange interaction is not enough to decrease the Néel temperature from 118 K to about 56 K. Secondly the defects of Li in the host matrix decreases the intralayer exchange interactions and induces magnetic disordering. Since the Raman intensity by the vibrations of defect in the host matrix is strong, considerable quantity of Li atoms probably was inserted into the host matrix. The random distribution of these defects would give rise to magnetic disordering. But it is not easy to estimate the quantitative effect of the defects on the Néel temperature. Thirdly we must consider the increase of free carriers. It is known that the charge transfer from the

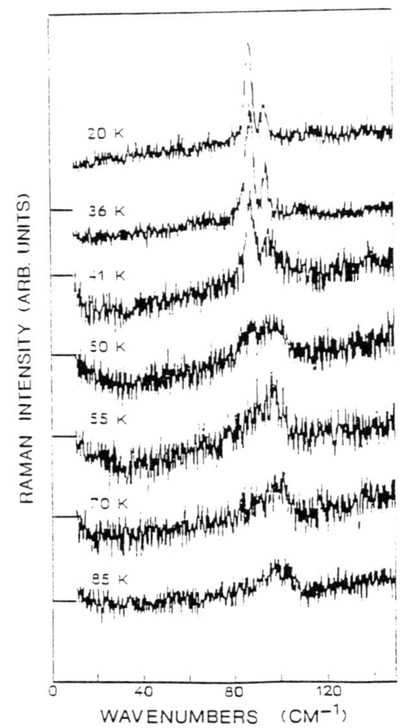

Fig. 8. Temperature dependence of the low-frequency Raman spectrum in Li_xFePS_3. This sample was intercalated by n-butyllithium method.

intercalants to the host lattice occurs. Brec *et al.* [22] observed a drastic increase of the conductivity in $FePS_3$ after intercalation with Li which is due to the increase of free carrier concentration upon intercalation. Mauger and coworkers [23] calculated the indirect exchange due to the conduction electrons in heavily doped europium chalcogenides by taking into account the finite band width of the conduction band and carrier statistics. Their calculation well explains the influence of free carrier concentration on the paramagnetic Curie temperature and the ordering temperature. We think that the indirect exchange due to the conduction electrons is also important in the present case.

5. Conclusion

The temperature dependence of the Raman integrated intensity of phonon bands which have become Raman active after magnetic ordering has occurred below the Néel temperature show the behaviour predicted by the Suzuki and Kamimura theory on spin dependent phonon Raman scattering.

The macroscopic aspect of this theory predicts a temperature dependence of the integrated Raman intensity which is proportional to the nearest neighbor spin correlation function $< S_0 \cdot S_1 >$.

Closer examination of the experimental results shows that the temperature dependence of the magnetic moment has a more squared shape, departing from the squares of Brillouin functions, which points to the possibility of attributing the enhancement of Raman intensity in the magnetically ordered phase to the magnetostriction effect.

Insertion of lithium in the 2D magnetically ordered intercalation compounds like $FePS_3$ lowers the Néel temperature. This effect requires further theoretical attention.

References

1. N. Suzuki and H. Kamimura: 1972, *Solid State Commun.* **11**, 1603; 1973, *J. Phys. Soc. Jpn.* **35**, 985.
2. M. Scagliotti, M. Jouanne, M. Balkanski and G. Ouvrard: 1985, *Solid State Commun.* **54**, 291.
3. M. Scagliotti, M. Jouanne, M. Balkanski, G. Ouvrard and G. Benedek: 1987, *Phys. Rev. B* **35**, 7097.
4. M. Balkanski, M. Jouanne, G. Ouvrard and M. Scagliotti: 1987, *J. Phys. C.: Solid State Phys.* **20**, 4397.
5. E.F. Steigmeier and G. Harbeke: 1970, *Phys. Kondens. Materie* **12**, 1.
6. N. Koshizuki, Y. Yokoyama and T. Tsushima: 1977, *Solid State Commun.* **23**, 967; 1977, *Physica B* **89**, 214; N. Koshizuka, S. Ushioda and T. Tsushima: 1980, *Phys. Rev. B* **21**, 1316.
7. R. Merlin, R. Zeyher and G. Güntherodt: 1977, *Phys. Rev. Lett.* **39**, 1215.
8. R.P. Silberstein, L.E. Schurek, V.J. Tekippe, M.S. Dresselhaus and R.L. Agrawal: 1976, *Solid State Commun.* **18**, 1173; G. Güntherodt: 1979, *J. Magn. Mater.* **11**, 394.
9. G Güntherodt, W. Banhofer and G. Benedek: 1979, *Phys. Rev. Lett.* **43**, 1427.
10. P.A. Fleury S.P.S. Porto, L. Cheesman and H.J. Guggenheim: 1966, *Phys. Rev. Lett.* **17**, 84.
11. M. Balkanski, M. Jouanne and M. Scagliotti: 1987, *J. Pure Appl. Chem.* **59**, 1247.
12. M. Balkanski, K.P. Jain, R. Beserman and M. Jouanne: 1975, *Phys. Rev. B* **12**, 4328.
13. I. Samaras, S.I. Saikh, C. Julien and M. Balkanski: 1989, *Mat. Sc. Engin. B* **3**, 209.
14. T. Sekine, M. Jouanne, C. Julien and M. Balkanski: 1989, *Mat. Sc. Engin. B* **3**, 91.
15. T. Moriya: 1967, *J. Phys. Soc. Jpn.* **23**, 490.
16. See, for example: 1986, *Many Body Aspects of Solid State Spectroscopy* by R.F. Wallis and M. Balkanski, North Holland.
17. K. Kurosawa, S. Saito and Y. Yamaguchi: 1983, *J. Phys. Soc. Jpn.* **52**, 3919.
18. R. Loudon: 1964, *Adv. Physics* **13**, 423.
19. C. Sourisseau, J.P. Forgerit and Y. Mathey: 1983, *J. Solid State Chem.* **49**, 134.
20. R. Brec: 1986, *Solid State Ionics* **22**, 3.

21. P. da Costa, R.F. Wallis and M. Balkanski: 1991, *Phys. Rev. B* **43**, 7066.
22. R. Brec, M. Schleich, G. Ouvrard, L. Louisy and J. Rouxel: 1979, *Inorg. Chem.* **18**, 1814.
23. A. Mauger: 1977, *Phys. Stat. Sol. (b)* **84**, 761; A. Mauger, C. Godart, M. Escome, J.C. Achard and J.P. Desfours: 1978, *J. Phys. (Paris)* **39**, 1125; A. Mauger and C. Godart: 1986, *Phys. Rep.* **141**, 51.

A NEW METHOD TO CREATE ARTIFICIALLY STACKED LAYERED MATERIALS: VAN DER WAALS EPITAXY

ATSUSHI KOMA
Department of Chemistry, University of Tokyo, Bunkyo-ku, Tokyo 113, Japan.

Abstract. The lattice matching condition usually encountered in heteroepitaxial growth has been a bottleneck in creating artificially stacked layered structures. It has been proved that this difficulty can be overcome when the interface between constituent materials has a van der Waals nature and forms no direct chemical bonds. Layered transition metal dichalcogenides are typical materials having that nature, and a variety of heterostructures can be grown by using them. This kind of approach has also been shown to be applicable to heteroepitaxial growth between such quasi-one dimensional materials as tellurium and selenium, onto dangling-bond terminated three dimensional material substrates, and to organic materials which form van der Waals-type crystals.

1. Introduction

Molecular beam epitaxy has opened a new way to create such artificially stacked layered materials as semiconductor superlattices. But so far only very limited combinations of materials can be used for that purpose. This stems from the fact that there are usually dangling bonds on the surface of a substrate material and that those dangling bonds cannot be connected to the atoms of grown materials without good lattice matching between the substrate and the grown materials (see Figure 1a). The lattice matching condition, however, has been found to be greatly relaxed when the heteroepitaxial growth proceeds with van der Waals interaction. A typical example is the growth of a layered material onto a cleaved face of the other layered material having no dangling bonds (see Figure 1b). We have called this type of epitaxy 'van der Waals epitaxy' [1–4]. It has been proved that an ultrathin film with its own lattice constant can be grown by the method of van der Waals epitaxy, even under the existence of a lattice mismatch as large as 50%. It has also been shown that a very abrupt interface with small amounts of defects can be fabricated by van der Waals epitaxy because of the nonexistence of the dangling bonds [5]. As will be shown in the following, the idea of van der Waals epitaxy can be applied to rather a wide choice of materials, indicating its usefulness for creating various kinds of artificially stacked layered materials.

2. Materials to which van der Waals Epitaxy is Applicable

Table I shows a variety of materials which we have shown may be used in van der Waals epitaxy. The materials extend from quasi-one dimensional ones to three dimensional ones.

H. Aoki et al. (eds.), New Horizons in Low-Dimensional Electron Systems, 85–95.

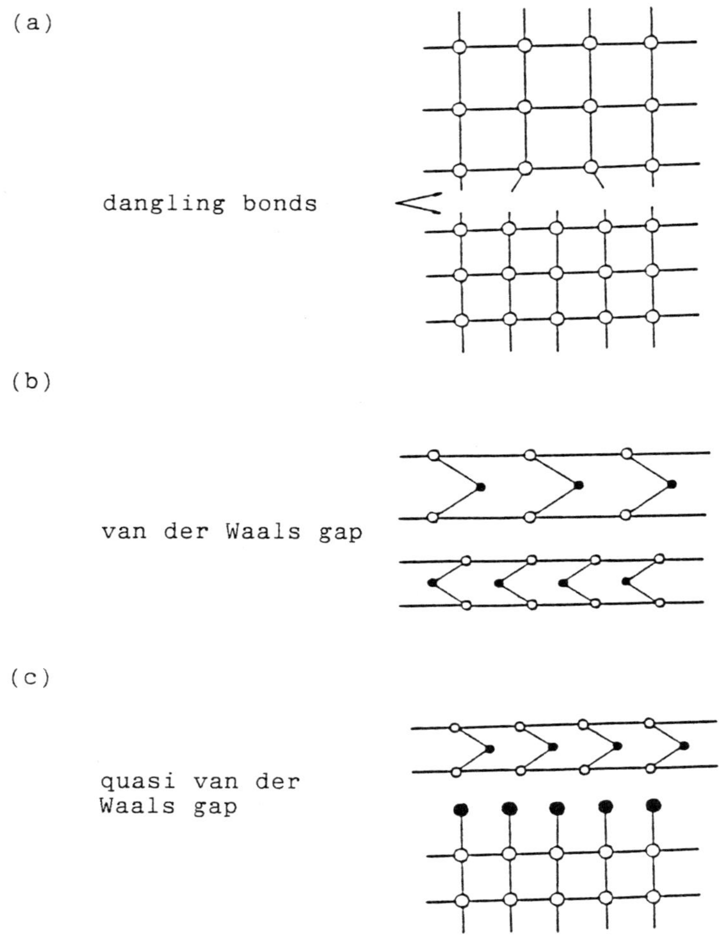

Fig. 1. Interfaces connected by (a) active bonds, (b) van der Waals gap, and (c) quasi-van der Waals gap.

2.1. Quasi-one-dimensional materials

Selenium and tellurium are Group VI element semiconductors and have peculiar crystal structures consisting of atoms in spiral chains as is seen in Figure 2. The atoms in a chain are bound to each other with strong covalent forces, whereas the chains are held together only via weak van der Waals forces. Thus they are considered as quasi-one-dimensional materials in the sense that the atoms are strongly bound to each other only in one direction. The crystal can be cleaved easily along the chains without producing dangling bonds on their surfaces, thus admitting the possibility of

TABLE I
Materials grown with van der Waals epitaxy. Here TX_2 denotes a transition metal dichalcogenide.

Material Group	Materials Grown with VdWe	References
Quasi-2D	Se/Te Te/Se/Te	[6]
Quasi-2D	TX_2/TX_2 TX_2/SnS_2 TX_2/Mica	[1–4] [7] [8,9]
Quasi-2D on 3D	TX_2/S-GaAs (111) TX_2/CaF_2 (111)	[11] [14]
Organic	Phthalocyanines/TX_2	

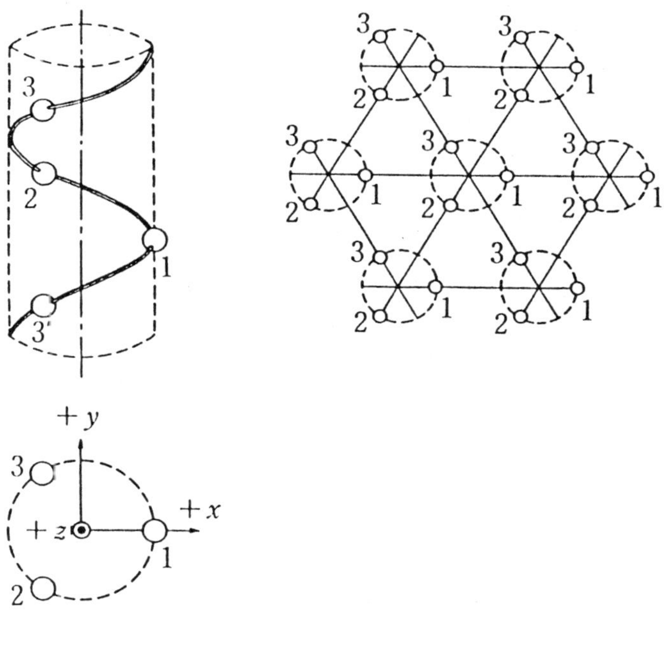

Fig. 2. Crystal structure of selenium and tellurium.

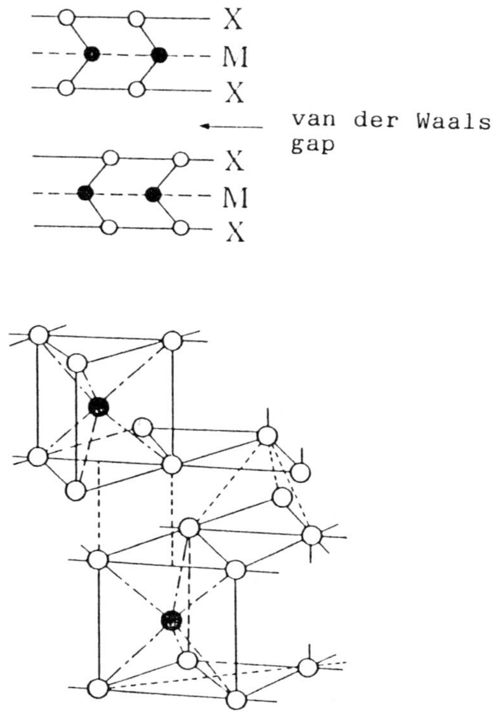

Fig. 3. Crystal structure of transition metal dichalcogenides.

van der Waals epitaxy. Although the lattice mismatch along the *c*-axis (the chain axis) is as large as 20%, ultrathin selenium film of good quality has been proved to grow on a cleaved face of tellurium [6]. Recently we have also succeeded in growing heterostructures of Te/Se/Te.

2.2. Quasi-two-dimensional materials (layered materials)

Transition metal dichalcogenides (TX_2, T = transition metal, X = chalcogen) are the most suitable materials for the van der Waals epitaxy. They have layered crystal structures, which are formed of unit layers consisting of transition metal atoms sandwiched by chalcogen atoms (Figure 3). Atoms in a unit layer are bound to each other by strong covalent bonds, whereas the layers are held together only by weak van der Waals forces. Thus they are easily cleaved parallel to the layers, and no dangling appear on their cleaved surface. Therefore it is expected that the growth of a layered material onto such surfaces proceeds via van der Waals forces, resulting in the relaxation of the lattice matching condition. Since there are such insulators as HfS_2, such semiconductors as MoS_2 and $MoSe_2$, and such superconducting metals as NbS_2 and $NbSe_2$ among the TX_2 compounds, the fabrication of various kinds of heterostructures using these as constituent materials has become a distinct possibility.

Recently van der Waals epitaxial growth of $MoSe_2$ has been succesfully achieved on a cleaved face of SnS_2 [7], which is also a layered material but different from the TX_2 type. Moreover it has been proved that $MoSe_2$ and $NbSe_2$ can be grown on a cleaved face of muscovite, a kind of mica [8,9]. Although muscovite is a layered material, it has a much more complicated crystal structure than TX_2 and its lattice constant is larger than that of TX_2 by 50%. It is expected from these facts that van der Waals epitaxy is possible between any layered materials, regardless of their crystal structures and their lattice constants.

2.3. Van der Waals epitaxy onto dangling-bond-terminated surfaces

So far van der Waals epitaxy has been limited to the heteroepitaxial growth between quasi-one or quasi-two-dimensional materials. If it can be applied to widely used three-dimensional materials, its field of application will be greatly extended. As has been mentioned above, dangling bonds appear on a clean surface of an ordinary material, which prevents good heteroepitaxial growth of a layered material on it. But growth with van der Waals forces becomes possible if regular termination of the surface dangling bonds is accomplished, as is shown in Figure 1(c) and if it is kept stable, even at the high temperatures required for good epitaxial growth. Sulfide treated GaAs (111) surface seems to be the most suitable candidate for this purpose, since it was found that the surface dangling bonds are regularly terminated with sulfur atoms up to 520 °C [10]. Epitaxial growth of semiconducting $MoSe_2$ and superconducting $NbSe_2$ on $(NH_4)_2S_x$-treated GaAs (111) surfaces have been tried successfully [11].

Another promising three-dimensional substrate material is CaF_2. It has been found that the (111) surface of CaF_2 (cleaved face) is covered with F atoms, resulting in a very inactive nature [12]. This inactive nature causes difficulty in achieving good heteroepitaxial growth of such a typical three-dimensional material as GaAs on it [13], but it is favorable to van der Waals epitaxy of a layered material on it. We have succeeded in growing an ultra-thin film of $MoSe_2$ on a cleaved face of CaF_2 [14]. CaF_2 is a well-known insulating material, lattice-matched to silicon, and good heteroepitaxial films can be grown on Si [13]. Thus the present method makes it possible to grow heteroepitaxial layers of various transition metal dichalcogenides on a Si (111) substrate by putting an intermediate CaF_2 layer between them.

2.4. Organic materials

Van der Waals epitaxy seems to have potential application to heteroepitaxial growth of organic material thin films, because many organic material crystallize with van der Waals force. Very recently we have succeeded in growing Pb-, Cu- and VO-phthalocyanine films on a cleaved face of MoS_2. Cu- and VO-phthalocyanine molecules have been found to form square lattice on a cleaved face of MoS_2, whereas Pb-phthalocyanine forms rectangular lattices. The axes of those lattices align with the hexagonal lattice of MoS_2, which results in the formation of three equivalent domains. This is our first application of van der Waals epitaxy to organic materials, and this type of approach will be extended to the growth of thin films of a vast range organic materials.

3. Characterization of Ultrathin Heterostructures Grown with van der Waals Epitaxy

In situ observations of reflection high energy electron diffraction (RHEED) have been performed during the growth of all of the materials mentioned above. A typical example is seen in Figure 4 which shows the evolution of the RHEED patterns during the growth of $MoSe_2$ on a mica substrate. Figure 4a is the RHEED pattern of a mica substrate. Figures 4b–4e show RHEED patterns taken after the deposition of 1/5, 1/2, 1 and 2 unit layers of $MoSe_2$, respectively. Here 1/5 unit layer means that a unit layer having 0.65 nm thickness covers 1/5 of the substrate face. As the amount deposited increased, the intensity of streaks from the mica substrate became weak. Instead, sharp streaks with a line interval corresponding to the lattice constant of a bulk $2H$-$MoSe_2$ appeared clearly. After two unit layers of $MoSe_2$ were grown, the streak pattern of the mica completely disappeared. The change of RHEED patterns indicates that a smooth and uniform $MoSe_2$ film with its own lattice constant grows epitaxially even from the initial stage. This is very different from the case of a strained superlattice, in which the lattice constant is forced to be the same as that of a substrate [15]. Nevertheless, the crystal axes of the grown film and the substrate align well to each other. Such features have also been confirmed by recent transmission microscope observation of a $MoSe_2/SnS_2$ heterostructure [7]. Thus the substrate and the film grown on it with van der Waals epitaxy are usually incommensurate but they are rotationally-commensurate. This is the most characteristic feature of van der Waals epitaxy, which induces a drastic relaxation of the lattice matching condition.

The abruptness of the interfaces of heterostructures grown with van der Waals epitaxy has been measured by nondestructive in-depth profiling using low-energy electron energy loss spectroscopy (LEELS) [5]. Figure 5 shows

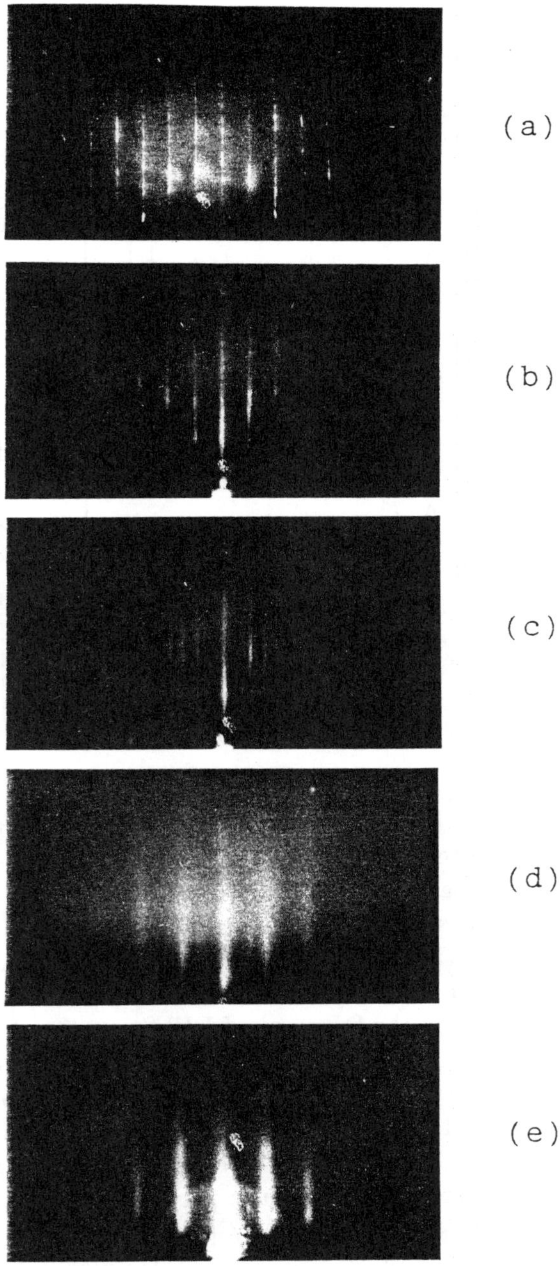

Fig. 4. Evolution of RHEED patterns during the growth of $MoSe_2$ on a mica substrate.

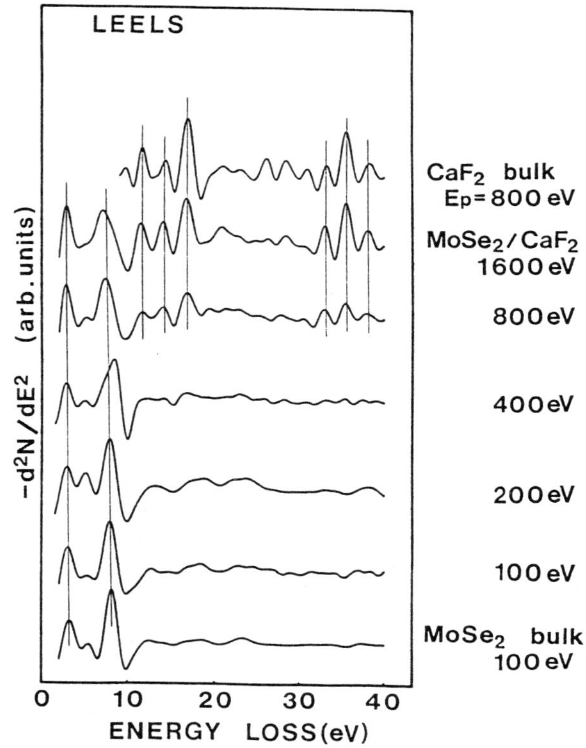

Fig. 5. In-depth profiling of a $MoSe_2/CaF_2$ interface using LEELS.

the result of LEELS in-depth profiling of a unit layer of $MoSe_2$ grown on a CaF_2 (111) substrate. Spectra for a cleaved face of a single crystal of $2H$-$MoSe_2$ and for a clean CaF_2 (111) surface are also shown in the figure for comparison. The loss peaks below 10 eV arise from electronic transitions from valence band states to conduction band states of $MoSe_2$, and the loss peaks above 10 eV are due to interband transitions or plasmon excitations in CaF_2. As can be seen in the figure, the LEELS spectra of the ultrathin $MoSe_2$ film for a primary electron energy less than about 400 eV are very similar to that of a single crystal of $2H$-$MoSe_2$, whereas loss peaks corresponding to CaF_2 are seen in the spectra for a primary electron energy of 800 eV or more. This clearly indicates that there exists a sharp interface at a depth of 0.6 to 0.8 nm, corresponding to the probing depth at a primary electron energy ranging from 400 eV to 800 eV [6], which agrees quite well with the thickness of a unit layer of $MoSe_2$. The spectra also indicate that the grown $MoSe_2$ layer has the same electronic structure as the bulk $MoSe_2$, reflecting a good quality of the grown film. The interfaces between $MoSe_2/MoS_2$ [5] and $NbSe_2$/S-GaAs [11] have also been found to have an abruptness as sharp as 0.1 nm.

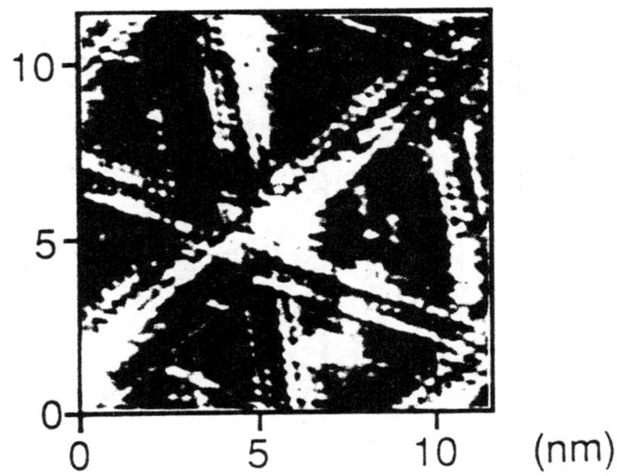

Fig. 6. A constant height STM image of a 2 monolayer coverage of $MoSe_2$ on MoS_2.

The top surfaces of heterostructures grown with van der Waals epitaxy are almost free from contamination, even under atmospheric pressure, because of the nonexistence of dangling bonds on them. This feature is especially suitable for the observation of their surfaces by a scanning tunneling microscope (STM). Recently we have succeeded in obtaining clear STM images of ultrathin $MoSe_2$ films on an SnS_2 [7] or a MoS_2 [16] substrate. Figure 6 shows and STM image of a two-monolayer coverage of $MoSe_2$ on a MoS_2 substrate. A Moiré-type modulation structure is seen in the figure, in addition to well-resolved atom images. This structure can be well explained if a very abrupt change exists in the lattice constant at the interface, as is shown in Figure 7. It has been concluded from the detailed analysis of the observed modulation structure that the lattice constant of the grown film agrees with the accepted value of the film material (in bulk) within an error of 0.1% and that the crystal axis of the grown film aligns with that of the substrate within 0.1 degree. In the next step the STM observation is expected to disclose the details of van der Waals epitaxial growth at atomic resolutions.

4. Summary

In conclusion we have proved that ultrathin heterostructures of good quality can be fabricated by van der Waals epitaxy by using various kinds of materials, ranging from quasi-one-dimensional to three-dimensional ones, as constituents. The present technique has opened a new way to prepare many kinds of heterostructures consisting of superconductor, metal, semiconductor or insulator films with any thickness from subnanometer upwards by using a variety of transition metal dichalcogenide materials and other mate-

Fig. 7. A top view of the interface two unrotated hexagonal closest packed lattices with a 4% lattice mismatch showing the Moiré pattern.

rials having no dangling bonds on their surfaces. This method has also been found to be useful in the preparation of organic thin films. It is expected that van der Waals epitaxy will be especially useful for creating new man-made materials formed of stacked ultrathin films of many kinds of substance.

Acknowledgements

The authors would like to thank Drs. B.A. Parkinson and F.S. Ohuchi for the STM experiments and stimulating discussions. The present work is supported by Grant-in-Aid for Special Distinguished Research (No. 02102002) from the Ministry of Education, Science and Culture of Japan.

References

1. A. Koma, K. Sunouchi and T. Miyajima: 1984, *Microelectronic Engng.* **2**, 129.
2. A. Koma, K. Sunouchi and T. Miyajima: 1985, *J. Vac. Sci. Technol. B* **3**, 724.
3. A. Koma, K. Sunouchi and T. Miyajima: 1985, *Proc. 17th Intern. Conf. Phys. Semiconductors*, San Fransisco, 1984, Springer-Verlag, New York, p. 1465.
4. A. Koma: 1989, *Proc. 1st Intern. Conf. Electronic Materials*, Tokyo, 1988, Materials Research Society, Pittsburgh, p. 105.
5. A. Koma and K. Yoshimura: 1986, *Surf. Sci.* **174**, 556.
6. N. Aoto, K. Sunouchi and A. Koma: 1983, *Extended Abstracts of the 15th Conference of Solid State Devices and Materials*, Tokyo, p. 309.
7. F.S. Ohuchi, B.A. Parkinson, K. Ueno and A. Koma: 1990, *J. Appl. Phys.* **68**, 2168.
8. K. Saiki, K. Ueno, T. Shimada and A. Koma: 1989, *J. Crystal Growth* **95**, 603.
9. K. Ueno, K. Saiki, T. Shimada and A. Koma: 1990, *J. Vac. Sci. Technol. A* **8**, 68.
10. H. Oigawa, J.-F. Fan, Y. Nannichi, K. Ando, K. Saiki and A. Koma: 1989, *Jpn. J. Appl. Phys.* **28**, L340.
11. K. Ueno, T. Shimada, K. Saiki and A. Koma: 1990, *Appl. Phys. Lett.* **56**, 327.
12. K. Saiki, Y. Sato and A. Koma: 1989, *Jpn. J. Appl. Phys.* **28**, L134.
13. H.C. Lee, H. Ishiwara, S. Kanemaru and S. Furukawa: 1987, *Jpn. J. Appl. Phys.* **26**, L1734.
14. A. Koma, K. Saiki and Y. Sato: 1989, *Appl. Surface Sci.* **41/42**, 451.
15. G.C. Osbourn: 1982, *J. Appl. Phys.* **53**, 1586.
16. B.A. Parkinson, F.S. Ohuchi, K. Ueno and A. Koma: 1991, *Appl. Phys. Lett.* **58**, 472.

PART II

Two-Dimensional Systems and Superlattices

TWENTY YEARS OF SEMICONDUCTOR SUPERLATTICE RESEARCH

L. ESAKI
IBM Thomas J. Watson Research Center,
Yorktown Heights, New York 10598, U.S.A.

Abstract. Following the past two-decade evolutionary path in the interdisciplinary research of semiconductor superlattices and quantum wells, significant milestones are presented, with an emphasis on electric field-induced effects in the frontier of semiconductor physics associated with technological advances.

1. Introduction

In 1969, research on quantum structures originated from a proposal of an *engineered* semiconductor superlattice by Esaki and Tsu [1,2]. In anticipation of an advancement in epitaxy, two types of superlattices were envisioned, with alternative deposition of ultrathin layers: doping and compositional, as shown at the top and bottom of Figure 1, respectively.

This was, perhaps, the first proposal of *designed semiconductor quantum structures*: namely, we asserted that quantum states such as narrow bands or confined states could be artificially created if potential barriers and wells were fabricated by means of successive deposition of different semiconductor layers with thicknesses smaller than the phase-coherent length of electrons. Since the electronic characteristics of semiconductor structures are mainly governed by such quantum states, it was predicted that new electronic materials can be designed and engineered to obtain unprecedented transport and optical properties through tailoring band parameters with the control of the potential profile during the layer deposition.

Since the one-dimensional potential is introduced along with the superlattice (SL) axis (perpendicular to the deposited plane layers), we thought that, if our attempt was successful, elegantly simple examples in one-dimensional quantum physics, for instance, resonant electron tunneling [3] or Kronig–Penney band model, which had remained textbook exercises, could, for the first time, be practiced in a laboratory. Perhaps, *do-it-yourself quantum mechanics* would be possible, since the principles of quantum theory dictate the design of semiconductor structures.

At the inception of the SL idea, it was recognized that the long, tailored lattice period provided the unique opportunity to exploit effects induced by applying a modestly high electric field along with the SL axis. Our early analysis of electron dynamics in an electric field led to the prediction of the occurrence of a negative differential resistance which could be a precursor

H. Aoki et al. (eds.), New Horizons in Low-Dimensional Electron Systems, 99–121.

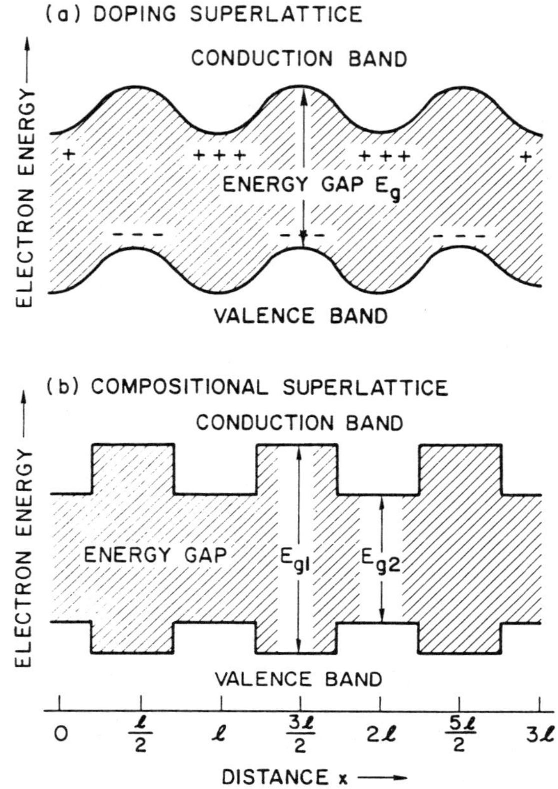

Fig. 1. Spacial variations of the conduction and valence bandedges in two types of superlattices: doping (top) and compositional.

of the Bloch oscillation. The introduction of the SL, apparently, allows us to explore the regime of electric field-induced quantization such as Stark ladders, which is not readily accessible in *natural* crystals.

Before reaching the SL concept, we had explored the feasibility of structural formation by epitaxy of ultrathin barriers and wells which might exhibit resonant tunneling. It was thought that semiconductors and technologies developed with them, might be called for in order to demonstrate the quantum wave nature of electrons associated with the interference phenomena, since their small Fermi energies due to low carrier densities help make the de Broglie wavelength relatively large. Namely, the Fermi wavelength $\lambda_1 = 2\pi/k_f$, where k_f is the magnitude of the Fermi wavevector, is given as a function of the carrier density n, as follows: $\lambda_f = (8\pi/3n)^{1/3}$ for a three-dimensional (3D) system; $(2\pi/n)^{1/2}$ for a two-dimensional (2D) system; $(4/n)$ for a one-dimensional (1D) system.

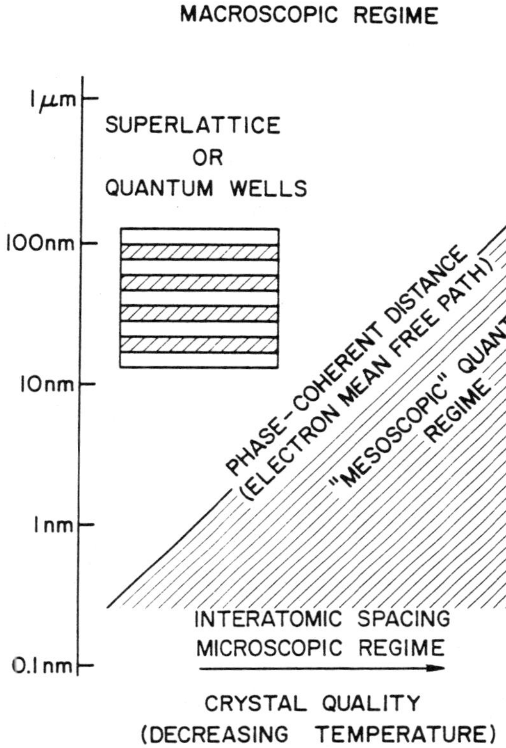

Fig. 2. Schematic illustration of a mesoscopic quantum regime (hatched) with a superlattice or quantum wells in the insert.

The idea of the SL occurred to us as a natural extension of double- and multi-barrier structures: namely, the SL is a series of quantum wells (QWs) coupled by resonant tunneling, where quantum effects are expected to prevail. An important parameter relevant to the observation of such effects is the phase-coherent length or, roughly, the electron inelastic mean free path, which depends heavily on bulk as well as interface quality of crystals and also on temperatures and values of the effective mass. As schematically illustrated in Figure 2, if characteristic dimensions such as SL periods and QW widths are reduced to less than the phase-coherent length, the entire electron system will enter a *mesoscopic* quantum regime of reduced dimensionality, being placed in the scale between the macroscopic and the microscopic.

It was theoretically shown that the introduction of the SL potential perturbs the band structure of the host materials, yielding unprecedented electronic properties of quasi-two-dimensional character [1, 2]. Figure 3 shows the density of states $\rho(E)$ for electrons with $m^* = 0.067\ m_0$ in an SL with a well width of 100 Å and the same barrier width, where the first three

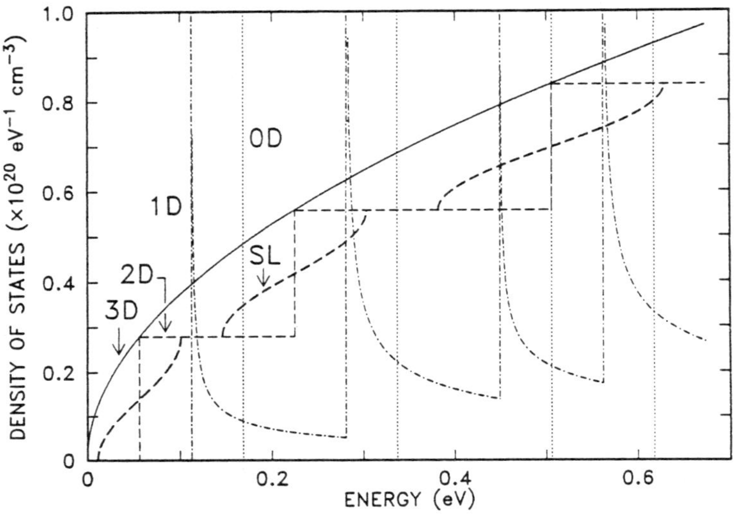

Fig. 3. Comparison of density of states in the three-dimensional (3D) electron system with those of a superlattice, and the two-dimensional (2D), one-dimensional (1D), and zero-dimensional (0D) electron systems.

subbands are indicated with dashed curves. The figure also includes, for comparison, a parabolic curve $E^{1/2}$ for 3D, a steplike density of states for 2D (quantum well), a curve $\sum (E-E_m-E_n)^{-1/2}$ for 1D (quantum wire), and a delta function $\sum \delta(E-E_1-E_m-E_n)$ for 0D system (quantum dot) where the quantum unit is taken to be 100 Å for all cases and the barrier height is assumed to be infinite in obtaining the quantized energy levels, E_1, E_m and E_n. Notice that the ground state energy increases with decrease in dimensionality if the quantum unit is kept constant. Each quantized energy level in 2D, 1D and 0D is identified with the one, two and three quantum numbers, respectively. The unit for the density of states here is normalized to $\mathrm{eV^{-1}cm^{-3}}$ for all the dimensions, although $\mathrm{eV^{-1}cm^{-2}}$ $\mathrm{eV^{-1}cm^{-1}}$ and $\mathrm{eV^{-1}}$ may be commonly used for 2D, 1D and 0D, respectively.

2. Epitaxy and Superlattice Growth

In the early 1970s, we initiated the seemingly formidable task of engineering nanostructures in the search for novel quantum phenomena [4]. Heteroepitaxy is of fundamental interest for the SL and QW growth. Innovations and improvements in growth techniques such as MBE [5], MOCVD (metalorganic chemical vapor deposition) [6] and MOMBE or CBE during the last decade have made possible high-quality heterostructures. Such structures possess predesigned potential profiles and impurity distributions with di-

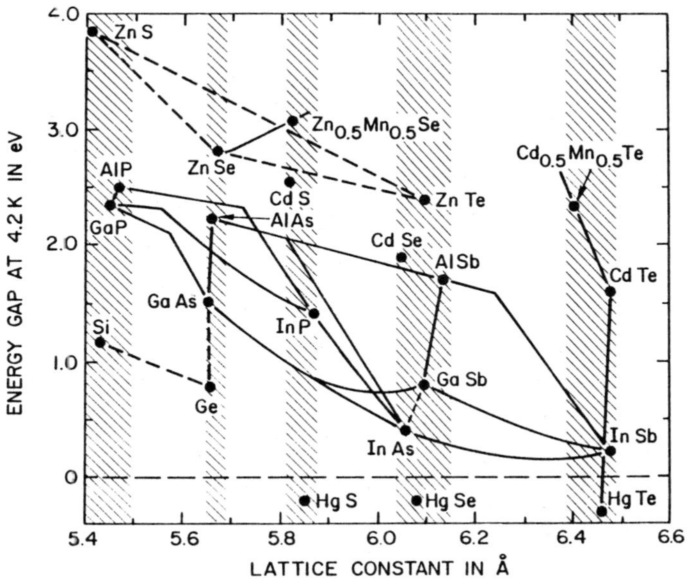

Fig. 4. Plot of energy gaps at 4.2 K versus lattice constants.

mensional control close to interatomic spacing and with virtually defect-free interfaces, particularly, in a lattice-matched case such as GaAs–$Ga_{1-x}Al_xAs$. This great precision has cleared access to a *mesoscopic* quantum regime.

The semiconductor SL structures have been grown with III–V, II–VI and IV–VI compounds, as well as elemental semiconductors. Figure 4 shows the plot of energy gaps at 4.2 K versus lattice constants for zinc-blende semiconductors together with Si and Ge. Joining lines represent ternary alloys except for Si–Ge, GaAs–Ge and InAs–GaSb. The introduction of II–VI compounds apparently extended the available range of energy gaps in both the high and low directions: that of ZnS is as high as 3.8 eV and all the Hg compounds have a negative energy gap or can be called zero-gap semiconductors. The magnetic compounds, CdMnTe and ZnMnSe are relative newcomers in the SL and QW arena.

Semiconductor hetero-interfaces exhibit a band offset, usually associated with a gradual band-bending in its neighborhood which reflects space-charge effects. According to the character of such discontinuity, known hetero-interfaces can be classified into four kinds: Type I, type II-staggered, type II-misaligned, and type III, as illustrated in Figure 5(a) (b) (c) (d): band offsets (left), band bending and carrier confinement (middle), and SLs (right).

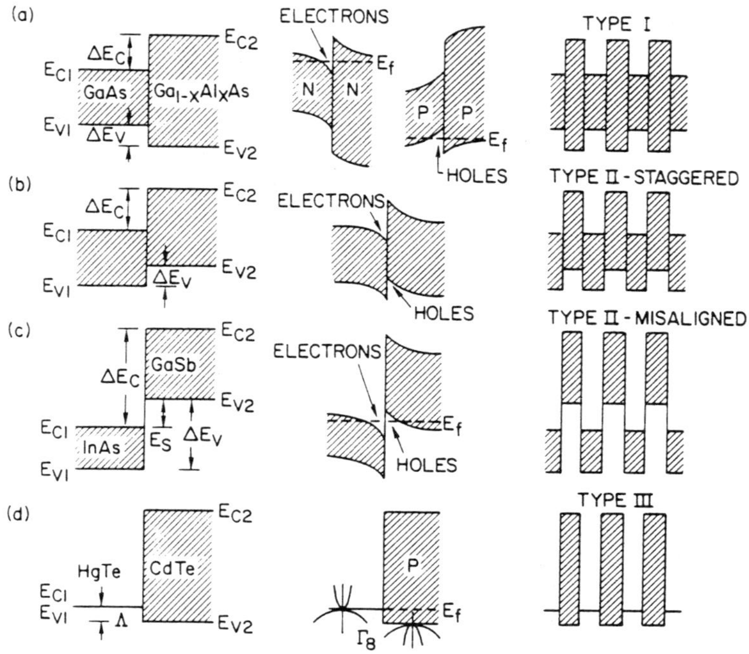

Fig. 5. Discontinuities of bandedge energies at four types of hetero-interfaces: band offsets (left), band bending and carrier confinement (middle), and superlattice (right).

3. Resonant Tunneling in an Electric Field

Resonant tunneling in multi-barrier structures can be observed with the application of an electric field perpendicular to the layer planes. In this context, Tsu and Esaki [7] derived the voltage dependence of the tunneling current in such structures with the transmission coefficient computed as a function of electron energy (Figure 6), in assuming the conservation of the total electron energy as well as its momentum transverse to the direction of tunneling. Chang, Esaki and Tsu [8] measured the tunneling current and conductance as a function of applied voltage in GaAs–GaAlAs double barriers, as shown in Figure 7, which exhibited a large component of the excess current primarily arising from imperfections in the structures, but agreed qualitatively with the Tsu–Esaki formula [7]. The energy diagram is shown in the inset where resonance occurs at such applied voltages as to align the Fermi level of the electrode with the bound states, as shown in cases (a) and (c) of Figure 7. This experimental result probably constitutes the first observation of manmade bound states in single QWs.

The technological advance in MBE has resulted in remarkably improved characteristics in resonant tunneling, as shown in Figure 8, which spurred

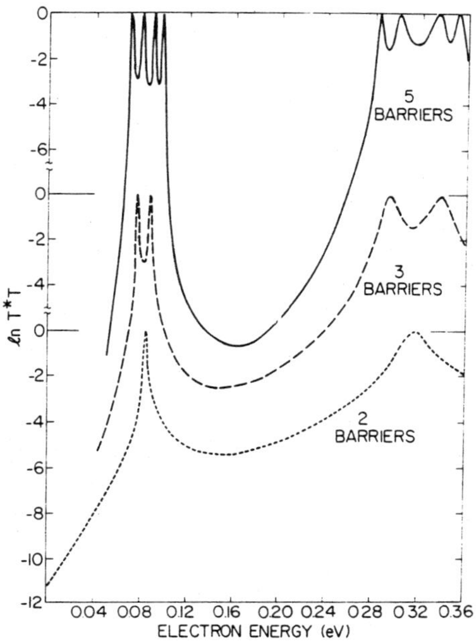

Fig. 6. Natural log of the transmission coefficient vs. electron energy in eV for cases of a double-, a triple-, and a quintuple-barrier. The peaks show resonance. The barrier and well widths are 20 and 50 Å, respectively. The barrier height is 0.5 eV.

renewed interest in such structures [9]. Recently, high-quality resonant-tunneling double-barrier junctions, exhibiting not only large values of the peak-to-valley ratio but also fine structure, have been the focus of intense experimental and theoretical investigations. The observation of resonant tunneling of holes reveals structure corresponding to respective bound states of both heavy and light holes [10]. The existence of optical phonon-assisted resonant tunneling was found in the valley region [11] and analyzed [12]. Resonant magneto-tunneling measurements manifested the energy levels quantized into Landau levels of index n [13]. Later, in magneto-tunneling, LO phonon (magnetopolaron)-assisted process was clearly resolved [14, 15]. Charge accumulation in a double-barrier resonant tunneling structure was studied by photoluminescence [16] and photoluminescence-excitation spectroscopy [17].

A variety of tunneling heterostructures have been developed with a combination of materials other than GaAs/AlAs, such as InGaAs/InAlAs [18] and InAs/AlSb [19]. Research in polytype heterostructures of InAs/AlSb/GaSb [20, 21], has been rejuvenated with the observation of a large negative resistance based on resonant interband tunneling in GaSb/AlSb/InAs/AlSb/GaSb [22] and InAs/AlSb/GaSb/AlSb/InAs [23].

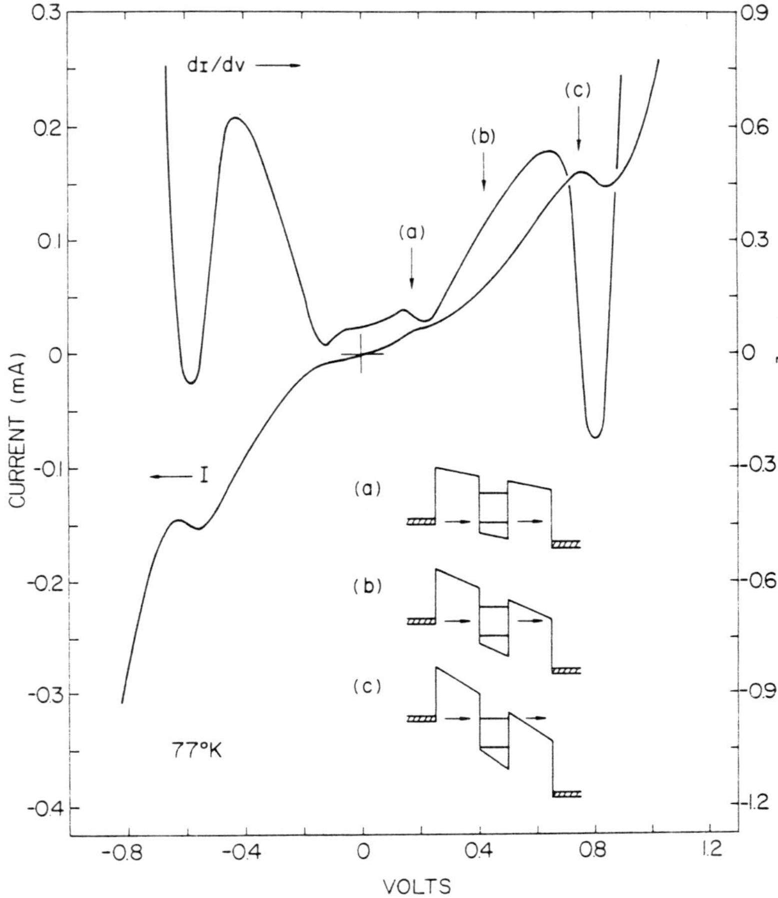

Fig. 7. Current- and conductance-voltage characteristics of a GaAs–GaAlAs double-barrier structure. Conditions at resonance (a) and (c), and at off-resonance (b), are indicated by arrows in the insert.

Those multi-heterojunctions, indeed, correspond to a portion of the proposed polytype superlattices [20]. A great deal of engineering interest in resonant tunneling, obviously, has arisen from its potential application to high-speed semiconductor devices. Picosecond bistable operation has been observed in a double-barrier resonant tunneling diode, where a rise time of 2 ps is said to be the fastest switching yet measured for an electronic device [24]. Oscillations up to frequencies of 420 GHz were reported in a GaAs resonant tunneling diode at room temperature [25].

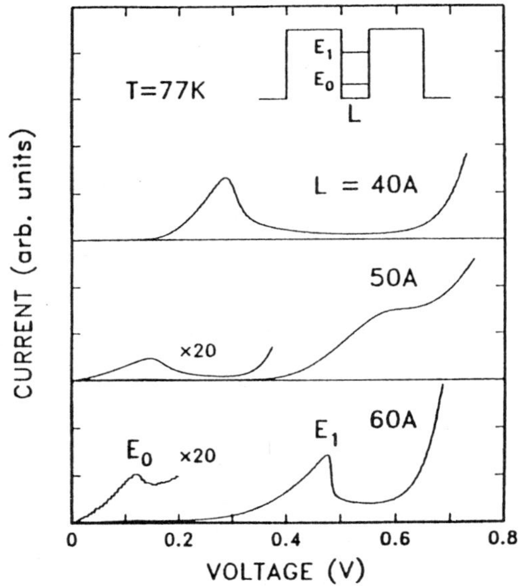

Fig. 8. Current-voltage curves for three double-barriers with well widths, L, of 40 Å, 50 Å, and 60 Å. The bound energies, E_1 and E_2, are shown in the energy diagram of the upper part.

4. Transport and Optical Properties in Superlattices and Quantum Wells Subject to an Applied Electric Field

4.1. Theoretical background

The introduction of the SL potential, as shown in Figure 1, perturbs the band structure of the host materials: The degree of perturbation obviously depends on its potential profile. With the SL period, ℓ, of the order of 100 Å, the Brillouin zone is divided into a series of minizones, as shown in the bottom of Figure 9, giving rise to the first and second subbands, E_1 and E_2, separated by forbidden regions, as shown in the top.

With application of an electric field in the SL direction, x, the electron dynamics in a narrow subband has been analyzed in the framework of a simplified path integration method to obtain a relationship between the applied field F and the average drift velocity v_d. The equations of electron motion are

$$\hbar\,\partial k_x/\partial t = eF \quad \text{and} \quad v_x = \hbar^{-1}\,\partial E_x/\partial k_x \;,$$

then, the velocity increment in a time interval dt is

$$\mathrm{d}v_x = eF\hbar^{-2}(\partial^2 E_x/\partial k_x^2)\,\mathrm{d}t \;.$$

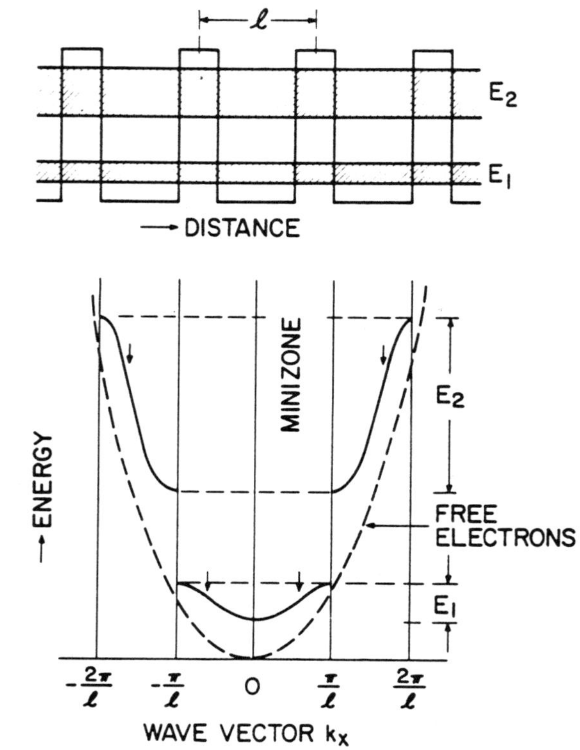

Fig. 9. (a) Potential profile of a superlattice, and (b) electron energy vs. wave vector in the minizones.

The average drift velocity, taking into account the electron scattering time τ, is written as

$$v_d = \int_0^\infty \exp(-t/\tau)\,\mathrm{d}v_x = eF\hbar^{-2} \int_0^\infty (\partial^2 E_x/\partial k_x^2)\ \exp(-t/\tau)\,\mathrm{d}t\ .$$

If a sinusoidal relationship, $E_x = \frac{1}{2}(1 - \cos k_x\ell)$, is assumed, we obtain

$$v_d = eF\tau/m^*(0)\ [1 + (eF\ell\tau/\hbar)^2]^{-1}\ ,$$

where $m^*(0)$ is determined by $\hbar^2(\partial^2 E_x/2k_x^2)^{-1}$ at $k_x = 0$. Thus, the current-field curve is given by

$$J = env_d$$

where n is the electron concentration. The current J, plotted as a function of F in Figure 10, has the maximum at $eF\ell\tau/\hbar = 1$ (indicated by an arrow) and thereafter decreases, giving rise to a differential negative resistance.

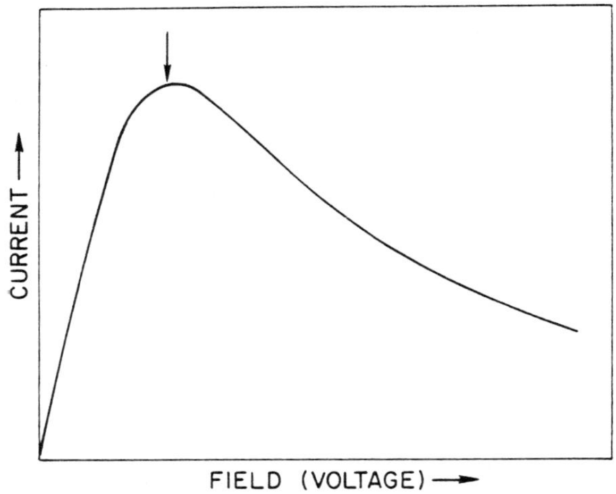

Fig. 10. Current versus electric field curve.

The threshold field for the negative resistance is 10^3 V/cm if $\ell = 100$ Å and $\tau = 0.66$ ps. This result indicates that, if, in applying modestly high electric fields, electrons gain enough energy to reach beyond the inflection points (indicated by arrows in Figure 9), they will be decelerated rather than accelerated by such electric fields.

If, in further increasing electric fields, electrons reach the mini-Brillouin zone boundary ($k_x = \pi/\ell$ in Figure 10), an umklapp process (or Bragg scattering) will occur, that is, the electron wave vector, k_x, is changed by a reciprocal lattice vector, $2\pi/\ell$. If this happens, electrons reappear at the opposite point of the Brillouin zone, $k_x = -\pi/\ell$, and the process will be repeated. This is the physical picture of the so-called Bloch oscillation (the centroid of the electron wave packet moves back and forth, steadily, in the k-space), whose frequency is determined by the time period T required to cross the Brillouin zone, namely, $T = 2\pi\hbar/eF\ell$. Thus, the Bloch angular frequency ω_B is given by $\omega_B = 2\pi/T = eF\ell/\hbar$. In order to realize this oscillation, the scattering time τ should be greater than T, that is, $\omega_B\tau > 2\pi$. A precursor of this oscillation is the above-mentioned differential negative resistance which starts at $\omega_B\tau > 1$. The frequency of the Bloch oscillator is as high as 1 THz for $F = 4 \times 10^3$ V/cm and $\ell = 100$ Å.

The Bloch oscillation involves resonating transitions between equally spaced and localized energy levels called Stark ladders which are the consequence of electric field-induced quantization. Figure 11 shows schematically electronic wave functions of Stark ladders formed in a superlattice under low and high electric fields; at the low field, optical transitions may be possible because of sufficient overlap of wave functions between adjacent Stark

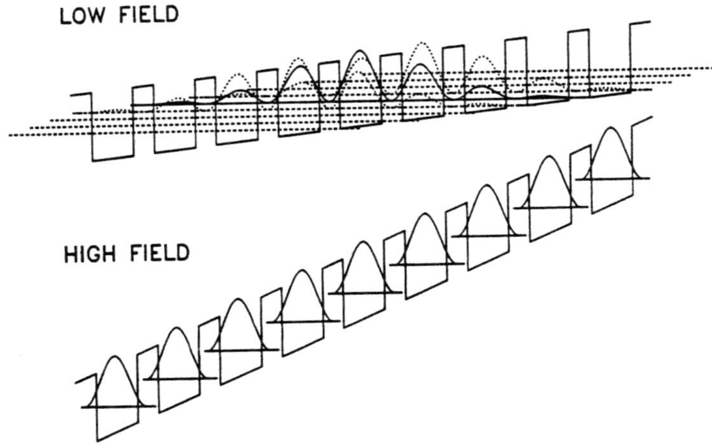

Fig. 11. Schematic potential profiles with electronic wave functions of Stark ladders formed in a superlattice under low and high electric fields.

states; the energy separation of these levels ΔE is equal to $\hbar\omega_B = eF\ell$.

Before the introduction by Wannier [26] of the term Stark ladder, James [27] heuristically reached the basic concept of Stark localization. The explicit wave function was first derived by Kane [28], presenting a clear picture of localization under the application of an electric field. The existence and real nature of these localized levels in bulk crystals, however, were quite controversial [29, 30]. Experimental evidences supporting their existence are rather tenuous [31–33], because the observation, firstly, requires enough high fields to produce the energy separation greater than the collision broadening, that is,

$$\Delta E = eFa > \hbar/\tau \ , \quad \text{then} \quad \omega_B\tau > 1 \ ,$$

where a is the lattice constant which is typically 2 to 3 Å. If $\tau = 0.66$ ps, $\hbar/\tau = 1$ meV. Therefore, F should be of the order of 10^5 V/cm. On the other hand, the introduction of the SL structures with large lattice constants ℓ provides easy access to the electric field-induced quantization regime with a few tens of kV/cm.

Stark ladder states are localized over the length λ given by E_1/eF, where E_1 is the subband width. The k-space approach presented here is justified when $\Delta E = eF\ell \ll E_1$ or $\ell \ll \lambda$. If, however, the value of $eF\ell$ approaches to $E_1(\lambda \to \ell)$ because of high fields or narrow bandwidths (tight binding SL), a hopping model by Tsu and Döhler [34] becomes valid. In this case, although a negative differential resistance may arise, the Bloch oscillation is unlikely to occur because of insufficient spacial overlap between adjacent Stark states.

Artaki and Hess [35] carried out Monte Carlo simulations for electron transport in SLs, taking into account scattering by acoustic and optical phonons and impurities; the collision broadening effect on SL transport associated with a negative resistance was assessed.

Incidentally, it should be appreciated that the quantum theory for the electric field effects in an SL presented here is quite different from the traditional approach to a bulk semiconductor with E_c and E_v bandwidths of a few electron-volts: The electro-optical properties (a tail below the energy gap and an oscillatory behavior above it) have been treated with the effective mass approach, where the envelope eigenfunctions under an external electric field are Airy functions: The observation is consistent with the resultant Franz–Keldish effect.

4.2. Early attempts in search for the quantum effects

Esaki *et al.* [36] found a weak negative resistance resembling Figure 10, starting at $F \sim 2 \times 10^4$ V/cm in an MBE-grown GaAs–GaAlAs SL with $E_1 \sim 40$ meV. Later, Esaki and Chang [37] studied transport properties in an SL of a tight-binding potential with $E_1 = 5$ meV, which exhibited an oscillatory behavior: the current and conductance as a function of applied voltage are shown in Figure 12. A marginal band-conduction mechanism (case (a) in the inset) is broken down, as $eF\ell$ approaches E_1, at relatively low voltages, resulting in a negative resistance with creation of a localized high-field domain. The current recovers due to resonant tunneling between the first miniband and the second miniband in the adjacent QWs (case (b) in the set) when their energies are aligned with each other. The oscillatory behavior in the conductance reflects the repetition of this resonance with increase in applied voltage, whose period, 0.24 V, agreed with the calculated value.

Tsu *et al.* [38] performed photocurrent measurements on GaAs–GaAlAs SLs subject to an electric field perpendicular to the plane layers with the use of a semitransparent Schottky contact. The photocurrents as a function of incident photon energy are shown in Figure 13 for samples of three different configurations, designated A (35 ÅGaAs–35 ÅGaAlAs), B (50 ÅGaAs–50 ÅGaAlAs), and C (110 ÅGaAs–110 ÅGaAlAs). The energy diagram is shown schematically in the upper part of the figure where minibands created by the SL potential are labeled E_1 and E_2. In sample C, the minibands are essentially discrete states. Calculated energy locations and bandwidths shown in the figure are found to be in satisfactory agreement with the observation. The photocurrent, as a function of applied voltage, exhibited a pronounced negative resistance which was interpreted with the hopping model [34] since it occured in the range of $eF\ell > E_1$.

Since the mid-80s, advanced growth facilities at many institutions have

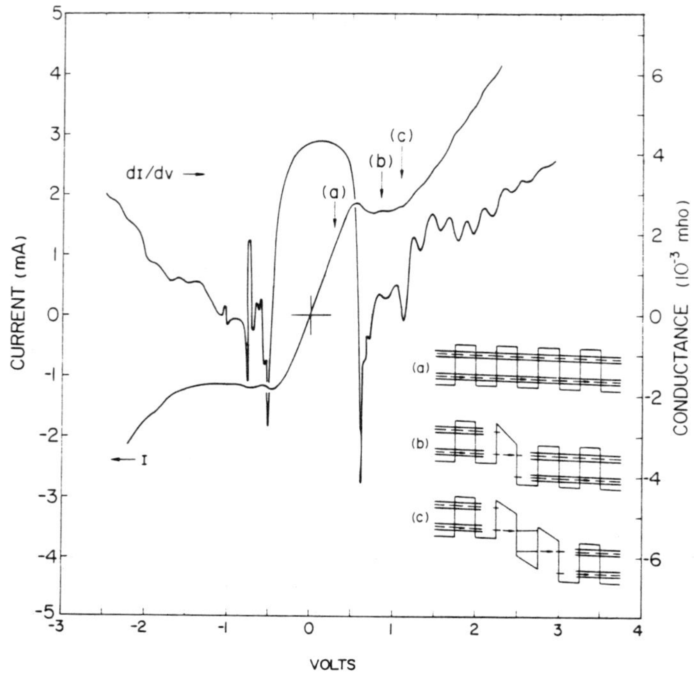

Fig. 12. Current- and conductance-voltage characteristics of a GaAs–AlAs superlattice. Band-type conduction (a) and resonant tunneling at an expanding high-field domain (b) and (c), are indicated in the insert.

begun to provide high-quality SLs and QWs, meeting stringent requirements in the structural design. As a result, investigations into such quantum structures have been intensified and have proliferated extensively. We present here only a limited number of reports, selected rather arbitrarily by the author.

4.3. Recent transport experiments

Deveaud *et al.* [39] carried out transport measurements of 2D carriers in minibands using subpicosecond luminescence spectroscopy. Schneider *et al.* [40] studied the dynamics of electron transport along the growth direction of tight-binding GaAs–AlAs SLs by electrical and optical time-of-flight experiments. Helm *et al.* [41] observed infrared light (intersubband) emission from SLs excited by sequential resonant tunneling. Tatham *et al.* [42] made time-resolved Raman measurements of intersubband relaxation in GaAs QWs. Sibille *et al.* [43] studied transport properties in a number of GaAs–AlAs SLs with relatively wide miniband widths: The data were analyzed in terms of the original theory [1, 2]. Grahn *et al.* [44] reported the observation of

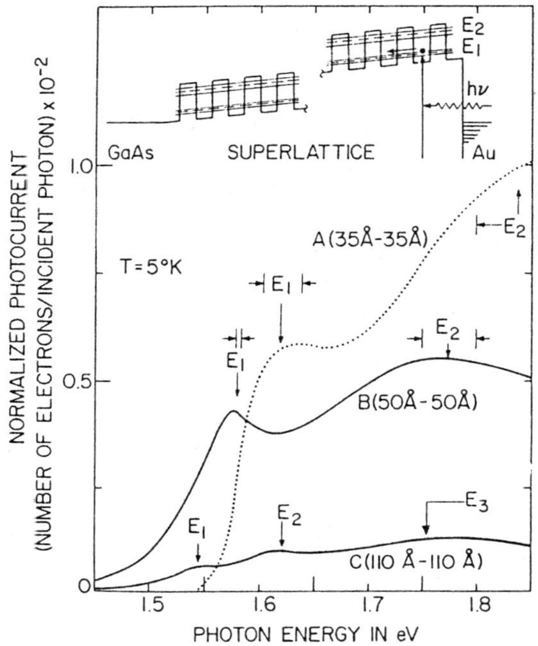

Fig. 13. Photocurrent versus photon energy for three superlattices samples A, B, and C. Calculated energies and bandwidths are indicated. The energy diagram of a Schottky-barrier structure is shown in the upper part.

photoluminescence from the second and third subbands in GaAs–AlAs SLs with an applied electric field: The occupation of the higher subbands was achieved by sequential resonant tunneling of electrons. Brozak *et al.* [45] observed thermal saturation of transport in a narrow band of an SL: the conductivity was quenched as the temperature was raised. Beltram *et al.* [46] studied transport properties in a transistor with an SL structure built between the base and the collector, observing negative differential conductance and scattering-controlled transmission resonances.

4.4. Stark shifts in QWs

Mendez *et al.* [47] studied the field-induced effect on the photoluminescence in GaAs–GaAlAs QWs, where the main peak is attributed to the excitonic transition between 2D electrons and heavy holes: when the electric field was applied perpendicular to the plane layers, pronounced field-effects, Stark shifts, were found. More recently, Vina *et al.* [48] studied the excitonic coupling as indicated in Figure 14, which shows Stark shifts of the light-hole exciton (open circles) and excited states of the heavy-hole exciton (open triangles), exhibiting strong coupling (solid circles) between them. Chemla *et al.* [49] observed Stark shifts of the excitonic absorption peak, even at

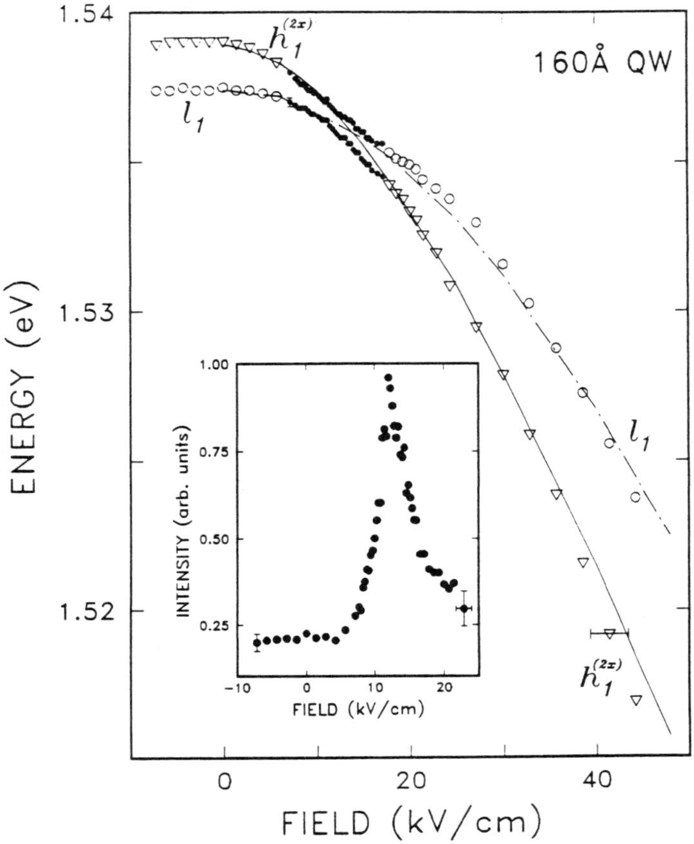

Fig. 14. Stark shifts of the light-hole exciton (open circles) and excited states of the heavy-hole exiton (open triangles), exhibiting strong coupling (solid circles) between them. The integrated intensity of $h^{(2x)}$ normalized to l_1 is shown as a function of electric field in the insert.

room temperature; Miller *et al.* [50] analyzed such electroabsorption spectra and demonstrated optical bistability, level shifting and modulation.

4.5. Formation of Stark ladders in SLs with the application of an electric field

Bleuse *et al.* [51] calculated field-induced Stark localization of the eigenstates in an SL, predicting a blue shift of the optical-absorption edge and the presence of oscillations in F^{-1}. Mendez *et al.* [52] observed the formation of Stark ladders in an SL by field-induced movement of the peak positions in photoluminescence and photocurrent spectra with applied electric fields. Figure 15 illustrates the conduction- and valence-band potential profiles,

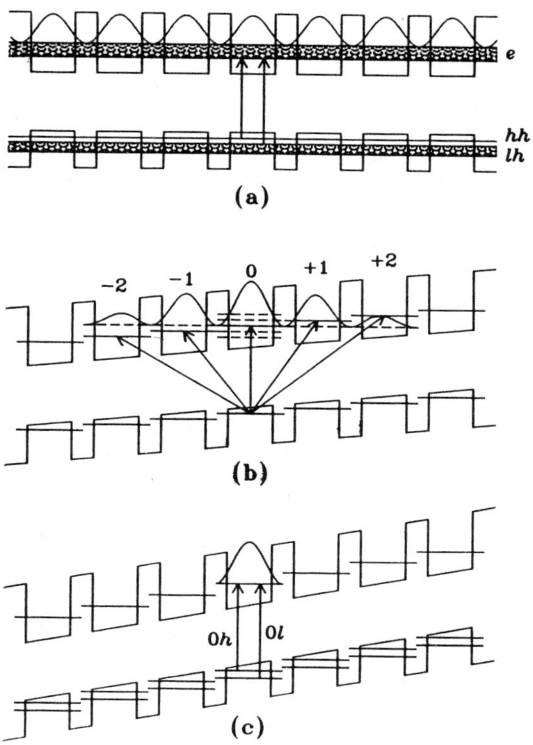

Fig. 15. Conduction- and valence-band potential profiles, including the Stark ladder formation, for a 40 Å–20 Å GaAs–GaAlAs superlattice at zero field, (a), at a moderate field (2×10^4 V/cm), (b), and at a high field (10^5 V/cm), (c).

including the Stark ladder formation, for a 40 Å–20 Å GaAs–GaAlAs SL in the absence, (a), and in the presence of an electric field, (b) and (c). At $F = 0$, the electron and hole wavefunctions are delocalized and interband transitions take place between SL minibands. With the application of an electric field, the localization of wavefunctions begin, forming the ladder whose energy spectrum is given by $E_0 + neF\ell$ $(n = 0, \pm 1, \pm 2, \ldots)$, where E_0 is the eigenenergy of an isolated QW. At a moderate field (2×10^4 V/cm), (b), electron and light-hole states are still extended over several SL periods whereas heavy holes are fully localized. Interband transitions (indicated by $-2, -1, 0, +1, +2$) involve Stark ladder states like the ones represented here for the conduction band. For simplicity, the light-hole ladder states are not included. At a high field (10^5 V/cm), (c), all states are completely localized and transitions take place between levels corresponding to isolated QWs. Figure 16 shows the interband transition energies as a function of field for the SL in Figure 15, determined from the peak positions in pho-

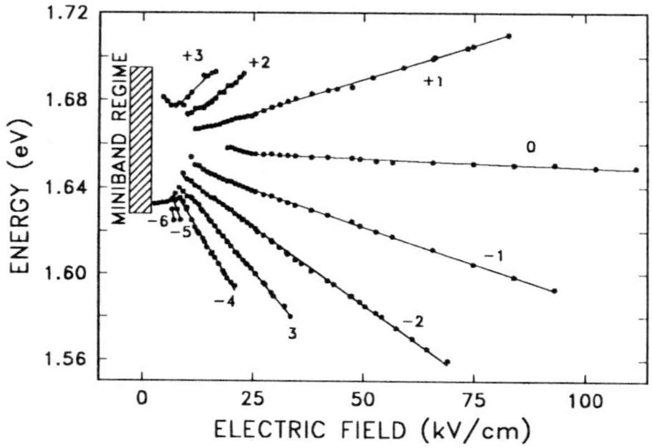

Fig. 16. Interband transition energies as a function of electric field for the superlattice in Figure 14. Each branch corresponds to a member of the Stark ladder, labeled by an index, n.

tocurrent spectra. Each branch corresponds to a member of the Stark ladder, labeled by an index, n. With the same technique, Agullo-Rueda *et al.* [53] determined the coherence length of electrons from the number of observed interband transitions between the valence- and conduction-band Stark ladders, which was found at least ten periods when $\ell = 60$ Å and increased with decreasing ℓ (or increasing the miniband width). Later, it was shown that the coherence length did not depend strongly on temperature in the range 5–292 K [54]. Voisin *et al.* [55] confirmed the predicted electro-optical effect [51] by an electroreflectance study for an SL. Dignam and Sipe [56] presented calculations of the energies and absorption spectra of GaAs–GaAlAs SL excitons in the presence of an applied electric field, which were in good agreement with the experimental photocurrent spectra [53]. Ohno *et al.* [57] observed localized surface states (Tamm states) intentionally introduced in GaAs–GaAlAs SLs by excitonic interband transitions in photoluminescence excitation spectra. Photocurrent experiments under an electric field showed additional transitions as well as anticrossing interactions between the Tamm states and the Stark ladder states associated with the SL.

5. Conclusion

Our original proposal and early works apparently triggered a wide spectrum of experimental and theoretical investigations in the new domain of semiconductor science and engineering. A variety of *engineered quantum structures* exhibited extraordinary transport and optical properties which may not even exist in any *natural* crystal. Thus, this new degree of freedom offered in semiconductor research *through advanced material engineering* has

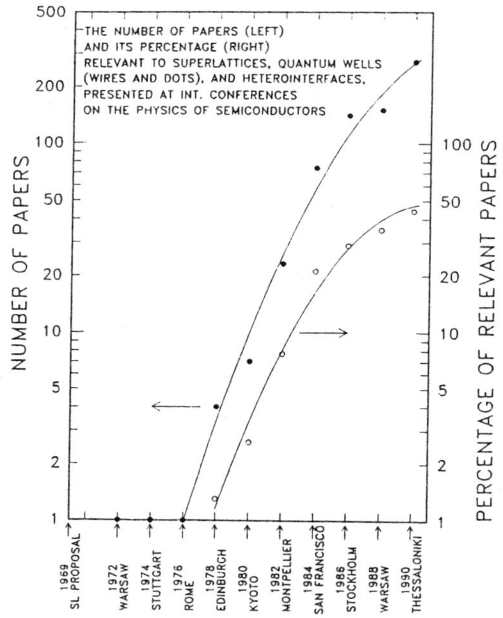

Fig. 17. Growth of papers at ICPS.

inspired many ingenious experiments, resulting in observations of not only predicted effects but also totally unknown phenomena. The growth of papers published on the subject for the last two decades is indeed phenomenal. For instance, the number of papers and the percentage of those to the total presented at the International Conference on the Physics of Semiconductors (ICPS) being held every other year, increased as shown in Figure 17. After the incubation period of several years, it really took off, with a rapid annual growth, around ten years ago. It says that, currently, nearly one half of semiconductor physicists in the world are working in this field. Activity in this new frontier of semiconductor physics has in turn given immeasurable stimulus to device physics, provoking new ideas for applications. Thus, unprecedented transport and opto-electronic devices have emerged. In this interdisciplinary research, beneficial cross-fertilizations are prevalent.

In this review, we have primarily focussed on intriguing phenomena which occur with the application of an electric field in the direction of the SL axis. The significance of such electric field-induced effects was recognized at the inception. The observation of Stark ladders probably constitutes one of the milestones in quantum structure research. It should be appreciated that an energy separation of 10 meV between adjacent Stark ladders was achieved only with 20 kV/cm, while bulk semiconductors would require far greater electric fields for such electric quantization. It is interesting to compare this with magnetic quantization of the orbital motion of electrons, in which an energy separation, $\hbar\omega_c$, of 10 meV between Landau levels for electrons with a small effective mass, is readily obtainable with a magnetic flux of 10 T or so. The electric counterparts of the Landau level and cyclotron resonance are considered to be the Stark ladder and Bloch oscillation, respectively. "Whether the Bloch oscillation in superlattices is really observable or not" is, undoubtedly, one of the most fascinating unanswered questions.

References

1. L. Esaki and R. Tsu, Superlattice and negative conductivity in semiconductors: 1969, *IBM Research Note*, RC-2418.
2. L. Esaki and R. Tsu, Superlattice and negative differential conductivity in semiconductors: 1970, *IBM J. Res. Develop.* **14**, 61.
3. D. Bohm: 1951, *Quantum Theory*, Prentice Hall, Englewood Cliffs, N.J., p. 283.
4. L. Esaki, A bird's-eye view on the evolution of semiconductor superlattices and quantum wells: 1986, *IEEE J. Quantum Electron. QU* **22**, 1611.
5. A.C. Gossard, Growth of microstructures by molecular beam epitaxy: 1986, *IEEE J. Quantum Electron. QU* **22**, 1969.
6. M. Razeghi: 1989, *The MOCVD Challenge*, Adam Hilger, Bristol and Philadelphia.
7. R. Tsu and L. Esaki, Tunneling in a finite superlattice: 1973, *Appl. Phys. Lett.* **22**, 562.
8. L.L. Chang, L. Esaki, and R. Tsu, Resonant tunneling in semiconductor double barriers: 1974, *Appl. Phys. Lett.* **24**, 593.
9. F. Capasso, K. Mohammed and A.Y. Cho, Resonant tunneling through double barriers: 1986, *IEEE J. Quantum Electron. QE* **22**, 1853.
10. E.E. Mendez, W.I. Wang, B. Ricco, and L. Esaki, Resonant tunneling of holes in AlAs–GaAs–AlAs heterostructures: 1985, *Appl. Phys. Lett.* **47**, 415.
11. V.J. Goldman, D.C. Tsui and J.E Cunningham, Evidence for LO-phonon-emission-assisted tunneling in double-barrier heterostructures: 1987, *Phys. Rev. B* **36**, 7635.
12. W. Cai, T.F. Zheng, P. Hu, B. Yudanin and M. Lax, Model of phonon-associated electron tunneling through a semiconductor double barrier: 1989, *Phys. Rev. Lett.* **63**, 418.
13. E.E. Mendez, L. Esaki and W.I. Wang, Resonant magneto-tunneling in GaAlAs–GaAs–GaAlAs heterostructures: 1986, *Phys. Rev. B* **33**, 2893.
14. M.L. Leadbeater, E.S. Alves, L. Eaves, M. Henini, O.H. Hughes, A. Celeste, J.C. Portal, G. Hill and M.A. Pate, Magnetic field studies of elastic scattering and optic-phonon emission in resonant-tunneling devices: 1989, *Phys. Rev. B* **39**, 3438.
15. G.S. Boelinger, A.F. Levi, S. Schmitt-Rink, A. Passner, L.N. Pfeiffer and K.W. West, Direct observation of two-dimensional magnetopolarons in a resonant tunnel junction: 1990, *Phys. Rev. Lett.* **65**, 235.
16. J.F. Young, B.M. Wood, G.C. Aers, R.L. Devine, H.C. Liu, D. Landheer and M. Buchanan, Determination of charge accumulation and its characteristic time in double-barrier resonant tunneling structures using steady-state photoluminescence: 1988, *Phys. Rev. Lett.* **60**, 2085.
17. H. Yoshimura, J.N. Schulman and H. Sakaki, Charge accumulation in a double-barrier resonant-tunneling structure studied by photoluminescence and photoluminescence-excitation spectroscopy: 1990, *Phys. Rev. Lett.* **64**, 2422.

18. A. Tackeuchi, T. Inata, S. Muto and E. Miyauchi, Picosecond characterization of InGaAs/InAlAs resonant tunneling barrier diode by electro-optic sampling: 1989, *Jpn. J. Appl. Phys.* **28**, L750.
19. L.F. Luo, R Beresford and W.I. Wang, Resonant tunneling in AlSb/InAs/AlSb double-barrier heterostructures: 1988, *Appl. Phys. Lett.* **53**, 2320.
20. L. Esaki, L.L. Chang and E.E. Mendez, Polytype superlattices and multi-heterojunctions: 1981, *Jpn. J. Appl. Phys.* **20**, L529.
21. H. Takaoka, Chin-An Chang, E.E. Mendez, L.L. Chang and L. Esaki, GaSb–AlSb–InAs multi-heterojunctions: 1983, *Physica* **117B** & **118B**, 741.
22. L.F. Luo, R. Beresford and W.I. Wang, Interband tunneling in polytype GaSb/AlSb/InAs heterostructures: 1989, *Appl. Phys. Lett.* **55**, 2023; R. Beresford, L.F. Luo, K.F. Longenbach and W.I. Wang, Resonant interband tunneling through a 110 nm InAs quantum well: 1990, *Appl. Phys. Lett.* **56**, 551; Interband tunneling in single-barrier InAs/AlSb/GaSb heterostructures: 1990, *Appl. Phys. Lett.* **56**, 952; E.E. Mendez, H. Ohno, L. Esaki and W.I. Wang, Resonant interband tunneling via Landau levels in polytype heterostructures: 1991, *Phys. Rev. B.* **43**, 5196.
23. J.R. Soderstrom, D.H. Chow and T.C. McGill, New negative differential resistance device based on resonant interband tunneling: 1989, *Appl. Phys. Lett.* **55**, 1094.
24. J.F. Whitaker, G.A. Mourou, T.C.L.G. Sollner and W.D. Goodhue, Picosecond switching time measurement of a resonant tunnel diode: 1988, *Appl. Phys. Lett.* **53**, 385.
25. E.R. Brown, T.C.L.G. Sollner, C.D. Parker, W.D. Goodhue and C.L. Chen, Oscillations up to 420 GHz in GaAs/AlAs resonant tunneling diodes: 1989, *Appl. Phys. Lett.* **55**, 1777.
26. G.H. Wannier, Wave functions and effective Hamiltonian for Bloch electrons in an electric field: 1960, *Phys. Rev.* **117**, 432.
27. H.M. James, Electronic states in perturbed periodic systems: 1949, *Phys. Rev.* **76**, 1611.
28. E.O. Kane, Zener tunneling in semiconductors: 1959, *J. Phys. Chem. Solids* **28**, 181.
29. A. Rabinovitch and J. Zac, Electrons in crystals in a finite-range electric field: 1971, *Phys. Rev. B* **4**, 2358.
30. W. Shockley, Stark ladders for finite, one-dimensional models of crystals: 1972, *Phys. Rev. Lett.* **28**, 349.
31. A.G. Chynoweth, G.H. Wannier, R.A. Logan and D.E. Thomas, Observation of Stark splitting of energy bands by means of tunneling transitions: 1960, *Phys. Rev. Lett.* **5**, 57.
32. S. Maekawa, Nonlinear conduction of ZnS in strong electric fields: 1970, *Phys. Rev. Lett.* **24**, 1175.
33. R.W. Koss, Experimental observation of Wannier levels in semi-insulating GaAs: 1972, *Phys. Rev. B* **5**, 1479.
34. R. Tsu and G. Döhler, Hopping conduction in a superlattice: 1975, *Phys. Rev. B* **12**, 680.
35. M. Artaki and K. Hess, Monte Carlo calculations of electron transport in GaAs/AlGaAs superlattices: 1985, *Superlatt. Microstruct.* **1**, 489.
36. L. Esaki, L.L. Chang, W.E. Howard, and V.L. Rideout, Transport properties of a GaAs-GaAlAs superlattice: 1972, *Proc. 11th Int. Conf. on the Physics of Semiconductors*, Warsaw, Poland, PWN-Polish Scientific Publishers, Warsaw, Poland, p. 431.

37. L. Esaki and L.L. Chang, New transport phenomenon in a semiconductor superlattice: 1974, *Phys. Rev. Lett.* **33**, 495.
38. R. Tsu, L.L. Chang, G.A. Sai-Halasz, and L. Esaki, Effects of quantum states on the photocurrent in a superlattice: 1975, *Phys. Rev. Lett.* **34**, 1509.
39. B. Deveaud, J. Shah, T.C. Damen, B. Lambert and A. Regreny, Bloch transport of electrons and holes in superlattice minibands: direct measurement by subpicosecond luminescence spectroscopy: 1987, *Phys. Rev. Lett.* **58**, 2582.
40. H. Schneider, W.W. Rühle, K. von Klitzing and K. Ploog, Electrical and optical time-of-flight experiments in GaAs/AlAs superlattices: 1989, *Appl. Phys. Lett.* **54**, 2656.
41. M. Helm, P. England, E. Colas, F. DeRosa, and S.J. Allen, Jr., Intersubband emission from semiconductor superlattices excited by sequential resonant tunneling: 1989, *Phys. Rev. Lett.* **63**, 74.
42. M.C. Tatham and J.F. Ryan, Time-resolved Raman measurements intersubband relaxation in GaAs quantum wells: 1989, *Phys. Rev. Lett.* **63**, 1637.
43. A. Sibille, J.F. Palmier, H. Wang, and F. Mollot, Observation of Esaki–Tsu negative differential velocity in GaAs/AlAs superlattices: 1990, *Phys. Rev. Lett.* **64**, 52.
44. H.T. Grahn, H. Schneider, W.W. Rühle, K. von Klitzing and K. Ploog, Nonthermal occupation of higher subbands in semiconductor superlattices via sequential resonant tunneling: 1990, *Phys. Rev. Lett.* **64**, 2426.
45. G. Brozak, M. Helm, F. DeRosa, C.H. Perry, M. Koza, R. Bhat and S.J. Allen, Jr., Thermal saturation of band transport in a superlattice: 1990, *Phys. Rev. Lett.* **64**, 3163.
46. F. Beltram, F. Capasso, D.L. Sivco, A.L. Hutchinson, S.N.G. Chu, and A. Cho, Scattering-controlled transmission resonances and negative differential conductance by field-induced localization in superlattices: 1990, *Phys. Rev. Lett.* **64**, 3167.
47. E.E. Mendez, G. Bastard, L.L. Chang, and L. Esaki, Effect of an electric field on the luminescence of GaAs quantum wells: 1982, *Phys. Rev. B* **26**, 7101.
48. L. Vina, R.T. Collins, E.E. Mendez and W.I. Wang, Excitonic coupling in GaAs/GaAlAs quantum wells in an electric field: 1987, *Phys. Rev. Lett.* **58**, 832.
49. D.S. Chemla, T.C. Damen, D.A.B. Miller, A.C. Gossard, and W. Wiegmann, Electroabsorption by Stark effect on room-temperature excitons in GaAs/GaAlAs multiple quantum well structures: 1983, *Appl. Phys. Lett.* **42**, 864.
50. D.A.B. Miller, J.S. Weiner, and D.S. Chemla, Electric-field dependence of linear optical properties in quantum well structure: 1987, *IEEE J. Quantum Electron. QE* **22**, 1816.
51. J. Bleuse, G. Bastard and P. Voison, Electric-field-induced localization and oscillatory electro-optical properties of semiconductor superlattices: 1988, *Phys. Rev. Lett.* **60**, 220.
52. E.E. Mendez, F. Agullo-Rueda and J.M. Hong, Stark localization in GaAs–GaAlAs superlattices under an electric field: 1988, *Phys. Rev. Lett.* **60**, 2426.
53. F. Agullo-Rueda, E.E. Mendez and J.M. Hong, Quantum coherence in semiconductor superlattices: 1989, *Phys. Rev. B* **40**, 1357.
54. E.E. Mendez, F. Agullo-Rueda and J.M. Hong, Temperature dependence of the electronic coherence of GaAs–GaAlAs superlattices: 1990, *Appl. Phys. Lett.* **56**, 2545.

55. P. Voisin and J. Bleuse, Observation of the Wannier–Stark quantization in a semiconductor superlattice: 1988, *Phys. Rev. Lett.* **61**, 1639.
56. M.M. Dignam and J.E. Sipe, Exciton Stark ladder in $GaAs/Ga_{1-x}Al_xAs$ superlattices 1990, *Phys. Rev. Lett.* **64**, 1797.
57. H. Ohno, E.E. Mendez, J.A. Brum, J.M. Hong, F. Agullo-Rueda, L.L. Chang and L. Esaki, Observation of 'Tamm States' in superlattices: 1990, *Phys. Rev. Lett.* **64**, 2555.

A RECENT STUDY OF ELECTRONIC STRUCTURES OF SEMICONDUCTOR SUPERLATTICES

T. NAKAYAMA
Department of Physics, Chiba University, Yayoi, Chiba 260, Japan.

Abstract. Crystal structures and electronic structures of zinc-compound strained superlattice and GaP/ZnS heterovalent superlattices are calculated by the first-principles pseudopotential method. It is shown that zinc-compound strained superlattices are good quantum well systems and that offset can drastically change under the variation of the strain environment, mechanism of which is analyzed in order to predict the general feature of the variation of offset in strained superlattices. As for the GaP/ZnS heterovalent superlattice, it is clarified that large lattice relaxation occurs at the interface and that the appearance of interface states affects electronic properties of these systems.

1. Introduction

It has been possible by the recent crystal growth technique to synthesize superlattices (SLs) of new category without lattice imperfection; (a) strained superlattices with various types of lattice deformation, and (b) heterovalent superlattices made of materials with different valency. Among such strained superlattices, II–VI zinc-compound (ZnS, ZnSe, ZnTe) superlattices have been extensively studied because they have the potentiality for new kinds of opto-electronic devices. On the other hand, heterovalent superlattices such as GaAs(III–V)/ZnSe(II–VI) have begun to be synthesized in the last few years [1].

In view of this, we have calculated band structures of above mentioned types of short period superlattices from the first principles. By the theoretical calculation, we aim to clarify (1) properties intrinsic to short-period superlattices and (2) properties of the hetero-structure which are common to superlattices of arbitrary layer-thickness. The characteristics obtained from the present theoretical calculation will help the evaluation of the quality of crystal growth and the challenges to produce new types of devices.

Throughout this article, we have adopted the *ab initio* pseudopotential total energy method within the Local Density Approximation (LDA), and investigated electronic structures of short-period superlattices of various pairs of materials and of various layer-thickness.

2. Zinc-Compound Strained Superlattices

2.1. Crystal structures [2]

In this subsection, we comment briefly on crystal structures of strained SLs. Lattice constants of ZnS, ZnSe and ZnTe are 5.41, 5.66 and 6.09 Å, respec-

H. Aoki et al. (eds.), New Horizons in Low-Dimensional Electron Systems, 123–135.

tively. When one uses these materials as constituents of a SL, a strain appears in the SL. In order to determine the crystal structure which includes strain, the valence force field model and the first-principles pseudopotential total energy calculation are adopted. In every SL of $(ZnS)_m/(ZnSe)_n$, $(ZnSe)_m/(ZnTe)_n$ and $(ZnS)_m/(ZnTe)_n$, the following crystal structures are found.

In case of the free-standing SL which is free from the restraint of the substrate, the distribution of strain is uniform apart from the interface by monolayer and the magnitude of strain in inner layers is linear to the ratio $x = n/(m+n)$ of constituent materials. This result indicates that crystal structures are well simulated by the macroscopic elastic theory. On the other hand, in case of the SL on the GaAs substrate, the crystal structure is similar to that of the free-standing SL whose lattice constant parallel to the interface equals to that of bulk GaAs. Furthermore, strain does not depend on the ratio of constituents. In experiments, a free-standing SL is synthesized by epitaxializing on the alloying-substrate with constituents of moderate concentration. Thus we would like to emphasize here that we can easily produce a lot of strained SLs with various strain configurations.

2.2. Electronic structures [2,3]

As for the effect of strain on electronic structures, the following characteristics have been clarified. (1) Energies of degenerate states, for example the energy level at X_z and the averaged energy level between two inequivalent X_{xy} points in the bulk Brillouin zone, split widely. This is because the magnitude of strain parallel to the interface is different from that perpendicular to the interface. (2) In case of the SL ZnS/ZnSe on the GaAs substrate, there is the possibility that, because strain forces the barrier height for the light-hole state lower than that for the heavy-hole, the light-hole state becomes higher in energy than the heavy-hole state as the layer-thickness decreases. This system is the only one among such types of strained SLs, where the transition of the character of the fundamental band gap between the heavy-hole and the light-hole types occurs as the layer-thickness is varied. It should be noted here that this transition is caused not by the applied pressure but by the superstructure itself. When this transition happens, physical properties such as the polarization of the optical spectrum from the fundamental gap and the conductive character change.

As for the characteristics intrinsic to the short-period SL, we have clarified the following; (1) band gap energies of representative lowest conduction band states at Γ, X_{xy}, X_z and L points decrease as the layer-thickness increases. In Table I, for example, we show minimum band gap energies of SLs and bulks on the GaAs substrate, where every band gap energy corresponds to the energy between the top of the valence band state (heavy-hole) and the

TABLE I

Band gap energies and oscillator strengths for the optical transition of minimum energy (at point Γ) of various kinds of superlattices and bulks on the GaAs substrate. Considering the correction to LDA, all the calculated gap energies are shifted by +0.4 eV, 2/2 means the A_2/B_2 superlattice, where A and B are ZnS, ZnSe and ZnTe. The values of calculated (experimentally observed) gap energies (eV) and oscillator strengths (*arb.unit*) of free-standing bulks ZnS, ZnSe and ZnTe are 3.783(3.76) 0.2600, 2531(2.60) 0.2408 and 2.321(2.38) 0.2262, respectively.

bulk	ZnS	ZnSe	ZnTe	ZnS/ZnSe	2/2	3/3	4/4
gap	3.629	2.533	2.298	gap	2.922	2.870	2.829
oscil.	0.2650	0.2405	0.1993	oscil.	0.2147	0.1945	0.1852
ZnSe/ZnTe	2/2	3/3	4/4	ZnS/ZnTe	2/2	3/3	4/4
gap	2.010	1.928	1.872	gap	2.151	2.034	2.035
oscil.	0.1279	0.1026	0.0829	oscil.	0.0957	0.0677	0.0717

bottom of the lowest conduction band state at point Γ. This feature indicates that the wavefunctions of various states are localized in one material even in ultrathin SLs and easily continue to the confined ones in a *quantum well picture*. In fact, the wavefunctions of the heavy-hole, the light-hole and the electron states are more localized in one material with smooth envelopes: the character of each wavefunction is monochromatic, which can be seen from the charge densities in Figure 1. (2) Degenerate states in bulk, such as states at X_{xy} and L points in the Bulk Brillouin zone, split or coincide in energy depending on the even-odd nature of the layer-thickness. This oscillation (split- and stick-together phenomena) is caused by the symmetry of the superstructure.

As opto-electronic devices, it is important to estimate the range of gap energy of zinc-compound SLs. Using the Kronig–Penney model for energy levels of thicker-layer SLs and the calculated band gap energies of short-period SLs, the ranges of band gap energy of SLs on the GaAs substrate are 2.5 $\sim$ 2.9 eV for ZnS/ZnSe, 1.2 $\sim$ 2.0 eV for ZnSe/ZnTe and 1.5 $\sim$ 2.2 eV for ZnS/ZnTe SLs, respectively.

2.3. Valence band offset

Among the physical quantities which characterize the hetero-interface, band offset is the most important one. From the calculated results of the electronic structures of short-period SLs, we obtained band offsets for various strained structures. Since the heavy-hole state often appears as a band edge state, we consider the band offset for the heavy-hole state $E_{v,hh}$ in the following [4]. The method for obtaining the band offset of the heavy-hole state is roughly shown in Ref. [2].

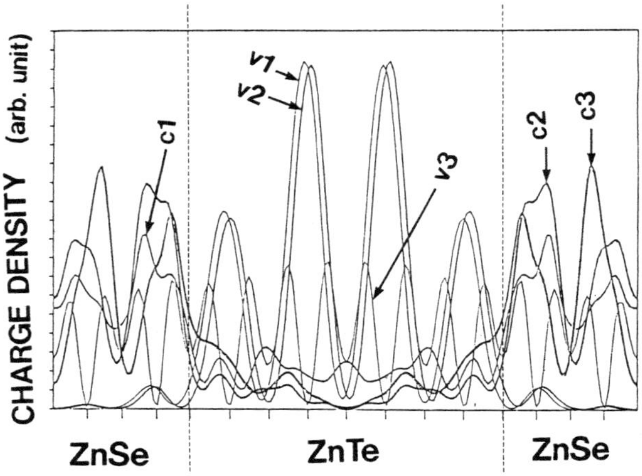

Fig. 1. Charge densities of the higher valence band states and the lower conduction band states at Γ of $(ZnSe)_3/(ZnTe)_3$ on the GaAs substrate. The charge density is averaged parallel to the interface and displayed along the SL direction. n in cn or vn is numbered in order from the band gap. Especially $c1$ is the lowest conduction band state at Γ, $v1$ and $v2$ the heavy-hole states at Γ and $v3$ the light-hole state at Γ.

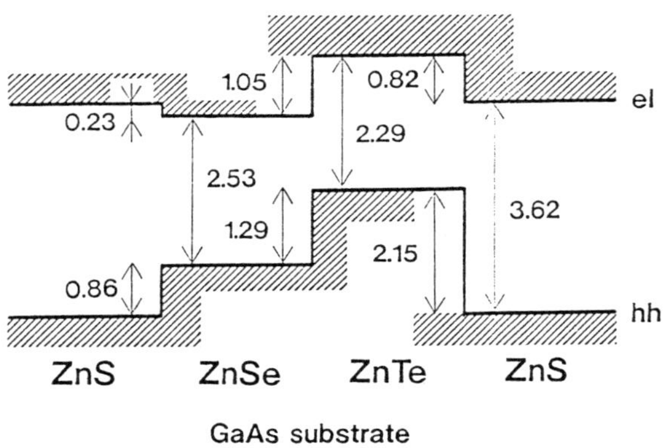

Fig. 2. Band offsets of various zinc-compound hetero-structures on the GaAs substrate. All the values are in eV.

2.3.1. Offset on the GaAs substrate

First, we consider the variation of band offset when the constituent materials are changed. Figure 2 shows band offsets of three kinds of heterostructures. Here we consider SLs on the GaAs substrate. (1) Among the present SLs, *transitivity* of band offset is very well satisfied; $\Delta E_{v,hh}(C-A) = \Delta E_{v,hh}(C-B) + \Delta E_{v,hh}(B-A)$. This is because the quantity of charge transfer through the interface is additive, corresponding to the combination of materials. This result indicates that under the similar strained configuration there exists *some* midgap energy, as pointed out by Tersoff [5] or Cardona and Christensen [6] who discussed the natural (strain-free) offset or the one under the hydrostatic deformation, while in the present case the shear strain also exists. (2) The direction of charge transfer is ZnS $\Rightarrow$ ZnSe, ZnSe $\Rightarrow$ ZnTe and ZnS $\Rightarrow$ ZnTe, respectively. This tendency also coincides with their natural predictions. (3) Using the calculated heavy-hole offsets and the experimentally observed band gap energies of bulk ZnS, ZnSe and ZnTe, we evaluate the conduction band offsets, which are also shown in Figure 2. It is concluded that ZnS/ZnSe SL is type-I while ZnSe/ZnTe and ZnS/ZnTe are type-II, which is consistent with the experimental results [7]. This type-II band alignment can also be seen in small oscillator strengths of the latter SLs in Table I. The calculated heavy-hole offset of ZnS/ZnSe system (0.86 eV) is in good agreement with the values determined experimentally (0.75–0.90 eV) [7].

2.3.2. Offset vs. strain

Next, we discuss the relation between band offset and the magnitude of strain. Figure 3 shows offsets ($\Delta E_{v,hh}$) of $(ZnS)_m/(ZnSe)_n$, $(ZnSe)_m/(ZnTe)_n$ and $(ZnS)_m/(ZnTe)_n$ systems as a function of $x = n/(m+n)$, which is linearly dependent on strain. Offset in the strained crystal structure is determined mainly by the following two contributions: (1) the strain-induced energy-level-shift of the heavy-hole state $\Delta\varepsilon$, which is related to the deformation potential and is purely the bulk property. (2) The height of dipole-field (total potential) ΔV at the interface caused by charge transfer $\Delta\rho$, which is intrinsic to the hetero-interface. In all of the three systems, with increasing x, the former decreases offset, while the latter increases offset. The mechanism of this variation can be explained qualitatively by the following two-level tight-binding model at the interface (Figure 4), where ε_s and ε_p are energies representative to the conduction and the valence bands of constituent, various t's are transfer energies. When only $\varepsilon_{p'}$ of material A' decreases due to the change of strain-configuration in A' (contribution (1)), the charge transfer from A to A' is induced (contribution (2)) because the charge transfer from A to A' depends on $\varepsilon_{p'}$ as $-|t_{sp'}|^2/(\varepsilon_s - \varepsilon_{p'})$. Thus these two contri-

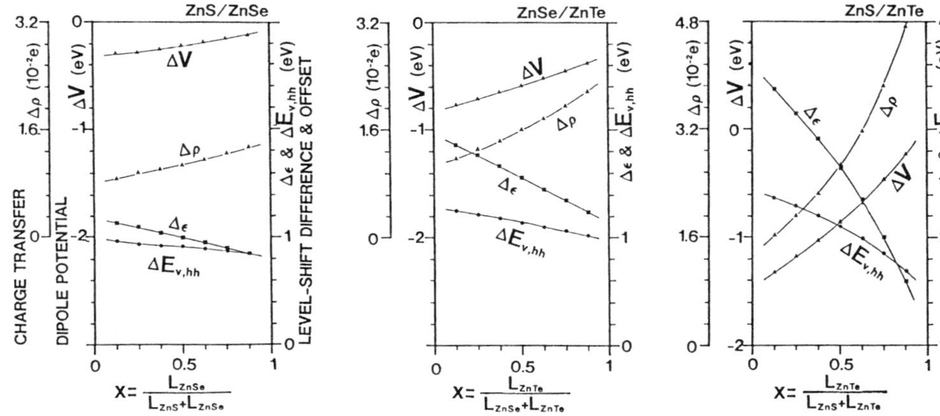

Fig. 3. Band offsets ($\Delta E_{v,hh}$) for the heavy-hole states of zinc-compound superlattices as a function of the ratio (x) of constituent materials; i.e. as a function of the magnitude of strain. $\Delta\rho$: the charge transfer, ΔV: the dipole-field at the interface, $\Delta\varepsilon$: the difference of the strain-induced energy-shift of the heavy-hole state between constituents.

butions cancel each other: the valence-band offset does not change as large as the anticipated value from the difference of the deformation potential between constituents, due to the accompanying charge transfer.

3. GaP/ZnS Heterovalent Superlattices

3.1. Crystal structures

Heterovalent interfaces have been constructed so far, for example, when ZnSe is epitaxialized on the GaAs substrate, and heterovalent SLs have begun to be synthesized. Among various heterovalent SLs, we pick up GaP/ZnS SLs, which are lattice-matched but valency mismatched systems. (Lattice constants of GaP and ZnS are 5.15 and 5.11 a.u.) Corresponding to the growth process, many crystal structures are considered to be realized in GaP/ZnS SLs. For example, the crystal structure of (110) nonpolar interface is considered to be quite different from that of (001) polar interface [8]. Moreover, there is a possibility of the existence of two kinds of interfaces in GaP/ZnS SLs; P–Zn (H) interface and P–Ga (E) interface. The former lacks one electron (hole-like) and the latter possesses excess one electron (electron-like) in a sense of ordinary covalent-bond picture. Then in GaP/ZnS SLs there are three types: (H & H), (E & E) and (H & E). Here we suppose only the following crystal structure. We concentrate on SLs with (001)–(H & E) interfaces on the GaP substrate. Moreover, we assume no additional superstructure along the interface [8].

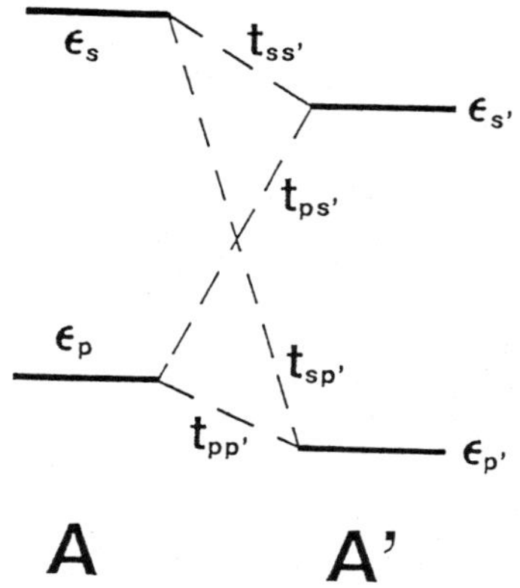

Fig. 4. Two-level tight-binding model at the interface.

As the candidates of crystal structures of GaP/ZnS SLs, we consider two cases (see Table II). Case A is the intermediate-bond-length case, where bond lengths in inner regions of SL layers equal to those of bulk materials on the GaP substrate and bond lengths at two interfaces are taken as the averaged value of those of bulk materials. Case B is the relaxed case, where, considering the tendency of the optimized structure within the pseudopotential total-energy method, the bond length (P–Zn) is decreased and the bond length (S–Ga) is increased compared to case A. As seen in the calculated results from the first-principles in Table II, we expect that the bond length at the heterovalent interface is drastically changed from that of the bulk materials.

3.2. Electronic structures

First we consider the charge distribution in the heterovalent SL. How does the charge redistribute in H & E interfaces? Table III shows the integrated charge between the nearest neighboring atoms along the SL direction. In the bulk and homovalent SL, the value of integrated charge becomes 4. As can be seen from this table, the integrated charge around the H(P–Zn)-interface lacks about one electron and that around the E(S–Ga)-interface possesses excess one electron with little relation to the bond-relaxation at the interface. In other words, excess charge at the E-interface hardly transfers to the H-interface. The net charge transfer from the E- to the H-interface is at most 2%. Then where is the excess electron (hole) situated? Analyzing total valence charge density, the charge is mainly accumulated around atoms, and extend its hands somewhat towards the neighboring atoms, Moreover, the charge distribution around each atom is very similar to those in bulk materials. Thus we had better try to understand the bonding at the interface

TABLE II

Interatomic distances of hypothetical crystal structures (case A and B) for investigating electronic structures of heterovalent SLs, and those optimized by the total energy calculation, in a.u. All these SLs are assumed on the GaP substrate. Thus lattice constants parallel to the interface equal to that of GaP. In the present calculation, 3d electrons of Zn atoms are not included as valence electrons. Then the optimized interatomic distance in bulk ZnS is fairly short. Relaxed-C corresponds to the case where only the interactomic distance at the interface is allowed to relax, while in Relaxed-D all the distances along the SL direction are relaxed. Both the relaxed cases C and D are for $(GaP)_1/(ZnS)_1$.

structure	Ga–P inner	P–Zn interface	Zn–S inner	S–Ga interface
case A	2.576	2.552	2.528	2.552
case B	2.576	2.212	2.528	2.898
bulk GaP	2.499			
bulk ZnS			1.695	
Relaxed-C	2.499	1.953	1.695	2.579
Relaxed-D	2.25	1.85	1.74	2.44

TABLE III

Interatomic charge in GaP/ZnS SL. n/n corresponds to the superlattice $(GaP)_n/(ZnS)_n$.

Ga	P	Ga	P	Ga	P	Zn	S	Zn	S	Zn	S	Ga
case A												
1/1					3.986	3.588	4.023					4.402
2/2			4.006	3.963	4.003	3.578	3.998	4.034	3.999			4.415
3/3	4.005	3.975	4.023	3.963	4.003	3.575	4.000	4.026	3.974	4.032	3.999	4.418
case B												
1/1					3.975	3.418	4.088					4.517
2/2			3.950	3.965	4.050	3.419	4.105	4.030	3.957			4.521
3/3	3.950	3.971	4.024	3.967	4.049	3.417	4.106	4.028	3.973	4.027	3.958	4.523

from the ionic bonding picture than from the covalent bonding picture.

Although the charge transfer from the E to the H-interface is small, this transfer strongly influences the electronic structures. Figures 5(a) and (b) show total self-consistent potentials along the SL direction for cases A and B, which are averaged parallel to the interface. As seen in these figures, the above mentioned charge transfer causes the large electric field (polarization). Comparing potentials for cases A and B, the lattice relaxation decreases the strength of the electric field. Thus we may expect that in a heterovalent SL the lattice relaxation occurs around the interface, and that the magnitude of the relaxation is determined from the balance of energies of both polarization and elasticity.

The calculated band structure of $(GaP)_3/(ZnS)_3$ in case B is shown in

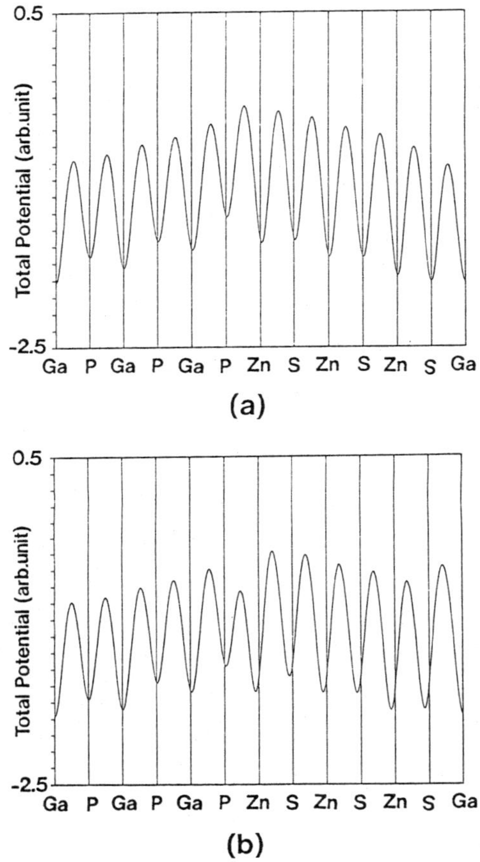

Fig. 5. Total self-consistent potential along the SL direction. (a): case A, (b): case B.

Figure 6. Around the band gap interface states appear. In order to analyze the origin of interface states, we show the charge density of these states in Figure 7, where the charge density is averaged along the interface. At first, the higher valence band states are localized near the H-interface, while the lower conduction states near the E-interface. This is easily understood from the aligned potential profile shown in Figure 5. In this way, as for the states around the fundamental gap, the electronic structure of the (H & E) heterovalent SL is very similar to that of the modulation doped SL (i.e. dipole layered SL).

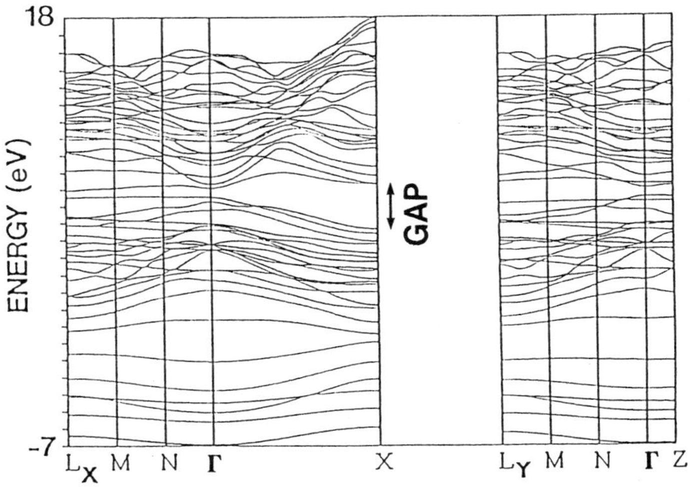

Fig. 6. Band structure of case B $(GaP)_3/(ZnS)_3$. Dispersion is displayed along ΓL_x, ΓL_y and ΓX lines corresponding to the bulk ΓL and ΓX lines.

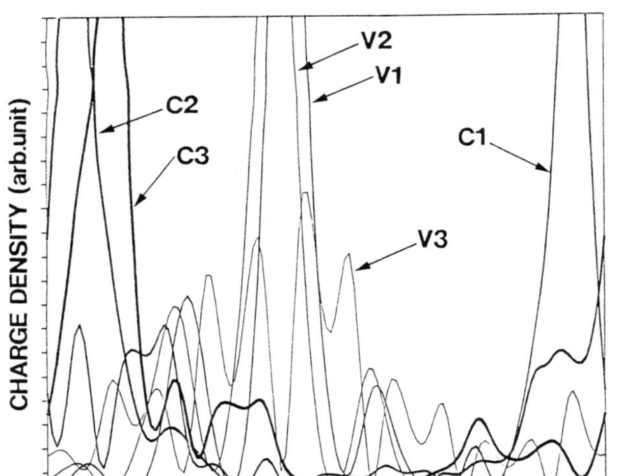

Fig. 7. Charge densities of states around the band gap at Γ point of case B $(GaP)_3/(ZnS)_3$ along the SL direction. cn is the n-th lowest conduction band state and vn the n-th highest valence band state. Especially $c1$ is the lowest conduction band state at Γ, $v1$ and $v2$ the heavy-hole states at Γ and $v3$ the light-hole state at Γ.

TABLE IV

Band gap energies (eV) of SLs of various layer-thickness and oscillator strengths (10^{-6}a.u.$^{-2}$) for various optical transitions at Γ point. n/n corresponds to $(GaP)_n/(ZnS)_n$. $v2 \rightarrow c1$ corresponds to the oscillator strength of the interband transition from the second highest valence band state to the lowest conduction band state. Symbcls in parentheses are polarizations of light. All the band gaps of SLs are direct.

	1/1	2/2	3/3	GaP(bulk)	ZnS(bulk)
case A					
band gap	0.988	0.729	≤ 0	2.02 (X) (indirect)	3.72 (direct)
$v1 \rightarrow c1(y)$	0.0849	0.0302		1.0404	0.9077
$v2 \rightarrow c1(x)$	0.2255	0.0699		1.0404	0.9077
$v3 \rightarrow c1(z)$	0.4862	0.0983		1.0415	0.8785
case B					
band gap	1.620	1.336	0.514		
$v1 \rightarrow c1(y)$	0.1896	0.0519	0.0088		
$v2 \rightarrow c1(x)$	0.4559	0.1205	0.0163		
$v3 \rightarrow c1(z)$	0.5437	0.1457	0.0035		

3.3. Optical properties [9]

In a sense of SL, this system is a spatially type-II SL, because the hole and the electron states are localized separately. As the charge transfer from the E- to the H-interface is almost the same in every SL in Table III, the magnitude of the internal electric field is considered to be constant in SLs with various layer thickness. Thus, as the layer thickness increases, the band gap sharply decreases, which can be also seen from the calculated band gap energies shown in Table IV. Moreover, as the layer thickness increases more than a few layers, there is a possibility that the system becomes semi-metallic or that the additional large lattice relaxation occurs.

The highest valence band state ($v1$) is the heavy-hole state with p_y atomic character, which shows little dispersion along the ΓL_x direction, while the second highest one ($v2$) is the heavy-hole state with p_x character, which shows little dispersion along the ΓL_y. The third highest valence one ($v3$) is the light-hole state with p_z character. In the present calculation, the spin-orbit interaction is not included. However, the degeneracy of the valence band top is completely solved due to the crystal symmetry. This crystal symmetry of GaP/ZnS SL also splits the E_1 shoulder of the optical transition of bulk semicconductor into two shoulders.

As seen in Figure 6, the effective masses of these three highest valence band states are strongly anisotropic, and there appears a strong trichroism in the transitions from these higher valence band states to the lowest conduction band state due to the difference of atomic character between these states. (See the oscillator strength in Table IV.) As the layer thickness increases, all the strengths of these interband transitions decrease. This is because these systems are spatially type-II systems.

4. Conclusion

In the present article we have introduced our recent study of the electronic structures of new category of semiconductor superlattices. As for the strained superlattice, we have selected zinc-compound (001) strained superlattices and clarified that strain causes large level splitting and that the strain-induced light-hole gap appears in some strained superlattices. These superlattices become good quantum well systems. With varying the strain environment, band offset drastically changes due to the energy-level shift by strain and charge transfer at the interface, mechanism of which is analyzed in order to predict the general feature of the variation of offset in strained superlattices. On the other hand, as for the heterovalent superlattice, we have selected GaP/ZnS superlattices with (001) polar interfaces and clarified that the bond lengths at the interface change drastically from those of the bulk materials in order to decrease the polarization field in the superlattice. Around the fundamental gap, there appear resonant states localized at the interfaces, which make GaP/ZnS superlattice a direct-gap type-II semiconductor and show large anisotropy in the conductive and optical properties. Experimental investigations for these systems have just begun. We hope that more intensive studies for these kinds of unexplored superlattices will be done in the near future.

This work is supported by a Grant-in-Aid from the Ministry of Education, Science and Culture, Japan.

References

1. N. Kobayashi: 1989, *Appl. Phys. Lett.* **55**, 1235; S. Ramesh, N. Kobayashi and Y. Horikoshi: 1990, *Appl. Phys. Lett.* **57**, 1102.
2. T. Nakayama: 1990, *J. Phys. Soc. Jpn.* **59**, 1029.
3. T. Nakayama: 1990, *Proc. 20th Int. Conf. Phys. Semiconductors*, Greece.
4. In the present *ab initio* calculations, we have not included the effect of spin-orbit interaction. Thus we include such effect *a posteriori* and obtain the heavy-hole state offset as $\Delta E_{v,hh} = \Delta\varepsilon + \Delta V + \Delta(\Delta_0)/3$.
5. J. Tersoff: 1986, *Phys. Rev. Lett.* **56**, 2755.
6. M. Cardona and N.E. Christensen: 1987, *Phys. Rev. B* **35**, 6182.
7. H. Fujiyasu, K. Mochizuki, Y. Yamazaki, M. Aoki, A. Sasaki, H. Kuwabara, Y. Nakanishi and G. Shimaoka: 1985, *2nd Int. Conf. on Modulated Semiconductor Structures*, Kyoto; T. Taguchi, S. Hayamizu, H. Fujiyasu, M. Satoh, T. Yao and A. Hiraki: 1987, *Proc. 12th Int. Symp. Application of Ion Beams in Material*

Sciences, Hosei University, Tokyo, Japan, T. Sebe and Y. Yamamoto (eds.), Hosei University Press, p. 511; Y. Kawakami, T. Taguchi and A. Hiraki: 1988, *J. Cryst. Growth* **93**, 714; N. Teraguchi, Y. Takemura, R. Kimura, M. Konagai and K. Takahashi: 1988, *J. Cryst. Growth* **93**, 720.

8. G.A. Baraff, J.A. Appelbaum and D.R. Hamann: 1977, *Phys. Rev. Lett.* **38**, 237; W.A. Harrison, E.A. Kraut, J.R. Waldrop and R.W. Grant: 1976, *Phys. Rev. B* **18**, 4402; J. Pollmann and S.T. Pantelides: 1980, *Phys. Rev. B* **21**, 709; R. Eppenga: 1989, *Phys. Rev. B* **40**, 10402; A. Munoz, N. Chetty and R.M. Martin: 1990, *Phys. Rev. B* **41**, 2976; R.G. Dandrea, S. Froyen and A. Zunger: 1990, *Phys. Rev. B* **42**, 3213; D.M. Bylander and L. Kleinman: 1990, *Phys. Rev. B* **41**, 3509.
9. T. Nakayama, to be published.

QUANTUM INTERFERENCE AND MAGNETO-RESISTANCE IN GALLIUM ARSENIDE STRUCTURES

D.J. NEWSON[1] and M. PEPPER[2]
Cavendish Laboratory, Madingley Road, Cambridge CB3 0HE, U.K.

Abstract. A brief account is presented on the application of Gallium Arsenide structures to studies of quantum interference and its suppression by a magnetic field, leading to a negative magneto-resistance. The MESFET structure allows the 3D to 2D transition to be studied. It is shown that various regions of magneto-resistance can be encountered depending on the strength and direction of the magnetic field. In particular, a parallel field can be used to convert two dimensional behaviour into three. Excellent agreement is found with theoretical predictions for both dimensions. We conclude with a brief account of results with a narrow Gallium Arsenide heterojunction in which a 2D electron gas exhibits one dimensional quantum interference, and, possible, flux cancellation effects may be removed by small angle scattering.

1. Introduction

Since the work of Abrahams *et al.* [1] and Gorkov *et al.* [2] it has been known that quantum interference plays a major role in determining the conductivity of metallic systems at low temperatures. The Boltzmann conductivity,

$$\sigma_B = \frac{ne^2\tau}{m} \tag{1}$$

(where n is the carrier concentration, τ is the elastic scattering time and m is the effective mass), is subject to a correction which varies with dimensionality and is given by:

$$\text{3D} \qquad \sigma = \sigma_B - \frac{e^2}{2\pi^2\hbar}\left[\frac{1}{l} - \frac{1}{L_\varphi}\right] \tag{2}$$

$$\text{2D} \qquad \sigma = \sigma_B - \frac{e^2}{2\pi^2\hbar}\ \ln(L_\varphi/l) \tag{3}$$

$$\text{1D} \qquad \sigma = \sigma_B - \frac{e^2 L_\varphi}{\pi\,\hbar}\ . \tag{4}$$

[1] Current address: British Telecom Research Laboratories, Martlesham Heath, Ipswich, Suffolk IP5 7RE, U.K.
[2] Also at the Toshiba Cambridge Research Centre, 260 Cambridge Science Park, Milton Road, Cambridge CB4 4WE, U.K.

H. Aoki et al. (eds.), New Horizons in Low-Dimensional Electron Systems, 137–153.

In the above expressions l and L_φ are the elastic and phase coherence mean free paths, respectively. As $T \to 0$ so $L_\varphi \to \infty$, or, more realistically, the sample length, and, as seen, the 3D correction saturates but in 2D and 1D the correction becomes increasingly large and is the precursor to the localisation of all states. In general, quantum interference becomes increasingly significant as $k_F l$ approaches unity where k_F is the Fermi wave vector.

The physical origin of the effect lies in the interference between two partial electron waves which have completed identical, but time reversed, loops. This gives rise to an enhanced probability of an electron returning to the origin, and a reduced diffusion coefficient. The loops increase in size as the temperature drops and L_φ, which cuts off the interference, increases. All the other diffusion loops which produce interference will average to zero in a macroscopic sample. However, in a small sample the averaging does not occur, resulting in the observation of universal conductance fluctuations [3], which will not be discussed here.

Application of a magnetic field results in a change of phase between the two partial waves, so decreasing magneto-resistance has the following forms for a degenerate electron gas in the regime $k_F l > 1$:

3D Kawabata [4] has found the following asymptotic limits depending on the values of L_c, the cyclotron length, which determines the scaling length cutting off the interference, and L_φ.

$$L_c \gg L_\varphi; \ \delta\sigma(B) = \frac{e^2}{12\pi^2\hbar}\left[\frac{e\,B}{\hbar}\right]^{1/2}\left[\frac{De\,B\tau_\varphi}{\hbar}\right]^{3/2} \tag{5}$$

$$L_c \ll L_\varphi; \ \delta\sigma(B) = \frac{e^2}{12\pi^2\hbar}\left[\frac{e\,B}{\hbar}\right]^{1/2} \tag{6}$$

$$= 289.3\,\sqrt{B}\ (\Omega m)^{-1}\ (\text{Tesla})^{1/2}\ .$$

Here B is the magnetic field, τ_φ is the phase relaxation time and D is the diffusion coefficient of electrons at the Fermi energy.

2D Hikami *et al.* [5] find that for a transverse field

$$\delta\sigma(B) = \frac{\alpha e^2}{12\pi^2\hbar}\, f\left(\frac{4e\,BD\,\tau_\varphi}{\hbar}\right) \tag{7}$$

α is a constant of order unity and

$$f(x) = \ln(x) + \psi\left(\frac{1}{2} + \frac{1}{x}\right)$$

ψ is the digamma function, $f(x)$ has asymptotic forms,

$$\text{for } x \ll 1\ ,\ (L_c > L_\varphi)\ ,\ f(x) = x^2/24$$

$$x \gg 1\ ,\ (L_c < L_\varphi)\ ,\ f(x) = \ln x\ .$$

For a parallel field, Altshuler and Aronov [6] find that for $L_c \gg L_\varphi$

$$\delta\sigma(B) = \frac{e^2}{2\pi^2\hbar} \ln\left(\frac{l + t^2 L_{\varphi^2} B^2 e^2}{3\,\hbar^2}\right)\ . \tag{8}$$

For $L_c \ll t$, where t is the thickness of the layer, the field effectively converts 2D behaviour to 3D and the high field 3D formula should apply.

The case of one dimension will be discussed later in this article.

We propose to discuss results which were obtained on two different types of device. The first is the GaAs MESFET in which a thin (10^{-4}cm) layer of GaAs doped above the critical concentration for the metal-insulator transition forms the active layer. By application of a voltage to the Schottky gate, it is possible to decrease the thickness of the conducting layer to zero [7], with consequent observation of a reduction in the dimensionality of transport. The second device is the modulation doped GaAs–AlGaAs heterojunction [8] which will produce a 2D electron gas, the mobility of which can be sufficiently high as to give a value of elastic mean free path of $\approx 10^{-3}$cm.

While much rigorous work has been carried out to successfully explain the effects of both temperature and magnetic fields on the conductivity in truly 2D or 2D systems, such as the Si-MOSFET and bulk semiconductors, much less effort has been applied to the case of intermediate dimensionality [9–11]. Here many subbands may be occupied but certain effects are 2D because the phase relaxation length (L_φ) is greater than the channel thickness. The authors extended the physically intuitive model [12,13] of Bergmann to account for the observed effects.

2. Model for Calculating Magneto-Resistance of a Thin 3D Conductor ($L_\varphi > a$)

Bergmann pointed out that weak localisation is caused by self-intersecting diffusive paths. While electrons transversing such paths have undergone many elastic scattering events the paths are coherent. Such coherent back-scattering effectively reduces the diffusion constant, and gives rise to a conductivity correction dependent on phase relaxation time, and, consequently, temperature and magnetic field. The model requires the proportion of available states which contribute to back-scattering to be integrated between appropriate limits to obtain the correction to the electrical conductivity. In a 2D electron gas, the available area for conduction in reciprocal space is $2\pi^2 k_f/l$. In the absence of magnetic field the available area for coherent scattering is $\frac{1}{Dt}$ integrated between the limits τ and τ_φ. Magnetic fields

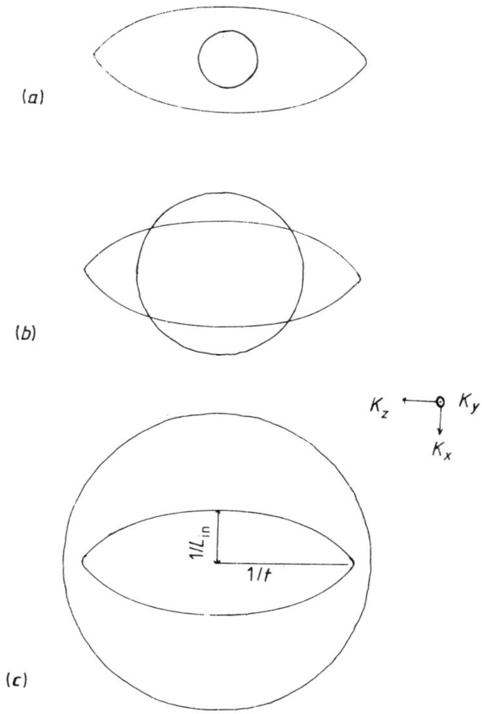

Fig. 1. Illustrates in k space the coherently backscattered spot at $-k$ for a conducting channel in the presence of a parallel magnetic field. L_c; L_{in} and t represent, respectively, the cyclotron length, the inelastic scattering length (equivalent to L_φ) and the thickness. (a) Weak magnetic field. (b) Strong magnetic field. (c) Intermediate situation.

introduce a phase difference between the time reversed paths with a consequent reduction in the area of coherent backscattering. A characteristic length $L_c = \sqrt{(\hbar/eB)}$ is introduced, in which an electron wave changes phase by 2π. The following discussion is found in References [10,11 and 14].

In the presence of a magnetic field the integration limit is no longer necessarily isotropic: it depends on the intersection of two ellipsoids and in general there are three distinct regions of conductivity correction, Figure 1. These will now be discussed.

(i) Weak Parallel Magnetic Fields

The magnetic spheroid is totally enclosed by the inelastic spheroid and hence the magnetic field has comparatively little effect on quantum interference.

(ii) Strong Parallel Magnetic Fields

The magnetic sphere is totally outside the inelastic spheroid and the inelastic effects have little effect on quantum interference. The magnetoresistance is temperature independent.

(iii) Intermediate Parallel Magnetic Fields
The magnetic and inelastic spheroids intersect and the conductivity correction has a complicated form, dependent on both magnetic field and temperature.

The effects will now be modelled quantitatively. In the following discussion we stress that weak localisation is an interference effect, with magnitude determined by L_c and L_φ.

2.1. Weak parallel magnetic fields

Magnetic fields reduce the number of coherently back scattered states. The magnetic and inelastic effects are on very different length scales and the system behaves as an interferometer with two inputs separated by more than the coherence length. Consequently intensities must be subtracted.

The amplitude is determined by the volume of the relevant spot. Thus intensity varies as (volume)^2. Hence, the effective volume in k-space

$$= \frac{1}{\sqrt{D\tau}} \left[\frac{1}{D\tau a^2} - \frac{1}{4L_c^4}\right]^{1/2}$$

$$= \frac{1}{aD\tau} \left[1 - \frac{a^2 D\tau}{8L_c^4}\right]$$

$$I_{\text{coh}} = \frac{al}{2\pi^2 k_f} \times \frac{1}{aD\tau} \left[1 - \frac{a^2 D\tau}{8L_c^4}\right] .$$

Integrating I_{coh} to obtain $\delta\sigma$

$$\delta\sigma = \frac{e^2}{2\pi^2\hbar} \left[-\ln(\tau_\varphi/\tau) + \frac{a^2 L_{\text{in}}^2}{8L_c}\right] .$$

Hence,

$$\delta\sigma(B) = \frac{e^4 a^2 L_{\text{in}}^2 B^2}{16\pi^2\hbar^3} . \tag{9}$$

This is compared with the rigorous low-field result of Altshuler and Aronov [6],

$$\delta\sigma(B) = \frac{e^2}{24\pi^2\hbar} \ln\left[1 + \frac{a^2 L_{\text{in}}^2 4e^2 B^2}{\hbar^2}\right] . \tag{10}$$

Approximating $\ln(1+x) = x$, the approaches give agreement to within a numerical factor of 2.5.

2.2. High parallel magnetic fields

In this case, $L_c < t < L_\varphi$, so the relevant cut-off spot is a sphere of radius $1/L_c$. Hence,

$$I_{\rm coh} = \frac{1}{L_c^3} \frac{\pi a l}{2\pi^2 k_f} \ .$$

Integration to obtain $\delta\sigma$ is now between τ and τ_B

$$\delta\sigma_{\rm 2D} = \frac{e^2}{2\pi^2\hbar} \, a(1/L_c - 1/1) \ . \tag{11}$$

To turn this into an equivalent 3D conductivity, Equation 11 is divided by thickness, a

$$\delta\sigma_{\rm 3D}(B) = \frac{e^2}{2\pi^2\hbar L_c} \ . \tag{12}$$

This should be compared with the high field result of Kawabata [4], namely,

$$\delta\sigma_{\rm 3D}(B) = \frac{0.605 e^2}{2\pi^2\hbar L_c}$$

giving

$$\delta\sigma_{\rm 3D}(B) = 289.3 \sqrt{B} \ (\Omega m)^{-1} \ {\rm Tesla}^{-1/2} \ . \tag{13}$$

In strong parallel magnetic fields, the magnetoresistance can be said to undergo a 2D to 3D transition.

2.3. Intermediate parallel magnetic fields

This is the most interesting case of three outlined above. Here the two spheroids intersect. Hence, the exterior surface is not solely governed by inelastic or magnetic effects but by both.

The easiest way of regarding this is to imagine the magnetic spheroid as cutting out part of the elastic spheroid. Consider the plane perpendicular to the magnetic field. Quantum interference can proceed in that area of the ellipse not incorporated in a circle of radius $1/L_c$.

To a first approximation this area is calculated to be equal to A given by

$$A = \left[\frac{1}{a} - \frac{1}{L_c}\right]^2 \frac{a}{L_\varphi} \ .$$

The relevant volume, V, in k-space is then given by multiplying by $1/L_\varphi$.

$$V = \left[\frac{1}{a} - \frac{1}{L_c}\right]^2 \frac{a}{L_{\varphi^2}} \ .$$

So

$$I_{\rm coh} = \frac{1}{2\pi^2 k_f} \, (1 - 2a/L_c) \, \frac{1}{Dt}$$

to a first approximation. The dummy variable, t, runs between τ and τ_φ, giving the conductivity correction as

$$\delta\sigma_{\rm 2D} = \frac{e^2}{2\pi^2 \hbar} \left[1 - \frac{2a}{L_c}\right] \ln T \tag{14}$$

or

$$\delta\sigma_{\rm 2D}(B) = -a \sqrt{B} \, \ln T \; . \tag{15}$$

As a tends to zero $\delta\sigma(B) \to 0$: agreeing with rigorous theory and experimental results.

3. Experimental Results

To check the parallel field predictions, measurements were made on GaAs MESFETs, these were first published in References [10] and [11]. Under certain conditions the channel acts as a doped 2DEG of variable thickness. It was found that a doping level of $7 \times 10^{16}\text{cm}^{-3}$ was optimal for these measurements, being sufficiently close to the metal-insulator transition for quantum interference effects to be important but the diffusion constant and phase relaxation time were both much larger than in samples doped just above the metal-insulator transition, giving 2D quantum interference effects over an enhanced range of channel thickness. The thickness of the electron gas, a, could be varied between zero and 2000 Å by applying a gate voltage in the normal way. The experiments were carried out under conditions of temperature and magnetic field such that interaction [15] effects were not significant.

Figure 2 shows the perpendicular magnetoresistance for $V_g = 0.0$ V (corresponding to a channel thickness of a = 1300 Å), plotted against ln B, showing logarithmic behaviour at high fields, indicative of 2D weak localisation (quantum interference) effects. Fitting these data to the theory of Hikami *et al.* [5] gave a phase relaxation length of 0.4 μm at 1.3 K. As $L_\varphi > a$, quantum interference effects were two dimensional.

In Figure 3 the parallel magnetoresistance is plotted against $B^{1/2}$ with the device under the same conditions as in Figure 2. At high fields a temperature-independent term is observed, varying as $B^{1/2}$. The resistance change per square in 1.0 Tesla$^{1/2}$, is 126.0Ω. Using a = 1300 Å and R, resistance per square, of 1867Ω at 4.2 K, gives a 3D conductivity increase of $\delta\sigma_{\rm 3D}$ = 277.0 $(\Omega\text{m})^{-1}$ $(\text{Tesla})^{-1/2}$, i.e. agreement to within 5% of the theoretical prediction fo Kawabata [6].

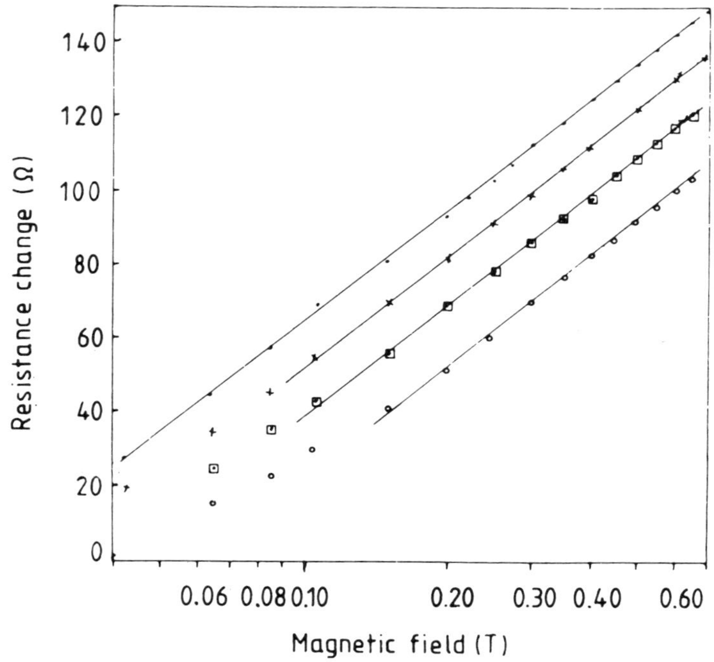

Fig. 2. The resistance change in a transverse magnetic field (in Tesla) is plotted against ln B, showing 2D behaviour for different temperatures. The channel thickness is 1300 Å and the doping is 7×10^{16} donors cm^{-3}: • 1.3 K; × 1.75 K; ◦ 2.25 K; ◦ 0.3 K.

The crossover to 3D behaviour in the presence of a parallel field occurs at a field of about 0.4 Tesla, when $L_c = 400$ Å, $= 0.3$ of the channel thickness. The field increases at lower temperatures, as expected from an effective length-scale argument. In a high parallel magnetic field, the resistance change becomes larger than in a perpendicular field, because of 3D behaviour.

Figure 4 shows the results at intermediate fields for the same channel thickness as in Figure 3. The change in resistance is plotted; this only depends on the localisation correction as interaction effects are not enhanced.

A temperature dependent magneto-resistance is illustrated in Figure 5, where the results of Figure 4 are plotted against ln T. This implies that $\delta\sigma = B^{1/2}$ ln T, as predicted in Equation 15.

In Figure 6, the very low field data is plotted against B^2: showing that the behaviour is obeying Equation 9. Hence, at one value of thickness, all three postulated regimes of conduction in a parallel magnetic field have been observed.

To find the dependence on channel thickness, similar measurements were repeated at various gate voltages and hence channel thicknesses.

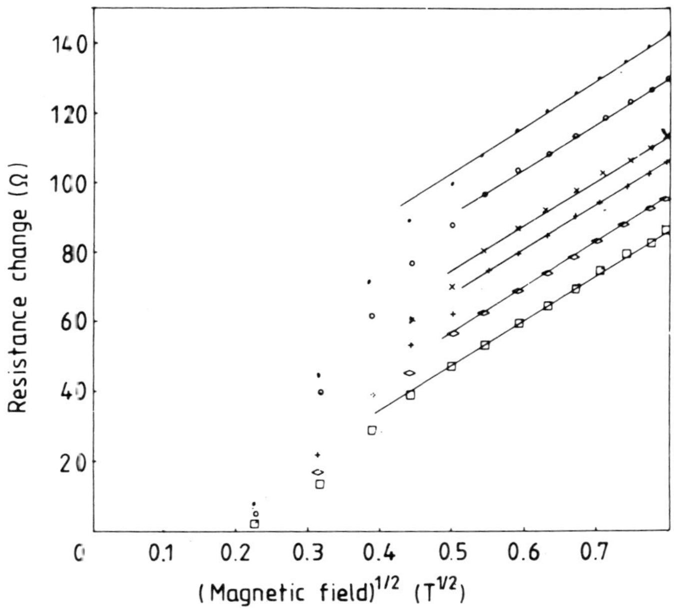

Fig. 3. The resistance change of the channel used for Figure 2 is plotted against $B^{1/2}$ for a parallel magnetic field (in Tesla, T): • 1.3 K; 01.6 K; × 2.15 K; + 2.8 K; ◇ 3.8 K; ▫ 4.2 K.

In Figure 7 $\delta\sigma/\delta(\ln T)$ is plotted against $B^{1/2}$ for different channel thicknesses. At very small values of field, the B^2 regime is observed, and there is consequently little dependence on field. However, once $L_c^4 \leq a\, L_\varphi^3$, a linear dependence on $B^{1/2}$ is seen, in agreement with Equation 15. At higher fields still, a flattening of $\delta\sigma$ is seen, probably due to a combination of interaction effects and a breakdown of the model as l (the elastic length) becomes comparable L_c.

In Figure 8 $(\delta\sigma/\Delta \ln T)/\Delta \sqrt{B}$, the gradients of the temperature dependent terms in Figure 7, are plotted against channel thickness. A reasonably linear fit is observed, the line passing close to the origin. The breakdown at small channel thickness occurs when l is comparable to the thickness, so that the diffusive motion is modified.

To summarise, measurements of magnetoresistance have been made on a GaAs MESFET and good agreement has been found with the model we have proposed. A particularly noteworthy feature of the results is that at low temperatures the quantum interference is clearly 2D, as L_φ exceeds the thickness, with the appropriate magnetoresistance in a transverse field. Yet, when the field is parallel, the predicted conversion to 3D magnetoresistance occurs with excellent agreement between experiment and theory.

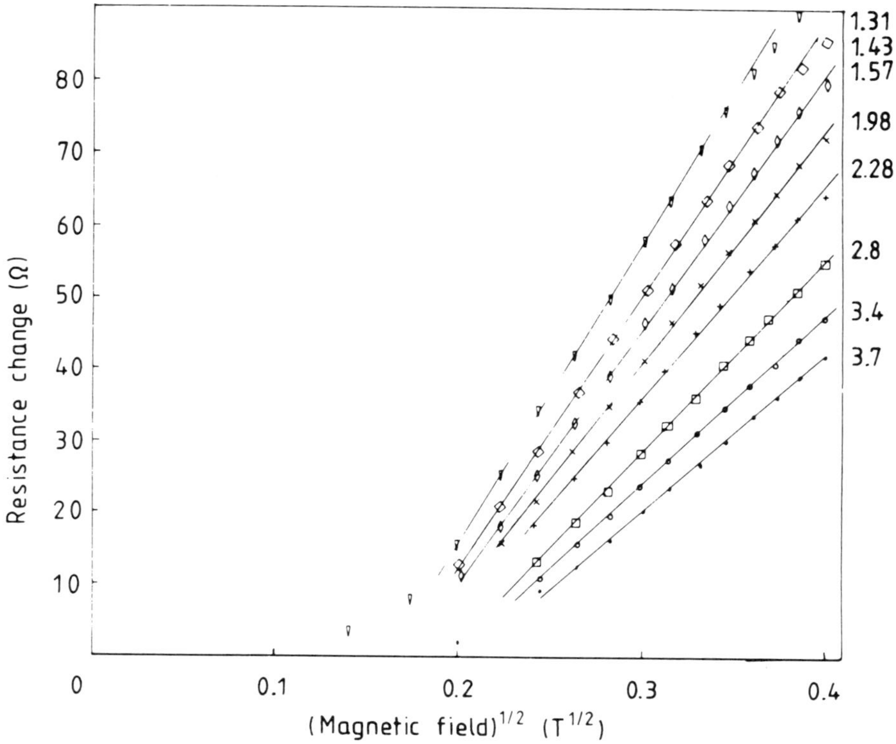

Fig. 4. The resistance change for the 1300 Å channel illustrated in Figures 2 and 3 in intermediate parallel magnetic fields, is plotted agains $B^{1/2}$; the temperatures in Kelvin are indicated on the figure.

4. One Dimension

One dimensional quantum interference is observed when L_φ exceeds two of the sample dimensions. The first separation of quantum interference and interaction [15] effects was achieved by Thornton *et al.* [16] using the split gate technique of electrostatically narrowing the high mobility 2D electron gas formed at the heterojunction of a modulation doped AlGaAs–GaAs structure. This method subsequently resulted in the observation of the quantized 1D ballistic resistance.

Altshuler and Aronov [18] found that for a narrow system with L_φ greater than the width, a 2D density of states and elastic mean free path less than the width.

$$\Delta g = \frac{e^2}{\pi\hbar}\,\frac{\sqrt{D}}{L}\left(\frac{1}{\tau_\varphi}+\frac{1}{\tau_B}\right)^{-1/2} \tag{16}$$

with

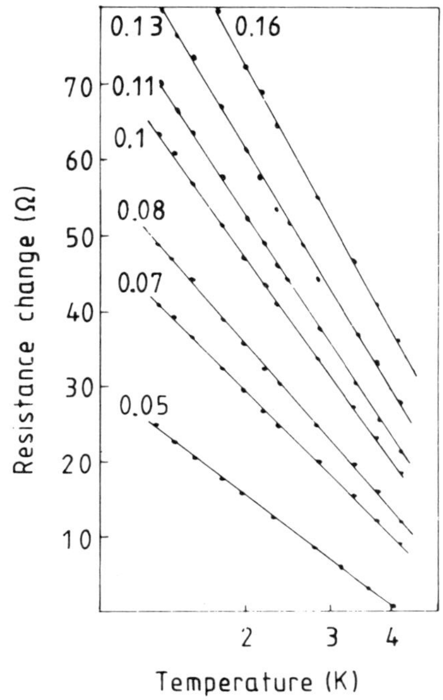

Fig. 5. The resistance change in different values of parallel magnetic fields (Figure 4) is plotted against ln T. The magnetic field, in Tesla, is shown on the figure.

$$\tau_B = \frac{3L_B^4}{DW^2} ,$$

where W is the width, D the diffusion constant, L the length, τ_φ the phase breaking time, τ the scattering time and the 2D magnetic length is given by $L_B = \sqrt{\hbar/eB}$.

The diffusion approach is invalid when the elastic mean free path is longer than the width of the channel and scattering from the channel walls is specular. Dugaev and Khmelnitskii [19] first theoretically investigated weak localisation in a film with specular scattering from the walls, i.e. the mean free path is longer than the thickness. Beenakker and van Houten [20] subsequently adapted this approach to the quasi one dimensional situation, and also computed the results for the intermediate case in which the width is of the same order of magnitude as the mean free path. Often it is not straightforward to measure the width of the conducting layer. This can be different from the physical width due to sidewall depletion or the use of a split gate technique to narrow the electron gas. The method adopted is that first proposed by Berggren *et al.* [21,22], in which the periodicity of the Shubnikov–de Haas oscillations is used to estimate the width. The period deviates from a $1/B$ relation due to the formation of magneto-electric subbands.

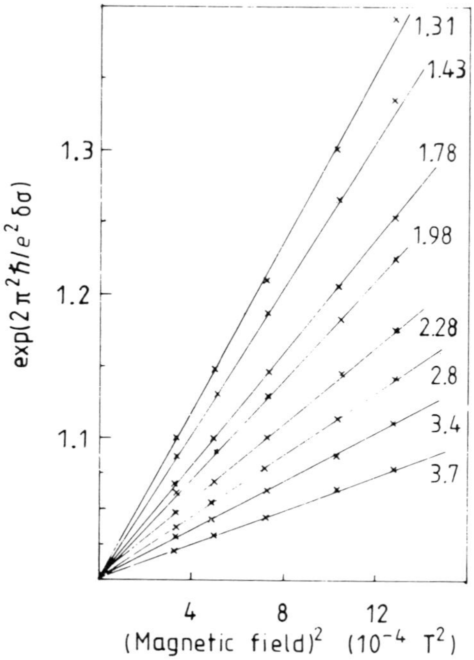

Fig. 6. The resistance change at very low values of parallel magnetic field (from Figure 4) is plotted against B^2 for different values of temperature in Kelvin as indicated on the diagram.

The important feature, as pointed out by Dugaev and Khmelnitskii, which occurs when the mean free path is much longer than the width, is a flux cancellation. Those areas where the flux makes a positive contribution to the loop are cancelled out by areas which make a negative contribution. The exact cancellation is removed by the phase change in the immediate vicinity of an impurity scattering event and the net effect is a reduction in the phase change compared to a sample where all scattering is diffuse. In the case of the narrow channel, when the mean free path $l \gg W$, Beenakker and van Houten [20] give as a replacement to τ_B in Equation 15.

$$\tau_B = \frac{C L_B^4}{W^3 v_f}$$

The constant C is in the range 3–10, the exact value depending on the scattering mechanism and, weakly, on the field regime, v_f is the Fermi velocity.

It is clear from this result that application of the Altshuler-Aronov [18] formula when the mean free path exceeds the width will result in an estimation of L_φ which is incorrect by a numerical factor. This explains why Thornton *et al.* were able to extract the correct 1D temperature dependence

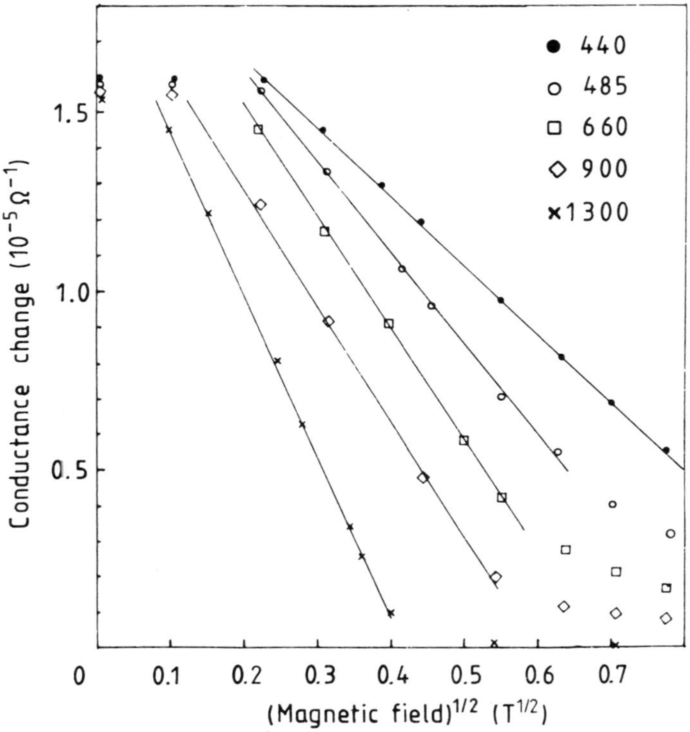

Fig. 7. The ratio ($\delta\sigma/\ln T$) is plotted against $B^{1/2}$ for a parallel magnetic field for different values of thickness of conducting channel, these are marked on the diagram and are in Å.

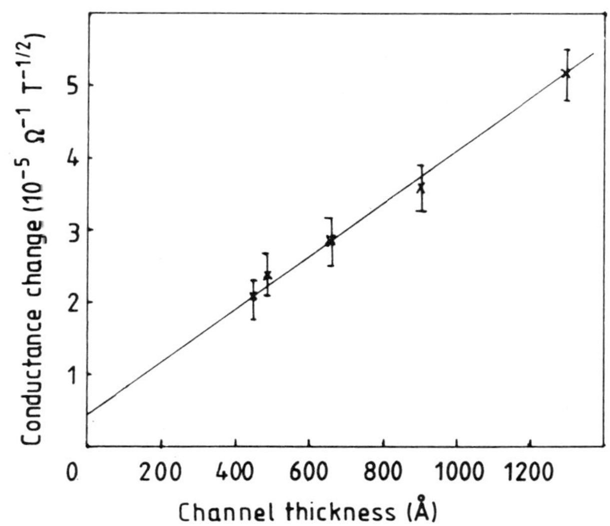

Fig. 8. The gradient of each line in Figure 7 is plotted against the channel thickness.

from their results by fitting this formula even though it was inapplicable due to $l \gg W$. Beenakker and van Houten find

$$\Delta g = \frac{e^2}{\pi \hbar} \frac{\sqrt{D}}{L} \left[\left(\frac{1}{\tau_\varphi} + \frac{1}{\tau_B} \right)^{-1/2} - \left(\frac{1}{\tau_\varphi} + \frac{1}{\tau_B} + \frac{1}{\tau} \right)^{-1/2} \right] . \tag{17}$$

In deriving Equation 17, the role, and cutoff, of small loops is important as the phase coherence length is often comparable with the mean free path.

Van Houten *et al.* [23] and Hiramoto *et al.* [24] found that both the original Altshuler and Aronov theory and the modified theory could fit their data. However, the original theory required a value of the mean free path greater than the width to fit their data, which is out of the range of the validity of that theory implying a proof of the cancellation theory. Green *et al.* [25] have investigated the two theories in GaAs–AlGaAs heterojunctions where the mobility is sufficient high that the flux cancellation should change the magnetic field required to quench weak localisation by a factor of five. However, the results show that the field scale remains unchanged from the Altshuler and Aronov approach.

Figure 9 shows the low field magnetoresistance obtained in a heterojunction with a value of elastic mean free path (4.5 ± 0.5) μm and an estimated width of (0.8 ± 0.2) μm. The negative magnetoresistance shown is temperature dependent and saturates at about 5×10^{-4} Tesla, characteristic of quantum interference. The agreement between experiment and theory has τ_φ as the only variable parameter, and, as seen, agreement is much better using the original formula, Equation (16). Here it is possible to match with only one parameter both the curvature near the conductance minimum and the field at which the weak localisation is roughly quenched, $\approx 5 \times 10^{-4}$T. It appears that in this sample the flux cancellation mechanism does not occur, even though the mean free path as obtained from the mobility is much greater than the width.

In samples with lower mobilities both Equations 16 and 17 could fit the data, although the mean free path was twice the width.

Greene *et al.* [25] suggested that the most likely reason for the failure of the mechanism of flux cancellation was the presence of small angle scattering. This will remove the role of the equal areas producing flux cancellation, particularly as such scattering will have a major effect on electron trajectories at a glancing angle to the sides of the sample, such trajectories tending to dominate the negative magnetoresistance. Evidence for a small angle scattering time less than the elastic scattering time was provided by the Shubnikov–de Haas oscillations. Analysis produced a mean free path for small angle scattering about a factor of ten smaller than the sample width.

The phase coherence length from the results of Figure 10 was extracted as a function of temperature. The best fit was to the Nyquist behaviour

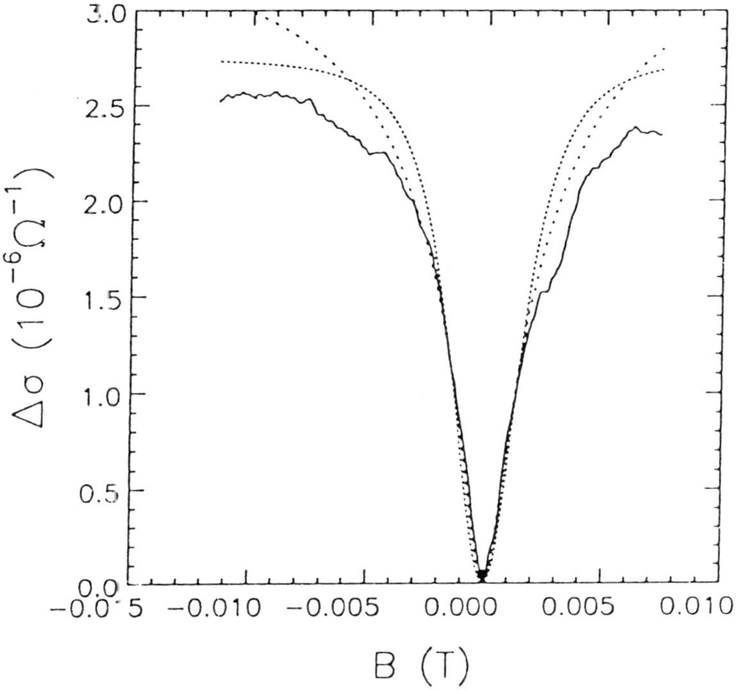

Fig. 9. Low field magnetoresistance of a sample which processes an elastic scattering length 1 = 4.5 ± 0.5 μm, width of 0.8 ± 0.2 μm, at a temperature of 0.1 K. The fits are to Equation 15 with the dense dotted line and Equation 16 with the loose dotted line.

in 1D previously observed in narrow Si channels [26] and theoretically predicted [27]. Agreement with the proposed $T^{-2/3}$ behaviour of τ_φ is found and the absolute value of τ_φ agrees with theory. The fit breaks down when L_φ approaches the elastic length, and for temperatures below 0.5 K the temperature dependence decreases. However, as the elastic scattering time τ is now less than $\frac{\hbar}{kT}$ it is expected that lifetime broadening will invalidate the predicted behaviour.

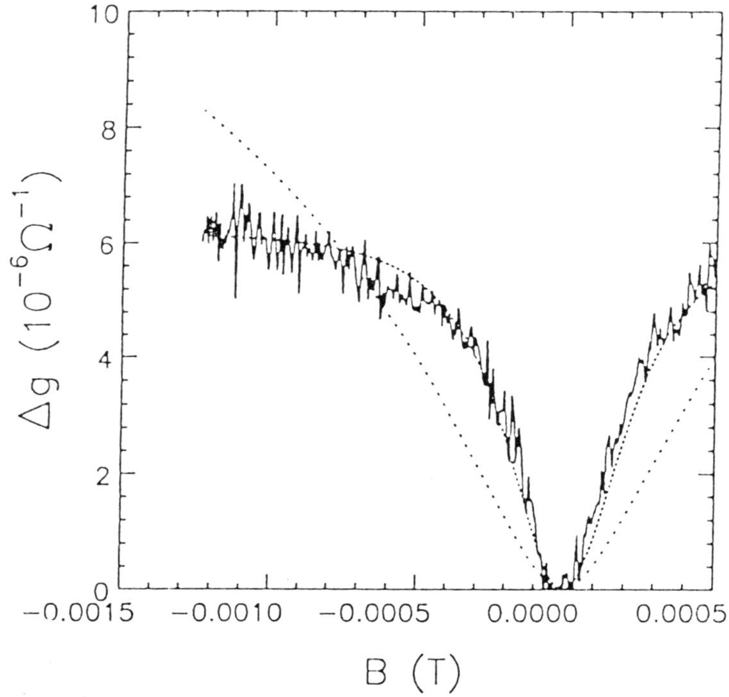

Fig. 10. The low field magnetoresistance exhibited at 0.8 K by a sample with mean free path of ~ 1 μm, which is comparable with the width. The dense dotted line is the fit with Equation 15 and the loose dotted line is with Equation 16.

Acknowledgements

We are indebted to many colleagues for discussions, in particular our co-workers in the original work upon which this article is based. It is a pleasure to acknowledge the many contributions of Professor H. Kamimura to this area of solid state physics, we have always derived a great deal of insight from discussions with him.

This work was supported by SERC, EC under the auspices of basic research action 3043 and, in part, the European Research Office of the U.S. Army.

References

1. E. Abrahams, P.W. Anderson, D.C. Licciardello and T.V. Ramakrishnan: 1979, *Phys. Rev. Lett.* **42**, 673.
2. L.P. Gorkov, A.P. Larkin and D.E. Khmelnitskii: 1979, *JETP Lett.* **30**, 249.
3. B.L. Altshuler and D.E. Khnelnitskii: 1986, *JETP Lett.* **42**, 359.
4. A. Kawabata: 1984, *J. Phys. Soc. Japan* **53**, 318; also 1981, *Solid State Comm.* **38**, 823.

5. S. Hikami, A.I. Larkin and Y. Nagaoka: 1980, *Progress Theor. Phys.* **63**, 707.
6. B.L. Altshuler and A.G. Aronov: 1981, *JETP Lett.* **31**, 445; also 1979, *JETP Lett.* **30**, 482.
7. D.A. Poole, M. Pepper, K.F. Berggren, G. Hill and H.W. Myron: 1982, *J. Phys. C* **15**, L21; also D.A. Poole, M. Pepper and H.W. Myron: 1983, *Physica* **117B**, 697.
8. R. Dingle: 1975, *Festkorperprobleme* **15**, 21.
9. D.J. Newson, C. McFadden and M. Pepper: 1985, *Phil. Mag. B* **52**, 437.
10. C. McFadden, D.J. Newson, M. Pepper and N.J. Mason: 1985, *J. Phys. C* **18**, L383.
11. D.J. Newson and M. Pepper: 1985, *J. Phys. C* **18**, L1049.
12. G. Bergmann: 1982, *Phys. Rev. B* **25**, 2937.
13. G. Bergmann: 1984, *Physics Reports* **107**, 1.
14. D.J. Newson: 1986, Ph.D. Thesis, Cambridge.
15. B.L. Altshuler and A.G. Aronov: 1985, *Electron-Electron Interactions in Disordered Systems*, A.L. Efros and M. Pollak (eds.), North-Holland.
16. T.J. Thornton, M. Pepper, H. Ahmed, D. Andrews and G.J. Davies: 1984, *Phys. Rev. Lett.* **56**, 1198.
17. D.A. Wharam, T.J. Thornton, R. Newbury, M. Pepper, H. Ahmed, J.E.F. Frost, D.G. Hasko, D.C. Peacock, D.P. Ritchie and G.A.C. Jones: 1988, *J. Phys. C* **21**, L209; also B.J. van Wees, H. van Houten, C.W.J. Beenakker, J.G. Williamson, C.E. Timmering, M.E.J. Broekaart, C.T. Foxon and J.J. Harris: 1984, *Phys. Rev. Lett.* **62**, 2523.
18. B.L. Altshuler and A.G. Aronov: 1981, *Soviet Phys. JETP* **54**, 411.
19. V.K. Dugaev and D.E. Khmelnitskii: 1984, *Soviet Phys. JETP* **59**, 1038.
20. C.W.J. Beenakker and H. van Houten: 1988, *Phys. Rev. B* **38**, 3232.
21. K.F. Berggren, T.J. Thornton, D.J. Newson and M. Pepper: 1986, *Phys. Rev. Lett.* **56**, 1198.
22. K.F. Berggren and D.J. Newson: 1986, *Semiconductor Sci. Technol.* **1**, 327.
23. H. van Houten, C.W.J. Beenakker, B.J. van Wees and J.E. Mooij: 1988, *Surface Science* **196**, 144.
24. T. Hiramoto, K. Hirakawa, Y. Iye and T. Ikoma: 1989, *Appl. Phys. Lett.* **54**, 2103.
25. S.K. Greene, M. Pepper, D.C. Peacock, D.A. Ritchie, V.J. Law, R. Newbury, J.E.F. Frost, G.A.C. Jones, R.J. Brown, H. Ahmed and D. Hasko, to be published in *J. Phys. Condensed Matter.*
26. D.M. Pooke, N. Paquin, M. Pepper and A. Gunlach: 1989, *J. Phys. Condensed Matter* **1**, 3289.
27. B.L. Altshuler, A.G. Aronov and D.E. Khmelnitskii: 1982, *J. Phys. C* **15**, 7367.

ELECTRODE SCREENING OF TWO DIMENSIONAL DISORDERED SYSTEMS[1,2]

M. GREEN and M. POLLAK
Department of Physics, Univ. of California at Riverside, Riverside CA 92521, U.S.A.

Abstract. Experiments on two dimensional disordered systems often use devices with an electrode near the two dimensional system. We use computer simulation to explore the screening effects of such an electrode. The main focus is on elastic tunneling through mesoscopic MOSFETs.

1. Introduction

There have been numerous experiments reported in the literature which utilize two dimensional disordered systems and attempt to evaluate the effects of electron-electron interactions. Most of these experiments fall into two categories – tunneling experiments [1,2], which attempts to evaluate the one-particle density of states, and transport experiments in MOSFETs [3–5]. The results are often interpreted in terms of theoretical predictions [6] on the Coulomb gap [7]. However, many of these predictions are strongly tied to the long range nature of Coulomb interactions, while in the experimental systems such interactions are screened by a metallic electrode placed near the disordered electrons, so the range of interaction between them is limited to a distance of the order of the thickness of the dielectric between the system and the electrode. This is particularly important in the tunnel experiments, where this distance is of the order of 1 nm. (Another problem with the tunnel experiments is a non-negligible voltage drop due to the hopping transport, but this problem is not investigated here.) The purpose of this work is to investigate the effect that screening by an electrode has on the Coulomb gap.

For future use we introduce here expressions for the Hamiltonian H, and for the energy E_i of a carrier on site i, in a form convenient for the screened interaction, in the so-called classical approximation.

$$E_i = \varepsilon_i + \sum_j q_i \varphi_j(\mathbf{r}_i) \ , \quad H = \sum_i [\varepsilon_i + \sum_{j<i} q_i \varphi_j(\mathbf{r}_i)] \ . \tag{1}$$

Here ε_i is a random energy at site i, $\varphi_j(\mathbf{r})$ is the (screened) potential at $\mathbf{r}$ from a charge q_j at site j, and $\mathbf{r}_i$ is the position of site i.

[1] Computational work supported in part by the San Diego Supercomputer Center.
[2] Most of this work is based on M. Green's Ph.D. dissertation (Ref.14).

H. Aoki et al. (eds.), New Horizons in Low-Dimensional Electron Systems, 155–167.

In the next section we report on a computer simulation of screening by an electrode in a simple model – electrons are placed on sites in a simple, square, two-dimensional lattice, where they interact by a Coulomb interaction, among themselves and with image charges placed at twice the distance of the simulated electrode. The one-particle density of states is computed for various distances of the electrode from the electrons. This part of the work is mainly relevant to the tunneling experiment, which report on the one-particle density of states of disordered systems.

In the third section we report on simulation of electrode screening in a more complex system. The location of electron sites is placed at random on a surface, imbedded in a dielectric at a certain distance from an interface with another dielectric. The second dielectric has a smaller dielectric constant, and a certain thickness, beyond which there is a metallic electrode. The model is designed to simulate a MOSFET structure, as proposed by the IBM group [5], the first dielectric being the silicon, the second dielectric of thickness λ_2 the silicon oxide, the distance of the electrons from the interface being as proposed by the IBM group. The work here is designed to relate to experiments [5] at IBM on gate-voltage dependent current through submicron MOSFETs, and to examine their proposal that the current is due to elastic tunneling between source and drain, via a localized state in the silicon, and is affected by Coulomb interactions with other localized electrons. The section briefly describes the IBM experiments, then presents a theory of interaction between electrons in the above model, and finally presents results of simulation of the changes in the potential of a site at the Fermi level induced by hopping of electrons among sites of the system, and between an electrode and a site of the system.

The fourth section briefly discusses inelastic, phonon assisted tunneling (or hopping) in MOSFETs at low temperature, and relates to some older experiments by the IBM group and by others [8].

2. A Simple Model

2.1. Description of the model

The model used here is essentially the two-dimensional version of the lattice model introduced by Efros and by Baranovskii *et al.* [10], which in turn is an extension of Anderson's models, to include effects due to Coulomb interactions. It consists of sites i, placed on a square lattice, with random energies ε_i distributed uniformly between $-W/2$ and $W/2$, and with a number of electrons which is a fraction K (here 1/2) of the number of sites. The electrons occupy the sites at most singly. The Hamiltonian includes the Coulomb interaction energy and the random energy, but the resonance energy between sites is neglected. For charge neutrality, a charge $+K\,|e|$ is placed on each site, so the charge q_i is $-|e|/2$ on occupied, and $+|e|/2$ on unoccupied sites. Here the model was extended to include the screening effect of an electrode a distance d from the system, by placing image charges $q_{i'} = -q_i$, shifted a distance 2λ from q_i along a normal to the plane of the q_i's. The effective interaction of q_i with q_j is then taken to include interaction with $q_{j'}$.

2.2. Description of the Algorithm

The main purpose was to study the one-particle density of state $N(E_i)$, as defined by Efros, and the one-particle excitations from ground state (sometimes called electron-hole excitations), using the interaction described above.

The first step in the procedure was to place the system in a ground state (or rather a pseudoground state, stable with respect to one-particle excitations), following basically the method of Baranovskii *et al.* [10]. In this method, a pseudoground state is generated in two steps – in the first step, the system is relaxed from some initial state until a chemical potential μ is established. In a second step, the energy of the system is reduced by one-particle relaxation until a minimum is reached. If a Fermi level still exists, the final state is a pseudoground state. The procedure was repeated many times on the same 'sample', generating a set of pseudoground states. The lowest of these is then taken to be the ground state. In this pseudoground state configuration the energy $E_i = \varepsilon_i + \sum_j q_j/r_{ij} + q_{j'}/r_{ij'}$ is calculated for each site i. Here r_{ij} is the distance between i and j, and $r^2_{ij'} = r^2_{ij} + \lambda^2$. (The charge of the electron and the dielectric constant are taken here to be unity.) In addition, the energy $E_{km} = E_k - E_m - (1/r_{km} - 1/r_{km'})$ is calculated for each pair with k occupied and m unoccupied. The whole process is repeated with many 'samples', differing from each other in random site energies. The energies E_i and E_{km} are then binned, and their histograms provide the $N(E_i)$, and the one-particle excitation spectrum $N(E_{km})$, respectively. When binning E_i, a certain problem arises with the chemical potential μ; due to the finite size of the systems, μ fluctuates from 'sample' to 'sample', and it is not obvious how to proceed in determining $N(E_i)$ by averaging over the 'samples'. The Leningrad group used two different ways [10,11], but here a method different from both is used. The symmetry of the system makes $\mu = 0$ in the thermodynamic limit. We therefore keep only systems for which $E_i < 0$ for the highest occupied site, and $E_i > 0$ for the lowest unoccupied site. Then all E_i in these systems can be binned in a straightforward way.

2.3. Results

The principal results of the computer simulation are shown in Figure 1 which shows $N(E_i)$, and in Figure 2 which shows $N(E_{km})$. The distance of the electrons from the plane of the sites is used as a parameter in both figures.

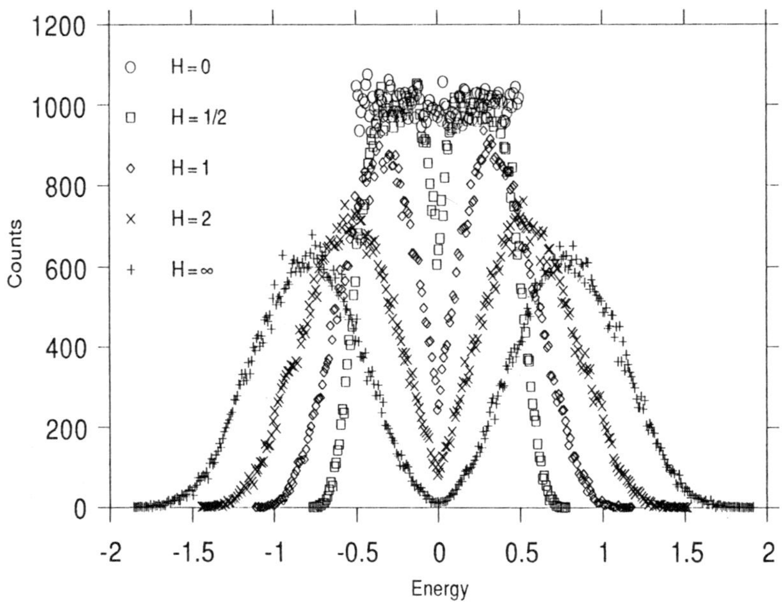

Fig. 1. The one-particle density of states. The separation distance between the surface and the image charges is given by $H = 2\lambda$, measured in units of the lattice constant. The energy is in units of the random energy W.

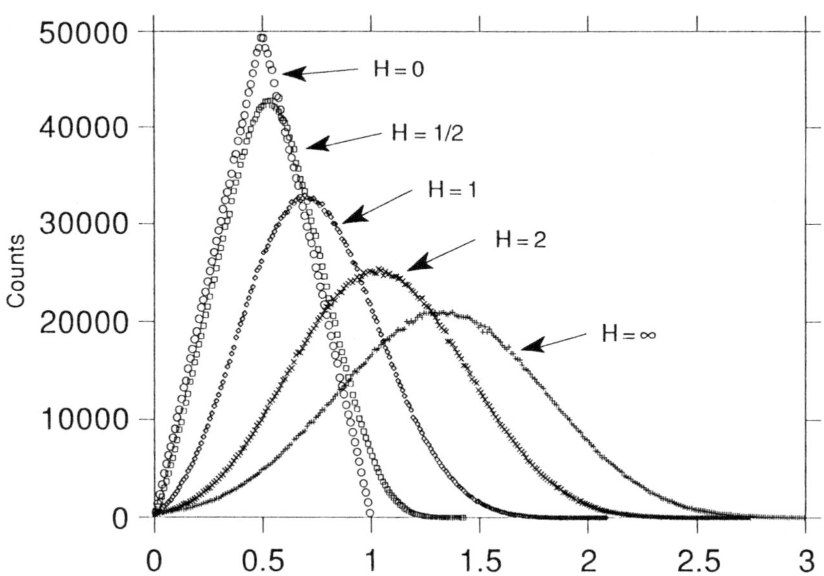

Fig. 2. The one-particle excitation spectrum. The separation distance between the surface and the image charges is given by $H = 2\lambda$, measured in units of the lattice constant. The energy is in units of the random energy W.

As expected. $N(E_i)$ goes to zero at the chemical potential only in the limit of λ becoming infinite, i.e. when the screening disappears. For finite λs, $N(\mu)$, the one-particle density of states at μ, is nonvanishing, and increases with decreasing λ, as expected. Interestingly, a distinct dip in $N(E_i)$ still exists at rather small λ.

It is interesting to note that in the limit of infinite λ we find that $N(E_i)$ increases superlinearly with $|E_i - \mu|$ (perhaps quadratically) rather than showing the linear dependence predicted by Efros [12], and observed in computer simulations [10,11]. The reason for this difference is most likely due to our discarding systems where μ is far from its value in the thermodynamic limit, as opposed to the procedures used in the other works. It seems to us that our procedure is more appropriate, but there is probably no way to state with certainty whether any method of averaging over relatively small systems can correctly give the thermodynamic $N(E_i)$ in the vicinity of μ. It is also interesting to note that $N(E_i)$ does go linearly with $|E_i - \mu|$ for smaller λ.

The one-particle excitation spectrum $N(E_{km})$ becomes, as expected, broadened with decreasing screening, from the theoretically predictable spectrum without interactions ($\lambda = 0$), to one of roughly 2.5 times that width for the totally unscreened system.

2.4. Discussion and Relevance to Experiments

We think that this work relates rather directly to several tunneling experiments [1,2]. The experimental data were processed to obtain $N(E_i)$ in the conventional way. On the insulating side of the disordered system $N(E_i)$ was stated to be compatible with $(E - \mu)^2$, as Efros predicted.

A few comments about the experiments are in order.

(i) The tunneling current needs to flow through the disordered system, where the resistivity is substantial. It is then probably impossible to assume, as is normally done in the analysis of such experiments, that the entire applied voltage is applied across the tunneling junction.

(ii) It should be noted that, while the system used in the experiment was nominally three dimensional, the tunneling must occur to sites very near the junction, so the system of relevance is probably closer to being two dimensional.

(iii) The oxide layer between the metallic electrode and the disordered system is very thin, of the order of 1 nm. Although the exact parameters for the system are not known, it seems clear that $\lambda = 1$ nm is not likely to approximate the infinite limit, so that $N(\mu)$ should not be zero, as reported. Thus, we think that the interpretation in terms of a quadratic Coulomb gap needs to be reexamined.

3. Elastic Hopping in Mesoscopic MOSFETs

3.1. Experimental background

Experiments [5] on submicron MOSFETs at dilution refrigerator temperatures show that the source-drain current exhibits, as a function of gate voltage, a sequence of pronounced peaks, with no measurable current between the peaks. The IBM group proposes that the peaks correspond to the gate voltages where the chemical potential is at the level of a localized site which is in a position to mediate tunneling between the source and the drain. They estimate that such sites should lie, within a localization radius, midway between them. The characteristic features of the peaks are non-Lorentzian, and they also have discernible subpeaks. Their width turns out to be about ten times broader than the estimated lifetime. The authors propose that the structure and the non-Lorentzian character may be due to changes of the Coulomb potential at the tunnel mediating site, from carriers undergoing thermally induced jumps. In the absence of Coulomb interactions, the peak would be a simple Lorentzian. We developed an algorithm to simulate the behaviour of the proposed model, to see whether we can obtain peaks which resemble, in structure and width, the observed peaks.

It should be mentioned that time dependent effects have been observed for very long durations [5]. Such effects are indeed quite common in disordered localized systems. So the rate of change of the gate voltage may be an important parameter, and it seems in fact unclear whether the gate voltage could be changed slowly enough for near-equilibrium conditions to prevail. This problem will be addressed again in the discussion of the algorithm.

3.2. The theoretical model

Since in this simulation we attempt to test quantitative features of experiments, we use a more realistic model than that described in Section 2, where we were interested mainly in qualitative features. Furthermore, we use values of parameters as close as possible to those believed to prevail in the experimental system [5,13]. The model consists of a semi-infinite medium of dielectric constant κ_1 (simulating the silicon), a layer of thickness λ_2 and dielectric constant κ_2 adjacent to it (simulating the oxide), and a metal adjacent to that (simulating the gate electrode). Sites are placed in the κ_1 medium, at random positions $\mathbf{r}_i \equiv (x_i, y_i, z_i = \lambda + \lambda_2)$ in a plane a distance λ from the κ_1/κ_2 interface, with a density n, and random energies ε_i distributed according to some distribution $N(\varepsilon_i)$. The main idealizations in this model are perfect planes for the interfaces and for the placement of sites, and a point charge approximation for the carriers.

The evaluation of the screened interaction between carriers is clearly more difficult here than in the previous model, but can still be handled by use of image charges. Each localized carrier is now associated with an infinite number of these, placed on a line crossing the carrier and normal to the interfaces. The procedure to obtain the image charges is outlined below.

The interaction energy between two point charges q_i, q_j at $\mathbf{r}_i, \mathbf{r}_j$ can be written as $e_{ij} = q_i\varphi_j(\mathbf{r}_i) = q_j\varphi_i(\mathbf{r}_j)$, where the notation is the same as for Equation (1). The coordinate z will be normal to the interfaces, with $0 < z < \lambda_2$ in κ_2, $\lambda_2 < z$ in κ_1. $\varphi(\mathbf{r})$ can be evaluated by an expansion in k into cylindrical Bessel functions $J_o(k\rho)$ and matching of boundary conditions at interfaces, as follows,

$$\varphi_i(\mathbf{r}) = q_i \int_0^\infty \exp[k(z-\lambda-\lambda_2)]\, J_o(k\rho)\, \mathrm{d}k/\kappa_1$$

$$+ \int_0^\infty f(k)\, \exp(-kz)\, J_o(k\rho)\, \mathrm{d}k\ , \quad 0 < z < \lambda_2$$

$$\varphi_i(\mathbf{r}) = \int_0^\infty g(k)\, \sinh(kz)\, J_o(k\rho)\, \mathrm{d}k\ , \quad \lambda_2 < z \leq \lambda + \lambda_2 \tag{2}$$

where $\rho = (x^2+y^2)^{1/2}$, and $f(k)$ and $g(k)$ are functions to be solved for by matching boundary conditions. The first term in the upper equation comes directly from the point charge q_i and the identity

$$|\mathbf{r}|^{-1} \equiv (\rho^2+z^2)^{-1/2} \equiv \int \exp(kz)\, J_o(k\rho)\, \mathrm{d}k\ , \tag{3}$$

the other terms are the potentials due to the interfaces. After considerable algebra, the desired result for $\varphi(\mathbf{r})$ in the plane of the charges becomes [14]

$$\begin{aligned}\varphi_i(\mathbf{r}) = (q_i/\kappa)\ \{&[\rho^2 + (z-\lambda-\lambda_2)^2]^{-1/2} \\ &+[e^{-2b}(\rho^2 + (z+\lambda-\lambda_2)^2)]^{-1/2} \\ &-(1-e^{-4b}) \sum_\nu [e^{-2\nu b}\,(\rho^2 + (z+2\nu\lambda_2+\lambda+\lambda_2)^2)]^{-1/2}\}\end{aligned} \tag{4}$$

where $b = \tanh^{-1}(\kappa_2/\kappa_1)$. The summation comes from an expansion of hyperbolic functions, made to bring the expression of the potential into a form which conforms to a potential from a series of point charges. The first term in Equation (4) corresponds to the potential from the original charge. The second term corresponds to a charge in the insulator layer which has the same sign as the point charge, expressing that the self-energy of q_i is larger when $\kappa_2 < \kappa_1$ than when κ_2 is the same as κ_1. The terms which follow correspond to charges with an opposite sign, expressing that screening by the electrode decreases the self-energy. Notice that the value of the image charges decreases with increasing ν, i.e. as the image charges get further from the electrode. This allows for an accurate evaluation of the potential when truncating ν at a relatively small value. In this work we used $\nu \leq 100$.

The experimentally controlled gate voltage V_g alters the number of carriers in the system, and thus translates into a chemical potential μ. It should be remembered that such a translation is possible only for a very slow variation of the gate voltage, since μ is well defined only if near equilibrium conditions can be maintained. When this is so, in a mesoscopic MOSFET at very low T, a smooth change of V_g does not correspond to a smooth change of μ. Rather, μ varies in steps, being very close to E_i while the occupation probability of i changes between 0 and 1, and jumping to the next site in energy thereafter.

3.3. Description of the algorithm

In the first step, a random generator is used to assign positions ρ_i within a certain area, and energies ε_i within a range W, to N sites. Any such set of (ρ, ε) will be referred to as a 'sample'. Each sample will have a group (say G) of sites in the strip where tunneling can presumably occur. The distribution $N(\varepsilon_i)$ was chosen as either uniform or exponential, but no important differences between the two were obtained in terms of results. For a varying number of N_e electrons ($0 \leq N_e/N \equiv K \leq 1$), a pseudoground state for the system is found, using the algorithm originally used by Baranovskii *et al.* [10], and the Hamiltonian $\sum(\varepsilon_i + \varphi_i(\mathbf{r}_j)\, q_j$. If the highest occupied site or the lowest unoccupied site in in the group G, a current peak can occur, since μ (at very low T) is bound to coincide with one of these states. In such a situation, the excitation energies E_{kj} are computed for all possible one-particle transitions $k \to j$ within the N-site system, and between the system and the electrodes (the latter is simply simulated by inserting or deleting a carrier). When $E_{kj}/kT < 5$, the probability $P(E_{kj}/kT)$ for the excited state is computed in a straightforward way, as well as the shift ΔE_i in the E_i of the tunneling site due to the change of the Coulomb energy at i, induced by the transition. The structure of a skeleton peak is then constructed by assigning an amplitude $1 - \sum P(E_{kj}/kT)$ to a central line, and amplitudes $P(E_{kj}/kT)$ to lines shifted by $\Delta E_i(kj)$ from the central line. The actual peak is then taken to be the skeleton peak, after broadening each line into a Lorentzian with the width Γ appropriate to the tunneling rate.

We now turn to the rate of change of V_g. The current peaks were simulated for two different such rates – when equilibrium can be maintained continuously while the occupation of the tunneling site changes from zero to one, and when the rate is too fast for that. For brevity these will be

TABLE I
The values used in the simulation of current peaks, as obtained from Reference 5, and from personal communications from Dr. A.B. Fowler.

Quantity	Value
κ_1	11.7
κ_2	3.95
λ_2	10 nm
λ	4 nm
T	22 mK – 150 mK
Γ	8.6 μeV
N	780
W	4 mev
system size	500 × 1000 nm

referred to as 'slow' and 'fast' cases. Even for the 'fast' case it is assumed that equilibrium is reestablished by the time μ reaches the next tunneling site. In the algorithm for the 'fast' case the system does not respond to the changing occupation of the tunneling site, while for the 'slow' case it does.

For the fast case one expects a hysteresis effect, since the state of the system will generally be different at a given V_g, when V_g is shifted downward or upward. This might be a good experimental way to see whether V_g is changed slowly enough to maintain equilibrium. However, such experiments were not reported, so the simulation was done only for an upward change. For slow cases, a rearrangement of electrons can occur in response to the change of occupation of the tunneling site – the back-action of the rearrangement on the potential of the tunneling site then may turn off the tunneling. Only cases (say type A) when this does not happen were explored; the cases where it does (say type B) would require a study of the equilibrium configuration for intermediate occupations, between 0 and 1, of the tunneling site.

3.4. Results

Table I lists the values of the various parameters [5,13]. The main results of our work are the width and the structure of the simulated current peaks. A few such typical peaks are shown in Figure 3. The general features can be summarized as follows.

1. The width of the peak is an order of magnitude larger than the life-time broadening. This result is in good agreement with the experimentally observed peaks [5].

2. The number of lines in the peak is relatively small, so the individual subpeaks could be resolved. This result does not agree with the exper-

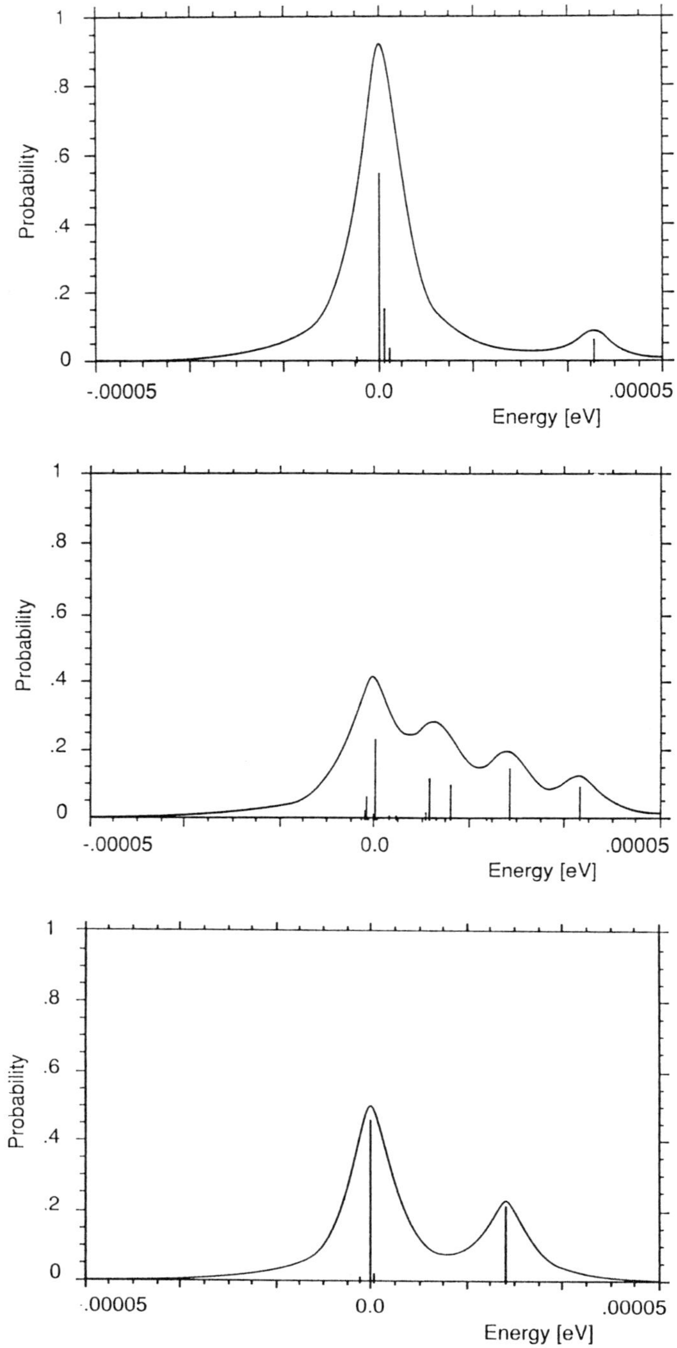

Fig. 3. Three examples of simulated current peaks, at a temperature of 22 mK.

imental observations [5], where the resolution of subpeaks is much less pronounced.

3.5. Discussion and relevance to experiments

Although we did not evaluate type B processes, it is of interest to discuss qualitatively the expected behavior in order to ascertain whether observation of such behavior was reported. Assume that at some occupation between 0 and 1 the system responds by changing configuration. Such a change must imply a lowering of energy, so the average occupation near the tunneling site diminishes, thus lowering the interaction energy with the tunneling site, and hence its energy below μ. If such lowering is by much more than kT, tunneling is turned off. Such a behavior was not reported, which may be an indication that the rate of change of V_g in the experiments was not slow enough to maintain equilibrium while the tunneling site changes occupation.

The processes of type A can be expected to behave similarly to the fast processes, except for the hysteresis in the latter case, not experimentally explored, and therefore also not in our simulations. We believe, however, that such experiments would be of considerable interest, as one could obtain from them good information about the rate of the electronic processes in the system.

It is noted that the simulation included only one-particle transitions, either within the system or between an electrode and the system. A question of much interest is whether many-particle transitions in the system are sufficiently numerous to fill in the gaps between the subpeaks, so the simulated peaks would better resemble the experimental peaks. A procedure which allows examination of all the low energy many-particle excitations has been developed [15], and we are currently in the process of applying it to this problem.

As a final remark, we address the question of localization. Although the authors of Reference [5] believe that the system is strongly localized [13], we estimated the localization problem using the parameters of Table I. Our results suggested that it is a borderline case, so this question should be addressed in more detail.

4. Inelastic Hopping in MOSFETs

At temperatures above those where elastic tunneling is observed, the current, as is now quite well established, is due to inelastic tunneling between sites. Experiments in macroscopic [3,8] as well as mesoscopic [4] MOSFETs were reported in that regime. In both cases, the interpretation offered was in terms of the usual percolative transport by hopping of noninteracting carriers. The fact that the data could be fitted with such a theory, but could not be fitted with the accepted theory for hopping in the presence of Coulomb interactions [6,12], led to the conclusion that interactions are unimportant in these systems. This conclusion has been puzzling, since simple calculations show [16] that the Coulomb interactions in these systems should be of the same order or larger than other important energies of the system. We cannot shed much light on this puzzle here, but it is worthwhile putting the various considerations regarding this problem into perspective, and clarifying how the present work bears on the problem.

First, it should be mentioned that some difficulty exists with the interpretation of experiments by hopping of non-interacting carriers, namely the fact that in the activated regime the activation energy decreases with increasing concentration [8]. This is contrary to what is predicted by simple theory [17]. It was proposed [18] that the effect is due to collective hopping effects produced by interactions, as has also been often proposed for impurity conduction [16]. However, this view has not been generally accepted. For example, quantum effects [19] were put forward as an alternative explanation for a decrease in the activation energy with increasing site concentration.

As a possible reason why the expected effects of Coulomb interactions failed to be seen, screening by the gate electrode was proposed [16]. However, the results of the second section of this paper argue against such an explanation. For the thickness of the dielectric layer in the MOSFETs used, the results of Section 2 indicate a rather deep gap. Even if it could be argued (and this is controversial) that usual variable range hopping theory can be used, the relevant density of states would have to be the one at μ, which is considerably reduced by the interaction (see Figure 1), so a disagreement with the theory for hopping without interaction is still expected. Thus, screening by the gate electrode does not resolve the puzzle.

The work reported in the third section tentatively supports interaction effects as a valid explanation of transport through MOSFETs at millikelvin temperatures reported in Reference [5]. If future work confirms this, it will be difficult to see why the Coulomb effects should be observable in the regime of elastic hopping, and not in the regime of inelastic hopping, where the interaction energies are still considerably larger than kT.

The problem thus remains unresolved. It should be mentioned, however, that the case against interactions is based at least in part on a transport theory for systems with Coulomb interaction which is controversial [16,20]. Interesting effects, not discussed here, associated with magnetic fields were observed [3,4]. Such effects will hopefully contribute to a full understanding of the problem.

Acknowledgements

We would like to thank Dr. A.B. Fowler for a number of communications regarding the work of his group, in particular the experiments of Reference [5]. We are grateful to the editors for inviting us to contribute to this volume honoring Professor Kamimura, an inspiration and a long-time friend to one of us (M.P.). The stimulating months spent with Professor Kamimura and his group in Tokyo will always be remembered with great pleasure.

References

1. W. McMillan and J. Mochel: 1981, *Phys. Rev. Lett.* **46**, 556.
2. R.C. Dynes and J.P. Garno: 1981, *Phys. Rev. Lett.* **46**, 371.
3. For reviews see H. Kamimura and H. Aoki: 1989, *The Physics of Interacting Electrons in Disordered Systems*, Clarendon Press; T. Ando, A.B. Fowler and F. Stern: 1982, *Mod. Phys.* **54**, 54.
4. For review on mezoscopic MOSFETs see A.B. Fowler, J.J. Wainer and R.A. Webb, *Hopping Transport in Solids*, M. Pollak and B. Shklovskii (eds.), N. Holland Physics, in press (Jan. 1991).
5. A.B. Fowler, G.L. Timp and J.J. Wainer and R.A. Webb: 1986, *Phil. Mag. B* **42**; *Phys. Rev. Lett.* **57**, 138.
6. A.L. Efros and B.I. Shklovskii: 1975, *J. Phys. C* **8**, L49; S.D. Baranovskii, A.L. Efros, B.L. Gelmont and B.I. Shklovskii; 1979, *J. Phys. C* **12**, 1023; S.D. Baranovskii, A.A. Uzakov and A.L. Efros: 1982, *Zh. Exp. Teor. Fiz.* **83**, 756; 1982, *Sov. Phys. JETP* **56**, 442.
7. M. Pollak: 1970, *Disc. Faraday Soc.* **50**, 11; G. Srinivasan: 1971, *Phys. Rev. B* **4**, 2581.
8. A.B. Fowler and A. Hartstein: 1980, *Phil. Mag. B* **42**, 949; G.C. Timp, A.B. Fowler and A. Hartstein, *Physics of Semiconductors*, J.D. Chadi and W.A. Harrison (eds.), Springer Verlag 1985, p. 381; G. Timp, A.B. Fowler, A. Hartstein and P.N. Butcher: 1986, *Phys. Rev. B* **34**, 8771; K.T. Hayden and P.N. Butcher: 1979, *Phil. Mag. B* **138**, 607.
9. M. Pepper and M.J. Uren: 1982, *J. Phys. C* **15**, L617; Thornton *et al.*: 1986, *Phys. Rev. Lett.* **56**, 1198.
10. Second paper in Reference 6.
11. B.I. Shklovskii and A.L. Efros: 1984, *Electronic Properties of Doped Semiconductors*, Springer Verlag.
12. A.L. Efros: 1976, *J. Phys. C* **9**, 2021.
13. A.B. Fowler, personal communication.
14. For a detailed derivation see M. Green, Ph.D. dissertation, University of California, Riverside, 1990.
15. M. Mochena and M. Pollak, *Hopping and Related Phenomena*, H. Fritzche and M. Pollak (eds.), Worlk Scientific, 1990, p. 129, and *J. Non-Crystalline Solids*, in press.
16. Chapter 4, *Electron-Electron Interactions in Disordered Systems*, A.L. Efros and M. Pollak (eds.), N. Holland Physics, 1985.
17. M.L. Miller and E. Abrahams: 1960, *Phys. Rev.* **120**, 745 and many later papers.
18. M. Pollak: 1982, *Surface Physics* **113**, 170.
19. R.N. Bhatt and T.V. Ramakrishnan: 1984, *J. Phys. C* **17**, L639.
20. Chapter 4, *Electron-Electron Interactions in Disordered Systems*, A.L. Efros and M. Pollak (eds.), N. Holland Physics, 1985.

MULTIFRACTAL PROPERTIES OF EIGENSTATES IN WEAKLY DISORDERED TWO-DIMENSIONAL SYSTEMS WITHOUT MAGNETIC FIELD

MICHAEL SCHREIBER
Institut für Physikalische Chemie, Johannes-Gutenberg-Universität Mainz Jakob-Welder-Weg 11, 6500 Mainz, Germany.

Abstract. In order to investigate the electronic states in weakly disordered 2D samples very large (up to 180 000 * 180 000) secular matrices corresponding to the Anderson Hamiltonian are diagonalized. The analysis of the resulting wave functions shows multifractal fluctuations on all length scales in the considered systems. The set of generalized (fractal) dimensions and the singularity spectrum of the fractal measure are determined in order to completely characterize the eigenfunctions.

1. Introduction

The properties of the electronic wavefunctions are of special interest in the investigation of disordered systems, because they determine important features, such as the transport in amorphous semiconductors or the dynamics of excitons in molecular crystals. The general behaviour of the electronic states in these systems can be studied within a simple tight-binding model like the Anderson Hamiltonian which describes a regular lattice with site-diagonal disorder only.

In three dimensions (3D), as well as in 2D under the influence of a strong magnetic field, extended states exist for weak disorder, separated from the localized states by a mobility edge E_c. The localized states are generally believed to fall off exponentially in space, a result which can be proved for 1D systems of any disorder [1] and one which has been explicitly assumed in the scaling hypothesis of localization [2]. However, this decay might be adopted only asymptotically so that the short range behaviour of localized wavefunctions may be completely different. Approaching the mobility edge from the localized regime the localization length (i.e. the exponential decay constant) will diverge so that the different behaviour characterizes the wavefunction throughout.

But even arbitrarily close to the mobility edge the localized state should still occupy only an infinitesimal fraction of the total volume available. On the other side of the mobility edge, states should extend over the whole sample. Without implying a drastic transition in the nature of the wavefunction, these two characteristics can be reconciled at E_c if one assumes a fractal wavefunction with a filamentary structure like a net over the whole sample. This notion was originally suggested by Aoki [3,4], and exploited by

H. Aoki et al. (eds.), New Horizons in Low-Dimensional Electron Systems, 169–178.

several authors to determine a fractal dimension either from the density–density correlation function [5] or the participation number [6–9] or directly from the amplitude of the wavefunction [4]. These investigations showed that not only the states at the mobility edge are fractal objects, but also off the edge the fluctuations of the wavefunction can be characterized by a fractal dimension. Of course, this behaviour cannot be expected to persist on length scales of the order of or larger than the correlation length or the localization length, for the extended and the localized states, respectively.

The use of different quantities for the derivation of a fractal dimension [4–6] as well as the observation of anomalous scaling properties [10] suggested that the more general concept of multifractality involving a set of generalized dimensions has to be employed for a complete characterization of the wavefunctions. A few of these dimensions have been computed [11–13] to corroborate the multifractal picture. A complete multifractal analysis has recently been given for extended states close to the critical energy for a 2D disordered system in a high magnetic field [4]. In these systems a metal-insulator transition exists at which a fractal behaviour is no longer surprising. In the following sections it will be shown that even a wavefunction in a 2D sample without magnetic field, i.e. without a metal-insulator transition, shows multifractal fluctuations. This observation is in contrast to an analysis of the same model for 1D systems which disproved self-similar spatial fluctuations of the wavefunction [15]. However, the 1D results should not be simply carried over to 2D samples, because 2D systems are marginal [2] with respect to the existence of extended states (and thus the existence of a mobility edge). From the results presented below we conclude that for the parameters considered the system is close enough to criticality so that the multifractal properties persist over appreciable length scales.

2. Wavefunctions and Participation Ratio

Our investigation is based on the Anderson Hamiltonian which is commonly used to study the localization of electronic states in disordered systems [5,6,9,11–13,15]. Straightforward diagonalization of the $N \times N$ secular matrix by means of the Lanczos algorithm without reorthogonalization [9] yields the amplitudes e_{in} of the wave function φ_i at site n at an eigenenergy E_i for a square lattice of N sites with periodic boundary conditions. (The energy E and the width W of the box distribution of the randomly chosen site energies are measured as usual in units of the constant nearest neighbour transfer matrix element.) For a thorough analysis of the multifractal properties a system as large as possible should be treated. Figure 1 shows an example of the largest system (with $N = 180\ 000$ sites) that we have so far been able to investigate. Due to the numerical instability inherent in the Lanczos algorithm one has to check carefully for the occurrence of ghosts

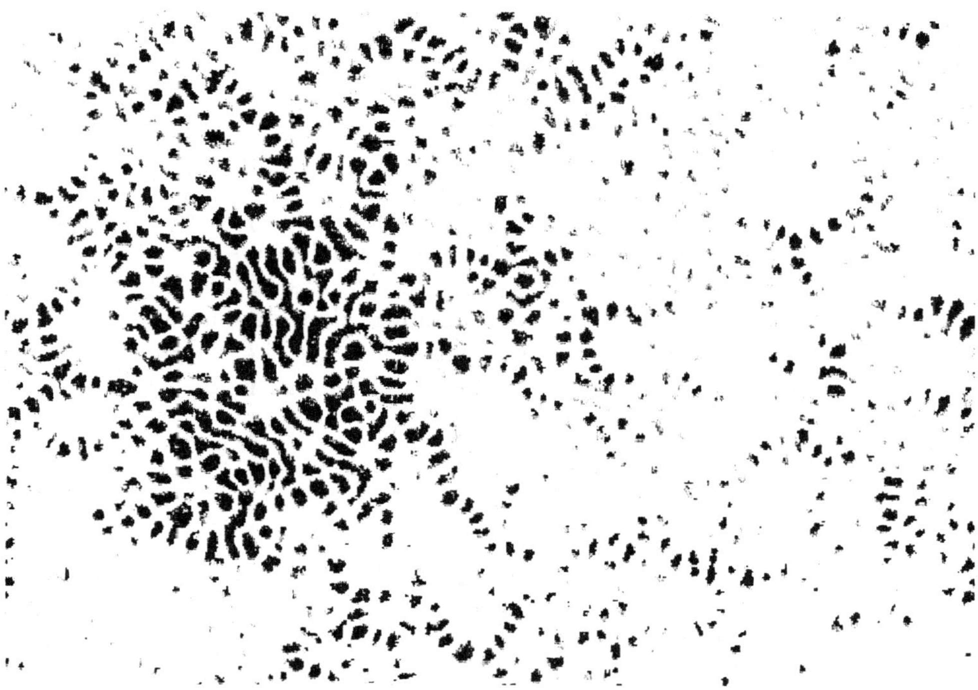

Fig. 1. Wavefunction on a 2D square lattice with $N = 360 * 500$ sites for disorder $W = 1$ at the eigenenergy $E_i = 3.91$. Every site with an amplitude larger than average is shown, i.e. $|e_{in}| > N^{-1/2}$. Four different grey levels ($j = 0, 1, 2, 3$) distinguish whether $e_{in}^2 > 2^j/N$.

in the eigenvalue spectrum [9]. In the present case we ascertained that our solution fulfills $||H\varphi_i - E_i\varphi_i|| \leq 10^{-6}$.

The filamentary structure of the wavefunction in Figure 1 is obvious. The speckle pattern in some regions is reminiscent of the Bloch oscillations, but the fragmented character of the state clearly suggests a fractal behaviour. It is interesting to note that from the strip and bar method [16] an exponential decay length larger than 1000 lattice constants can be estimated for the considered parameters [17]. This means an extension of the state which cannot be imagined from Figure 1.

In the present case it therefore appears to be better to use the participation number P_i instead of the localization length to measure the size of a state. This is given by

$$P_i^{-1} = \sum_n e_{in}^4 \tag{1}$$

and reflects the number of sites which contribute to a given eigenfunction. For the example in Figure 1 we obtain $P_i = 28\,084$, confirming that only a small fraction of the sites contributes.

3. Derivation of the Multifractal Properties

A suitable average of P_i^{-1} over the eigenstates in a (small) energy range and over different configurations of the random potential yields $P^{-1}(E, W)$. Its proportionality to some power of the system size has been used to define a fractal dimension D_2 [6–9]. It can be shown that the same result can be obtained by the standard box counting procedure, i.e. subdividing the system into N_L boxes of linear size L and determining the box probability of the wavefunction in the k-th box

$$\mu_k(L) = \sum_{n=1}^{L*L} |e_{in}|^2 \; ; \quad k = 1, \ldots, N(L) \, . \tag{2}$$

Then the (inverse) participation number for the boxes is analyzed with respect to L:

$$P^{-1}(L) = \; < \sum_{k=1}^{N_L} \mu_k^2(L) > \; \sim L^{D_2} \, . \tag{3}$$

Here the brackets denote the abovementioned average over energy and random potential.

This definition can easily be generalized to

$$\chi_q(L) = \; < \sum_k \mu_k^q(L) > \; \sim L^{-\tau(q)} \, . \tag{4}$$

Thus χ_q describes the q-th moment of the probability μ_k which is used as a suitable *measure* for the wave function. If the proportionality in Equation (4) holds, multifractal behaviour may be derived.

In the present case, of course, the discrete lattice restricts the possible range of (multi)fractal behaviour to scales above the lattice constant; the results presented below have been obtained by a least square fit for box sizes down to $L = 2$. We note that a good accuracy can be achieved by averaging over all possible choices of the origin of the partitioning; this procedure also avoids the restriction that $N_L = N/L^2$ has to be an integer. Moreover, it is also profitable to include (slightly) rectangular boxes with $L_x = L_y \pm 1$.

The typical dependence of the resulting $\tau(q)$-curve is shown in Figure 2. For a homogeneous fractal $\tau(q)$ would be a linear curve, so that only two

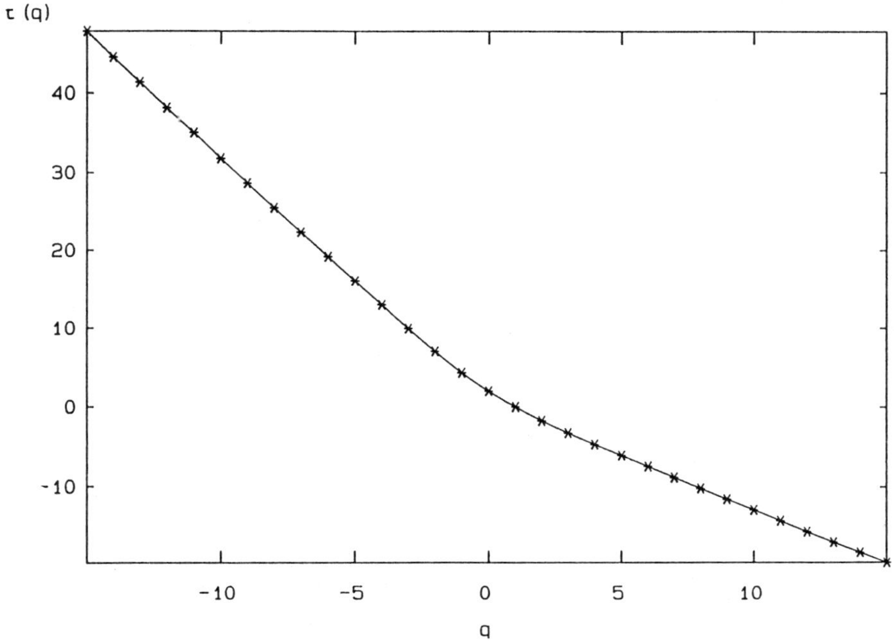

Fig. 2. Mass exponents $\tau(q)$ of the measure $\mu_k(L)$ for the wavefunction in Figure 1. Integer values cf q are marked by symbols.

independent moments would suffice to specify $\tau(q)$. If in addition the steepness of this straight line would agree with $-D_E$, the Euclidean dimension, then no fractal behaviour would be obtained at all.

Therefore Figure 2 demonstrates the multifractality of the wavefunction. The physical interpretation of this result is that intertwined regions of the state distinguished by their measure scale in a different way as determined by the scaling exponents $\tau(q)$. Each subset of the measure characterizes a fractal with its own fractal dimension. As a consequence, no self-similar picture of the complete state can be obtained.

From $\tau(q)$ the generalized dimensions D_q can be derived, defined [9] by

$$-\tau(q) = (q-1)\ D_q = \lim_{L\to 0}\ \ln\ \chi_q(L)/\ln\ L\ . \tag{5}$$

These generalized dimensions D_q are plotted in Figure 3. Typically the attention is focused [20] on the values for $q = 0,\ 1$, and 2, which are labeled the similarity, information, and correlation dimension, respectively.

In the present case, all boxes contribute to the summation (4) because the wavefunction is nowhere zero exactly, in contrast, for example, to a percolating network or for diffusion limited aggregation. Accordingly, the support of the wavefunction and thus the support of the measure is given by the total number of boxes in the whole plane, i.e. $N(L) = N/L^2$. A

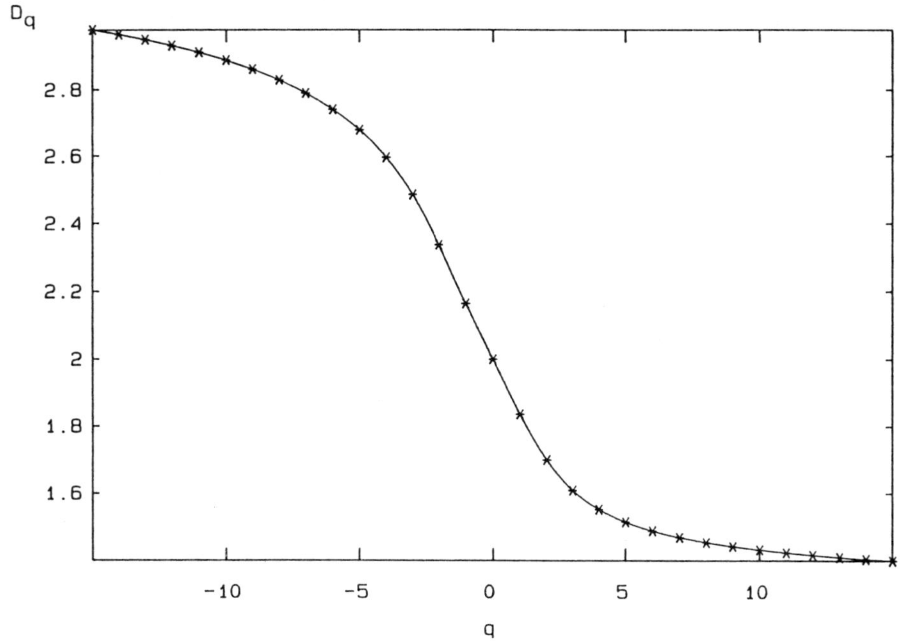

Fig. 3. Generalized dimensions D_q of the q-th moment of the measure $\mu_k(L)$ for the wavefunction in Figure 1. Integer values of q are marked by symbols.

comparison with Equation 4 shows that this box counting of the support corresponds to the evaluation of χ at $q = 0$ yielding $\tau(0) = -2$, so that the similarity dimension $D_0 = 2$ equals the Euclidean dimension.

For $q = 1$, the normalization of the wave function gives $\chi_1(L) = 1$, therefore $\tau(1) = 0$. To determine the information dimension D_1 from Equation 5, one has to compute $\chi(q)$ numerically in the limit $q \to 1$, or one can develop μ_k^q around $q = 1$ to obtain

$$D_1 = \lim_{L \to 0} \sum_k \mu_k \ln \mu_k / \ln L \ . \tag{6}$$

This formula shows that D_1 describes the scaling of the entropy S of the measure, which is defined as usual in statistical physics from the box probabilities as

$$S = -\sum_k \mu_k \ln \mu_k \ . \tag{7}$$

To illustrate the meaning of D_1, we note that the measure concentrates on a subset with this fractal dimension [21], which means that this subset contains a fraction of the measure arbitrarily close to unity, and thus contains (nearly) all the information.

The correlation dimension D_2 obtained from Equation (5) for $q = 2$ describes the scaling of the participation number P in Equation (3), or, equivalently, the scaling of the density–density correlation function. Higher correlations are characterized by larger q.

Finally, the limiting values of D_q for $q \to +\infty$ and $q \to -\infty$ are of special interest, because they reflect the scaling of the subsets where the measure (and thus the wavefunction) is most concentrated and most rarefied, respectively. A precise numerical evaluation of the latter, however, is difficult, because the small amplitudes of the wavefunction are most strongly blurred by numerical inaccuracies.

Another exponent commonly used in the multifractal analysis is the Lipschitz–Hölder exponent α, which describes the singularity strength of a fractal [21]. Again the corresponding singularity spectrum $f(\alpha)$ gives the Hausdorff dimensions of the intertwined subsets, on which the measure displays those singularities. Considering the k-th box the singularity strength α_k is determined from

$$\mu_k(L) \sim L^{\alpha_k} \tag{8}$$

and then the number of boxes $N(\alpha)$ in which the probability shows this strength scales with the fractal dimension $f(\alpha)$:

$$N(\alpha) \sim L^{-f(\alpha)} . \tag{9}$$

The relations $f(\alpha)$ and $\tau(q)$ are two equivalent ways to describe the multifractal properties completely. They are simply related to each other by a Legendre transformation [19,21]. In particular, one obtains $\tau(q)$ from

$$\tau(q) = f(\alpha) - \alpha q \tag{10}$$

and α is given by

$$\alpha(q) = -\frac{\mathrm{d}\tau}{\mathrm{d}q} . \tag{11}$$

For an accurate computation of the $f(\alpha)$-spectrum the Legendre-transformation is widely used, but is impractical due to numerical inaccuracies [22]. Instead it is better to consider Equation (10) as a definition of $f(\alpha)$ which, together with Equation (11), gives a parametric representation of $f(\alpha)$. We have used a simple method suggested by Chhabra and Jensen [22] to calculate $f(q)$ and $\alpha(q)$ directly from the q-th moment of the box probabilities for the separate boxes

$$f(q) = \lim_{L\to 0} \sum_k \mu_k(q,L) \ln \mu_k(q,L) \,/\, \ln L \tag{12}$$

$$\alpha(q) = \lim_{L\to 0} \sum_k \mu_k(q,L) \ln \mu_k(1,L) \,/\, \ln L$$

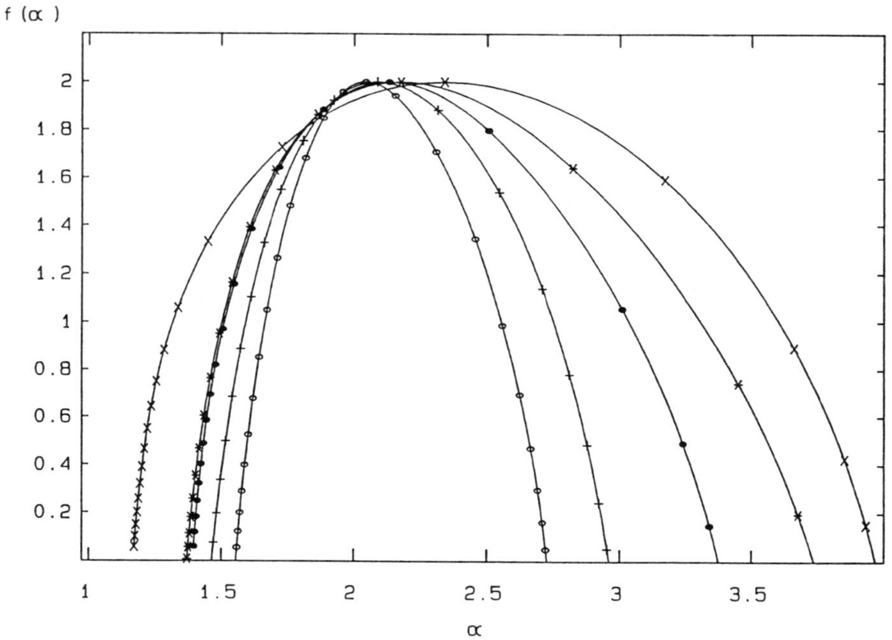

Fig. 4. Singularity spectrum $f(\alpha)$ for wavefunctions on a 2D square lattice with $N = 100 * 100$ sites for disorder $W = 1$ at different eigenenergies E_i. Integer values of the implicit parameter q are marked by symbols for $E_i = 2$ (○), 3.5 (+), 3.75 (•), 3.9 (∗), and 4.0 (×).

where $\mu_k(q, L)$ are the generalized measures

$$\mu_k(q, L) = \mu_k^q(L) \,/ \sum_k \mu_k^q(L) \tag{13}$$

so that $\mu_k(1, L) = \mu_k(L)$. The limits are again performed by least square fits for box sizes down to $L = 2$, averaging over all possible origins of the boxes and including rectangular boxes.

Typical results are displayed in Figure 4. From the discussion of $\tau(q)$ and D_q it is obvious that $q = 0$ corresponds to $f(\alpha) = 2$. This is the maximum of the spectrum, because it is the fractal dimension of the support of the measure. For $q = 1$ one obtains $f(\alpha) = \alpha = S$ as given by Equation (7). The minimum value of α is obtained from $\mathrm{d}\tau/\mathrm{d}q$ for $q \to +\infty$, as can be imagined from the monotonous behaviour of $\tau(q)$ in Figure 2. This value corresponds to the subset which contains the box with the largest measure. Similarly the maximum value of α projects out the singularity associated with the smallest box probability. In the limit $q \to \pm\infty$ the singularity spectrum $f(\alpha(q))$ approaches zero. Due to numerical limitations the plots in Figure 4 are restricted to $|q| \leq 15$ so that $f(\alpha_{\min}) = 0$ is not reached in all cases. The accuracy of $f(q)$ and $\alpha(q)$ was tested in all cases by computing the linear correlation coefficient. Only for the smallest q, hence for the largest values of α in Figure 4, are the fits of Equation (12) to our data points inaccurate. This is also the explanation of why the $f(\alpha)$-curves do not have infinite slope for $\alpha \to \alpha_{\max}$, i.e. $q \to -\infty$.

It is obvious from Figure 4 that the wavefunctions investigated are multifractal entities, because for homogeneous fractals the $f(\alpha)$-relation would be given by a single point. The important observation from Figure 4 is the strong change of the $f(\alpha)$ spectrum with energy. The spectrum widens with energy so that an increasing number of values of the singularity strength α become relevant. We note that this effect is even more drastic with respect to constant q, showing that the system increasingly deviates from the homogeneous situation.

For smaller energies the $f(\alpha)$-curves in Figure 4 can be approximated by a parabola. This observation, which shows that the box probability follows a log-normal distribution, is in agreement with the multifractal analysis of the states in disordered 2D systems with a strong magnetic field [14].

4. Concluding Remarks

We have shown that the spatial fluctuations of the eigenfunctions of the Anderson Hamiltonian display multifractal properties, in contrast to previous claims [15]. The singularity spectra for several wavefunctions have been determined. Either these spectra or the complete set of generalized dimensions should be taken into account when seeking to compare or to distinguish the states in disordered systems. It is obvious that the essential features of these systems, such as the transport properties, will be drastically influenced by the fractal behaviour, as has already been mentioned by Aoki [4]. It will be particularly interesting to analyze the wavefunctions close to or at the mobility edge which appears in 3D systems. However, the numerical effort is obviously much larger in these systems if one wants to investigate 3D samples of reasonably large linear size for the multifractal analysis [23]. From the present investigation, which demonstrates multifractal behaviour in 2D (i.e. not at a mobility edge), one can expect a similar behaviour in 3D, in contrast to the 1D results [15]. The question of whether it appears on an even larger length scale and whether it persists in infinite samples remains open to further study.

References

1. N.F. Mott and W.D. Twose: 1961, *Adv. Phys.* **10**, 107.
2. E. Abrahams, P.W. Anderson, D.C. Licciardello, T.V. Ramakrishan: 1979, *Phys. Rev. Lett.* **42**, 673.
3. H. Aoki: 1983, *J. Phys. C* **16**, L205.
4. H. Aoki: 1986, *Phys. Rev. B* **33**, 7310.

5. C.M. Soukoulis and E.N. Economou: 1984, *Phys. Rev. Lett.* **52**, 565.
6. M. Schreiber: 1985, *Phys. Rev. B* **31**, 6146.
7. B. Kramer, Y. Ono and T. Ohtsuki: 1988, *Surf. Sci.* **196**, 127.
8. Y. Ono, T. Ohtsuki and B. Kramer: 1989, *J. Phys. Soc. Jpn.* **58**, 1705.
9. M. Schreiber: 1990, *Physica A* **167**, 188.
10. C. Castellani and L. Peliti: 1986, *J. Phys. A* **19**, L429.
11. S. Evangelou: 1990, *Physica A* **167**, 199.
12. J. Bauer, T.M. Chang and J.L. Skinner: 1990, *Phys. Rev. B* **42**, 8121.
13. T.M. Chang, J. Bauer, J.L. Skinner: 1991, *J. Chem. Phys.* **93**, 8973.
14. W. Pook and M. Janssen: 1991, *Z. Phys. B* **82**, 295.
15. L. Pietronero, A.P. Siebesma, E. Tosatti and M. Zannetti: 1986, *Phys. Rev. B* **36**, 5635.
16. A. MacKinnon and B. Kramer: 1981, *Phys. Rev. Lett.* **47**, 1546.
17. M. Schreiber and M. Ottomeier, unpublished.
18. F. Wegner: 1980, *Z. Phys. B* **36**, 209.
19. T.C. Halsey, M.H. Jensen, L.P. Kadanoff, I. Procaccia and B.I. Shraiman: 1986, *Phys. Rev. A* **33**, 1141.
20. H.G.E. Hentschel and I. Procaccia: 1983, *Physica D* **8**, 435.
21. J. Feder: 1988, *Fractals*, Plenum, New York.
22. A. Chhabra and R.V. Jensen: 1989, *Phys. Rev. Lett.* **62**, 1327.
23. M. Schreiber and H. Grußbach: 1991, *Phys. Rev. Lett.* **67**, 607.

PART III

High T_c Compounds

SOME THEORETICAL IDEAS ON HIGH T_c SUPERCONDUCTORS

NEVILL MOTT
Cavendish Laboratory, Cambridge, England

Abstract. An account is given of a model for high temperature superconductors developed by the author in a number of recent papers. The carriers are holes in the $2P$ band, hybridised with Cu $3d^8$. A form of spin polaron is assumed, similar to the tJ model with influence on the orientation of surrounding copper spins. A fraction, 10–20%, of these form spin bipolarons, and there is some polarisation of the lattice by the charges, which can give rise to an isotope effect. A discussion of the electrical properties above T_c is given; a mixture of fermions and bosons is envisaged, the former having higher effective mass. The model differs from a strong interaction BCS model, mainly in the assumption that the current above T_c is carried by bosons.

1. Introduction

Kamimura and co-workers have proposed in a number of papers [1–4] that the carriers in the new superconductors are spin bipolarons. Their work is based on detailed calculations for $La_{2-x}Sr_xCuO_4$. The present author made the same proposal in a paper [5] published in 1989 in a Festschrift dedicated to Jacques Friedel, and has developed it in a number of further papers [6–13]. This model differs in several respects from that of Kamimura, being based on experiment rather than calculations, and so has been subject to frequent changes as experimental results have become available. The most important difference is that, in Kamimura's model, as in the BCS theory, the critical temperature T_c is that at which the bosons dissociate; we propose that for the copper oxide materials and possibly for the cubic bismuth oxide compounds too [12], T_c is the temperature at which the Bose gas becomes non-degenerate, as is, of course, the case for the critical temperature in superfluid liquid He^4. That this might be so for some superconductors was proposed by Alexandrov [13] in 1986, and prior to BCS by Shafroth [14,15]. This paper gives an outline of this model, as it appears in the autumn of 1990, in the full realisation of how provisional it must be.

The assumptions made for instance for $La_{2-x}Sr_xCuO_4$ and $YBa_2Cu_3O_{7-\delta}$ are as follows:

(a) The materials are quasi-two-dimensional, and can be considered as doped antiferromagnetic (Mott) insulators.

(b) The carriers are oxygen $2p$ holes hybridised with copper $3d^8$ states moving through the antiferromagnetic array of copper $3d^9$ ions; their number can be deduced from the chemical formula; according to Kamimura, their symmetry is dz^2.

H. Aoki et al. (eds.), New Horizons in Low-Dimensional Electron Systems, 181–190.

(c) They form some kind of mixed spin and dielectric polaron, which has a large effective mass.

(d) Some, but perhaps not all of them form bipolarons (bosons).

(e) T_c is the temperature at which the boson gas becomes non-degenerate.

(f) Above T_c the remaining fermions (80–90% of the carriers) form a Fermi liquid, but the states are Anderson [16] localized, and so can contribute to the specific heat and (Pauli) paramagnetism,[1] but not to the transport properties; these are determined by the non-degenerate gas of bosons.

2. Spin Polarons

As emphasized by Anderson [18], the properties of a doped Mott insulator are imperfectly understood. The case of $Gd_{3-x}V_xS_4$ (V denotes a vacancy), investigated by von Molnar *et al.* [19], has, however, been fully discussed. Each electron in the conduction band orients perhaps 10–20 gadolinium moments antiparallel to itself, forming a polaron with heavy effective mass which can move through the lattice and breaks up the antiferromagnetic order, the orientations of its moment being determined by the conduction electron. Though in principle mobile, at low temperature they are Anderson localized by the random field of the vacancies, so that the conductivity tends to zero with T. A strong magnetic field breaks up the (antiferromagnetic) polaron (cf. Mott [8]), so that the carriers are no longer heavy, Anderson localization no longer occurs and metallic conduction sets in.

In several early papers, I took this for a model in the superconductors, but have now modified it in the light of the experimental evidence in a way that I shall discuss in Section 6. Whatever the form of spin polaron, however, I make the following assertion about it:

(a) Two such polarons can under certain conditions, particularly in 2D structures, combine together to form a spinless boson.

(b) This charge carrier, with charge $q(= +2e)$ must polarize the surrounding medium, giving a phonon-dependent enhancement of the effective mass and so some isotope effect. Some authors (Emin and Hillery [20], Alexandrov and Ray [21]) propose that dielectric polarons are the sole mechanism of bonding, but we reject such a model in view of point (c) below. We follow de Jongh [22] in believing that both play a role.

(c) As shown by Vigren [23] in 1973, the diffusion coefficient D of a spin polaron is roughly independent of temperature, being of the form

[1] See the discussion by Kamimura and Aoki [17] of the phenomenon for impurity bands.

$$D \sim \omega a^2 \tag{1}$$

where $\hbar\omega = kT_N$, T_N is the two-dimensional Néel temperature and a the distance diffused by the polaron as a result of one spin flip.

3. Superconductivity

If superconductivity is due to the presence of a degenerate gas of bosons, which do not dissociate, the critical temperature is given in three dimensions by

$$k\,T_c = 3.3\;\hbar^2\,n^{2/3}\,/\,m \tag{2}$$

where n is the density of bosons and m their mean effective mass. This equation neglects interaction, which is known to be important in the helium problem, and indeed superconductivity could not exist without it (Mott [9]). In two dimensions, because the integral

$$\int N(E)\,/\,\{e^{E/kT} - 1\}\;\mathrm{d}E$$

diverges at $E = 0$, condensation is impossible without interaction, but there is evidence that truly two-dimensional conduction does exist (Lowndes *et al.* [24]). However, in this paper we shall treat these materials as three dimensional but highly anisotropic and following Prelovsek *et al.* [25], we write

$$m = (m_1\,m_2\,m_3)^{\frac{1}{3}}$$

with m_1 etc. referring to three directions in space.

The entropy S released at the transition is then (in 3D)

$$S = nk\;\ln 2\;. \tag{3}$$

Equation (2) indicates that, for varying levels of doping, T_c should vary as $n^{2/3}$. There is much evidence that, for low values of n_c, T_c varies as a power of n_c near to unity and consistent with 2/3 (Schafer *et al.* [26], Uemura *et al.* [27], de Jongh [22]). Here n is the concentration of carriers deduced from the composition. Our model, then is of a degenerate gas of bosons which do not overlap strongly (in contrast with BCS, where they do). If motion is truly two dimensional, $n^{2/3}$ should be replaced by n.

4. Conductivity above T_c

Above T_c good specimens show a resistivity proportional to T, extrapolating to zero, and extending [28] for LSCO to ca. 1000 K. Many theories have been proposed for this effect, some of which are discussed by the present author [9], who rules out scattering by phonons because the effect is observed even for a material for which T_c is only 10 K. In the similar (but three dimensional) material $La_{1-x}Sr_xVO_3$ for ranges of x where metallic conduction is observed $\sigma \sim \sigma_{\min}$ and is independent of T ($\sigma_{\min} = 0.03\ e^2/\hbar a$); this is discussed by Mott [8]. If above T_c the carriers are fermions, the very different behaviour must be caused by the quasi-two dimensional structure, and no theory to predict this without phonons has been proposed. If, on the other hand, the carriers are a non-degenerate gas of bosons, this behaviour follows from (1) using the Einstein relation

$$\sigma = nq^2\ D/kT \qquad q = 2e\ . \tag{4}$$

This is proposed in the authors papers.

If the movement of the carriers is of this diffusive type, we should not expect scattering by point defects; a residual resistance will only occur through the presence of rather large defects. A residual resistance is observed, for instance, through the substitution of Co for Cu in YBCO; this is believed to enter the chains, and may give the kind of obstacle required (Mott [9]).

It is this phenomenon which has lead the present author to suppose that a non-degenerate gas of some kind of *spin* polaron must determine the movement of carriers above T; we shall see in the next two sections why they must be bipolarons.

5. Thermopower

Evidence that the carriers are bosons comes from the thermopower S of YBCO above T_c. This is only weakly dependent on T, and is small ($\sim 4\,\mu V/K$). It can be negative or positive, changing sign with oxygen content. A degenerate gas of fermions cannot easily explain this behaviour.[2] The author's explanation [8] is as follows:

(a) In LSCO the carriers move in an impurity band, and the Hubbard U splits the states into an upper and lower Hubbard band. This splitting also occurs in YBCO (see Mott [8]).

(b) For diffusing polarons one can use the Heikes formula

$$S = (k/q)\ \ln\{(1-z)/z\} \tag{5}$$

where z is the ratio of carriers to sites.

[2] The change of sign should be caused by fermions if the Fermi surface consisted of n-type and p-type branches contributing equally to the current (a coincidence); but a linear increase of S with T would then be expected.

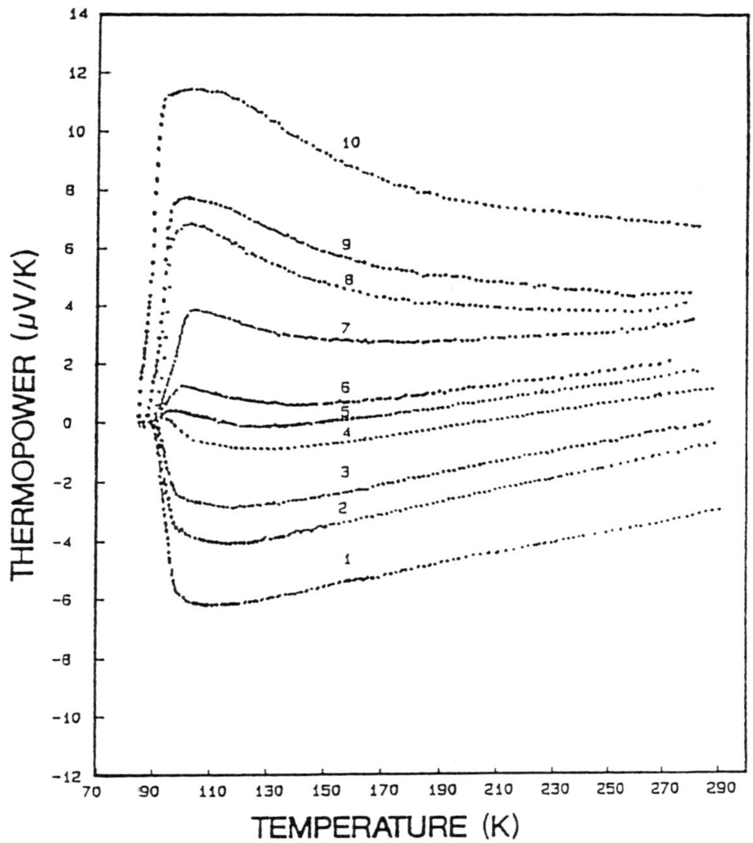

Fig. 1. Variation of thermopower of $YBa_2Cu_3O_7-\delta$ as a function of T with different oxygen content. Oxygen content $7 - \delta$ for curve 1 is 7, for curve 2, 6.98, curve 3, 6.94, curve 4, 6.93, curve 5, 6.91, curve 6, 6.90, curve 7, 6.86, curve 8, 6.84, curve 9, 6.83 and for curve 10, 6.81 (Ouseph and Bryan [43]).

For an impurity band, in the absence of compensation, $z = 1$, and no conduction is possible by fermions. If they form pairs (bosons), however, $z = \frac{1}{2}$ and $S = 0$. It has been suggested on the basis of experiment, that for perfect stoichiometry S should *vanish* (Yu *et al.* [29]). Figure 1 shows the behaviour of S for varying oxygen content, which may be expected to change the compensation and thus the value of z from small positive to negative values. Similar results have recently been obtained by the Cambridge group.

Above about 300 K, S is observed to increase; we propose that at these temperatures the carriers are no longer confined to the lower Hubbard band, so z deviates from $\frac{1}{2}$.

Other investigations showing this behaviour of S are Howson *et al.* [30], Wang *et al.* [31] and Crommie *et al.* [32].

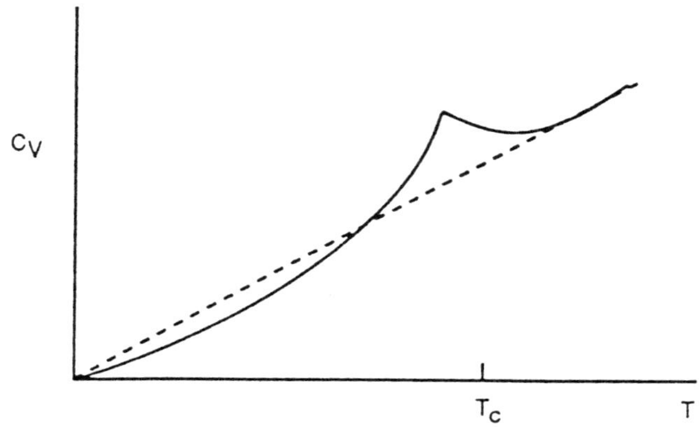

Fig. 2. Electronic specific heat of YBCO (schematic).

The present author [19] has given a theory of the Hall effect in YBCO based on this model; the Hall constant cannot be used directly to determine the number of of carriers. Recent results in Cambridge show that it also changes sign when S does so.

6. The Concentration of Bosons

In the author's earlier papers it was suggested that *all* the fermions would form bosons. This in our view is ruled out[3] by the observations of the electronic specific heat of YBCO and LSCO by Loram *et al.* [36]. These are illustrated schematically in Figure 2. The entropy in the peak is found to have values from 0.1 to 0.2 of the quantity nk ln 2 (Equation 3). Above T_c a linear specific heat is observed suggesting a degenerate gas of fermions with degeneracy temperature $\sim$ 1200 K – if nearly *all* the carriers predicted by the chemical formula form fermions. These authors point out that their results are compatible with a strong interaction BCS theory, with a small fraction of the fermions condensing into bosons, and disappearing at T_c. Conduction above T_c would then be by fermions.

We reject the latter hypothesis in view of the evidence of the last two sections, particularly that relating to the thermopower, which in our view could not be caused by a degenerate gas of fermions. Also, the sharp transition observed in normal semiconductors in the BSC theory is essentially caused by the interaction resulting from the strong overlap of the Cooper paris, and it is not clear that it would result from non-overlapping bosons. The results of Loram *et al.* [33,37] and the electrical properties are, in our view, compatible with the following hypotheses:

[3] See, however, p. 189.

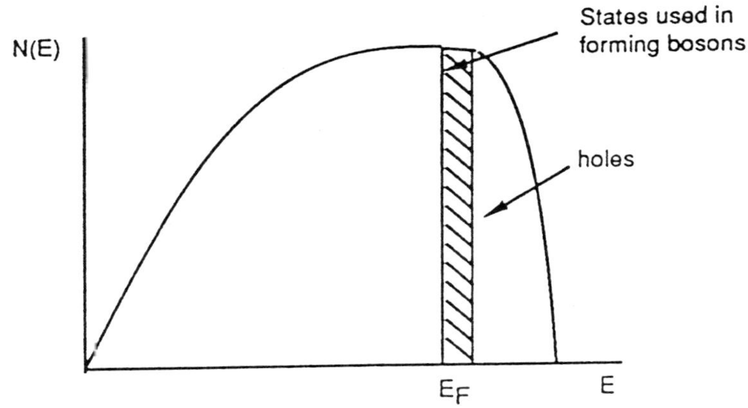

Fig. 3. Density of electron states. The shaded region represents the energy range of fermion holes which form bosons at zero temperature.

(a) Only a definite fraction of the fermions, independent of temperature, form bosons, this fraction being of the order 10–20%.

(b) The fermions are Anderson localized by the disorder (as in gadolinium sulphide, but here at high temperature), so the bosons carry the current. They are probably lighter than the fermions, as is suggested by Kamimura.

If this is so, we have to ask why only a fraction form bosons. It happens in BCS theory, because the phonon-induced binding energy is small compared to the Fermi energy, but this is unlikely to be the case there if bosons are stable up to ca. 1000 K. I have suggested [12] that the 'overcrowding effect' of Ruscher *et al.* [35] and Salje [36] may be responsible; there is not enough space for all the polarons to form bipolarons – but this hypothesis seems incompatible with the relationship $T \propto n_c^{2/3}$ (or n_c as observed), n_c being the concentration of carriers given by the chemical formula. I now think we may find an explanation by comparing a superconductor with superfluid He^4, in both of which there is strong repulsive interaction (see Feynman [39]). It is known that in the many-particle wavefunction of the latter only a weight of about 10% corresponds to zero wavenumber (Svensson [38]). If the same is true for the superconductor, only these may contribute to the entropy S according to Equation (3) while the rest transform into localized fermions. We hope to give a fuller account of these ideas in a forthcoming paper (see p. 189).

We thus envisage a density of states as envisaged in Figure 3. A proportion of the holes above the Fermi energy form pairs, and at $T = 0$ condense in the shaded area just above E_F; this is predicted by Wheatley [37] if the

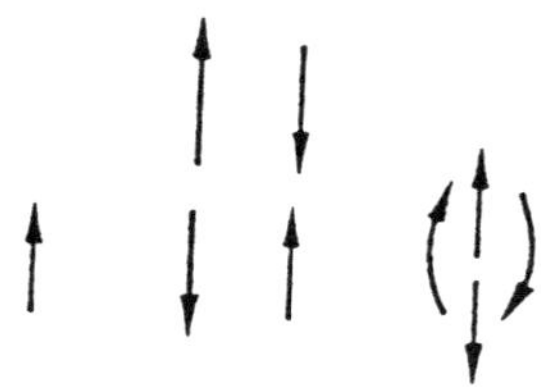

Fig. 4. Model of the spin bipolaron. Above are two oxygen holes, below the copper spins, that on the right resonating.

bosons are lighter than the fermions. The specific heat at low T is due to these bosons and (with interaction) should vary as T^3. But above T_c the bosons are distributed over the whole range up to E_o in Figure 3, so a Fermi surface exists with specific heat as shown in Figure 2. Also angle-resolved photoemission enables the density of states to be explored as for a Fermi liquid.

The situation, then, is not dissimilar to that of the BCS theory, the major difference that the bosons do not overlap (large Hubbard U) and do not disappear at T_c. We think that, since the London equations and the Meissner effect follow from BCS, they should obtain here, though the critical magnetic fields and penetration defects have yet to be deduced from the model.

7. Nature of the Pairing

The (Pauli) paramagnetism κ above T_c in the range of the linear specific heat has been measured for YBCO by Cooper [40]. If it is assumed to be given by

$$\kappa = n_F \; \mu^2 / E_F$$

with n_F some 80–90% of the total number of carriers, μ for each carrier is of the order of one Bohr magneton. This is not compatible with the kind of spin polaron found in gadolinium sulphide, where μ must be much larger. We think that a correct description may be as follows (Mott [13]). We start with the so-called tJ model of Zhang and Rice [41] in which one oxygen $2p$ hole and one copper $3d^9$ form a positively charged singlet fermion. The surrounding copper ions retain their short range AF order, but since the polarons are moving, at long range the order breaks down. A paper by von Szczepanski *et al.* [42] gives a model of this kind.

The bipolaron is then envisaged as in Figure 4. The antiferromagnetic forces keep it together. Clearly two charged particles will repel at small distances, and any attraction at larger distances may or may not give rise to a bound state. The importance of the quasi two-dimensional structure is, we believe, that bound states are more easily formed in two than in three dimensions.

Note Added in Proof

A discussion with Professor Alexandrov (Moscow) has lead the author to return to his original concept, namely that all the fermions in conducting copper oxide layers form bosons. If fermions were present above T_c, an EPR signal would be observed, which is normally not the case. The specific heat and paramagnetism shown in Figure 2 are thus due to bosons, a spin bipolaron showing a paramagnetic susceptibility independent of T. Details have recently been published in *Phil. Mag. Letters* **63**, 319 (1991) and depend on the concept outlined on p. 187.

References

1. H. Kamimura, *Proc. Adriatico Conf. on Theoretical Understanding of High Temperature Superconductivity*, Y. Lu, E. Tosati (eds.), World Scientific, Singapore, 1988, p. 163.
2. H. Kamimura, S. Matsuno, R. Saito: 1988, *Solid State Commun.* **67**, 363.
3. H. Kamimura, S. Matsuno and R. Saito, *Mechanisms of High Temperature Superconductivity*, H. Kamimura and A. Oshiyama (eds.), 1989, p. 8.
4. H. Kamimura, M. Eto, K. Ishida, S. Matsuno, T. Schimizu and H. Ushio, *Supercond. Sci. Technology* **4**, 5121 (1991).
5. N.F. Mott: 1989, *J. Phys. (Paris)* **50**, 2811.
6. N.F. Mott: 1990, *Phil. Mag. Lett.* **61**, 217.
7. N.F. Mott: 1990, *Phil. Mag. Lett.* **62**, 37.
8. N.F. Mott: 1990, *Adv. Phys.* **39**, 55.
9. N.F. Mott: 1990, *Phil. Mag. Lett.* **62**, 273.
10. N.F. Mott: 1990, *Contemp Phys.* **31**, 373.
11. N.F. Mott: 1990, *Physica A* **168**, 221.
12. N.F. Mott: 1991, *Supercond. Sci. Technology* **4**, 559.
13. A.S. Alexandrov, J. Ranninger and S. Robaszkiewicz: 1986, *Phys. Rev. B* **33**, 4526.
14. M.R. Shafroth: 1954, *Phys. Rev.* **96**, 1149; 1955, *ibid.* **100**, 463.
15. M.R. Shafroth, S. Butler and J. Blott: 1957, *Hel. Physica Acta* **30**, 93.
16. P.W. Anderson: 1958, *Phys. Rev.* **109**, 1492.
17. H. Kamimura and H. Aoki, *Physics of Interacting Electrons in Disordered Systems*, Oxford, 1989.
18. P.W. Anderson, Chap. II of a book to be published.
19. S. von Molnar and T. Penney: 1985, in *A Festschrift in Honor of Sir N. Mott*, Vol. 3, p. 183; H. Fritzsche and D. Adler (eds.), Plenum Press, N.Y.; see also S. Washburn, R.A. Webb, S. von Molnar, F. Holtzberg, J. Flouquet and G. Remenyi: 1984, *Phys. Rev. B* **30**, 6224.

20. D. Emin and M.S. Hillery: 1989, *Phys. Rev. B* **39**, 6575.
21. A. Alexandrov and D.K. Ray: 1991, *Phil. Mag. Lett.* **63**, 295.
22. L.J. de Jongh: 1988, *Solid State Commun.* **65**, 961; 1988, *Physica C* **152**, 171.
23. P.T. Vigren: 1973, *J. Phys. C* **6**, 967.
24. D.A. Lowndes, D.P. Norton and J.D. Badai: 1990, *Phys. Rev. Lett.* **65**, 1160.
25. P. Prelovsek, T.M. Rice and F.C. Zhang: 1987, *J. Phys. C* **20**, L229.
26. M.W. Schafer, T. Penney and B.L. Olson: 1987, *Phys. Rev. B* **30**, 1047.
27. Y.J. Uemura: 1988, *Phys. Rev. B* **38**, 902.
28. L.P. Gor'kov and N.B. Kopnin: 1987, *Soviet Phys. Usp.* **156**, 117 (English translation, 1988, *Soviet Phys. Usp.* **31**, 850).
29. R.C. Yu. M.J. Naughton, X. Yan, F. Chaikin, P. Holtzberg, R.L. Green, J. Stuart and P. Davies: 1988, *Phys. Rev. B* **37**, 7963.
30. M.A. Howson, M.B. Salamon, T.A. Friedmann, S.E. Inderlees, J.P. Rice, D.M. Ginsberg and K.M. Ghoran: 1989, *J. Phys.* **1**, 465.
31. Z.Z. Wang and P. Ong: 1988, *Phys. Rev. B* **301**, 7160.
32. M.F. Crommie, A. Zettle, T.W. Barber and M.L. Cohen: 1988, *Phys. Rev. B* **37**, 9734.
33. J. Loram, K.A. Mirza, K.A. Liang and J. Osborne, *Physica C* **162–164**, 4981.
34. J. Loram and K.A. Mirza, 1990: *Cambridge Conf.*
35. C. Ruscher, E. Salje and A. Hussain: 1988, *J. Phys. C* **21**, 3737.
36. E. Salje: 1991, *Phil. Mag. Lett.* (in press).
37. J.M. Wheatley: 1990, *Phys. Rev. B* **42**, 946.
38. E.C. Svensson: 1984, in *75th Jubilee Conf.*, J.G.M. Armirage (ed.), World Scientific.
39. R.P. Feynman: 1955, *Progress in Low Temperature Physics* **1**, 17.
40. J. Cooper (to be published).
41. F.C. Zhang and T.M. Rice: 1988, *Phys. Rev. B* **37**, 3759.
42. K.J. Von Szczapanski, P. Horsch, W. Stephan and M. Ziegler: 1990, *Phys. Rev. B* **41**, 2017.
43. D.J. Ouseph and M.R. Bryan: 1990, *Phys. Rev. B* **41**, 4123.

LIMITATIONS ON THE THEORIES OF HIGH TEMPERATURE SUPERCONDUCTORS

MARVIN L. COHEN

Department of Physics, University of California and Materials Sciences Division, Lawrence Berkelely Laboratory, Berkeley, California 94720, U.S.A.

Abstract. Experimental data and consideration of theoretical constraints put limitations on the applicability of Boson-exchange BCS-like theoretical proposals to explain high temperature superconductivity in the copper oxide systems.

1. Introduction

It is a pleasure to contribute to this volume honoring Professor Hiroshi Kamimura. Along with his many other contributions, his recent work on superconductivity in the high-T_c oxides illustrates his creativeness and scholarly approach to important problems in physics. The Kamimura 'two storey house model', his studies of phase diagrams for hole- and electron-doped systems, and his proposal of a spin polaron mechanism are some examples of his research in this area. We look forward to continued stimulation from Professor Kamimura in the high-T_c area and in other fields of condensed matter theory.

The approach taken in this paper is a modest one since only a fraction of the various types of proposed mechanisms for high-T_c superconductivity will be examined. In particular, only theories that are BCS-like and involve an exchange Boson will be considered. The central idea in most of these theories is to start with a BCS-Eliashberg formalism which is appropriate for conventional superconductors and explore the simplest changes one can make in this approach to explain properties of high-T_c oxides. Many theoretical proposals of this type have been suggested and there is no consensus at this time on which of these are viable. Here we discuss some of the possible schemes and use experimental data to restrict the domain over which these theories can be used.

Before the BCS theory successfully passed almost every experimental test on the then known superconductors, many researchers felt there was merit to Felix Bloch's view that "all theories of superconductivity can be shown to be wrong". For the past 30 years since the publication of the BCS theory successful explanations of experimental data have been so numerous that proposals of non-BCS superconductivity theories were very rare. However, modified pairing mechanisms were considered and the idea that Bosons other than phonons can cause pairing was proposed often. Plasmons, excitons, and

H. Aoki et al. (eds.), New Horizons in Low-Dimensional Electron Systems, 191–199.

demons (acoustic plasmons) were the most common exchange Bosons suggested. Usually these were introduced into a BCS-like theory and properties such as the transition temperature T_c were predicted.

After the discovery [1] of high-T_c copper oxide superconductors, several of these proposals experienced a rebirth and formed the basis for explanations of the properties of the high-T_c oxides. New theories such as the ones mentioned above by Professor Kamimura were also introduced. Other 'new' theories include resonating valence bonds (RVB), spin-bags, anyons, ... A majority of these theories were developed around the observation that the normal state properties, such as the magnetic properties, of the Cu oxides were unusual. Early on it was suggested that these highly correlated systems require non-standard explanations for their normal state properties. Hence it should not be surprising if their superconducting properties would require non-standard theories too. These 'non-standard' theories cannot be easily grouped and examined as a class; the predictions of each should be compared with experiment to test their validity. In contrast, many of the theories involving BCS-like Boson exchange mechanisms can be examined as a group, and experimental data can be used as constraints on these theories. This is the major motivation of this paper.

2. Experimental Data

There are a large number of experiments on the high-T_c oxides which bear on the microscopic pairing mechanism. The discovery of a reduced isotope effect [2–5] had profound effect on motivating the search for non-phonon mechanisms. The importance of determining whether these systems are best represented by weak or strong coupling theories was pointed out [6] at an early stage, but at present the experimental situation is still unclear. Tunneling measurements were also urged and now it appears [7] that phonon structure in the tunneling spectra is being observed albeit only at low energies.

Two conventional experimental measures of weak versus strong coupling have been the gap to transition temperature ratio, $2\Delta/kT_c$, and the ratio of the heat capacity jump at T_c to the normal state heat capacity, $\Delta C/C$. The weak coupling BCS values are 3.52 and 1.43, respectively. Measurements of $2\Delta/kT_c$ and $\Delta C/C$ have tended to give values well above 3.52 and 1.43. However, recent laser studies [8] on a Bi–Sr–CA–Cu–O system yield values corresponding to weak coupling, $2\Delta/kT_c = 3.3$. In contrast $\Delta C/C$ is close to 6. One way to reconcile these results [9] is to consider the measured non-BCS-like temperature dependence of the superconducting energy gap, $\Delta(T)$. The optical measurements [8] of $\Delta(T)$ suggest that $\Delta(T)$ remains larger than what is expected from a BCS $\Delta(T)$ as T approaches T_c. One possible microscopic explanation for this effect is that the pairing interaction itself is temperature dependent, but others are possible. Irrespective of the microscopic origins, the observed $\Delta(T)$ can be used to explain the large $\Delta C/C$ and this may be a common feature of many of the high-T_c oxides.

It is unfortunate that the issue of strong versus weak coupling has not been settled. In particular, for the Boson-exchange models, a high T_c and weak coupling would favor high frequency Bosons such as those associated with electronic mechanisms. A high T_c and strong coupling would favor low frequency exchange Bosons such as phonons. Since phonon mediated pairing is likely to be present whether it is the dominant mechanism or not, even without a determination of strong versus weak coupling, it is possible to use experimental data and theoretical limits to examine the conditions under which a non-phonon Boson exchange in addition to phonon exchange can contribute in a significant way to electron pairing.

3. Pairing via Phonons Plus Other Bosons

Phonon mediated pairing can yield values of T_c in the range of and higher than those which have been observed. However, it is difficult to reconcile [10] a high T_c and a low isotope parameter α. Although this is possible for low T_c transition metals [11], the restriction of $T_c \sim 95$ K and $\alpha \sim 0.02$ yields unphysically large electron–phonon coupling parameters λ. Some additional assumptions which may be capable of giving the values are two dimensionality [12], material inhomogeneities [13], and anharmonic phonons [14]. In fact, in the latter case, it may be possible to explain [14] measurements [15] which give a $\alpha > 1/2$. This is outside of the range of the usual harmonic Eliashberg theory. Each of these explanations will require more experimental studies before they are generally accepted. The 2D models lead to Van Hove singularities in the density of states which may not be realistic for these systems because of their 3D nature [16], scattering by imperfections, and screening of the pairing interaction. The suggestions related to material properties will require more characterization of the systems studied and there is only marginal evidence [17] of anharmonicity with the needed characteristics.

A popular suggestion for obtaining a high T_c and a low α is to add a pairing mechanism which involves an exchange Boson with frequency ω_e which is independent of ionic mass. The electronic mechanisms mentioned earlier are of this type. The motivation of this model is that a mass independent mechanism could yield a high T_c with $\alpha = 0$. At the same time the phonons would contribute only slightly to T_c and this should lead to the observed small values of α. A model of this kind was examined [18] and the results were not consistent with these expectations. Some features of this calculations are described briefly below.

An electron–phonon spectral function $\alpha^2 F(\omega)$ is used as input. The peaks in $\alpha^2 F(\omega)$ are taken to be at the dominant structure in the phonon density of states. These peaks are assumed to move with changes in ionic mass in the conventional way. An additional peak is introduced at ω_e corresponding to a mass independent electronic contribution to $\alpha^2 F(\omega)$. The frequency ω_e is taken to be a characteristic frequency of the exchange Boson for the electronic pairing mechanism. Limits on the parameters representing the coupling to phonons λ_p and the electronic Boson λ_e can be evaluated for a chosen ω_e by the constraints on T_c and α. In practice, λ_p and λ_e are varied until the experimental T_c and α are found. The gap to T_c ratio is also evaluated as a function of ω_e.

The results of this calculation for particular case of Y–Ba–Cu–O are the most striking. With the above assumptions, it is found that constraints of $T_c = 93$ K, $\alpha = 0.01$, and $\omega > 100$ meV yield $0.05 < \lambda_p < 0.10$ which is an order of magnitude below most estimates based on experiment [19] and theory [20]. For $10 < \omega_e < 100$ meV, $\lambda_p < 0.3$. For $\omega_e < 10$ meV, the total coupling λ becomes unphysically large. The situation is less restrictive for the La–Sr–Cu–O ($T_c \sim 40$ K, $\alpha = 0.14$) and the Ba–K–Bi–O ($T_c \sim 27$ K, $\alpha = 0.40$) systems because of the larger values of α and lower values of T_c involved.

If the above conclusions related to the phonon only or phonon plus electronic pairing mechanisms are considered as rigid, then the restrictions on λ_{ph} would rule out these mechanisms for the Y–Ba–Cu–O systems. However, as discussed earlier, there are simplifying assumptions in these calculations which need to be explored before considering conclusions of this kind as fixed. Three examples were given before. Two of these, anharmonic phonons and reduced dimensionality, are discussed in more detail in the next section.

4. Anharmonic Phonons

The Eliashberg theory [21] and variations based on it for calculating T_c is the state-of-the-art theory in this area [22]. However, the standard formulation assumes that the phonons causing pairing are harmonic. Some formal modifications have been explored [23], but the usual approach is to assume that the theory is appropriate and that the major changes can be introduced into the kernel of the equation. Using the method introduced by Hui and Allen [24], Hardy and Flocken [25] suggested that large values of λ could be obtained when anharmonic effects are included. For strong coupling this does not yield an extremely large T_c because $T_c \sim \sqrt{\lambda_p \omega_{ph}^2}$ in this limit and the phonon frequency ω_{ph} decreases as λ_p increases. In addition, there is the question of the reduced α.

In the standard two-square well model the isotope parameter α is reduced from 1/2 because of the difference in the frequency dependence of the phonon and Coulomb interactions

$$\alpha = \tfrac{1}{2}\left[1 - \left(\frac{\mu^*}{\lambda^* - \mu^*}\right)^2\right] \tag{1}$$

where λ^* and μ^* are the renormalized phonon and Coulomb couplings. The above model is only appropriate for weak coupling theory, but it does illustrate some general features of α. In this model, a reduction in α and even $\alpha < 0$ are possible. However, $\alpha > 1/2$ is not possible within this theory. If the extreme strong coupling theory is investigated, then α approaches 1/2.

In contrast, values of $\alpha > 1/2$ can be obtained [14] when anharmonic effects are included. In the weak coupling limit,

$$\alpha = \tfrac{1}{2}\left[1 - \left(\frac{\mu^*}{\lambda^* - \mu^*}\right)^2\right](1 + \lambda_M) - \frac{\lambda_M}{\lambda}\left(\frac{\lambda^*}{\lambda^* - \mu^*}\right)^2 \tag{2}$$

where $\lambda_M = \mathrm{d}\ \log \lambda / \mathrm{d}\ \log M$. For strong coupling and $\mu^* = 0$

$$\alpha = \tfrac{1}{2} - \frac{\lambda_M}{2\lambda}\,. \tag{3}$$

Hence a $\lambda_M < 0$ can give rise to $\alpha > 1/2$. By considering various forms of the potentials it is possible to generate curves which reproduce the measured [15] dependence of T_c and α on doping for the La–Sr–Cu–O system, including the values of $\alpha > 1/2$.

5. Reduced Dimensionality

As discussed earlier, the latest energy band calculations [16] for the oxides do indicate that they are not 2D systems. Overlap of wavefunctions along the 'low conductivity' axis leads to dispersion and itinerant behavior for electron motion. However, the Cu–O planes are prevalent and 2D theories are therefore interesting candidates. The work by Labbe and Bok [12] gives some of the main arguments which have been repeated in a number of subsequent papers by other authors. Several of the ideas have their roots in earlier studies of A15 superconductors and in model studies of the influence of electronic density of states structure on T_c and α.

In particular, there are a finite number of topological Van Hove singularities for 3D and 2D systems. Criticisms of the appropriateness of tight-binding band calculations which were used in some of the studies of the role of 2D behavior are not relevant. The singularities in the density of states must be present. Their positions and strengths will depend on the methods and potentials used to compute the band structures. Given that Van Hove singularities will exist in periodic clean systems, then it is relevant to explore their role.

For 2D systems, one expects a density of states function near a critical point of the form

$$D(\varepsilon) = \frac{A}{b} \ln \left(\frac{b}{\varepsilon}\right) \tag{4}$$

where A and b are constants and ε is the energy variable. Some of the physical consequences of the logarithmic peak can be seen by using a BCS model for a constant interaction V. The gap equation simplifies to the condition

$$1 = \tfrac{1}{2}V \int_{-E_D}^{E_D} \frac{D(\varepsilon)}{\varepsilon} \tanh\left(\frac{\varepsilon}{2k_B T_c}\right) d\varepsilon \tag{5}$$

where E_D is a characteristic phonon frequency and k_B is Boltzmann's constant. If a non-phonon interaction is considered then E_D would be the characteristic frequency of the exchange Boson. Solution of (5) for T_c gives the standard exponential form but a $\sqrt{\lambda}$ replaces λ and if $b < E_D$ then it determines the energy scale,

$$T_c = 1.13 b e^{-1/\sqrt{\lambda}} \tag{6}$$

where $\lambda = AV/2b$.

The square root dependence enhances the effect of the pairing interaction for $\lambda < 1$ and since b is independent of mass, the isotope effect vanishes. The case for $b > E_D$ can be solved and similar results are obtained since the log peak is so dominant. There is great sensitivity to A and b and the position of the peak. Also, it is critical that the log behaviour persist, which may be not the case for non-ideal systems.

Some of the possible problems with this approach were mentioned earlier. Overlap along the third dimension, broadening due to scattering and defects, the theory need normalizing, and so on. However, there are still attractive features to this proposal and further exploration is needed. One effect which should be explored is the possibility that the beneficial peaks in $D(\varepsilon)$ will be cancelled out by the screening function. Since the electron–electron pairing interactions are screened by an electronic dielectric function, the frequency and wavevector dependence of the interaction should be accounted for averaged in a appropriate way. The constant V model which has been used is particularly inappropriate in 2D with a rapidly varying $D(\varepsilon)$. When screening is included, the dielectric function will be approximately linearly dependent on the density of states. Hence the peaks in $D(\omega)$ arising from Van Hove singularities will appear in the screening function. Since this function appears in the denominator its peaks will tend to cancel the peaks in $D(\varepsilon)$ in the numerator. The degree of cancellation can be determined for specific models.

If electron–phonon or other coupling mechanisms becomes strong, then real processes can accompany the virtual exchange of the Bosons resulting in a lifetime or energy width ΔE. For phonons in 3D at high temperatures,

$$\Delta E = 2\pi\lambda_p k_B T \ . \tag{7}$$

In the A15 materials this was suggested [26] as a broadening mechanism for peaks in the electronic density of states. Similar considerations should be addressed for the 2D Van Hove singularities.

Because of the layered structural features of the high-T_c oxides, theories have been introduced which consider intraplane and interplane couplings. When groups of planes or blocks are considered, then inter- and intra-block couplings are used. The detailed computations of the coupling strength have not been done so they are treated as parameters. Two examples of these theories are those proposed by Wheatley, Hsu and Anderson (WHA) [27] and by Ihm and Yu (IY) [28]. In the WHA proposal, pairing is dominated by coherent hopping between planes within a block (intra-block coupling) with an enhancement from hopping between planes in adjacent blocks (inter-block coupling). For IY, which is based on a BCS-like formalism, intra-plane coupling is dominant, but intra-block coupling is also included.

Again, experiment may help determine the detailed role of the couplings. Recently, it has been shown [29] that iodine can be intercalated into the Bi–Sr–Ca–Cu–O systems. The iodine can act as spacers between Bi–O bilayers and hence reduce various couplings. At this point the iodine intercalation has been used to change the distance between CuO_2 layers and therefore the intercalation primarily affects interblock coupling. Changes in T_c up to 10 K for Bi–2212 and Bi–2223 intercalation compounds suggest that next block coupling contributes $\sim$ 10 K to T_c. More detailed studies are in progress, however based on studies up to now, it appears that elements of both the WHA and IY models are needed to explain the observations.

6. Conclusions

Although the most exotic theories were not considered here, it is apparent that even with BCS-like theories there are rich variations and possible theoretical concepts capable of explaining the superconducting properties of the copper oxide systems. Problems still exist for most of the proposed models, but some of them may be surmounted when the detailed materials properties related to crystal structure and electronic structure are included. This brings up the central problem in evaluating theoretical proposals for high-T_c oxides. Because of the complexity of most of the materials, variable parameters are inserted into the theoretical formulations. Hence most of the proposed theories can be forced to agree with most of the existing experiments. As material characterization improves, and measurements of normal state and superconducting properties become more definitive, the appropriateness of the many theoretical proposals will be judged more accurately.

Acknowledgements

This work was supported by NSF Grant No. DMR88-18404 and by the Director, Office of Energy Research, Office of Basic Energy Sciences, Materials Sciences Division of the Department of Energy under Contract No. DE-AC03-76SF00098.

References

1. J.G. Bednorz and K.A. Müller: 1986, *Z. Phys. B* **64**, 189.
2. B. Batlogg, R.J. Cava, A. Jayaraman, R.B. van Dover, G.A. Kourouklis, S. Sunshine, D.W. Murphy, L.W. Rupp, H.S. Chen, A. White, K.T. Short, A.f M. Mujsce and E.A. Rietmand: 1987, *Phys. Rev. Lett.* **58**, 2333.
3. L.C. Bourne, M.F. Crommie, A. Zettl, H-C. zur Loye, S.W. Keller, K.L. Leary, A.M. Stacy, K.J. Chang, M.L. Cohen and D.E. Morris: 1987, *Phys. Rev. Lett.* **58**, 2337.
4. B. Batlogg, G. Kourouklis, W. Weber, R.J. Cava, A. Jayaraman, A.E. White, K.T. Short, L.W. Rupp and E.A. Rietman: 1987, *Phys. Rev. Lett.* **59**, 912.
5. T.A. Faltens, W.K. Ham, S.W. Keller, K.J. Leary, J.N. Michaels, A.M. Stacy, H-C. zur Loye, D.E. Morris, T.W. Barbee III, L.C. Bourne, M.L. Cohen, S.Hoen and A. Zettl: 1987, *Phys. Rev. Lett.* **59**, 915.
6. M.L. Cohen: 1987, *Novel Superconductivity*, S.A. Wolf and V.Z. Kresin (eds.), Plenum, New York, p. 1095.
7. Q. Huang, J.F. Zasadzinski, N. Tralshawala, K.E. Gray, D.G. Hinks, J.L. Peng and R.L. Greene: 1990, *Nature* **347**, 369.
8. L.C. Brunel, S.G. Louie, G. Martinez, S. Labdi and H. Raffy: 1991, *Phys. Rev. Lett.* **66**, 1346.
9. M.L. Cohen and D.R. Penn: 1990, *Phys. Rev. B* **42**, 8702.
10. T.W. Barbee III, M.L. Cohen, L.C. Bourne and A. Zettl: 1988, *J. Phys. C* **21**, 5977.
11. J.W. Garland: 1963, *Phys. Rev. Lett.* **111**, 111 and 114.
12. J. Labbe and J. Bok: 1987, *Europhys. Lett.* **3**, 1225.
13. J.C. Phillips: 1990, *Phys. Rev. Lett.* **64**, 1605.
14. V.H. Crespi, M.L. Cohen and D.R. Penn: 1991, *Phys. Rev. B* **43**, 12921.
15. M.W. Crawford, M.N. Kunchur, W.E. Farneth, E.M. McCarron III and S.J. Poon: 1990, *Phys. Rev. B* **41**, 282.
16. W.E. Pickett, R.E. Cohen and H. Krakauer: 1990, *Phys. Rev. B* **42**, 8764.
17. R.E. Cohen, W.E. Pickett and H. Krakauer: 1989, *Phys. Rev. Lett.* **64**, 2575.
18. M.L. Cohen, D.R. Penn and T.W. Barbee III: 1990, *Solid State Commun.* **75**, 971.
19. S.D. Brorson, A. Kazeroonian, D.W. Face, T.K. Cheng, G.L. Doll, M.S. Dresselhaus, G. Dresselhaus, E.P. Ippen, T. Venkatesan, X.D. Wu and A. Inam: 1990, *Solid State Commun.* **74**, 1305.
20. W.E. Pickett: 1989, *Rev. Mod. Phys.* **61**, 433.
21. G.M. Eliashberg: 1960, *Soviet Phys. JETP* **11**, 696.
22. P.B. Allen and B. Mitrovic: 1982, *Solid State Phys.* **37**, 1.
23. A.E. Karakozov and E.G. Maksimov: 1978, *Soviet Phys. JETP* **47**, 358.
24. J.C.K. Hui and P.B. Allen: 1974, *J. Phys. F: Metal Phys.* **4**, L42.
25. J.R. Hardy and J.W. Flocken: 1988, *Phys. Rev. Lett.* **60**, 2191.

26. K.M. Ho, M.L. Cohen and W.E. Pickett: 1978, *Phys. Rev. Lett.* **41**, 815.
27. J.M. Wheatley, T.C. Hsu and P.W. Anderson: 1988, *Nature* **333**, 121.
28. J. Ihm and B.D. Yu: 1989, *Phys. Rev. B* **39**, 4760.
29. X.-D. Xiang, S. McKernan, W.A. Vareka, A. Zettl, J.L. Corkill, T.W. Barbee III and M.L. Cohen: 1990, *Nature* **348**, 145.

OBSERVATION OF MAGNETOOPTICAL EFFECTS IN SEVERAL HIGH T_c SUPERCONDUCTORS

J.F. DILLON, JR. and K.B. LYONS
AT & T Bell Laboratories, Murray Hill, NJ 07974, U.S.A.

Abstract. Recent so called 'anyon' theories of high temperature superconductivity in layer structure materials have suggested that at some temperature $T_{TP} \geq T_c$ there is a symmetry breaking transition below which these materials may be in either of two distinct states related to each other by time reversal. The study of magnetooptical effects in superconductors reviewed here was undertaken to explore the time reversal symmetry of these materials. Using a novel technique with a rotating $\lambda/2$ plate at 515 nm, we have observed 'circular dichroism' on reflection from epitaxial films and single crystals of cuprate superconductors with layer structures. The onset of the dichroism was at temperatures of $\approx$ 180 K to $\approx$ 300 K. These results appear to support the 'anyon' theories. However, circular dichroism was also seen in films and single crystals of bismuthate superconductors with cubic structure, to which the theories seem inapplicable. In sharp contrast, Spielman *et al.*, at Stanford in a very sensitive experiment at 1060 nm have seen no evidence of non-reciprocal circular birefringence in epitaxial cuprate superconducting films. Weber *et al.* at Dortmund have recently reported the observation at 633 nm of non-reciprocal magnetooptical effects on single crystals of cuprate superconductors, but none on films.

1. Introduction

The recent discussion of theoretical models [1] of high temperature superconductivity which imply spontaneous breaking of time reversal symmetry has led us to perform sensitive measurements of magnetooptical effects in several such superconductors and related materials. In this article we attempt a tutorial summary of our own experiments [2–4] and of the situation concerning related optical measurements. In this introduction we have drawn on Halperin's [5] recent review of theoretical proposals and the experimental search for anyon superconductivity. In these models, the charge carriers 'anyons' are confined to two dimensions and obey fractional statistics. Such systems cannot be invariant under time reversal (T) or two dimensional reflection (P). The theoretical idea is that below some spontaneous symmetry breaking temperature T_{TP} there appears a correlated state with broken symmetry among the copper spins in a layer, and that T_{TP} could be equal to or greater than the superconducting critical temperature T_c. The symmetry breaking can be of either sign, and there are magnetic fields associated with it. There might be a coupling between the layers which would tend to make the sign of the ordering the same in adjacent layers, 'ferromagnetic'. Alternatively antiferromagnetic coupling is also possible. In either case, on cooling a macroscopic sample through T_{TP} the ordering could be distributed in domains of indeterminate size. Since nucleation is likely to be an important consideration, differences in microstructure from sample to sample might

H. Aoki et al. (eds.), New Horizons in Low-Dimensional Electron Systems, 201–215.

play a large role. On the other hand, it is conceivable that the interlayer coupling could be so small as produce no long range order between layers. In any event, the presence of magnetic fields on cooling through T_{TP} might produce radically different distributions of the ordering.

March-Russell and Wilczek [6] pointed out that independent of any model, the symmetry properties under time reversal of the high temperature superconducting materials could be studied experimentally. A leading possibility is to look for the magnetooptical effects listed above. These non-reciprocal optical effects are of great use in the study of magnetic materials. In that case the primary property they probe is, in fact, time reversal asymmetry. Usually a detailed understanding of the magnitudes is lacking.

Thus we have provocative theoretical suggestions concerning the mechanism of high temperature superconductivity. Inherent in these is the lack of time reversal symmetry. The experiments described here were undertaken to explore whether there is a T violation.

2. Jones Vectors and Matrices

When a monochromatic linearly polarized wave traveling along the z-axis strikes the surface of a uniform material parallel to the $x-y$ plane, the material responds according to its dielectric tensor, and reradiates reflected and transmitted waves. For the simple case in which there are no linear birefringences or dichroisms, the appropriate Cartesian form of the tensor [7] is

$$\overleftrightarrow{\varepsilon}=\begin{bmatrix} \varepsilon_{11} & i\varepsilon_{12} & 0 \\ -i\varepsilon_{12} & \varepsilon_{11} & 0 \\ 0 & 0 & \varepsilon_{33} \end{bmatrix} .$$

Magnetooptical circular birefringence (CB) and circular dichroism (CD) arise from antisymmetric terms in the off-diagonal elements. For a magnetic material these are proportional to the z-component of the order parameter m_z. If the incident light is linearly polarized along the x-axis, the polarization states of both reflected and transmitted light experience a complex rotation. In the reflection case this is complex polar Kerr rotation, and is written

$$\hat{\Phi}_K = \frac{E_y^r}{E_x^r} = \varphi_K' + i\varphi_K''$$

where the first term is a rotation of the major axis, the second an ellipticity. Neglecting small interface contributions, the rotation of the transmitted beam is proportional to the thickness traversed, and has a parallel form

$$\hat{\Phi}_F = \frac{E_y^t}{t\,E_x^t} = \frac{1}{t}\,(\varphi_F' + i\varphi_K'') .$$

The two terms are simply magnetooptical (or Faraday) rotation and ellipticity. Solution of the boundary value problem shows that the normal modes of propagation in the material are the two circular polarizations (CP). The φ'_K and φ'_F represent a phase difference between the two CP after reflection or transmission. Similarly φ''_K and φ''_F correspond to a difference in the amplitudes after reflection or transmission. Following the analysis of magnetooptical effects given by Scott and Lacklison [8] we can write expressions for φ'_K and φ''_K in terms of ε'_{12} and ε''_{12}. These are frequency dependent in that they contain both components of the complex refractive index $N = n - ik$. Though they are Kramers–Kronig related, at a given photon energy one can be much larger than the other. These expressions and calculations of the relative contribution of the two components of $\hat{\varepsilon}_{12}$ to the various magnetooptical effects constitute too much detail to give here. However, starting with plots of ε'_{11} and ε''_{11} given by Bozovic [9], we find that at 500 nm ε''_{12} is weighted about 8× as heavily as ε'_{12} in determining φ''_K, the quantity measured in our experiments.

In their Cartesian form the 2×2 Jones matrices used for representing the effect of optical elements on polarized light are closely related to the 2×2 minor of the dielectric tensor given above. Here we are interested in CB and CD in an experiment in which light is reflected normally from a surface, so it seems natural to use *circular* Jones matrices and vectors [10] to represent the effect of our sample as well as that of the optical components on the light beam. In this formalism, the vector $\hat{e}_+ = \begin{bmatrix} 1 \\ 0 \end{bmatrix}$ is plus-polarized light, and $\begin{bmatrix} 0 \\ 1 \end{bmatrix}$ is minus-polarized light. The vector $\begin{bmatrix} e^{i\theta} \\ 1 \end{bmatrix}$ is linear polarization with a phase difference of θ between the two circular components. Note that we adopt a convention in which polarization states are referred to the same set of axes before and after reflection. Thus the sign of circular polarization does not change on reflection from a perfect mirror. The most general sample can be represented by the matrix

$$\mathbf{M}_s = \begin{bmatrix} A_+ & B_- \\ B_+ & A_- \end{bmatrix} ,$$

where the elements may be complex. Thus, plus polarized light becomes $A_+\hat{e}_+ + B_+\hat{e}_-$ on reflection. When the input is x-polarized light $\begin{bmatrix} 1 \\ 1 \end{bmatrix}$, after reflection the amplitude is $\begin{bmatrix} A_+ + B_- \\ A_- + B_+ \end{bmatrix}$. Insofar as $\mathbf{M}_s$ differs from a scalar multiple of the unit matrix, the polarization state will be changed. We note that by a suitable decomposition, it is possible to attribute these changes to various birefringences and dichroisms.

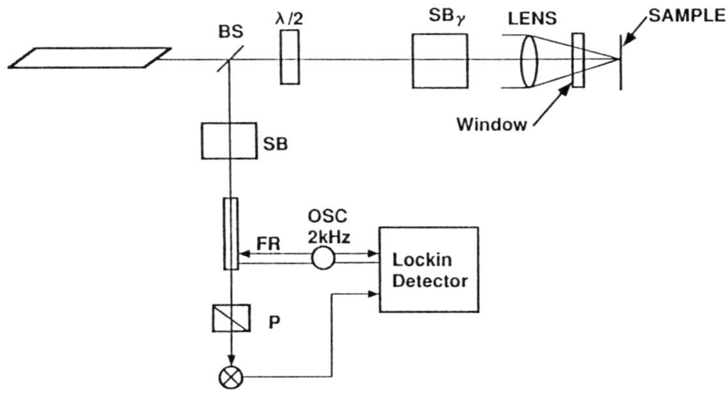

Fig. 1. Sketch of apparatus used to measure circular dichroism on reflection, the Kerr ellipticity. SB_γ and SB are Soleil–Babinet compensators; $\lambda/2$ is a rotating half wave plate; FR is a Faraday rotator. See text for further discussion.

The matrix $\mathbf{M}_s$ plays a central role in our view of these experiments. It is merely a statement of the most general changes in polarization state. Ordinary circular birefringence and circular dichroism are contained in the real and imaginary differences between A_+ and A_-. For a material which is invariant to time reversal, $A_+ = A_-$. On the other hand, as we shall see below, if $|B_+| \neq |B_-|$ the resultant change in polarization will be indistinguishable from a circular dichroism unless further polarization selection is employed. A striking example of this inequality in the cross channel terms lies in the reflection of one sense of circular polarization from a cholesteric liquid crystal [11] for the wavelength at which the helicity of the wave matches the helicity of the structure. Both incoming and reflected waves have the same helicity, and therefore opposite senses of rotation. Such inequalities in $|B_\pm|$ do not represent a violation of time reversal symmetry, and must arise from variations in the index ellipsoid along the propagation direction.

3. Experiment

Our search for evidence of time reversal asymmetry was carried out with the experimental arrangement sketched in Figure 1 [2,4]. This should be viewed in conjunction with Figure 2 in which we have unfolded the optical path by regarding the sample surface as a plane in which polarization changes take place. To the left of the schematic in the second figure are given the circular Jones matrices for each of the elements the beam encounters. Linearly polarized (515 nm) laser light which is accurately in the plane of incidence passes

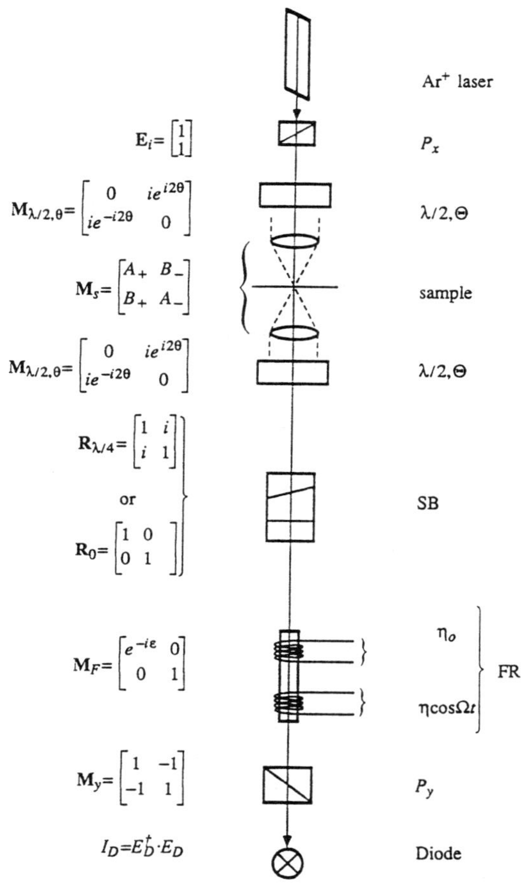

Fig. 2. Schematic of the optical components of Figure 1 unfolded at the sample reflection. The circular Jones matrices for the optical components are shown at left. P_x and P_y are x- and y-polarizers; SB is a Soleil–Babinet compensator; FR is a Faraday rotator with two windings. See text for further discussion.

through a beam splitter, then encounters a $\lambda/2$ retarder rotating at θ. Thus the light leaving the $\lambda/2$ plate is linearly polarized with its axis rotating at 2θ. The beam encounters a compensator, a lens and several Dewar windows both before and after reflection by the sample. The lens and windows may have linear stress birefringences with arbitrary azimuths, and those of the inner Dewar are likely to be temperature dependent. The azimuth and retardation of the Soleil–Babinet compensator SB_γ are adjusted to compensate the spurious birefringences of the elements as well as possible. Detection at 2θ makes it possible to do this with great sensitivity. Note particularly that in Figure 2 the birefringence of lens, windows and the compensating SB_γ are all subsumed into the matrix $\mathbf{M}_s$. The rotating half wave plate enables us to reject net linear birefringences and dichroisms. Since these effects will have opposite sign when the incoming linear polarization is rotated by 90°, a half wave retarder rotating at θ will produce a signal at 4θ, and our final detection at some other frequency will average this to zero. The time average of the reflected beam is restored to very nearly its original linear polarization on the second passage through the rotating $\lambda/2$ plate. However, it may have acquired a small ellipticity and rotation from the net circular effects. If the succeeding element, Soleil–Babinet compensator SB, is set at zero retardation, it does not change the polarization state. The Faraday rotator is driven by two coils which together rotate the polarization by $\varepsilon = \eta_0 + \eta \cos \Omega t$. This is followed by a y-polarizer. The $\cos \Omega t$ signal at the detector is proportional to the rotation minus η_0. It is readily seen that η_0 can be used to set a zero point and, quite separately, to calibrate the signal.

Analytically we see that the amplitude at the detector E_D is given by the product of the matrices listed in Figure 2 times the input vector E_i,

$$E_D = \mathbf{M}_y \, \mathbf{M}_F \, \mathbf{R} \, \mathbf{M}_{\lambda/2,\theta} \, \mathbf{M}_s \, \mathbf{M}_{\lambda/2,\theta} \, E_i \ .$$

The detector responds to the intensity $I_D = E_D^\dagger \cdot E_D$, and we use a lock-in detector to select the $\cos \Omega t$ component. There are many terms containing the rotational frequency of the half wave plate, and we must average over these. If SB is set at $\lambda/4$ with the fast axis along the x-axis, it converts the ellipticity of the beam into a rotation, and the $\cos \Omega t$ signal at the detector thus measures ellipticity. The expansion and subsequent manipulation of the expression for I_D by hand is straightforward but exceedingly tedious. We have found it convenient to handle these symbolic matrix problems on a small computer using the software package 'Derive – A Mathematical Assistant' [12]. If SB is set to zero, we use $\mathbf{R} = \mathbf{R}_0$, and measure circular birefringence or rotation. The full expression may be written

$$I_{\mathrm{CB}} \propto \cos \Omega t \left[\eta_0 + \frac{\mathrm{Im}\,(A_+^* \, A_-)}{\mathrm{Re}\,(A_+^* \, A_-)} \right] .$$

Note that the cross channel terms $B_\pm$ do not appear in this. A corresponding expression for circular dichroism is obtained by setting BS at $\lambda/4$ with its fast axis along the x-axis, and using the matrix $\mathbf{R} = \mathbf{R}_{\lambda/4}$. This rotates the ellipticity into a rotation to which our detection system with its rotation modulation is then sensitive. We find

$$I_{\mathrm{CD}} \propto \cos \Omega t \left[\eta_0 + \frac{|A_+|^2 - |A_-|^2 - |B_+|^2 + |B_-|^2}{\mathrm{Re}\,(A_+^* \, A_-)} \right] .$$

In surprising contrast to the preceding case, the $B_\pm$ do appear here. The A's are associated with time reversal symmetry; the B's might arise from inhomogeneity in the dielectric tensor along the propagation direction. In the CB measurement, the rotating half wave plate eliminated the effects of $B_\pm$, but they can appear in the CD measurements.

Although the expression for I_{CB} is unencumbered by the $B_\pm$ terms which appear in I_{CD} we find that the statistical variance in making CB measurements with this rotating plate technique is $\approx$ 200 μrad as compared with perhaps 20 μrad for the corresponding CD determinations. The possible sources of these errors is to be discussed in a forthcoming paper [13] on the technique of using the rotating $\lambda/2$ plate. This difference in residual error has led us to measure CD on reflection in spite of the appearance of cross channel terms.

The wavelength of the light was 515 nm. The half wave plate was rotated at 16 Hz, and the Faraday modulator frequency was $\approx$ 2000 Hz. The size of the focused spot on the sample was about 20 μm.

To measure the CD at a spot on the sample, the intensity was maximized by adjustment of the lens, and the azimuth and retardation of SB_γ were adjusted to cancel the birefringence of lens and windows. Then the steady current through the Faraday rotator was adjusted to null the cos (Ωt) signal at the detector. This constituted a measure of the polar Kerr ellipticity φ_{CD}, i.e. φ''_K in the definition of complex $\hat{\Phi}_K$ above. Finally the sample was moved about 100 μm. This process was repeated as many as 25 times at each temperature. From these data an average $\bar{\varphi}$ and a standard deviation σ_φ were extracted.

Consider briefly three possible experimental subjects. For a sample which showed no dichroism, we would expect $\bar{\varphi} = 0$ and σ_φ to be a measure instrumental noise, and this would not vary with temperature. Indeed measurements on a polished silicon sample and on a gold film (180 Å) deposited on an MgO substrate showed a temperature independent value of σ_φ of about 20 μrad. A one domain sample with CD would have a temperature dependent $\bar{\varphi} = \pm\varphi_0$, and σ_φ would be the instrumental value. Finally, for a sample with a dichroism of φ_0 which is broken up into an equal number of positive and negative domains smaller than the spot size, we would expect $\bar{\varphi} = 0$ and an added temperature dependent contribution to σ_φ of $\varphi_0/\sqrt{N}$. If r is the radius of the domains, and d the diameter of the laser spot, σ_φ would be $\varphi_0 r/d$. As will appear below, this last possibility corresponds to the experimental results.

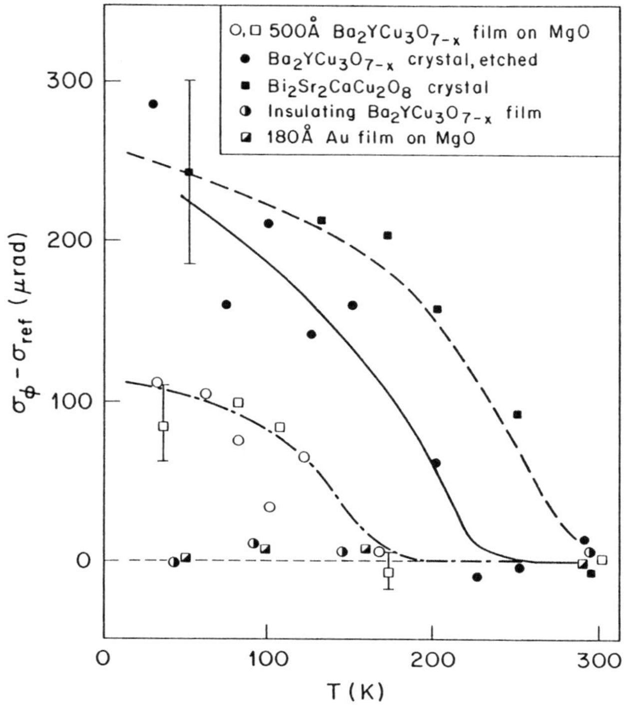

Fig. 3. Standard deviation (less a background value) of the measured CD on reflection for cuprate samples indicated. Note the difference in the onset temperatures for the two different single crystal compositions, and that both are higher than the films. The lines are merely guides to the eye.

4. Results

4.1. Cuprate superconductors

The CD measurements just described were first performed on a number of films and crystals of cuprate superconductors [2] as well as on control samples. These are layer structure compounds in which the conductivity is confined to sheets of copper ions, and to which the anyon theories are aimed. There has been a tremendous worldwide effect on these materials, and it is now possible to grow high quality epitaxial films as well as bulk single crystals. Figure 3, taken from Reference 2 shows some of these results. In it are plotted standard deviation of the $\bar{\varphi}_{CD}$ measured at a number of points against temperature. Error bars are shown for a few representative points. Though more complete details are given in the paper cited, the samples of the figure were:

a. Films of $YBa_2Cu_3O_{7-x}$. These were about 500 Å thick, and were epitaxially grown on MgO (100) substrates by molecular-beam epitaxy. X-ray diffraction showed them to be (001) oriented with a mosaic spread $\leq 0.3°$. The substrates had cleavage steps 50 – 100 μm wide. The transition temperatures were 86.5 to 88 K, with the width (10% to 90%) about 0.7 K.

For these films, $\sigma_\varphi - \sigma_{\text{ref}}$ is zero at high temperature. Just below 200 K it begins to increase, but there is no feature associated with T_c.

b. One of the $YBa_2Cu_3O_{7-x}$ films was reduced to insulating $YBa_2Cu_3O_6$ by annealing in N_2 at 450 C.

Though before the anneal it behaved like the films described above, afterwards there was no sign of CD.

c. Single crystals of $YBa_2Cu_3O_{7-x}$ and of $Bi_2Sr_2CaCu_8$. The first shows a sharp superconducting transition at 92 K; the second shows steps in the susceptibility vs temperature curve at 87 and 65 K.

For both crystals the behavior is qualitatively the same, except that the onset is at higher temperatures, and the extreme values are substantially higher. Again, nothing happens at T_c.

d. A 180 Å film of Au on MgO which has about the same reflectivity as the Y–Ba–Cu–O films, and thus a surface topography like that of the superconducting films. We also made measurements on a polished silicon surface.

The data for the gold film are shown on Figure 3. Those for the silicon surface are similar. No CD signal was seen for either.

An increasing σ_φ signal was seen below a clear onset temperature in every superconducting specimen studied with the exception of a single film. That film was less than an ideal superconductor in that it showed zero Meissner fraction. This contrasts with a film showing a finite σ_φ that exhibited a Meissner fraction of 0.8. Thus a variety of well behaved superconducting cuprate sampels with layer structures show a finite variance in the apparent circular dichroism. The average $\bar{\varphi}$ over a number of points does not differ significantly from zero, even when the sample is cooled in small magnetic fields (approx 20 Oe). No structure in the $\sigma_\varphi(T)$ plot is associated with T_c. These results seem consonant with 'anyon' theories, and suggest that the temperature T_{TP} corresponds to the 'onset temperature' seen here, and further that the time reversed domains are smaller than our light spot.

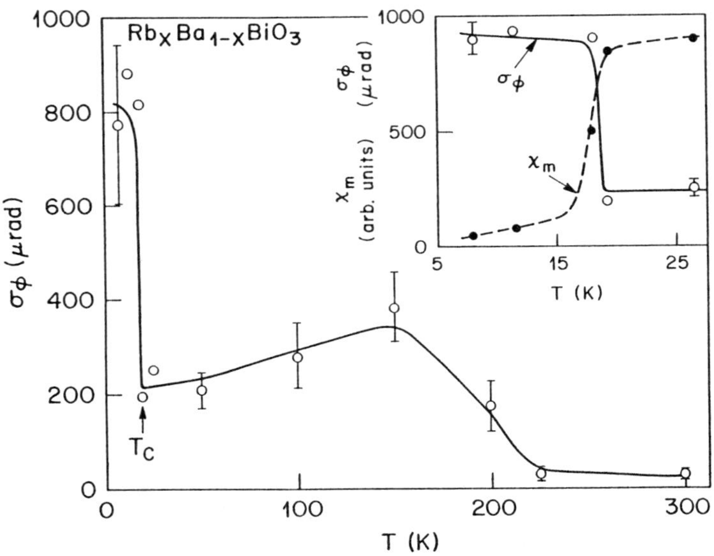

Fig. 4. Standard deviation of the circular dichroism on reflection σ_φ for a 2000 Å thick bismuthate superconducting film with cation ratio $Rb_{0.25}Ba_{0.23}Bi_{0.52}$. The inset shows data near T_c on an expanded temperature scale. The susceptibility data, solid points in the inset, were taken on a separate run. Note the abrupt change in σ_φ at T_c. The lines are merely guides to the eye.

4.2. Bismuthate superconductors

Circular dichroism has also been seen in films of $Ba_{1-x}Rb_xBiO_3$. These are cubic compounds for which the anyon models do not seem to pertain since the theories are built on the assumption of two dimensionality. More complete characterizations of the films are given in the original reference [3]. We note that the overall quality of these films is much lower than that of the cuprate compounds, and the superconducting properties are less ideal. Furthermore, there are significant differences between the results for two samples. Thus these were presented as preliminary and incomplete observations. Two superconducting samples and a control were studied. Again these were grown epitaxially on MgO substrates by molecular beam epitaxy.

a. The cation ratio of the sample of Figure 4 was measured by Rutherford backscattering to be $Rb_{0.25}Ba_{0.23}Bi_{0.52}$, and x-ray diffraction showed some RbO_2 phase. This impurity compound is an antiferromagnet with $T_N = 15$ K. The film was 2000 Å thick, and it had a resistive transition 3 K wide reaching zero at 18 K. In susceptibility, a smeared transition

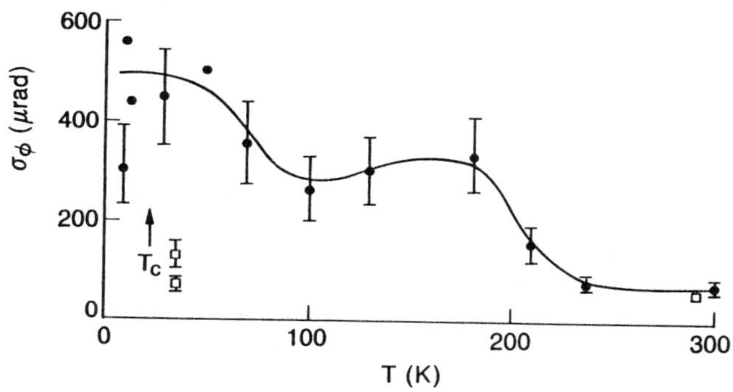

Fig. 5. Standard deviation of the circular dichroism on reflection σ_φ for a 1500 Å film with cation ratio $Rb_{0.21}Ba_{0.27}Bi_{0.52}$ (solid points). The solid line is merely a guide to the eye. Note that there is no distinctive structure at T_c. The two open points at 36 K are for an insulating film of $Rb_{0.3}Ba_{0.7}BiO_3$. It shows no variation with temperature.

started at 18 K. The normal state conductivity was metallic, and the sample showed no dc shielding or Meissner effect.

In Figure 4 we plot the standard deviation of the CD as measured on a set of spots. On cooling this deviates from the instrumental noise value at about 220 K, rises to a peak at about 150 K, then falls gradually. But then at T_c CD increases abruptly by a factor of four. The inset shows the sharpness of this transition on an expanded temperature scale. The dashed curve marked χ_m is from an ac susceptibility probe placed just behind the sample. The χ_m points were obtained in a separate run. The closest pair of points to the transition are only 1.5 K apart. This sharp structure at T_c is unlike anything seen in the cuprate samples, and the magnitude of σ_φ below T_c is substantially higher.

b. The sample of Figure 5 showed a cation ratio of $Rb_{0.21}Ba_{0.27}Bi_{0.52}$, and RbO_2 less than a tenth that of the previous specimen. It was 1500 Å thick, and had a semiconducting normal resistivity. On cooling, the resistivity started to drop near 30 K, and reached zero at 22 K. Susceptibility measurements showed a sharp transition at ≈ 20 K. Squid measurements showed dc shielding near 20 K, but no observable Meissner effect. On cooling this sample there is an increase in σ_φ beginning at about 220 K, and perhaps two broad peaks below this. In sharp contrast to the preceding sample, there is no distinctive feature at the superconducting transition.

c. To provide a non-superconducting control, σ_φ was measured on an insulating sample of $Rb_{0.7}Ba_{0.3}BiO_3$.

   After it was determined that there was no significant variation of the CD with temperature, only two points were measured. These are plotted as open squares on Figure 5.

As in the cuprate case there is a clear correlation between superconductivity and the appearance of CD on reflection below some clearly defined temperature. The variation between samples coupled with their poorer crystallographic and superconducting properties make it clear that these results must be viewed somewhat tentatively. However, the fact that CD on reflection is seen in materials with three dimensional structures clouds the connection between 'anyon' theory and magnetooptical effects.

5. Related Reports

Since the initial report on the observation of CD in high temperature superconductors, several other groups have begun to publish papers about closely related experiments. Given the theoretical background, the complexity of the material situation, and the difficulty of the experiments along with their interpretation, it is certain that there will be more work reported in the near future.

5.1. Sagnac interferometer – Spielman et al.

Very shortly after they had preprints of Reference [2], S. Spielman *et al.* [14] at Stanford University used a fiber optic Sagnac interferometer to seek non-reciprocal circular birefringence in high quality films of $YBa_2Cu_3O_7$. The technique is of great sensitivity, but they found no measurable CB. In their interferometer counterpropagating beams travel around a (1 km) loop of polarization maintaining fiber, and interfere at the output. If the system rotates, the path length is different for the two beams, and there is a phase shift when the beams are recombined. Thus it is a 'fiber optic gyroscope' delicately sensitive to rate of rotation. The earth's rotation is in fact used for calibration. The instrument is useful here because non-reciprocal CB in the optical path also caused a phase shift, whereas all reciprocal effects cancel out. Spielman *et al.* introduced somewhat transparent films of $YBa_2Cu_3O_7$ film into the interferometer loop. The film was placed between a pair of $\lambda/4$ plates arranged so that the transmitted light was circularly polarized, and the two beams had opposite senses of rotation at the sample. Detailed characterization of the films is given in the paper. On the face of it these films should be comparable to those of Lyons *et al.* The wavelength used was 1.06 μm, and the spot size at the sample was 15 μm. Films with thicknesses of 50 Å, 200 Å, 500 Å, and 800 Å were used. To an accuracy of 2 μrad the authors were unable to measure a non-reciprocal circular birefringence, a result apparently contravening that of Reference [2].

5.2. Reflection and transmission at Dortmund – Weber et al.

H.J. Weber *et al.* [15] at Dortmund used a complex technique to measure quantities proportional to CD on reflection from one orthorhombic domain crystal of $YBa_2Cu_3O_7$. The crystal had a $T_c = 92.5$ K. Light of $\lambda = 633$ nm was used. They found a CD which deviated from zero at ≈ 120 K, and there was no anomaly at T_c.

In a transmission experiment (at $\lambda = 633$ nm) they measured CB on a thin section ($t \approx 1000$ Å) of $Bi_2Sr_2CaCu_2O_8$ single crystal. The results began to deviate from zero at 140 K, and then increased to about 100 K where they leveled off. In both this and the reflection measurements the deviation versus temperature was qualitatively similar to that seen by Lyons *et al.* [2] except that the onset here was at substantially lower temperatures.

These authors also sought non-reciprocal effects in c-axis oriented films of $YBa_2Cu_3O_{7-\delta}$. They found no evidence of symmetry breaking.

Particularly interesting were the effects of cooling in a magnetic field. In both reflection and transmission experiments it was found that the CD or CB after cooling in the absence of a magnetic field varied statistically in sign and magnitude. In field cooling experiments, the samples were cooled from 180 K to a temperature below the onset temperature but above T_c in a field of 200–500 Oe. Then the field was switched off and the magnetooptical measurements made. It was found that the sign of the field on cooling had determined the sign of CD or CB. This was interpreted as indicating that there is a built in magnetic 'moment' associated with the broken symmetry state, and that a 'Zeeman' interaction between this and the applied field determines the low temperature state.

6. Discussion

The question of non-reciprocal optical effects in superconductors first arose as a possible test for anyon theories of high temperature superconductivity. Initial experimental results on layer structure cuprate showed a CD, but with it came the realization that cross channel terms in the reflection matrix might contribute to the apparent CD. The Dortmund reflection and transmission experiments on single crystals seemed generally consistent though the difference in onset temperatures needs exploration. Observation of a statistical variation in sign and magnitude of reflection CD and transmission CB as well as the ability to determine the sign of these effects by that of a magnetic field applied on cooling greatly expands the range of behavior to be studied. The beautiful fiber optic interferometer experiment from Stanford constitutes a major difficulty. It appears to conflict with the film measurements of Lyons *et al.* The negative film experiments of Weber *et al.* alongside their positive experiments on single crystals seems to certainly implies that some films do not behave in the same way as the corresponding bulk material. We note the following differences between the experiments of References [2] and [14]:

a. *Photon energy.* The Stanford experiment was done at $\lambda = 1060$ nm, $h\nu = 1.17$ eV, ours at 515 nm, $h\nu = 2.48$ eV. We simply do not know how important this might be. It is worth noting that magnetooptical spectra of magnetic solids often show much more structure than the simple absorption against photon energy plot.

b. ε'_K *and* ε''_K. Insofar as we can eliminate the effects of $B_\pm$, these magnetooptical experiments essentially measure combinations of the real and imaginary parts of the antisymmetric component of the dielectric tensor. The absolute and relative sensitivity varies dramatically with photon energy, and great caution is necessary in comparing the various different measurements especially at different frequencies.

c. *Sample differences.* It is noted above that if the light spot illuminates a number of symmetry broken domains, the standard deviation $\sigma_\varphi \propto \varphi_0 r/d$ where φ_0 is the single domain value, r is a domain radius, and d is the diameter of the light spot. Radical differences in domain size might be traceable to nucleation conditions and to some aspect of film microstructure not probed by present sample characterization.

7. Conclusion

In the wake of the interesting 'anyon' theories of high temperature superconductivity in layer structure materials, we and others have undertaken to explore broadly the time reversal symmetry of superconductors. Initial measurements on high quality layer structure cuprate compounds showed circular dichroism which correlated with superconductivity, and thus lend support to the theories. However, preliminary measurements on cubic bismuthate superconductors to which the theories are presumably inapplicable also showed CD, and complicated the situation somewhat. The observation of a zero CB by Spielman *et al.* is surprising, and must be considered very carefully. The single crystal work of Weber *et al.* seemed to show very much larger domains than we have yet encountered. Beyond that, their ability to control the sign of CD or CB with that of a cooling field is behavior the 'anyon' theories would lead us to expect.

Further experimental studies of the time reversal properties of superconductors are underway at a number of laboratories, and the field appears to be expanding rapidly.

References

1. R.B. Laughlin: 1988, *Science* **242**, 525.
2. K.B. Lyons, J.Kwo, J.F. Dillon, Jr., G.P. Espinosa, M. McGlashan-Powell, A.P. Ramirez and L.F. Schneemeyer: 1990, *Phys. Rev. Letters* **64**, 2949.
3. K.B. Lyons, J.F. Dillon, Jr., E.S. Hellman, E.H. Hartford and M. McGlashan-Powell: 1990, *Phys. Rev. B* **43**, 11408.
4. K.B. Lyons, J.F. Dillon, Jr. and M. McGlashan-Powell, *J. Appl. Phys.* (in press).
5. B.I. Halperin, Proceedings on the International Seminar on *Theory of High Temperature Superconductivity*, Dubna USSR, 3–6 July 1990.
6. J. March-Russell and F. Wilczek: 1988, *Phys. Rev. Letters* **61**, 2066.
7. e.g. L.D. Landau and E. Lifschitz: 1960, *Electrodynamics of Continuous Media*, Sections **58–60, 82**, Oxford.
8. G.B. Scott and D.E. Lacklison: 1976, *IEEE Trans. Mag.* **12**, 292.
9. I. Bozovic: 1990, *Phys. Rev. B* **42**, 1969.
10. e.g. R.M.A. Azzam and N.M. Bashara: 1977, *Ellipsometry and Polarized Light*, North Holland, Amsterdam.
11. P.G. de Gennes: 1974, *The Physics of Liquid Crystals*, Clarendon, Oxford.
12. *Derive, A Mathematical Program*, Soft Warehouse, Inc., Honolulu, 1990.
13. K.B. Lyons, J.F. Dillon, Jr. and M. McGlashan-Powell (to be published).
14. S. Spielman, K. Fesler, C.B. Eon, T.H. Geballe, M.M. Fejer and A. Kapitulnik: 1990, *Phys. Rev. Letters* **65**, 123.
15. H.J. Weber, D. Weitbrecht, D. Brach, A.L. Shelankov, H. Keiter, W. Weber, Th. Wolf, J. Geerk, G. Linker, G. Roth, P.C. Splittgerber-Huennekes and G. Guenterodt: 1990, *Solid State Comm.* **76**, 511.

ASPECTS OF THE GAUGE FIELD THEORY OF THE NORMAL STATE PROPERTIES OF HIGH-T_c SUPERCONDUCTORS

N. NAGAOSA

Department of Applied Physics, University of Tokyo, Tokyo 113, Japan.

and

P.A. LEE

Department of Physics, Massachusetts Institute of Technology, Cambridge, MA02139, U.S.A.

Abstract. The resonating valence bond (RVB) theories are reviewed with emphasis on their physical properties. It has turned out that the spin chirality is of particular importance in the RVB theories, which is expressed by the phase variables of the RVB order parameters. These phases, together with the Lagrange multiplier introduced to impose the local constraint, constitute the gauge field which ensures the local gauge symmetry of the theory. Each RVB state is characterized by (1) the statistics of the spinons and holons, (2) the volume in the k-space enclosed by the Fermi surface, (3) the existence or absence of the gap in the triplet excitations and (4) the dynamics of the gauge field. The normal state properties as well as the superconductivity are discussed for each of these states, and we propose the uniform RVB state as the most promising candidate to describe the high T_c superconductors.

Since the discovery of high T_c superconductivity in 1986, extensive studies have been done to clarify its mechanism. It has turned out experimentally that the normal state properties rather than the superconducting ones are unusual, which is attributed to the strong correlations between the electrons on the Cu d-orbital and the quasi-two dimensional structure of the system. Without the carrier doping (half filled), the system is insulating in contradiction to the band calculation and is believed to be a Mott insulator with spin $S = 1/2$ at each orbital. The antiferromagnetic (AF) exchange interaction J between the spins causes the AF long range ordering (LRO) at zero temperature even in 2D. With the doping of the holes, the AFLRO diminishes rapidly and disappears. As the concentration of the holes x is further increased, the superconductivity appears with high T_c. Anderson [1] proposed that the system without AFLRO is a liquid of spin singlet pairs which are resonating quantum mechanically (RVB state). When we take out an electron from the system, a vacant site as well as a spin which can not find its partner to form a singlet pair is created simultaneously. When regarded as quasi-particles, these two particles carry charge and spin and are called a holon and a spinon, respectively. The holons and spinons are

H. Aoki et al. (eds.), New Horizons in Low-Dimensional Electron Systems, 217–228.

expected to move somehow independently, and this phenomenon is called the spin-charge separation.

To describe this situation, we start with the $t-J$ model given by [2]

$$H = -\sum_{ij,\sigma} t_{ij}\, \tilde{C}^{\dagger}_{i,\sigma}\, \tilde{C}_{j,\sigma} + \sum_{\langle ij\rangle} J\mathbf{S}_i \cdot \mathbf{S}_j \ , \tag{1}$$

which can be derived from the $d-p$ model as well as the single band Hubbard model as the effective Hamiltonian in the large U limit. Hence, the Hilbert space is restricted to the subspace where no orbital is doubly occupied. This is expressed by

$$\tilde{C}_{i,\sigma} = P\, C_{i,\sigma}\, P \quad (\tilde{C}^{\dagger}_{i,\sigma} = P\, C^{\dagger}_{i,\sigma}\, P) \ , \tag{2}$$

where the projection operator $P = \Pi_i(1 - n_{i,\uparrow}\, n_{i,\downarrow})$ excludes the double occupation. This projection can be conveniently described by the slave boson (slave fermion) method where the projected electron operator $\tilde{C}^{\dagger}_{i,\sigma}$ is decomposed into $f^{\dagger}_{i,\sigma}\, b_i (b^{\dagger}_{i,\sigma}\, f_i)$ where $b_i, b^{\dagger}_i (f_i, f^{\dagger}_i)$ are the boson (fermion) operators which represents the holon while $f^{\dagger}_{i,\sigma}, f_{i,\sigma}(b^{\dagger}_{i,\sigma}, b_{i,\sigma})$ are the fermion (boson) operators for the spinon. When we annihilate a physical electron from the system by $\tilde{C}_{i,\sigma} = f_{i,\sigma}\, b^{\dagger}_i (\tilde{C}_{i,\sigma} = b_{i,\sigma}\, f^{\dagger}_i)$, it simultaneously creates a holon and destroys a spinon which results in a unpaired spin. Therefore the total number of the fermions and the bosons are conserved at each site i to be unity.

$$\begin{gathered} f^{\dagger}_{i,\uparrow}\, f_{i,\uparrow} + f^{\dagger}_{i,\downarrow}\, f_{i,\downarrow} + b^{\dagger}_i\, b_i = 1 \ , \\ (b^{\dagger}_{i,\uparrow}\, b_{i,\uparrow} + b^{\dagger}_{i,\downarrow}\, b_{i,\downarrow} + f^{\dagger}_i\, f_i = 1 \ ,) \end{gathered} \tag{3}$$

where the three terms on the lefthand side correspond to the three possible states for site i. The spin operator $\mathbf{S}_i$ is expressed as

$$\mathbf{S}_i = \tfrac{1}{2} \sum_{\alpha,\beta} \tilde{C}^{\dagger}_{i,\alpha}\, \boldsymbol{\sigma}_{\alpha,\beta}\, \tilde{C}_{i,\beta} \ , \tag{4}$$

where $\boldsymbol{\sigma}_{\alpha,\beta}$ is the α,β-component of the Pauli matrices. Putting $\tilde{C}_{i,\sigma} = f_{i,\sigma}\, b^{\dagger}_i (\tilde{C}_{i,\sigma} = b_{i,\sigma}\, f^{\dagger}_i)$ into Equation (1), we have the Hamiltonian expressed in terms of the fermions and bosons with the constraint Equation (3). RVB theories start with the mean field solution with the order parameters $\chi_{ij} = \langle f^{\dagger}_{i,\sigma}\, f_{j,\sigma}\rangle$, $B_{ij} = \langle b^{\dagger}_i\, b_j\rangle$ and $\Delta_{ij} = \langle f_{i,\uparrow}\, f_{j,\downarrow} - f_{i,\downarrow}\, f_{j,\uparrow}\rangle$ ($\chi_{ij} = \langle b^{\dagger}_{i,\sigma}\, b_{j,\sigma}\rangle$, $B_{ij} = \langle f^{\dagger}_i\, f_j\rangle$ and $\Delta_{ij} = \langle b_{i,\uparrow}\, b_{j,\downarrow} - b_{i,\downarrow}\, b_{j,\uparrow}\rangle$). Thus far there appeared many RVB theories depending on the choice of the order parameters as well as the formalism (slave boson or slave fermion) which are summarized in Table I. Each RVB state is characterized by (1) the statistic of the spinons and holons, (2) the volume in the **k**-space enclosed by the Fermi surface, (3)

TABLE I
Various RVB states and their properties.

statistics	slave-fermion spinon:boson	slave-fermion holon:fermion	spinon:semion holon:semion	slave-boson spinon:fermion	slave-boson holon:boson
state	magnetically ordered states	quantum spin liquid state	generalized flux state	uniform RVB state	Fermi liquid
characteristic	condensation of the spinon Neel or spiral state	spinon pairing staggered gauge field $\tilde{a}$	broken P,T symmetries nonzero spin chirality	composition laws Reizer's singularity	quasi particle Landau theory $1/N$ expansion
Fermi surface	small F.S. of holons near $(\pm\frac{\pi}{2}, \pm\frac{\pi}{2})$ $\Gamma \sim \epsilon^2$	small F.S. of holons near $(\pm\frac{\pi}{2}, \pm\frac{\pi}{2})$	no F.S.	large spinon F.S. satisfying Luttinger's th. $\Gamma \sim \epsilon^{2/3}$	large electron F.S. satisfying Luttinger's th. $\Gamma \sim \epsilon^2$
spin excitations	spin wave $\omega \propto q$	gap	gap	no gap Pauli para	no gap Pauli para
gauge field	Π_S:Meissner Π_H:normal	Π_S:Meissner Π_H:normal	Π_S:C.S.term $\Pi_H = -\Pi_S + Mq^2$	Π_S:normal Π_H:normal	Π_S:normal Π_H:Meissner
physical properties	$R_H = 1/x$ $\rho \propto T^2/x$ interaction with spin wave	$R_H = 1/x$ $\rho \propto T^{4/3}/x$ $\langle f_A f_B \rangle$ paring	anyon gas model G.S. is S.C. KT-transition	R_H: temp. dep. $\rho \propto T/x$ see text for details	$R_H = -1/(1-x)$ $\rho \propto T^2/(1-x)$ d-wave S.C.
references	[9–11]	[13–15]	[17–22]	[2,4,5]	[23]

the existence or absence of the gap in the triplet excitations and (4) the dynamics of the gauge field. Before discussing Table I, some remarks about the gauge field are in order. The gauge field is obtained as the fluctuations around the mean field solution. The amplitude fluctuations have a gap in the excitation energy, and we limit our attention to the low lying modes. There are symmetric and antisymmetric combinations of the two phases of the RVB order parameters χ_{ij} and B_{ij}. The symmetric one is lower in energy which is nothing but the space component a_{ij} of the gauge field [3–6]. The spin chirality $\mathbf{S}_1 \cdot \mathbf{S}_2 \times \mathbf{S}_3$ can be expressed in terms of these space-components of the gauge field as $\nabla \times \mathbf{a}$ in the continuum limit [7]. The time component of the gauge field a_0 is the Lagrange multiplier which has been put to be a constant in the mean field theory. Because the hopping Hamiltonian is decoupled as

$$\sum_{ij,\sigma} t_{ij}\, \tilde{C}^\dagger_{i\,\sigma}\, \tilde{C}_{j,\sigma} = \sum_{ij,\sigma} t_{ij}\, f^\dagger_{i,\sigma}\, f_{j,\sigma}\, b_i\, b^\dagger_j$$

$$\rightarrow \sum_{ij,\sigma} t_{ij}\, \chi_{ij}\, b_i\, b^\dagger_j + \sum_{ij,\sigma} t_{ij}\, B_{ij}\, f^\dagger_{j,\sigma}\, f_{i,\sigma}\ , \tag{5}$$

the phase of the order parameters χ_{ij} and B_{ij} enters into the phase of the quantum amplitude for the particle (fermion or boson) as a vector potential.

In the continuum approximation, this is expressed by the minimal coupling as

$$L = L_S(\partial_\mu + ia_\mu + iA_\mu) + L_H(\partial_\mu + ia_\mu)$$

$$= L_S(\partial_\mu) + (a_\mu + A_\mu)\, j_{S_\mu} + L_H(\partial_\mu) + a_\mu\, j_{H_\mu}\ , \tag{6}$$

where $L_S(L_H)$ and $j_{S_\mu}(j_{H_\mu})$ $(\mu = 0, x, y)$ are the Lagrangian and the current density of the spinon (holon) system. As is evident from Equation (6), the local gauge symmetry with respect to the gauge transformation $f_\sigma \to f_\sigma\, e^{i\theta}$, $b \to be^{i\theta}(b_\sigma \to b_\sigma\, e^{i\theta},\ f \to fe^{i\theta})$ and $a_\mu \to a_\mu + \partial_\mu\theta$, which is equivalent to the constraint Equation (3), is restored. The external electromagnetic field A_μ transforms as $A_\mu \to A_\mu + \partial_\mu\theta$ with the other gauge transformation $\tilde{C}_\sigma \to \tilde{C}_\sigma\, e^{i\theta}$. This gauge transformation is realized by $f_\sigma \to f_\sigma\, e^{i\alpha\theta}$, $b \to be^{i(\alpha-1)\theta}$ $(b_\sigma \to b_\sigma\, e^{i\alpha\theta},\ f \to fe^{i(\alpha-1)\theta})$ for arbitrary α. The coupling of the system with A_μ is specified for each α but the gauge invariant quantities are not affected by the choice of α. In Equation (6) we choose $\alpha = 1$.

If we integrate $\mathbf{a}$ and a_0 first in Equation (6), we obtain the constraint $j_{S_\mu} + j_{H_\mu} = 0$ because the coupling is written as $a_\mu(j_{S_\mu} + j_{H_\mu})$. Instead we integrate the fermions and bosons first to obtain the effective action S_g for the gauge field as

$$S_g = \sum_{\mu\nu} \sum_{q\omega_n} \Big[\Pi_{S_{\mu\nu}}(q,\omega_n)\, (a_\mu(q,\omega_n) + A_\mu(q,\omega_n))\, (a_\nu(-q,-\omega_n)$$
$$+A_\nu(-q,-\omega_n))$$
$$+\Pi_{H_{\mu\nu}}(q,\omega_n)\, a_\mu(q,\omega_n)\, a_\nu(-q,-\omega_n)\Big]\ , \tag{7}$$

where $\Pi_{S(H)}$ is the spinon (holon) current-current correlation function. In the transverse gauge, we have the time component (0) and the transverse spatial component ($\perp$), and Π_S and Π_H are reduced to 2×2 matrices. Equation (7) is transformed into

$$S_g = \sum_{\alpha,\beta=0,\perp} \sum_{q\omega_n} \Big[(\Pi_S + \Pi_H)_{\alpha\beta} \Big(a(q,\omega_n) + (\Pi_S + \Pi_H)^{-1}\, \Pi_S\, A(q,\omega_n)\Big)_\alpha$$
$$\times \Big(a(-q,-\omega_n) + (\Pi_S + \Pi_H)^{-1}\, \Pi_S\, A(-q,-\omega_n)\Big)_\beta$$
$$+(\Pi_S^{-1} + \Pi_H^{-1})_{\alpha\beta}\, A_\alpha(q,\omega_n)\, A_\beta(-q,-\omega_n)\Big]\ . \tag{8}$$

The coefficient $\Pi_S + \Pi_H$ of a^2 is the inverse of the propagator of the gauge field, which describes the fluctuation. $-(\Pi_S + \Pi_H)^{-1}\, \Pi_S\, A$ is the shift of the average of a, and the gauge field which the fermion (boson) system feels on the average is $(\Pi_S+\Pi_H)^{-1}\,\Pi_H\,A(-(\Pi_S+\Pi_H)^{-1}\,\Pi_S\,A)$. As will be discussed below, this screening effect restores the local constraint and leads to the composition laws for various physical quantities. An example is already seen in Equation (8). The electromagnetic response Π of the total system is equal to the coefficient $\Pi_S^{-1}+\Pi_H^{-1}$ of A^2 in Equation (8) [5]. This means that Π_S or Π_H whichever is smaller determines dominantly the total response Π. There are three types of the current-current correlation function $\Pi_{S,H}$. (1) When the system remains normal, the response to the gauge field is characterized by the Landau diamagnetism with the dissipation which is proportional to ω. (2) When the fermion pairing or the Bose condensation occurs, the gauge symmetry is broken and the transverse $\Pi_\perp$ remains constant in the small q,ω limit (Meissner effect). (3) If the P (parity) and T (time-reversal) symmetries are broken with the nonzero expectation value of the spin chirality $\langle \mathbf{S}_1\cdot\mathbf{S}_2\times \mathbf{S}_3\rangle$, the leading order term of the effective action for the gauge field becomes the Chern–Simons term expressed as [8]

$$\frac{1}{8\theta}\int d\tau\, d^2\,\mathbf{r}\,\varepsilon^{\mu\nu\lambda}\,a_\mu\,\partial_\nu\,a_\lambda\ , \tag{9}$$

where $\varepsilon^{\mu\nu\lambda}$ is the antisymmetric tensor. The identical particles interacting with the gauge field acquire an additional phase $e^{i\theta}$ when two of them are interchanged and the statistical transmutation occurs.

We now turn to Table I. The slave fermion formalism is reduced to the Schwinger boson method when there are no holes ($x=0$). The spinon pairing Δ_{ij} is nonzero while $\chi_{ij}=0$ in the mean field solution. Therefore, the spinon response Π_S belongs to the type (2) (Meissner), and the electromagnetic response Π of the total system is the holon response Π_H in the normal state. The antiferromagnetic fluctuations are lower in energy than the ferromagnetic ones, and by integrating out the latter the nonlinear (NL)-σ model can be derived as the effective theory which describes the low lying states [9]. The coupling constant appearing in the NL-σ model is larger for smaller spin S, and the quantum disordered spin liquid is realized for the coupling constant larger than a critical value (strong coupling region). The renormalization group analysis of this model has been done to compare with the experiments and it was concluded that the 2D AF Heisenberg system shows AFLRO at zero temperature (weak coupling region in the NL-σ model) [9]. Arovas–Auerbach [10] and Yoshioka [11] studied the Heisenberg model ($x=0$) in the mean field approximation and also obtained the AFLRO which is expressed as the condensation of the Schwinger bosons (spinons). When a hole is introduced, it causes the local distortion of the spin configuration around it and the minimum of the dispersion lies at $\mathbf{k}=(\pm\pi 2,\pm\pi/2)$ [12]. With the finite concentration of the holes, there appear small hole pockets around these minima. Shraiman and Siggia [12] found that the degeneracy between these two pockets is lifted in the spiral state where $\chi_{ij} \neq 0$. In these magnetically ordered states, the low lying excitation is the spin wave mode with the dispersion $\omega \propto q$.

In the concentration with high-T_c, the magnetic order has already disappeared though the AF correlations still remain. Several authors take into account this situation by assuming that the spin system is described by the NL-σ model in the strong coupling region (quantum spin liquid state) [13–15]. In this phase there is a gap in the triplet excitations, and the important low energy excitation is the staggered gauge field $\tilde{a}_\mu$. The effective action for this staggered gauge field is the usual $\tilde{f}^2_{\mu\nu}$ which is obtained by integrating the spinons. The holes, which are spinless fermions, on different sublattices are not hybridized when $\chi_{ij} = 0$, and we define two species of fermions f_A, $f_A^\dagger$ and f_B, $f_B^\dagger$ which are moving along A- and B-sublattices, respectively. These A- and B-fermions interact with the staggered gauge field $\tilde{a}$ with the charges of opposite sign which was discussed to be the origin of the attractive force between them and hence the superconductivity with the pairing $\langle f_A f_B \rangle$ [13–15]. The spectrum of this staggered gauge field $\tilde{a}$ has a large weight in the low frequency region which results in anomalous temperature dependence of the physical quantities (Reizer's singularity) [15]. For example, the resistivity behaves as $T^{4/3}$. The imaginary part of the self-energy of the holons is of order $\varepsilon^{2/3}$ at zero temperature with ε being the energy measured from the Fermi energy, while it diverges at finite temperatures. This does no harm because the Green function of the holons is not gauge invariant and has no physical meaning. The Green function of the phyiscal electron $\tilde{C}$, which is gauge invariant, is well defined and has no sharp peak of its imaginary part because of the convolution with the spinon Green function which has a gap in the spectrum.

The above discussion ignores the spiral order parameter χ_{ij}, whose nonzero expectation value results in the Meissner effect to the staggered gauge field together with the hybridization between A- and B-fermions. Hence the scenario of the quantum spin liquid state discussed above is completely altered in the presence of the spiral order which should manifest itself in the location of the peak in the neutron scattering spectra away from (π, π) [16].

The most exotic possibility among the RVB theories is the anyon model pioneered by Kalmeyer and Laughlin [17,18]. The anyon model can be derived from the generalized flux state where the average flux $\langle \nabla \times \mathbf{a} \rangle$ per plaquette is equal to the average number of holes per spin per site [19]. There occurs a gap in the spinon excitations, and by integrating them out, the Chern–Simons term Equation (9) with $\theta = \pi/2$ is obtained [20]. The holons, which are coupled with the gauge field, turn into semions. In this way, the anyon model becomes relevant. The RPA theory of the anyon gas model has been developed [21], and it is concluded that the leading order terms of the correlation function Π_H cancels the Chern–Simons term Π_S, and the next order term is a diagonal constant matrix M times q^2 in the static case ($\omega = 0$), which corresponds to the 'electric' energy $c_1 q^2 a_0^2$ and the 'magnetic' energy $c_2 q^2 a_\perp^2$. Therefore, the electromagnetic response $\Pi = \Pi_S(\Pi_S + \Pi_H)^{-1}\Pi_H$ of the total system shows Meissner effect and the ground state of the anyon gas is superconducting. At finite temperatures, the competition between the logarithmic divergence of the 'electric' energy and the entropy for an isolated vortex leads to the Kosterlitz–Thouless transition at $T_{KT} \sim h^2 x/m$ with m being the mass of the anyon [22]. Above T_{KT} a finite density of the free vortices are present which screen the 'electric' field, and we have the constant term in addition to Mq^2 for Π_H. Therefore the superconductivity disappears above T_{KT}, but the detailed analysis of the normal state properties of the anyon gas is left for future study.

The states discussed thus far have one common feature that the holons with concentration x determine the electromagnetic response of the total system and the Fermi surface is small or there is no Fermi surface. The angle-resolved photoemission experiments, which observe the spectrum function of the physical electron Green function G_σ, have revealed the 'quasi-particle' peak near the 'Fermi surface' satisfying the Luttinger's theorem. Therefore, we turn to the RVB state with a large spinon Fermi surface. This is realized by assuming spatially uniform χ_{ij} and B_{ij} in the slave boson formalism (uniform RVB state) [2,4,5]. In the normal state, it is assumed that there is neither spinon pairing ($\Delta_{ij} = 0$) nor the Bose condensation ($\langle b_i \rangle = 0$). Therefore, both Π_F and Π_B belong to type (1) (normal) and the physical quantities of the total system is expressed as the combination of those for the fermions (spinons) and bosons (holons) (combination law). (Hereafter we use Π_F and Π_B instead of Π_S and Π_H.) As will be discussed later, we have only one transition temperature T_c below which both Π_F and Π_B show the Meissner effect due to the fermion pairing and the Bose condensation, respectively. Therefore, T_c is the superconducting transition temperature. The expectation value of the spin chirality $\langle \nabla \times \mathbf{a} \rangle$ is equal to zero, but its fluctuations (gauge field fluctuations) are important and give anomalous temperature dependence of physical quantities as discussed below [4].

The effective Lagrangian which describes the low energy excitation of this state is given by [4,5]:

$$L = \sum_{i,\sigma} f_{i,\sigma}^*(\partial_\tau - a_0 - A_0 - \mu_F)\, f_{i,\sigma} + \sum_{\langle ij \rangle} J e^{i a_{ij}}\, f_{i,\sigma}^*\, f_{j,\sigma} + c.c.$$

$$+ \sum_i b_i^*(\partial_\tau - a_0 - \mu_B)\, b_i + \sum_{\langle ij \rangle} \tilde{t} e^{i a_{ij}}\, b_i^*\, b_j + c.c. \,, \tag{10}$$

where the chemical potentials μ_F, μ_B enforce $\langle b_i^\dagger b_i \rangle = x$ and $\sum_\sigma \langle f_{i,\sigma}^\dagger\, f_{i,\sigma} \rangle = 1 - x$. In the large N mean field theory $\tilde{t}$ is the original hopping matrix element t of the physical electron, and the Bose–Einstein (BE) condensation temperature $T_{BE}^{(0)} \approx 4\pi\tilde{t}x$ is very high. Below this $T_{BE}^{(0)}$, the system behaves as a Fermi liquid [23]. In reality, however, $N = 2$ and the fluctuations are important. It has been discussed that $\tilde{t}$ is subject to the downward renormalization from t to J (exchange interaction) because the hole distorts the spin configuration around itself considerably. Therefore T_{BE} becomes of order $4\pi Jx \sim 1000$ K which is still higher than the temperature of interest. In this paper, we discuss that the inelastic scattering of the bosons by the gauge field suppresses T_{BE} down to ~ 100 K which at the same time explains the T-linear resistivity and the temperature dependent Hall constant. The Lagrangian in the continuum limit corresponding to Equation (6) is

$$L = \sum_{\sigma} \int d^2\mathbf{r}\Big[f_\sigma^*(\mathbf{r})\,(\partial_\tau - a_0 - A_0 - \mu_F)\,f_\sigma(\mathbf{r})$$

$$+ f_\sigma^*(\mathbf{r})\,\frac{1}{2m_F}\,(i\nabla - \mathbf{a} - \mathbf{A})^2\,f_\sigma(\mathbf{r})\Big]$$

$$+ \int d^2\mathbf{r}\Big[b^*(\mathbf{r})\,(\partial_\tau - a_0 - \mu_B)\,b(\mathbf{r}) + b^*(\mathbf{r})\,\frac{1}{2m_B}\,(i\nabla - \mathbf{a})^2\,b(\mathbf{r})\Big]\,, \quad (11)$$

where the lattice constant is set to unity. The energy is measured in units of J because $1/m_B$ and $1/m_F$ are considered to be of order J from the above discussion. In the present case the time component (0) and the transverse spatial component ($\perp$) are decoupled, and the important low lying excitation comes from the transverse component whose propagator D^T is given by

$$D^T = (\Pi_{F\perp} + \Pi_{B\perp})^{-1} = \left(\frac{|\omega_n|}{q} + \chi_d\,q^2\right)^{-1}\,, \quad (12)$$

where $\chi_d = \chi_F + \chi_B$ is the sum of the Landau diamagnetic susceptibility of the fermions and bosons. The current–current correlation functions Π_F, Π_B should be determined self-consistently taking into account the interaction with the fluctuating gauge field. We employ Feynman's path integral method to treat this problem. The boson polarization $\Pi_B(\mathbf{r},\tau)$ is expressed by the integral over the closed path passing the two space-time points $(\mathbf{0},0)$ and $(\mathbf{r},\tau)$ as follows:

$$\Pi_B(\mathbf{r},\tau) = \int D\mathbf{r}_1(t)\,\left\langle \exp\left[-\frac{m_B}{2}\int_0^{2\tau} dt\,\dot{\mathbf{r}}_1(t)^2 + i\Phi[\mathbf{r}_1]\right]\right\rangle_{\mathbf{a}}\,. \quad (13)$$

Here $\mathbf{r}_1(0) = \mathbf{r}_1(2\tau) = \mathbf{0}$, $\mathbf{r}_1(\tau) = \mathbf{r}$, $\Phi[\mathbf{r}] = \oint \mathbf{a}(\mathbf{r}_1(t)) \cdot \dot{\mathbf{r}}_1(t)\,dt$ and $\langle\ \rangle_{\mathbf{a}}$ means the average over the gauge field fluctuations. Since the important contributions come from $\omega < T^{3/2} \ll T$ because $T \ll J$, we approximate the transverse gauge field by a static but spatially varying 'magnetic' field

$h = \nabla \times \mathbf{a}$ and $\Phi[\mathbf{r}_1]$ becomes $\Phi[\mathbf{r}_1] = \int\int h dA$ which is the flux penetrating the area A enclosed by the path $\mathbf{r}_1(t)$ projected onto the real space. The fluctuation of this 'magnetic' field h is given by

$$\langle h_q h_{-q} \rangle = \int d\omega \, \frac{1}{e^{\omega/T} - 1} \, q^2 \, \text{Im} \, D^T(q, \omega) \cong \frac{T}{\chi_d} \, , \tag{14}$$

for $q < q_0$ with $q_0 \approx (T/\chi_d)^{1/3}$. Thus we envision a spatially random h which varies on a scale q_0^{-1}. When the linear dimension of the area A for the integral Equation (13) is larger than q_0^{-1}, the integral for $\Phi[\mathbf{r}_1]$ is the sum of many independent flux fluctuations, i.e. $\Phi[\mathbf{r}_1] = \sum_{i=1}^{N} \Phi_i$. In this case, $\langle \Phi[\mathbf{r}_1]^2 \rangle = N \langle \Phi_i^2 \rangle$ where the number of the independent flux fluctuations N is roughly Aq_0^2 and $\langle \Phi_i^2 \rangle$ is estimated as

$$\langle \Phi_i^2 \rangle \cong q_0^{-4} \int_0^{q_0} q dq \, \frac{T}{\chi_d} \cong q_0^{-2} \, \frac{T}{\chi_d} \, . \tag{15}$$

Averaging $e^{i\Phi[\mathbf{r}_1]}$ over $\mathbf{a}$ gives the Debye–Waller factor $e^{-\langle \Phi[\mathbf{r}_1]^2 \rangle /2}$, and we obtain

$$\Pi_B(\mathbf{r}, \tau) = \int D\mathbf{r}_1(t) \, \exp\left[-\frac{m_B}{2} \int_0^{2\tau} dt \dot{\mathbf{r}}_1(t)^2 - \frac{gA}{\lambda^2} \right] \, , \tag{16}$$

where $g = 1/m_B \chi_d$ is the dimensionless coupling constant and $\lambda = (2\pi/m_B T)^{1/2}$ is the thermal length of the boson. For the unperturbed trajectories, the first term of the exponent in Equation (16) is of order of 1, and $A \sim A_0 = r(\tau/m_B)^{1/2} < \lambda^2$. Therefore, for $g \ll 1$ the first term dominates the second and we obtain

$$\Pi_B(\mathbf{r}, \tau) = \Pi_B^{(0)}(\mathbf{r}, \tau) \, \exp[-gTm_B r(\tau/m_B)^{1/2}] \, , \tag{17}$$

where $\Pi_B^{(0)}(\mathbf{r}, \tau) = \tau^{-2} \exp[-2m_B r^2/\tau]$. Thus putting the typical $r \sim (\tau/m_B)^{1/2}$ into Equation (17), the mean free time τ_B and the mean free path l_B of the boson are read as $\tau_B^{-1} \cong gT$ and $l_B \cong \lambda/g$. This result can also be obtained by the Boltzmann transport theory [4]. In the opposite limit $g \gg 1$, the loops with areas of order $A \sim \lambda^2/g$ contribute. The reduction in the area A means that the almost retracing paths give dominant contributions. Even if we restrict ourselves to the exactly retraceable paths so that $\mathbf{r}_1(\tau + t) = \mathbf{r}_1(\tau - t)$ for $0 < t < \tau$, we still have $\Pi_B(\mathbf{r}, \tau)$ which is the Green function G_B of the boson with the mass $2m_B$ and its thermal length $\lambda' = \lambda/\sqrt{2}$. In general, we can choose an arbitrary function $f(t)$ and construct a path by $\mathbf{r}_1(\tau + t) = \mathbf{r}_1(\tau - f(t))$ which is also retraceable when projected onto the real space. Therefore Π_B is not exactly G_B, but the characteristic length, i.e. the mean free path l_B, still remains of order $\lambda' \sim \lambda$. Interpolating these results for $g \ll 1$ and $g \gg 1$, we conclude

$l_B \sim \lambda / \min(1, g)$, which is the characteristic magnitude of $|\mathbf{r}|$, the width of the Feynman tube $\sim \lambda/(1+g)$ and the area $A \sim \lambda^2/(1+g)$. Therefore the boson resistivity $\rho_B \sim m_B T \min(1,g)/x$ which remains T-linear irrespective of the T-dependence of $g = 1/m_B \chi_d$ as long as $g > 1$. The response of the boson system to the magnetic field H, however, is reduced because H always appears in the form of the flux $\phi = HA$ and A is reduced. For example, since σ_{xy} is proportional to H, the Hall constant R_H^B is reduced from the weak coupling value x^{-1} to $x^{-1}(1+g)^{-1}$. The Landau diamagnetic susceptibility χ_B is related to H^2 and hence reduced by the factor $(1+g)^{-2}$. If we write $\chi_B = T_{BE}/m_B T$, T_{BE} is also reduced by $(1+g)^{-2}$ from $4\pi J x \sim 1000$ K.

The similar discussions can be done for the fermion polarization Π_F and the physical electron Green function G_σ. For Π_F the width of the Feynman tube is of order k_F^{-1} which is much smaller than q_0^{-1}, and the calculation of the Debye–Waller factor is modified from the boson case. We obtain the mean free time $\tau_F \cong (T/\chi_d)^{4/3}\, E_F$ and the mean free path $l_F \cong v_F\, \tau_F$. It should be noted that $l_F\, k_F \gg 1$ and the interaction between the fermion and the gauge field is always in the weak coupling regime in contrast to the boson case. For the physical electron Green function G_σ we obtain its lifetime τ_{in} near the quasi-particle peak as $\tau_{in}^{-1} \cong (T/\chi_d)^{2/3}\, (m_B m_F^2)^{-1/3}$ in the weak coupling limit $g \ll 1$.

We now discuss the physical quantities of the total system [4,5]. The dynamics of the fermions and bosons are correlated by the constraint or equivalently through the coupling with the gauge field. The observable current J of the electron is $J_F = -J_B$. The fermion (boson) current $J_F (J_B)$ is $\Pi_F (\Pi_B)$ times the average gauge field $\Pi_{B\beta}\, A_\beta (\Pi_{F\beta} + \Pi_{B\beta})$ $(-\Pi_{F\beta}\, A_\beta/(\Pi_{F\beta} + \Pi_{B\beta}))$ which the fermions (bosons) feel. Then we obtain $J = J_F = -J_B = \Pi A$ with $\Pi = \Pi_F\, \Pi_B/(\Pi_F + \Pi_B)$ which has already been derived below Equation (8). From this, the conductivity σ of the total system is $\sigma_F\, \sigma_B/(\sigma_F + \sigma_B)$ which means that the resistivity ρ is the sum $\rho_B + \rho_F$. From the above discussion, $\rho_B \gg \rho_F$ and $\rho \cong \rho_B \cong m_B T\, \min(1,g)/x$. The dimensionless coupling constant g decreases when approaching the superconducting T_c because $\chi_d(T)$ diverges toward T_c. Well above T_c in the normal phase, g is estimated to be of order unity from the theoretical point of view and $g > 1$ is suggested from the analysis of the Hall constant below. Therefore $\rho \cong T m_B/x$ which explains the experiment. When the magnetic field is applied, the fermions (bosons) feel the magnetic field $\chi_B H/(\chi_F + \chi_B)$ $(-\chi_F H/(\chi_F + \chi_B))$ and the Hall constant R_H is given by

$$R_H = \frac{\chi_F R_H^B + \chi_B R_H^F}{\chi_F + \chi_B}\,, \tag{18}$$

where R_H^B is given above. R_H^F is the Hall constant for the fermion system which depends on the details of the band structure but its magnitude is expected to be smaller than x^{-1} and we approximate R_H by $R_H \cong$

$x^{-1}\chi_F/(m_B^{-1}+\chi_F+\chi_B)$. Well above T_c, $\chi_F \gg \chi_B$ and the T-dependence of χ_F, which is expected from the pairing fluctuation of the fermions, can explain the decreasing R_H as a function of T. Experimentally, the high temperature limit of R_H is reduced from x^{-1} by a factor of 2 to 4 depending on the material which can be regarded as $1+g$. Another interpretation of this temperature dependent R_H is the precursor of the chiral spin state [24]. The Landau diamagnetic susceptibility χ is $\chi_B\chi_F/(\chi_F+\chi_B)$ which is approximately χ_B when $\chi_F \gg \chi_B$. Because the reduction factor for χ_B is the square of that for R_H, we expect an order of magnitude reduction of χ_B and hence T_{BE}, which results in $T_{BE} \sim 4\pi Jx/10 \sim 100$ K.

The physical electron Green function $G_\sigma(\mathbf{k},\Omega)$ is given by the convolution of the Green functions of the fermion and boson in the mean field approximation. Taking into account the interaction with the gauge field we obtain the following results. The spectral function Im $G_\sigma(\mathbf{k},\Omega)$ consists of a continuum for $\Omega < -|\mu_B|-(|\mathbf{k}|-k_F)^2/2m_F$ and a peak at $\Omega = \mathbf{k}^2/2m_F-\mu_F$. The width of the peak is equal to $\max(\Omega^{2/3}\,J^{1/3},\,(Tm_B)^{1/2}\,m_F^{-1})$. $(Tm_B)^{1/2}\,m_F^{-1}$ comes from the momentum distribution of the bosons while $\Omega^{2/3}\,J^{1/3}$ from the coupling with the gauge field. This width violates the Landau criterion for the quasi-particle. If the bosons condense, however, this width becomes $\max(\Omega^2,T^2)$ and the system is a Fermi liquid. On the other hand, the location of the peak in $\mathbf{k}$ space for small $|\Omega|$ is determined by the large 'spinon' Fermi surface which satisfies Luttinger's theorem.

We conclude by outlining a scenario whereby T_{BE} can be identified with the superconducting T_c. The inelastic scattering by the gauge field gives rise to the pair-breaking effect and the transition temperature of the fermion pairing is reduced from its mean field value $T_D^{(0)}$ to a lower value T_D as in the case of the Bose–Einstein temperature T_{BE}. When either Π_F or Π_B shows the Meissner effect, $\chi_d\,q^2$ in Equation (12) is replaced by the superfluidity density ρ_S of the fermions or the bosons. This stiffens up the gauge field and the reduction of the transition temperature disappears. Therefore Bose–Einstein condensation and the pairing of the fermions occur simultaneously at $T_c = \max(T_D, T_{BE})$ as long as T_c is below $\min(T_D^{(0)}, T_{BE}^{(0)})$. This explains why the non-Fermi liquid behaviour is seen in the normal phase of the concentration region with high-T_c. In the over-doped case, however, $T_{BE} > T_D^{(0)} = T_c$ and we have the Fermi liquid behavior, i.e. $\rho \propto T^2$ and negative R_H, between these two temperatures. When the fluctuations are taken into account, the transition at T_{BE} becomes only a cross-over with the exponential growth of the correlation length ξ and $\chi_B \sim x\xi^2$ of the boson as $e^{T_{BE}/T}$ below T_{BE}.

In conclusion, we have analyzed the physical properties of various RVB states from the viewpoint of the gauge field, which are summarized in Table I. We propose the uniform RVB state as the most promising candidate to describe the normal state properties of high-T_c superconductors because it can explain the unusual transport properties in consistent with the large Fermi surface observed in angle-resolved photoemission experiments.

Acknowledgements

We dedicate this paper to Prof. H. Kamimura on his sixtieth birthday. We also thank Prof. B. Altshuler, Dr. T.K. Ng, Dr. B. Shraiman and Dr. C.L. Kane for discussions. Work at MIT is supported by the National Science Foundation through the Materials Research Laboratory under Grant No. DMR 89–13624.

References

1. P.W. Anderson: 1987, *Science* **235**, 1196.
2. P.W. Anderson and Z. Zou: 1988, *Phys. Rev. Lett.* **60**, 132.
3. G. Baskaran and P.W. Anderson: 1988, *Phys. Rev. B* **37**, 580; G. Baskaran: 1989, *Phys. Scr. T* **27**, 53.
4. N. Nagaosa and P.A. Lee: 1990, *Phys. Rev. Lett.* **64**, 2450 and *Phys. Rev. B* **43**, 1233 (1991).
5. L.B. Ioffe and A.I. Larkin: 1989, *Phys. Rev. B* **39**, 8988; L.B. Ioffe and G. Kotliar: 1990, *Phys. Rev. B* **42**, 10360.
6. L.B. Ioffe and P.B. Wiegmann: 1990, *Phys. Rev. Lett.* **65**, 653.
7. X.G. Wen, F. Wilczek and A. Zee: 1989, *Phys. Rev. B* **39**, 11413.
8. F. Wilczek and A. Zee: 1983, *Phys. Rev. Lett.* **51**, 2250.
9. S. Chakravarty, B.I. Halperin and D. Nelson: 1988, *Phys. Rev. Lett.* **60**, 1057.
10. D.P. Arovas and A. Auerbach: 1988, *Phys. Rev. B* **38**, 316; A. Auerbach and D.P. Arovas: 1988, *Phys. Rev. Lett.* **61**, 617.
11. D. Yoshioka: 1989, *J. Phys. Soc. Jpn.* **58**, 32 and 1516.
12. B. Shraiman and E. Siggia: 1989, *Phys. Rev. Lett.* **62**, 1564.
13. P. B. Wiegmann: 1988, *Phys. Rev. Lett.* **60**, 821.
14. X.G. Wen: 1989, *Phys. Rev. B* **39**, 7223.
15. P.A. Lee: 1989, *Phys. Rev. Lett* **63**, 680.
16. C.L. Kane, P.A. Lee, T.K. Ng, B. Chakraborty and N. Read: 1990, *Phys. Rev. B* **41**, 2653.
17. V. Kalmeyer and R.B. Laughlin: 1987, *Phys. Rev. Lett.* **59**, 2095.
18. R.B. Laughlin: 1988, *Phys. Rev. Lett.* **60**, 2677; 1988, *Science* **242**, 525.
19. P. Lederer, P. Poilblanc and T.M. Rice: 1988, *Phys. Rev. Lett.* **63**, 1519.
20. Y. Hasegawa, O. Narikiyo, K. Kuboki and H. Fukuyama: 1990, *J. Phys. Soc. Jpn.* **59**, 822.
21. A.L. Fetter, C.B. Hanna and R.B. Laughlin: 1988, *Phys. Rev. B* **39**, 9679.
22. D.H. Lee and M.P.A. Fisher: 1989, *Phys. Rev. Lett.* **63**, 903.
23. M. Grilli and G. Kotliar: 1990, *Phys. Rev. Lett.* **64**, 1170.
24. L.B. Ioffe, V. Kalmeyer and P.B. Wiegmann: 1991, *Phys. Rev. B* **43**, 1219.

MODELS FOR THE ELECTRONIC STRUCTURE OF CUPRATES

MARK S. HYBERTSEN and MICHAEL SCHLÜTER
AT & T Bell Laboratories, 600 Mountain Avenue, Murray Hill, New Jersey 07974, U.S.A.

Abstract. The development and analysis of a sequence of specific models for the electronic structure of copper oxide materials is briefly reviewed. Starting with a first principles Density Functional description of the materials properties, two successive renormalization procedures yield simplified models and their parameters. First an extended three-band Hubbard model is obtained and then, for energies near the Fermi level, the final models are of the one-band, either Hubbard or '$t-t'-J$', type. At each stage of the renormalization process, observables are calculated and found in agreement with experiments. In particular, a detailed, quantitative theory of recent soft X-ray absorption measurements strongly supports a doped charge-transfer insulator picture for the low energy electronic states.

1. Introduction

The discovery of high temperature superconductivity in the copper oxide materials has stimulated much theoretical work on correlated electron systems [1–10]. Many different simplified models have been proposed for the electronic structure of the Cu–O planes. The trend has been to explore the phenomenology of a particular model as its parameters are varied with less emphasis on material-specific properties. We have followed the opposite approach selecting a generic electronic structure model and its parameters from explicit chemical considerations of the Cu oxide material properties. This approach [11,12] begins with an accurate quantum chemical description of a Cu–O material (i.e. La_2CuO_4) and then proceeds to renormalize down in energy considering a selected set of simplified models. In this fashion one arrives at a final 'fixed-point' Hamiltonian and its parameters which should explicitly reflect the properties of the materials. Our approach is checked at each stage by comparison to experiment, e.g. photoemission, optical reflectivity and Raman scattering. Recently, a comparison to soft X-ray absorption measurements has allowed detailed analysis of the low energy electronic states in the undoped parent materials as well as the doped cuprates. The validity of the final model Hamiltonian for the metallic state has to be tested by considering a variety of low energy scale physical properties, such as normal state resistivity $\rho(T)$, optical conductivity $\sigma(\omega)$, etc.

In this brief review, we follow the course of this development in its successive stages. The overall picture which emerges indicates that the electronic structure of the cuprates may be described as doped charge-transfer insulators well into the metallic (superconducting) regime. We conclude with

H. Aoki et al. (eds.), New Horizons in Low-Dimensional Electron Systems, 229–245.

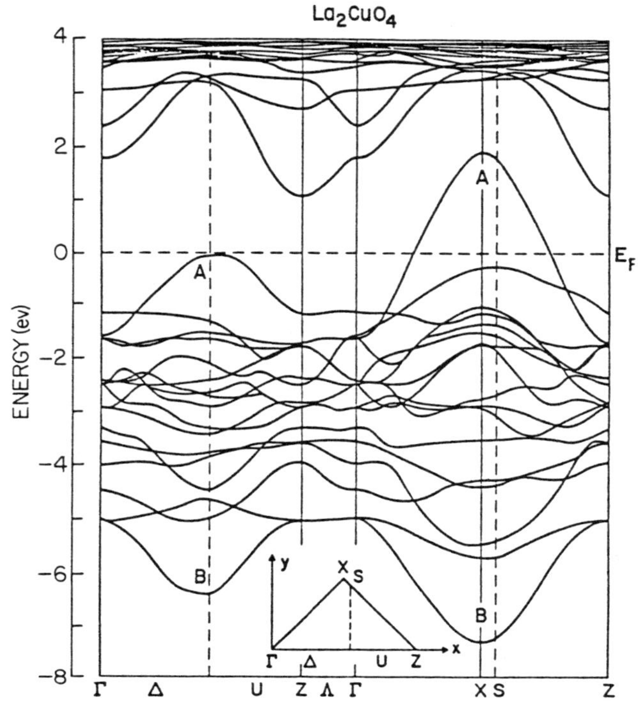

Fig. 1. Local Density Functional (LDA) band structure of La_2CuO_4. (From Reference [13].)

some remarks on the limitations and prospects for these models *vis–à–vis* high temperature superconductivity.

2. Density Functional Results

The strong covalency in the Cu–O materials, i.e. the strong mixing of TM d-states and oxygen p-states is clearly shown by Local Density Approximation (LDA)-type Density Functional results. Figure 1 shows a typical LDA band structure [13] of La_2CuO_4. A complex, about 10 eV wide, of 17 bands corresponding to mainly Cu d (5 states) and O p (4×3 states) is seen well separated from higher lying, mainly La d and f states. The large width of this complex derives from the strong hybridization. The band picture of La_2CuO_4 is metallic, in contrast to the observed AF insulating behavior. This failure signals the existence of strong electronic correlations among the electrons in the Cu–O planar complex which can not be described adequately within the LDA picture. As a consequence, electron dynamics on a low energy scale, such as magnetism and superconductivity and certain excitations on a high scale, such as localized photoemission shake-up processes are not accounted for in this approximation. However, the LDA results clearly support the point of view that in all these Cu–O materials superconductivity occurs in the strongly correlated Cu–O sheets when these are doped away from the half-filled band situation.

Inspite of the mentioned shortcomings, it has been found that the LDA gives a rather accurate description of the ground state electronic charge distribution, with similar quality as found for less correlated systems. For instance, cohesive properties, such as the equilibrium lattice structure and rather delicate quantities like phonon frequencies [14] have been well reproduced. This is the motivation and justification to use selected LDA results further, e.g. to select the relevant electronic degrees of freedom and to calculate the effective screening of the various bare parameters as they appear in the different model Hamiltonians we shall discuss.

3. Mapping onto a Generalized Hubbard Type Hamiltonian

A broadly useful model for correlated electron systems is based on a localized orbital representation of the Hamiltonian explicitly including the Coulomb interaction terms but limiting both the range of interactions as well as the number of electronic degrees of freedom. We retain only nearest (nn) and second nearest (nnn) neighbor terms. We select, based on the LDA results, the Cu $d_{x^2-y^2}$ and O p_σ planar orbitals as essential degrees of freedom. However, nearby in energy are the crystal field split Cu d-states, the planar and apical O p_π-states and the apical O p_σ-states. These will be eliminated at this stage, but we will discuss their role below. We implicitly assume the remaining interaction terms to be screened by the removed 'inactive' degrees of freedom in the solid and to become effective parameters. Thus we arrive at the following effective, multi-band Hubbard Hamiltonian (sum on sites implicit):

$$
\begin{aligned}
H = \sum_{\sigma} \Big\{ & \varepsilon_d \, n_{d\sigma} + \varepsilon_p \, n_{p\sigma} \\
& + t_{pd} \, c^{+}_{p\sigma} \, c_{d\sigma} + t_{pp'} \, c^{+}_{p\sigma} \, c_{p'\sigma} + h.c. \\
& + \tfrac{1}{2} \, U_d \, n_{d\sigma} \, n_{d-\sigma} + \tfrac{1}{2} \, U_p \, n_{p\sigma} \, n_{p-\sigma} \\
& + \sum_{\sigma'} \Big(U_{pd} \, n_{p\sigma} \, n_{d\sigma'} + U_{pp'} \, n_{p\sigma} \, n_{p'\sigma'} \\
& + K_{pd} c^{+}_{p\sigma} \, c_{p\sigma'} \, c^{+}_{d\sigma'} \, c_{d\sigma} \\
& + K_{pp'} \, c^{+}_{p\sigma} \, c_{p\sigma'} \, c^{+}_{p'\sigma'} \, c_{p'\sigma} \Big) \Big\}
\end{aligned}
\qquad (1)
$$

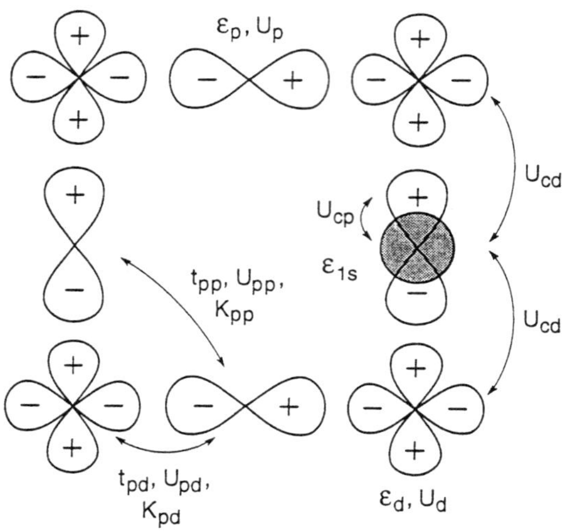

Fig. 2. Schematic representation of the orbitals and interactions included in the 3-band Hubbard model, Equation (1), as well as the additional interactions with an O 1*s* core-hole present.

where $c^+_{d\sigma}$ creates a *hole* in the Cu $d_{x^2-y^2}$ orbital of spin σ and $c^+_{p\sigma}$ creates an O p_x or p_y hole of spin σ as shown in Figure 2. A vacuum state $d^{10}p^6$ has been assumed. The retained terms are the on-site energies, one-electron hopping integrals and the direct as well as exchange Coulomb interaction integrals.

The approach for calculating the parameters in Equaton (1) is described in detail in Reference [11]. The complexity of the problem arises from the effective screening of the interaction parameters and the strong hybridization between the Cu and O orbitals. There are two classes of parameters: (i) the one-electron on-site and hopping energies, and (ii) the interaction parameters. They are determined by two different, but coupled procedures. The one-electron parameters are derived by assuming that the energy bands of the field solution of Equation (1) correspond to the $pd\sigma$-derived bands in the LDA bandstructure. This depends on the interaction parameters through the selfconsistency in the mean-field theory. One should note that the LDA approach contains interactions in mean-field somewhere in nature between a pure Hartree and Hartree–Fock (used here) treatment. The interaction parameters are determined using the so-called constrained density functional (CDF) approach [15]. For the purpose of calculating local interaction parameters like U_d one has to consider local d-charge 'fluctuations' (δn_d) away from this ground state density and calculate the associated energy change. This can be done variationally in DFT by constraining n_d to a chosen value N_{di} at an isolated site i. The local charge fluctuations (static) are optimally

TABLE I
Calculated set of parameters of the 3-band Hubbard model describing La_2CuO_4. Hole notation is used in connection with the phase conventions shown in Figure 2. (From References [11, 12 and 29].)

U_d	U_p	U_{pd}	$U_{pp'}$	U_{cp}	U_{cd}	K_{pd}	$K_{pp'}$	$\varepsilon_p - \varepsilon_d$	t_{pd}	t_{pp}
10.5	4	1.2	0	6	1.2	-0.18	-0.04	3.6	1.3	0.65

screened by the crystalline environment. The LDA calculations of this CDF energy surface are implemented using the linear muffin-tin orbital (LMTO) technique and up to (2×2) supercells to spatially isolate the 'frozen' charge fluctuations. The result of a sequence of such calculations is an energy surface as a function of local charge in the Cu d and O p channels. However, because of the strong hybridization, this energy includes both kinetic as well as Coulombic contributions. To obtain the U_{ij}, completely analogous calculations are also carried out for the MF-Hubbard model. The two energy surfaces $E^{LDA}[\delta n_i]$ and $E^{MF-H}[\delta n_i]$ are now mapped onto each other yielding the interaction parameters U_{ij} in the Hubbard model. These depend implicitly on the chosen one-electron parameters ε_{ij} so these must be chosen consistently. We have separately estimated the small exchange parameters (K_{pd} and $K_{pp'}$) from ionic cluster wavefunctions [7].

The final self-consistent set of parameters is listed in Table I. Notable are large onsite Coulomb energies, strong $p-d$ and $p-p$ hybridization terms and relatively small intersite Coulomb energies. The bare on-site energy difference $\varepsilon = (\varepsilon_p - \varepsilon_d)$ is larger than sometimes anticipated, though still smaller than U_d. A simple picture of the electronic structure of the Cu–O sheets is shown in Figure 3. An upper and a lower Hubbard band mainly Cu-derived are interleaved with the oxygen band complex. The Fermi level falls between the upper Cu Hubbard band and the oxygen band. Local configurations corresponding to the ground state, electron excitation and hole excitation are indicated (ligand hole denoted by $|\underline{L}>$ i.e. $|p^5>$), although hybridization mixes the wavefunction character. They illustrate that La_2CuO_4 is, according to the classification scheme proposed by Zaanen, Sawatzky and Allen [16] a charge transfer insulator. Note that this situation is different from that of an ordinary ionic insulator, since its insulating behavior depends on a large finite U_d.

The extended three-band Hubbard model derived here has been extensively discussed. An independent calculation by McMahan *et al.* [17], similar to ours in spirit but different in detail, comes to the same overall conclusions concerning the parameter values. Using analysis of small clusters, various spectroscopic data and broad experience with other Cu and O con-

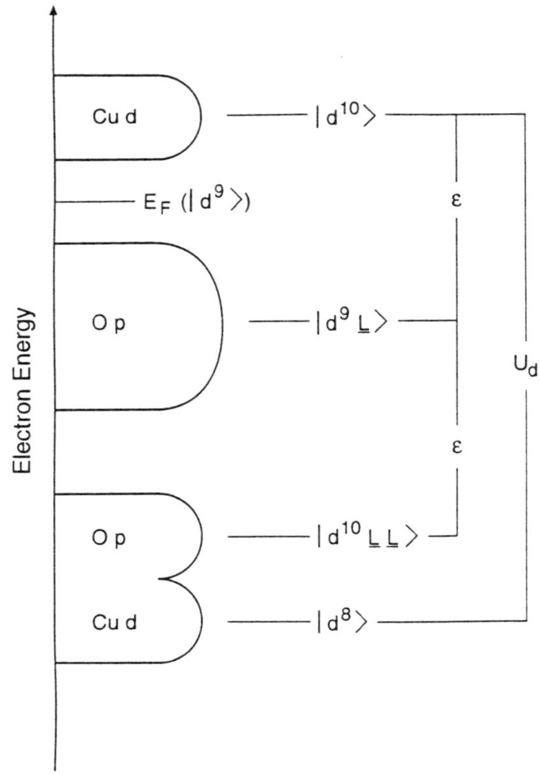

Fig. 3. Schematic diagram of the electronic structure of La_2CuO_4 as derived from the 3-band Hubbard model.

taining materials, Sawatzky and coworkers [18] obtain a Hubbard parameter set quite similar to ours. It should be noted that proper analysis of (for instance) photoemission spectroscopy including the satellite structure requires including all the Cu d-orbitals to properly account for the complex multiplet structure in the two hole states.

It remains controversial as to whether degrees of freedom outside the three-band model are required to describe low energy properties. In particular, Kamimura has early on proposed that upon doping the Cu $z^2 - r^2$ orbital enters directly, with the extra hole forming a triplet configuration stabilized by the large inter-orbital exchange energy [8]. This would suggest a rather different scenario for the doped materials than presented here, where the extra hole goes into an oxygen $p\sigma$ orbital and forms a singlet state. Cluster calculations favoring either the singlet or the triplet case have appeared [18,19]. However, these are difficult to assess because the competition between the singlet and triplet configurations only converges slowly with cluster size. We note that the LDA band structure shows only small admixture of the Cu $z^2 - r^2$ states at the Fermi level. We estimate that these states only enter our model as excited states above about 1 eV. This general point of view is further reinforced by the essential similarity of low energy properties in different materials independent of the number of apical oxygen atoms (two, one or zero). Since the apical p_z orbital strongly couples to the Cu $z^2 - r^2$ orbital, one would have expected strongly coordination dependent properties if the Cu $z^2 - r^2$ played an important role.

4. Exact Diagonalization Results

The generalized Hubbard Hamiltonian (Equation (1)) has been diagonalized exactly for a sequence of open clusters [12]: CuO_4, Cu_2O_7, Cu_3O_{10}, Cu_4O_{12} and Cu_5O_{16}. The clusters are embedded in an array of Cu d sites. The results are insensitive to details of this embedding procedure. The resulting sparse Hamiltonian matrices are numerically diagonalized using a block Lanczos technique. As expected from Table I and Figure 3, the chemical character of the holes in the insulating ground state (one hole per Cu) is copper dominated, Cu (80%) and O (20%). This is quite consistent with the observed magnetic moment in the AF Néel state of $\mu = 0.48 \pm 0.15\ \mu_B$ [20] which is reduced from the full moment of 1.14 μ_B for Cu^{2+} both by hybridization ($\sim$ factor of 0.8) and by quantum fluctuations ($\sim$ factor of 0.6) [21]. Drastic variations in the parameter set, such as a reduction in the intersite energy ε would clearly destroy this agreement. An added electron has 80% Cu character. However, results for the character of an added hole are essentially the reverse, Cu (20%) and O (80%), which is consistent with photoemission data [22] and other spectroscopic probes [23]. These results underline the essential *chemical* asymmetry between electron and hole doping.

Although detailed analysis of photoemission satellites requires the full multiplet structure, some basic insight follows from the three-band model. This was illustrated in Figure 3. Hybridization couples both the $|\underline{L}\,\underline{L}>$ and the $|d^8>$ configurations in the satellite region to the $|d^9\underline{L}>$ configuration in the main line. Due to this strong interaction, the satellites are not simply split off by U_d. Detailed calculations show that the parameters of Table I are consistent with observed satellite positions and resonance behavior [24].

In addition to photoemission, optical absorption or reflectivity can also be used to probe the validity of the derived parameter set. In particular the size of the optical gap, if non zero, is a delicate test for quantities like ε, t_{pd}, t_{pp} and U_{pd}. We have studied the total energy of clusters (with non-periodic boundary conditions) with N, $N-1$ and $N+1$ holes to determine the gap as:

$$\Delta = E(N+1) + E(N-1) - 2E(N) \ . \tag{2}$$

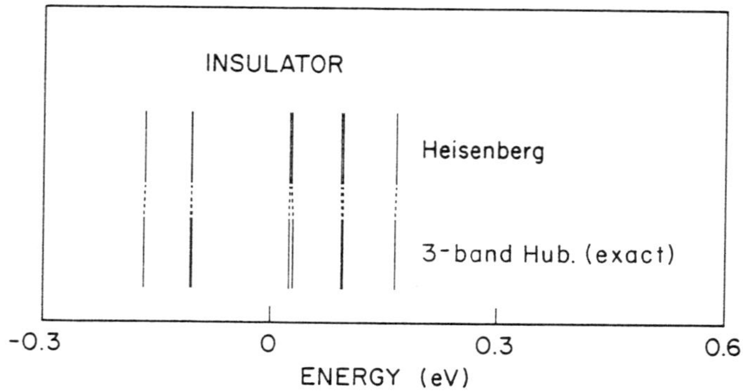

Fig. 4. The low energy spectrum of an insulating Cu_5 cluster calculated in the 3-band Hubbard model and compared to an effective 2D spin 1/2 Heisenberg Hamiltonian. (From Reference [12].)

Finite size effects are found to be important and scaling to the thermodynamic limit is necessary. The results extrapolated from Cu_1 to Cu_5 clusters give a charge transfer gap of $E_g = 2.4 \pm 0.3$ eV. This is in excellent agreement with ellipsometry data on $YBa_2Cu_3O_6$ which show a structure near 2.6 eV interpreted as the charge transfer absorption edge while a lower energy structure at 1.8 eV is assigned to the corresponding excitonic transition [25]. This, too, corresponds well with our parameters, since the excitonic binding energy should be on the scale of U_{pd}.

5. Mapping of Low Energy States

We now turn to the question of the nature of the low lying excitations. For the insulating phase, the calculated low energy cluster spectra with N holes show the correct number of lines for a spin 1/2 Heisenberg model separated by a large ($\sim$ 2.5 eV) gap from the charge transfer excitations, as already found in Reference [7]. Mapping these spectra to the Heisenberg Hamiltonian [12] for all clusters yields a nearest neighbor $J = 128 \pm 5$ meV and a next nearest neighbor $J' = 3 \pm 1$ meV, error bars indicating the variations with cluster size. The mapping is illustrated in Figure 4 where the RMS errors are 1 meV. Note that $J' \ll J$. Since the parameters in Table I have a precision of only about 10%, the calculated values of J, J' are uncertain within $\approx$ 50%. Experimentally, spin excitations in the AF have been measured by Raman scattering [26] and more recently also by neutron scattering [27]. The Raman spectra have been shown to arise from two-magnon type excitations of a simple 2D spin 1/2 Heisenberg AF on a square lattice, strongly influenced by quantum fluctuations. An excellent fit of the first three spectral moments was obtained for a value of $J = 128 \pm 6$ meV ($YBa_2Cu_3O_6$) [26] which is in

TABLE II
Calculated set of parameters of the one-band '$t-t'-J$' model describing electron and hole doped La_2CuO_4. Also shown are the parameters for a one- band Hubbard model. All energies in meV. (From Reference [12].)

	$t-t'-J$ electrons	$t-t'-J$ holes	Hubbard
t	410 ± 5	440 ± 5	430
t'	-70 ± 10	-60 ± 20	-70
J	128 ± 5	128 ± 5	–
J'	3 ± 1	3 ± 1	–
U_{eff}	–	–	~ 5400
			(~ 4100)

good (fortuitously close) agreement with the calculated value. The neutron data confirm this picture.

For the doped phases, represented by clusters with $N \pm 1$ holes, we again find for all clusters a spectral gap of $\gtrsim 2$ eV between the lowest manifold and higher lying states. This separation suggest the existence of a simplified Hamiltonian whose Hilbert space comprises only the lowest manifold. We find good mapping onto a generalized '$t-J$' type Hamiltonian [12] first proposed for these materials by Anderson [3] and by Zhang and Rice [6], which has the form:

$$H = \sum_{ij\sigma} t_{ij}\, c_{i\sigma}^{+}\, c_{j\sigma} + \sum_{\langle ij \rangle} J_{ij}\, \mathbf{S}_i \cdot \mathbf{S}_j \tag{3}$$

where $c_{j\sigma}$ creates an hole at an effective site j centered about a Cu in the plane and S_j is the corresponding spin. No double occupancy of these sites is admitted and the model reduces to the Heisenberg model at half filling. Carriers introduced by doping are represented by vacant sites.

The mapping consists of three steps. First the spectrum is mapped (see Figure 5) constraining J, J' to the values found for the insulator. Nearest neighbor (t) and second nearest neighbor (t') hopping is needed to obtain mappings with RMS error of ~ 60 meV (about 3% of the bandwidth). The values are given in Table II for both electron and hole doping. The error bars again reflect cluster to cluster consistency. The signs of t, t' refer to hole notation. This convention means that for the electron doped case, the effective carrier is a vacancy, while for hole doping it is a singlet bound state of two holes. Second, the subspace for the '$t-t'-J$' model has been identified. We used products of one-hole and two-hole wavefunction, individually derived from CuO_4 cluster calculations. The two-hole case for this cluster corresponds formally to the singlet proposed by Zhang and Rice [6], formed between two holes in the antibonding linear combination of the Cu

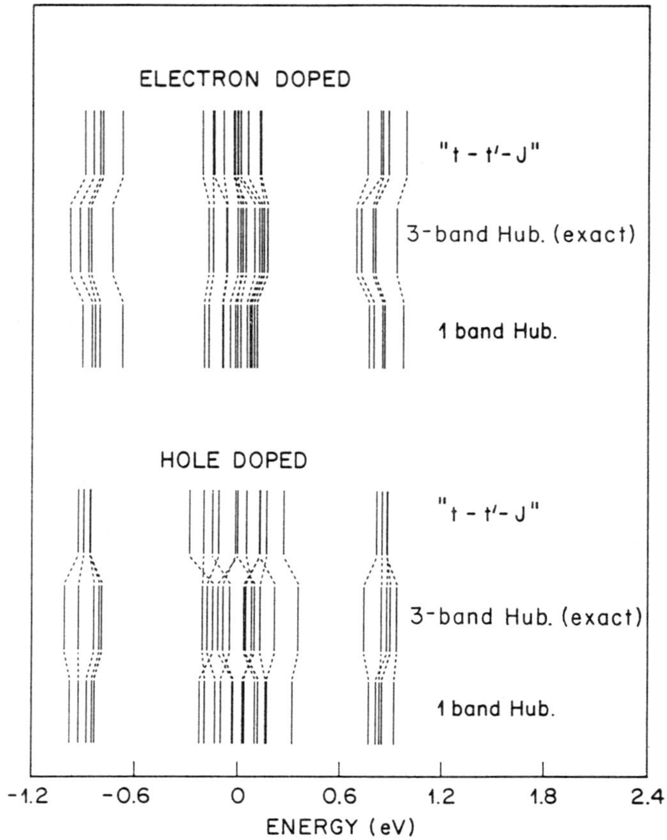

Fig. 5. The low energy spectrum of a doped Cu_5 cluster (electrons or holes) as calculated in the 3-band Hubbard model and compared to an effective one-band '$t - t' - J$' model or a one-band Hubbard model. (From Reference [12].)

$d_{x^2-y^2}$ and the O_{p_σ} orbitals, but contains full local correlations. The projection onto the space spanned by the low energy wavefunctions of the original three-band Hubbard model is at least 93% across the entire manifold. Third, we checked whether the internal dynamics in this subspace is correctly described by the '$t-t'-J$' wavefunctions. Again the mapping onto the low-lying Hubbard-model eigenfunctions was in the 80%–90% range. These are rather strong tests of the validity of the '$t - t' - J$' model for the low-energy electronic structure of the (doped) three-band Hubbard model with parameters appropriate for La_2CuO_4. Our numerical results thus confirm the original proposals by Anderson [3] and by Zhang and Rice [6].

The values in Table II show that the effective low energy carrier dynamics exhibits remarkable electron-hole symmetry, despite the underlying strong chemical asymmetry. We cannot identify any underlying symmetry for this, but rather believe it to be a consequence of the materials parameters. This symmetry has been noted in optical reflectivity experiments comparing electron and hole doped materials [28]. Although the spectral mapping is well defined for the clusters we studied, the large value of t suggests that in the limit of many sites the top of the '$t-t'-J$' band will merge with the higher energy continuum. Therefore, the '$t-t'-J$' band will strictly applies only to the region near the bottom of the band.

The '$t-t'-J$' model describes the dynamics of electrons or holes separately in single effective bands. These bands may be consistently placed relative to each other with a band-center separation found to be 4.1 eV, precisely the charge transfer gap for the CuO_4 cluster. A simple one-band Hubbard model with a single effective site per Cu-atom represents both bands in the same model and allows for dynamical interactions. All three low-energy spectra ($N-1, N, N+1$ holes) can be accurately reproduced with a single parameter set of $U_{eff} = 5.4$ eV, $t = 0.43$ eV and $t' = -0.07$ eV (Figure 5). This mapping places the upper Hubbard band so as to reproduce the magnetic energy scale as closely as possible. If instead, the value $U_{eff} = 4.1$ eV is used, the electron and hole bands are equally well represented individually, but the superexchange is about 30% larger in the insulator. It is important to note that exact diagonalization studies can be done for significantly larger clusters using the '$t-t'-J$' and one-band Hubbard models in comparison to using the three-band Hubbard models. These show charge transfer gaps of 2.5 and 2.7 eV respectively, quite consistent with the value extrapolated from the small open clusters treated in the three-band model.

6. Analysis of X-Ray Absorption Data

The validity of the models we have obtained for the low energy states must finally be assessed by making a detailed comparison to experimental data such as normal state resistivity, optical conductivity, etc. In each case, one must make a careful analysis of the appropriate response function, perhaps with the inclusion of additional scattering mechanisms. Each case constitutes a non-trivial problem and these have received on-going attention from various groups. Quite recently, Chen *et al.* have obtained high quality, bulk sensitive soft X-ray absorption data for $La_{2-x}Sr_xCuO_4$ which probes the O p-like states near the Fermi level for a range of Sr doping concentrations (Figure 6) [29]. For the insulator, a single 'pre-peak' occurs at ~ 530 eV, before the main K-edge. Upon doping with Sr, a new peak appears at lower energy, growing in with increasing x. This corresponds to new empty states available due to hole doping. At the same time the upper peak persists, but with decreasing intensity. The peak separation (~ 1.5 eV) is similar to the observed optical gap. This 'gap' persists well into the metallic (supercon-

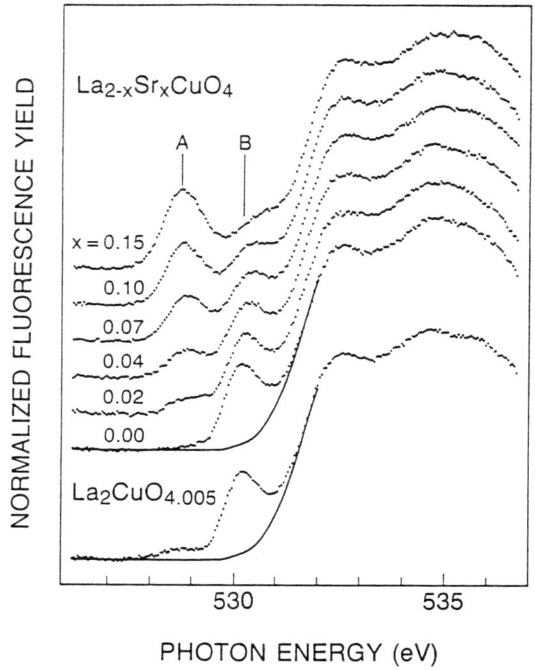

Fig. 6. Measured X-ray absorption spectra at the O K-edge of $La_{2-x}Sr_xCuO_4$. (From Reference [29].)

ducting) regime of $x = 0.15$. We have used the present models to give a detailed description of the pre- peak region which we summarize here [29].

For X-ray photon energies near the O K-edge, the excitation process promotes an O 1s electron into empty O p-like final states reflecting the dipole selection rules. The X-ray absorption is a local excitation whose analysis is complicated by the relatively long lifetime of the core-hole. We approach the problem using the so-called final state rule wherein the O 1s electron is removed to infinity and then returned to the O p-like states calculated in the presence of a static 1s core-hole. The extra local Coulomb interactions in the final state lead to well known excitonic features in the spectra which can be qualitatively quite important. Effects due to the finite core-hole lifetime are included together with other broadening mechanisms *post hoc* by convolving the final spectra with a Gaussian (FWHM 0.5 eV) reflecting instrumental resolution (0.3 eV), phonon broadening (0.3 eV) and free ion core-hole lifetime (0.2 eV).

The necessary Coulomb interaction parameters for the O 1s core-hole (illustrated in Figure 1) have been calculated using a procedure analogous to that described above. A frozen 1s hole is introduced in a LMTO LDA calculation and the valence electrons selfconsistently relax to screen it. From

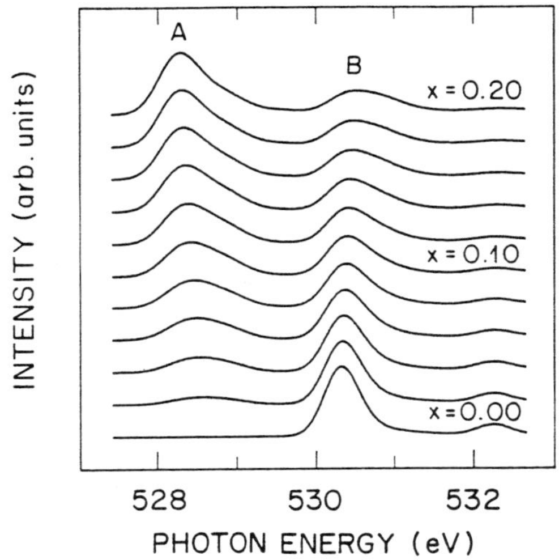

Fig. 7. Calculated X-ray absorption spectra as a function of carrier concentration x. (From Reference [29].)

the spectral features, the necessary Coulomb interaction parameters are derived: $U_{cp} = 6$ eV and $U_{cd} = 1.2$ eV. Using these results, qualitative features of the X-ray absorption can already be analyzed using the clusters previously studied. The spectral function may be obtained directly using Lanczos techniques. However, a more quantitative analysis requires study of larger systems. For that purpose, we map the X-ray absorption process including the core-hole potential into the one-band Hubbard model. In this way, both the Cu d-band and the charge transfer band (O p-like) are retained in the form of the lower and upper bands. We have extensively tested this mapping by comparing calculated absorption spectra in small finite clusters for the three and one band models.

The spectra are calculated for a 10 site cluster with periodic boundary conditions containing zero to four extra holes. Physically, the extra holes derive from Sr doping. We have extensively analyzed the role of a model Sr impurity potential. For a local probe with the resolution of X-ray absorption, it does not influence the net peak positions or intensities. However, fluctuations in the Sr distribution must be taken into account. To do this, the spectra for fixed hole number (n) are incoherently averaged according to the statistical occurrence of n Sr atoms in the 10 site cell for a given average concentration x. This procedure gives the family of spectra in Figure 7. These show the same features exhibited in the experiment. The peak position and intensity is compared (Figure 8) to the spectroscopic constants derived from the experiments. The quantitative agreement is quite good.

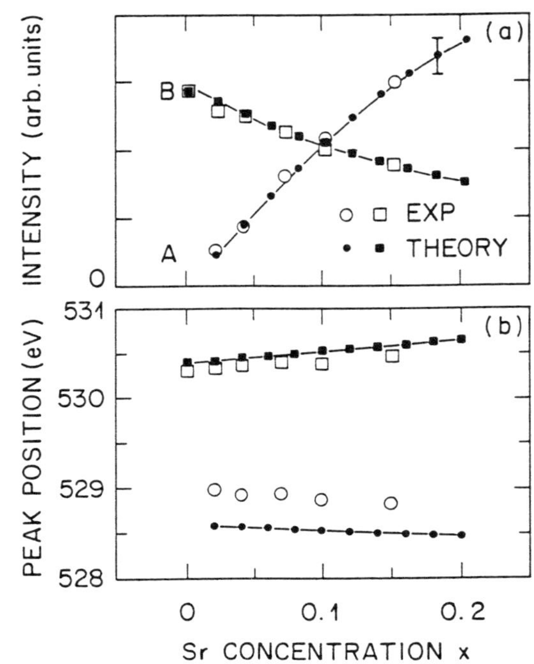

Fig. 8. Comparison of calculated peak intensity (a) and peak position (b) with experiment as a function of Sr concentration in X-ray absorption. (From Reference [29].)

Several points are important. First, the upper peak is a classic two-dimensional core-exciton. Second, the peak separation is precisely the charge transfer gap, renormalized by core-hole excitonic effects (about 0.8 eV). These derive, in the three band model, from U_{cd}. Third, there is a characteristic asymmetry in the lower peak which is the transfer of oscillator strength due to the well known 'singularity' at the Fermi level of the doped system. Fourth, the dispersion of the lower peak reflects, in part, motion of the Fermi level into the charge transfer band with increasing doping. Fifth, the dispersion of the upper peak derives from transfer of oscillator strength from the low energy side of the upper peak into the lower peak. Sixth, Figure 8a clearly shows the systematic shift of oscillator strength from the upper to the lower peak with increasing hole concentration. This may be traced, in a simple model, to final state interference effects [30]. It is a characteristic feature which points to the importance of strong correlation effects and has long been recognized in the one-band Hubbard model [31]. Recall that in an independent particle model (e.g. a semiconductor), the upper peak would have constant intensity. Finally, there is no strong signature of the insulator-metal transition apparent in the data consistent with recent EELS measurements [32]. One should note that in the theory, the incoherent average of spectra from finite cells can only yield a smooth interpolation and therefore does not address this issue.

7. Concluding Remarks

The consecutive mapping procedures described in this paper are neither exact nor unique. On the basis of our analysis, however, we come to the conclusion that the low energy dynamics of moderately electron or hole doped Cu–O planes should be describable by a *one-band* model, with additional degrees of freedom occurring at higher energies. This point of view is explicitly derived from materials properties and is supported by several experimental findings. In particular, the analysis of the X-ray absorption data gives strong evidence for the derivation of the metallic (superconducting) regime from the charge-transfer (Mott–Hubbard) insulator in a continuous fashion. This is to be contrasted with the viewpoint based on a phase transition upon doping leading to a more traditional Fermi liquid picture [33]. Further progress will require detailed analysis of other experimental probes. We can not claim on the basis of present work that other (neglected) interactions may not play an important role for the low energy dynamics, e.g. superconductivity. There may be interactions which do not trivially renormalize the effective parameters and therefore remain in addition to the terms in the one-band models. Nonetheless, the one-band model for electronic structure in the cuprates is a good starting point for further work.

Acknowledgements

We have benefited from collaborations with N.E. Christensen on the LDA part of the problem, with E.B Stechel and D.W. Jennison on the cluster part of the work and with W.M.C. Foulkes, E.B. Stechel, C.T. Chen and F. Sette on the analysis of the X-ray absorption experiment.

References

1. Proc. Intl. Conf. *High Temp. Supercond. Mat. Mechan. Supercond.*, J. Muller and J.L. Olsen (eds.), North-Holland, Amsterdam, 1988 (*Physica C*, 153–155, 1988).
2. Proc. Intl. Conf. *High Temp. Supercond. Mat. Mechan. Supercond.*, N.E. Phillips, R.N. Shelton and W.A. Harrison (eds.), North-Holland, Amsterdam, 1989 (*Physica C*, 162–164, 1989).
3. P.W. Anderson: 1987, *Science* **235**, 1196.
4. V.J. Emery: 1987, *Phys. Rev. Lett.* **58**, 3759.
5. C.M. Varma, S. Schmitt-Rink and E. Abrahams: 1987, *Solid Stat. Commun.* **62**, 681.
6. F.C. Zhang and T.M. Rice: 1988, *Phys. Rev. B* **37**, 3759.

7. E.B. Stechel and D.R. Jennison: 1988, *Phys. Rev. B* **38**; 1988, *Phys. Rev. B* **38**, 8873.
8. H. Kamimura: 1987, *Jpn. J. Appl. Phys.* **26**, L627; H. Kamimura, S. Matsuno and R. Saito, *Mech. High. Temp. Supercond.*, H. Kamimura and A. Oshiyama (eds.), Springer, 1989, Series Mat. Sci. 11, p. 8.
9. C.M. Varma, P.B. Littlewood, S. Schmitt-Rink, E. Abrahams and A.E. Ruckestein: 1989, *Phys. Rev. Lett.* **63**, 1996.
10. P.W. Anderson: 1990, *Phys. Rev. Lett.* **64**, 1839.
11. M.S. Hybertsen, M. Schlüter and N.E. Christensen: 1989, *Phys. Rev. B* **39**, 9028; Reference [1], p. 1217.
12. M.S. Hybertsen, E.B. Stechel, M. Schlüter and D.R. Jennison: 1990, *Phys. Rev. B* **41**, 11068; M. Schlüter and M.S. Hybertsen, Reference [2], p. 583; E.B. Stechel, in *Proceedings of High Temperature Superconductivity, Los Alamos, 1989*, K.S. Bedell, D. Coffey, D.F. Meltzer, D. Pines and J.R. Schrieffer (eds.), Addison-Wesley, New York, 1990, p. 377.
13. L.F. Mattheiss: 1987, *Phys. Rev. Lett.* **58**, 1028.
14. R.E. Cohen, W.E. Pickett and H. Krakauer: 1989, *Phys. Rev. Lett.* **62**, 831.
15. P.H. Dederichs, S. Blugel, R. Zeller and H. Akai: 1984, *Phys. Rev. Lett*, **53**, 2512.
16. J. Zaanen, G.A. Sawatzky and J.W. Allen: 1985, *Phys. Rev. Lett.* **55**, 418.
17. A.K. McMahan, R.M. Martin and S. Satpathy: 1988, *Phys. Rev. B* **38**, 6650; A. K. McMahan, J.F. Annett and R.M. Martin: 1990, *Phys. Rev. B* **42**, 6268.
18. H. Eskes and G.A. Sawatzky: 1988, *Phys. Rev. Lett.* **61**, 1415; H. Eskes, G.A. Sawatzky and L.F. Feiner: 1989, *Physica C* **160**, 424; H. Eskes, L.H. Tjeng and G.A. Sawatzky: 1990, *Phys. Rev. B* **41**, 288.
19. M. Eto, R. Saito and H. Kamimura: 1989, *Sol. State Commun.* **71**, 425; H. Kamimura and M. Eto: 1990, *J. Phys. Soc. Japan* **59**, 3053.
20. D. Vankin, S.K. Sinha, D.E. Moncton, D.C. Johnston, J.M. Newsam, C.R. Safinya and H.E. King Jr.: 1987, *Phys. Rev. Lett.* **58**, 2802.
21. J.D. Reger and A.P. Young: 1988, *Phys. Rev. B* **37**, 5978.
22. A.J. Arko, R.S. List, R.J. Bartlett, S.-W. Cheong, Z. Fisk, J.D. Thompson, C.G. Olson, A.-B. Yang, R. Liu, C. Gu, B.W. Veal, J.Z. Liu, A.P. Paulikas, K. Vandervoort, H. Claus, J.C. Campuzano, J.E. Schirber and N.D. Shinn: 1989, *Phys. Rev. B* **40**, 2268.
23. J.A. Yarmoff, D.R. Clarke, W. Drube, U.O. Karlsson, A. Taleb-Ibrahimi and F.J. Himpsel: 1987, *Phys. Rev. B* **36**, 3967; N. Nücker, J. Fink, C. Fuggle, P.J. Durham and W.M. Temmerman: 1988, *Phys. Rev. B* **37**, 5158.
24. E.B. Stechel, M.S. Hybertsen and M. Schlüter, unpublished.
25. J. Humlicek, M. Garriga and M. Cardona: 1988, *Sol. State Commun.* **7**, 589.
26. R.R.P. Singh, P.A. Fleury, V.B. Lyons and P.E. Sulewsky: 1989, *Phys. Rev. Lett.* **62**, 2736.
27. G. Aeppli, S.M. Hayden, H.A. Mook, Z. Fisk, S.-W. Cheong, D. Rytz, J.P. Remeika, G.P. Espinosa and A.S. Cooper: 1989, *Phys. Rev. Lett.* **62**, 2052.
28. S.L. Cooper, G.A. Thomas, D.H. Rapkine, A.J. Millis, S.-W. Cheong, A.S. Cooper and Z. Fisk: 1990, *Phys. Rev. B* **41**, 11605.
29. C.T. Chen, F. Sette, Y. Ma, M.S. Hybertsen, E.B. Stechel, W.M.C. Foulkes, M. Schlüter, S.-W. Cheong, A.S. Cooper, L.W. Rupp Jr., B. Batlogg, Y.L. Soo, Z.H. Ming, A. Krol and Y.H. Kao: 1991, *Phys. Rev. Lett.* **66**, 104.
30. P. Kuiper, G. Kruizinga, J. Ghijsen, G.A. Sawatzky and H. Verweij: 1989, *Phys. Rev. Lett.* **62**, 221; H. Eskes and G.A. Sawatzky: 1991, *Phys. Rev. B* **43**, 119.
31. A.B. Harris and R.V. Lange: 1967, *Phys. Rev.* **157**, 295.

32. H. Romberg, M. Alexander, N. Nücker, P. Adelmann and J. Fink: 1990, *Phys. Rev. B* **42**, 8768.
33. J.W. Allen, C.G. Olson, M.B. Maple, J.-S. Kang, L.Z. Liu, J.-H. Park, R.O. Anderson, W.P. Ellis, J.T. Markert, Y. Dalichaouch and R. Liu: 1990, *Phys. Rev. Lett* **64**, 595.

EXACT DYNAMICS OF HIGHLY CORRELATED ELECTRONS IN TWO DIMENSIONS

Y. KURAMOTO
Department of Applied Physics, Tohoku University, Sendai 980, Japan.

Abstract. This paper discusses dynamical property of highly correlated electrons in two dimensions on the basis of exact numerical results. The starting model is a multiband Hubbard-type model for cuprates which show high-temperature superconductivity. In the case of the Mott insulating state, an effective spin Hamiltonian is derived which involves a large 4-spin interaction. It is shown that the spectrum of the magnetic Raman scattering is much influenced by the 4-spin interaction. In the case of the metallic state with doped carriers, interaction between charge and spin degrees of freedom is almost absent if the transfer occurs only between the next-nearest neighbor sites and is smaller than the exchange interaction.

1. Introduction

Since the high T_c superconductivity in cuprates appears in the presence of strong electronic correlations, much effort has been directed to identifying the most essential element for its occurrence [1]. Despite the very intensive studies, however, no definite conclusion has so far been drawn on the mechanism. One of the reasons for the difficulty lies in the absence of a reliable theory for strong correlation effects in a macroscopic scale of the system. On the other hand, correlation effects in a single atom or a molecule have been understood much better by various quantum-chemical methods.

Evidently correlation effects in a macroscopic system have some features in common with those which appear already in atoms and molecules. Thus it seems that investigation of the size dependence of the correlation effect is useful to analyze step by step the more delicate effect in the condensed matter.

In the present paper we adopt an exact numerical method for small clusters, and discuss the dynamical property of the two-dimensional Mott insulating state as a model of the CuO_2 plane. The model we use for the Mott insulating state is not the conventional Heisenberg model, but includes a 4-spin exchange. It has been shown in [2] that for accurate description of the CuO_2 plane a considerable size of a 4-spin interaction term should be necessary. We discuss in detail how the effective spin model is derived and why there is so large 4-spin exchange. As a typical example of the dynamical property of the Mott insulating state we present exact numerical results [3] for Raman scattering, and show that the 4-spin interaction has a strong influence on the spectral line shape.

We also discuss how the doped carrier behaves and changes the spin dynamics with use of a simplified model. In particular we pay attention to

H. Aoki et al. (eds.), New Horizons in Low-Dimensional Electron Systems, 247–260.

possible separation of spin and charge degrees of freedom. The arguments of this paper are mostly based on our recent results which have been obtained in cooperation with H.J. Schmidt, Y. Honda and T. Watanabe at Tohoku University.

2. The Multiband Model and Its Low-Energy States

We consider a multiband Hubbard-type model with short-range Coulomb and exchange interactions as the starting effective Hamiltonian for the two-dimensional Cu–O lattice. Since we are mainly concerned with the low-energy properties of the Mott insulating state as in La_2CuO_4 we keep only those orbitals which constitute the covalent bond between Cu and O, namely the $x^2 - y^2$ orbital of Cu $3d$ and the $2p_\sigma$ orbital of O. We remark that according to band structure calculations [4] the $3z^2 - r^2$ orbital of Cu $3d$ is also populated by holes if Sr is sufficiently doped in La_2CuO_4. It has been discussed by Kamimura [5], Weber [6] and others that the $3z^2 - r^2$ orbital plays an essential role for the superconductivity. In the undoped case, however, we consider that the explicit account of the $3z^2 - r^2$ orbital is not necessary.

The multiband model in the hole picture is then given by

$$\mathcal{H} = \sum_{ij} [t_{ij}(\sum_{\sigma} c_{i\sigma}^{\dagger} c_{j\sigma}) + \frac{1}{2} U_{ij}\, n_i\, n_j + K_{ij}(\mathbf{S}_i \cdot \mathbf{S}_j + \frac{1}{4} n_i\, n_j)] \,, \quad (2.1)$$

where indices i and j denote either Cu or O sites and $c_{i\sigma}^{\dagger}$ and $c_{j\sigma}$ denote creation and annihilation operators of a hole with spin σ. We write the number operator as n_i and the spin operator as $\mathbf{S}_i$. The site diagonal part of t_{ij} is written as ε_d at the Cu site and ε_p at the O site. Then the charge transfer energy Δ is given by $\varepsilon_p - \varepsilon_d$. For the off-diagonal part we keep only the nearest neighbor Cu–O and O–O transfers which are written as $-t_{pd}$ and $-t_p$, respectively. Here we take a modified phase convention so that the bottom of the hole band in the absence of U_{ij} and K_{ij} is shifted to the origin of the Brillouin zone. The non-zero elements of the Coulomb repulsion U_{ij} are U_d at Cu and U_p at O sites, and U_{pd} for the nearest-neighbor Cu–O intersite interaction. Finally for K_{ij} we introduce $K_{pd}(< 0)$ as the ferromagnetic exchange between the nearest-neighbor p and d holes. Other exchanges are not included in the Hamiltonian but are generated from the present model. The final term $(1/4)\,n_i n_j$ in Equation (2.1) is included for comparison with the literature [7–9]. We follow [7,8] for the numerical values of the parameters and list them in Table I.

In a small system the nature of the state depends on the boundary condition. If one takes the 2×2 square Bravais lattice, for example, the periodic boundary condition counts the nearest-neighbor interaction twice, first

TABLE I
Numerical values of parameters (eV).

Δ	t_{pd}	t_p	U_d	U_p	U_{pd}	K_{pd}
3.6	1.3	0.65	10.5	4.0	1.2	−0.18

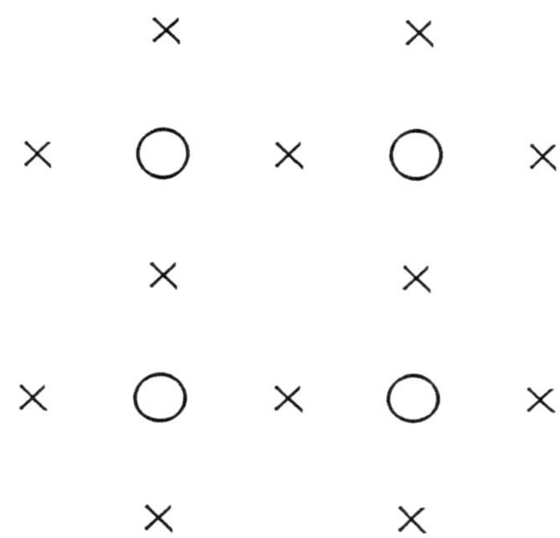

Fig. 1. The geometry of the Cu_4O_{12} cluster. The circle denotes a Cu site and the cross an O site.

inside the system and second by imposing the periodic image. Then the effective interaction strength becomes larger than in the macroscopic system. On the other hand, if one takes the free boundary condition there is no such double counting. However, in the latter case one cannot recover the energy band in the weak correlation limit. Thus it seems reasonable to take the free boundary condition in constructing an effective Hamiltonian which is characterized by tight-binding parameters, and then take the periodic boundary condition in calculating the response properties. We take a Cu_4O_{12} cluster as shown in Figure 1 with the free boundary condition. In order to check the sensitivity of the results on the geometry of the cluster, we have also taken a Cu_4O_8 cluster with the periodic boundary condition [2]. Related calculation has been done in [9] for a Cu_5O_{16} cluster with a more restricted set of basis functions.

The ground-state energy and the wavefunction are derived with the use of the modified Lanczos algorithm [10]. In this algorithm one improves the starting wavefunction by successive operation of the Hamiltonian. In our version we start from a normalized initial wavefunction $|1\rangle$ which is consistent with the symmetry of the Hamiltonian. Then another normalized wavefunction $|2\rangle$ orthogonal to $|1\rangle$ is constructed by

TABLE II
Energy levels and their fit by spin models.

	E_N	E_D	E_M	E_P
numerical	0.086	0.227	0.237	0.369
Heisenberg ($J = 0.086,\ J' = -0.027$)	0.086	0.227	0.172	0.258
4-spin ($J = 0.133,\ J' = 6 \times 10^{-4},\ J_c = 0.037$)	0.86	0.227	0.237	0.372

$$\mathcal{H}|1\rangle = E_1|1\rangle + V|2\rangle , \tag{2.2}$$

where $E_1 = \langle 1|\mathcal{H}|1\rangle$ and V is a c-number. We define $E_2 = \langle 2|\mathcal{H}|2\rangle$ and diagonalize the 2×2 matrix composed of E_1, E_2, V and V^*. Then the eigenfunction $|\tilde{1}\rangle$ corresponding to the lower eigenvalue constitutes an improved wavefunction over the initial one. It is evident that $|\tilde{1}\rangle$ has the same quantum number as $|1\rangle$. By repeating this procedure we can derive the lowest energy and the wavefunction for each set of quantum numbers.

3. Spin Dynamics in the Mott Insulating State

Let us first concentrate on the Mott insulating state where each Cu site is singly populated by a hole and every O site is empty. If the excitation energies are much smaller than Δ, the charge degrees of freedom need not be treated explicitly. Then a spin model with spin 1/2 at each Cu site emerges as the effective Hamiltonian.

The spin model in the 2×2 Cu lattice has five distinct energy levels [11]. The eigenfunctions are independent of the parameters in the spin Hamiltonian and are classified by the spatial symmetry and the total spin. We write E_R for the energy of the resonating valence bond (RVB) state with spin zero, E_N for the Néel-type state with spin 1, E_M for the other (doubly degenerate) spin 1 states, E_D for the singlet state with singlet bonds along the diagonals, and E_P for the polarized state with spin 2. These energies depend on the parameters in the effective spin model. Thus by comparing the energy levels with the corresponding ones in the model given by Equation (2.1) the parameters can be fixed.

Table II shows the energy levels (in units of eV) of the Cu_4O_{12} cluster with 4 holes. The free boundary condition is employed. The ground state corresponds to the RVB state whose energy is taken to be zero. If one adopts the Heisenberg model with the nearest-neighbor exchange J and the next-nearest one J', E_N and E_D are sufficient to fix J and J'. But the remaining levels E_M and E_P, which are predicted by the Heisenberg model, are not consistent with the numerical results as shown in Table II.

In order to improve the fit we have included the 4-spin term given by

$$\mathcal{H}_4 = J_c[(\mathbf{S}_1 \cdot \mathbf{S}_2)(\mathbf{S}_3 \cdot \mathbf{S}_4) + (\mathbf{S}_1 \cdot \mathbf{S}_4)(\mathbf{S}_2 \cdot \mathbf{S}_3) - (\mathbf{S}_1 \cdot \mathbf{S}_3)(\mathbf{S}_2 \cdot \mathbf{S}_4)] , \quad (3.1)$$

where the pair of sites 1, 3 or 2, 4 are along the diagonals of the square. $\mathcal{H}_4$ describes the cyclic exchange along the elementary plaquette [2]. The energy levels in the improved model are given by

$$\begin{pmatrix} 1 & -2 & 1/2 & 7/8 \\ 1 & -1 & 1/2 & -3/8 \\ 1 & 0 & -3/2 & -1/8 \\ 1 & 0 & 1/2 & 1/6 \end{pmatrix} \begin{pmatrix} E_0 \\ J \\ J' \\ J_c \end{pmatrix} = \begin{pmatrix} E_R \\ E_N \\ E_D \\ E_M \end{pmatrix} , \quad (3.2)$$

where E_0 is the origin of the energy in the spin model. The 4-spin exchange J_c thus determined is about 1/4 of J. As shown in Table II the remaining level E_P calculated in the spin model agrees well with that in the model of Equation (2.1). Thus the effective spin model turns out to be quite accurate for low-energy excitations below the charge-transfer gap provided that one includes the 4-spin exchange.

The magnitudes of exchange constants J and J' are consistent with those by Hybertsen *et al.* [9] who investigated various sizes of clusters up to Cu_5O_{16}. Therefore the small J' is not the artifact of the particular shape of the Cu_4O_{12} cluster taken here. The shape of the Cu_5O_{16} cluster taken in [9], however, does not contain the smallest plaquette of Cu_4. This is probably the reason why the 4-spin interaction is missed in [9]. As a possibility of other significant exchange interactions we first remark that in the square lattice there are no cyclic exchange interactions involving an odd number of sites [12]. Secondly, it has been shown in [2] that the next simplest cyclic exchange involving the diagonal path is negligibly small. In the present size of the cluster, however, we cannot evaluate the 6-spin [13] and more complex exchange interactions.

These exchange interactions are generated by combined action of t_{ij}, U_{ij} and K_{ij} terms in Equation (2.1). In the lowest-order perturbation theory with respect to the small parameter $t_{pd}/\Delta \equiv \alpha$ the nearest-neighbor exchange is given by

$$J = 4\left(\frac{t_{pd}^2}{\Delta + U_{pd} + K_{pd}}\right)^2 \left(\frac{1}{U_d} + \frac{2}{U_p + 2\Delta} + \frac{K_{pd}}{t_{pd}^2}\right) , \quad (3.3)$$

which is 0.09 eV with the parameters in Table I. Although Equation (3.3) is not accurate quantitatively, it shows clearly how, for example, U_{pd} and K_{pd} influence J. Namely the main effect of U_{pd} is to increase the effective charge transfer energy, and that of K_{pd} is to induce the ferromagnetic coupling between Cu spins via hybridization. Both of these act to reduce J.

A natural question is why the 4-spin interaction is so large in the present system. The main reason is the small charge transfer gap as compared with other transition-metal oxides. The small Δ and the strong covalency due to large t_{pd} lead to a larger α. We notice in this connection that the large 4-spin interaction as suggested from the magnetic structure in NiS_2 [14] should also be due to the small charge transfer gap between the Ni $3d$ and the S $3p$ orbitals. Although the perturbative order of the 4-spin interaction is higher than J by O (α^4) with neglect of K_{pd}, J_c can remain significant due to the moderate magnitude of $\alpha = 0.36$ and due to many possible paths of exchange which lead to a large numerical factor [12]. For a very rough estimate of the magnitude of the 6-spin interaction, we suggest an extrapolation from the ratio between J and J_c. Then the 6-spin interaction may be of the order of 10 meV.

4. Raman Spectrum and the Effect of 4-Spin Interactions

The dominant interaction responsible for the Raman scattering in antiferromagnetic insulators has been identified as such that involves the exchange interaction in an excited orbital and a final exchange of two neighboring spins [15]. The interaction is written as

$$\mathcal{H}_R = \tfrac{1}{2} C_R \sum_{ij} (\mathbf{E}_{in} \cdot \mathbf{d}_{ij})(\mathbf{E}_{sc} \cdot \mathbf{d}_{ij}) \mathbf{S}_i \cdot \mathbf{S}_j \ , \tag{4.1}$$

where $\mathbf{E}_{in}$ and $\mathbf{E}_{sc}$ are the electric fields of incident and scattered lights, respectively, and $\mathbf{d}_{ij}$ is the vector connecting the nearest-neighbor sites i and j. The constant C_R involves the matrix element of a virtual charge-transfer excitation from the ground-state orbital with a $3d$ character to an excited orbital with a $2p_\sigma$ character.

If the incident photon has an energy close to the charge-transfer excitation, C_R is resonantly enhanced. This is actually the situation in the cuprates where several experiments have been performed [16–18]. The immediate question then is whether other exchange processes involving the excited orbitals are relevant to the Raman scattering. For example the analogy of the 4-spin interaction in the ground state may also appear as a virtual process. Unfortunately very little is known on the magnitude of this interaction at present. In this paper, therefore, we would study the effect of the 4-spin interaction on the Raman scattering confining ourselves to the conventional Raman interaction given by Equation (4.1).

In the excitation energy range below the charge-transfer gap, it is reasonable to work with the effective spin Hamiltonian $\mathcal{H}_S$ with the nearest-neighbor exchange J and the 4-spin exchange J_c. We take a 4×4 square lattice and employ the periodic boundary condition. The ground state $|g\rangle$ of the system and its energy E_0 are derived by the modified Lanzcos algorithm as explained in Section 2. Then we define an unnormalized state vector

$$|0) = \mathcal{H}_R |g\rangle - |g\rangle \langle g|\mathcal{H}_R|g\rangle \ , \tag{4.2}$$

which is orthogonal to $|g\rangle$. The spectral function $S_R(\omega)$ of the Raman scattering is obtained in terms of a resolvent by

$$S_R(\omega) = -(\pi N)^{-1} \ \text{Im} \ (0|(\omega - \mathcal{H}_S + E_0 + i\gamma)^{-1}|0) \ , \tag{4.3}$$

where N is the number of lattice sites and a small positive number γ is put in order to make the spectrum smooth. We employ the continued fraction representation of the resolvent [19] in deriving the Raman spectrum. Starting from $|0)$ we successively generate orthogonal state vectors as follows:

$$|i+1) = \mathcal{H}_S\,|i) - a_i|i) - b_i|i-1) \ , \tag{4.4}$$

where $a_i = (i|\mathcal{H}_S|i)\,/\,(i|i)$, and $b_i = (i-1|\mathcal{H}_S|i)\,/\,(i-1|i-1)$. Then the spectral function is represented in the form of a continued-fraction:

$$S_R(\omega) = -\frac{1}{\pi} \ \text{Im} \ \frac{(0|0)}{z - a_0 -} \ \frac{b_1}{z - a_1 -} \ \frac{b_2}{z - a_2 -} \ \cdots \tag{4.5}$$

where $z = \omega - E_0 + i\gamma$. The continued fraction is truncated after 100 iterations where good convergence of the spectrum is observed [3].

Let us define the average frequency M_1 of the spectrum by

$$M_1 = \int_0^\infty \omega\, S_R(\omega)\, \mathrm{d}\omega \Big/ \int_0^\infty S_R(\omega)\, \mathrm{d}\omega \equiv \langle\omega\rangle \ , \tag{4.6}$$

and the n-th moment M_n for $n \geq 2$ by $M_n^n = \langle(\omega - M_1)^n\rangle$. Som lower-order moments with infinitesimal γ are simply given in terms of a_i and b_i by

$$M_1 = a_0 - E_0 \ , \quad M_2^2 = b_1 \ , \quad M_3^3 = b_1(a_1 - a_0) \ . \tag{4.7}$$

We consider two different configurations for the scattering: (1) $\mathbf{E}_{in} \,//\, (1,1)$ and $\mathbf{E}_{sc} \,//\, (-1,1)$; (2) $\mathbf{E}_{in} \,//\mathbf{E}_{sc} \,//\, (1,1)$. The configurations (1) and (2) probe excited states with B_{1g} and A_{1g} symmetries, respectively. In the case of the Heisenberg model, the A_{1g} scattering is absent because $\mathcal{H}_R$ becomes proportional to the Hamiltonian of the system and then $|0)$ vanishes. This is no longer the case when the 4-spin interaction is present.

Figures 2 and 3 show the results for the B_{1g} scattering [3]. There are only a few excited states which have significant intensities. The scale for $S_R(\omega)$ is taken for the case where all of C_R, $|\mathbf{E}_{in}|$, $|\mathbf{E}_{sc}|$, $|\mathbf{d}_{ij}|$ are of unit magnitude. We note that only such B_{1g} states with spin 0 and momentum 0 contribute to $S_R(\omega)$. This is because the ground state of the effective spin model is a singlet in the 4×4 cluster, and because the Raman interaction conserves the total spin and momentum of the system. In the macroscopic system with the Néel order the selection rule is less stringent since the ground state does ot have a definite total spin because of the broken symmetry.

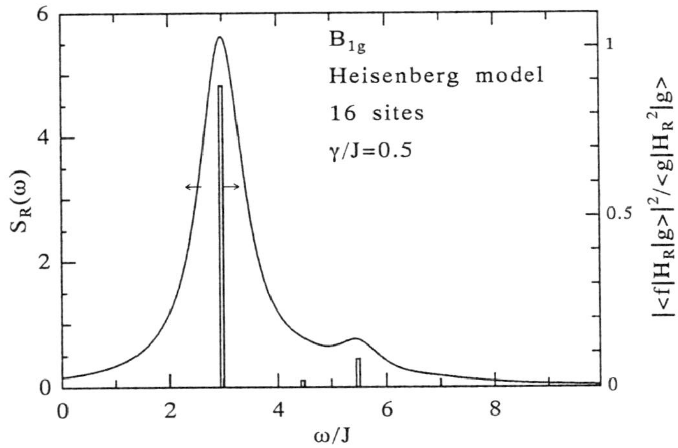

Fig. 2. The Raman spectrum for the Heisenberg model. The smooth curve is the result with a broadening factor $\gamma = 0.5J$. The scale to the right shows the relative intensity for each final state $|f\rangle$.

TABLE III
Spectral moments in the Heisenberg model and in the model with J_c.

	M_1/J	M_2/J	M_3/J
Heisenberg ($J_c = 0$)	3.24	0.80	1.14
4-spin ($J_c = J/4$)	3.08	0.87	1.16

In the absence of J_c as shown in Figure 2, related results consistent with ours have already been reported in [20,21]. With inclusion of J_c of $J/4$ as shown in Figure 3 the following changes are visible: First, the largest peak is shifted downwards. Secondly the peak at $\omega/J \sim 4.4$ is much enhanced and makes an appreciable shoulder in the broadened spectrum. These changes naturally appear in the spectral moments as summarized in Table III.

The largest peak in $S_R(\omega)$ is interpreted as due to a 2-magnon bound state. In the classical picture the state consists of a nearest-neighbor pair of reversed spins with respect to the Néel background. The number of broken exchange bonds is 6 which leads to $3J$ as the excitation energy. By analogy a 4-magnon bound state is pictured as such that consists of a smallest plaquette of reversed spins in the Néel background. The number 8 of broken bonds leads to $4J$ as the excitation energy, which is close to the second largest peak in Figure 3. In order to probe the 4-magnon bound state in the Raman scattering with Equation (4.1) as the interaction, there should be a configuration corresponding to the two-magnon bound state in the ground state. Thus a

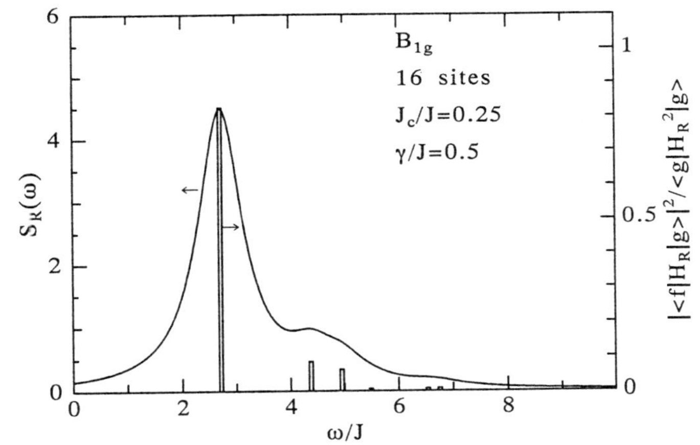

Fig. 3. The Raman spectrum with $J_c = 1/4$.

large quantum fluctuations is necessary for the 4-magnon bound state to be excited. We are currently analyzing the wavefunction of excited states if the simple picture described above is a relevant interpretation.

The intensity of the A_{1g} scattering activated by the 4-spin interaction is calculated to be smaller than the B_{1g} scattering by more than an order of magnitude. Thus the strong A_{1g} scattering observed experimentally [16,17] needs other sources for the activation such as the magnon-phonon interaction.

Let us briefly compare the present results with the results by Singh *et al.* [22]. In the method of [22] the moments of the spectrum are calculated by a perturbation theory from the Ising limit for a macroscopic system. They report $M_1 = 3.6J$, $M_2 = 0.8J$ and $M_3 = 1.0J$ for the Heisenberg model. The values for M_2 and M_3 are not far from the ones in Table III, but M_1 of [22] is much larger than our result. The most probable origin of the discrepancy is that we take a small cluster of 16 sites. It is necessary for a meaningful comparison to study the dependence of the result on the size of the cluster and to extrapolate to the macroscopic size. Preliminary study in this direction has been done and it is shown that the large discrepancy in M_1 is rather specific to the 4×4 lattice. More detailed discussion on this point will be reported elsewhere [3].

Experimentally Sulewski *et al.* [16] report that their Raman spectrum at room temperature is in good agreement with the theory of Singh. However the spectrum by Sugai *et al.* [17] at 30 K shows a distinct shoulder, which is not present in the data of [16], in the high energy side of the main peak. Sugai *et al.* ascribe the shoulder to the 4-magnon bound state. Whether there is indeed a large 4-spin exchange or not should be tested after careful inspection of the size dependence of the spectrum in the exact diagonalization method.

5. Charge and Spin Dynamics in the Doped State

Let us consider now the case where the number of holes is not the same as the number of Cu sites. We distinguish between the hole-doped case and the electron-doped one according to the direction of deviation from the Mott insulating state. In the doped case it is not obvious whether the charge degrees of freedom can be conveniently described by an effective Hamiltonian with a single energy band. Such description itself should be possible provided that a Fermi liquid state is realized and one is interested in the low-energy limit. Interband transitions have a finite threshold energy and their effect is put into renormalization of the parameters in the model. What is not obvious is the upper limit of the energy range to be described by a single-band model. If it is extremely small there is not much simplification to work with the single-band model.

From this standpoint Schmidt and Kuramoto [23] derived low-lying energy levels of the Cu_4O_{12} cluster with 3 and 5 holes by the modified Lanzcos algorithm. Each wavefunction corresponding to these energy levels has a counterpart with the same symmetry in the 2×2 lattice of Cu. The energy levels of the latter can be expressed analytically in terms of the parameters of the single-band model. Thus by comparing the energy levels of two models one can judge the valid range of the single-band model. As we have remarked in Section 2, however, we have to keep it in mind that high doping of carriers may make the starting model given by Equation (2.1) unrealistic since some carriers may enter the $3z^2 - r^2$ orbital.

It has been found in [23] that in the electron-doped case the so-called t–t'–J model with the nearest-neighbor transfer t and the next-nearest-neighbor transfer t' provides a fairly good fit to the original energy levels. However in the hole-doped case, the fit of the lowest four levels in terms of the t–t'–J model leads to a negative J. Moreover the higher levels deduced by the model do not agree at all with the original ones. It is necessary to introduce a spin-flip hopping term involving three sites for a satisfactory fit. The difference between the electron- and hole-doped cases can be interpreted in terms of the different radius of the local-singlet wavefunction [24]. More details of the fitting procedure are reported in [23].

If one adopts a single-band model the size of the cluster can be made much larger than 2×2. The largest cluster studied by the exact diagonalization in the literature consists of 20 sites [25]. It has been found [25] that the effective nearest-neighbor transfer of a hole is much suppressed by the antiferromagnetic correlation. This is easily understood because this transfer occurs between different sublattices with opposite spin directions. On the contrary the next-nearest-neighbor transfer occurs between the same sublattice. Therefore it is interesting to study whether the antiferromagnetic correlation has an important influence on the latter effective transfer.

Let us take the following single-band model written in terms of the $3d$-hole operators:

$$\mathcal{H}_{eff} = \sum_{ij\sigma} t_{ij}\, d^{\dagger}_{i\sigma}\, d_{j\sigma} + J \sum_{<ij>} \mathbf{S}_i \cdot \mathbf{S}_j \ , \tag{5.1}$$

where the second sum goes over the nearest-neighbor pair of sites. For the transfer t_{ij} we keep up to the third neighbor and use the notation $-t_1$, $-t_2$ and $-t_3$ in the order of increasing distance. We consider a situation where the one hole is removed from the Mott insulating state in the 4×4 cluster with the periodic boundary condition. Then each eigenstate is characterized by a crystal momentum of the total system. With the use of the modified Lanzcos algorithm we can derive the lowest energy and the wavefunction for each momentum and the total spin. In this paper we present recent results by Honda *et al.* [26] who consider the extreme situation with $t_1 = 0$.

Figure 4 shows the momentum dependence of the ground-state energy in the following two cases: $t_2 = 0.1J$, $t_3 = 0$ (circle); and $t_2 = t_3 = 0.1J$ (cross). These energies are remarkably well fitted by the equation:

$$E(k_x, k_y) = 4t_2 \ \cos\, k_x \ \cos\, k_y + 2t_3[\cos(2k_x) + \cos(2k_y)] - 16.88J \ , \tag{5.2}$$

where the lattice constant is taken to be unity and $-16.88J$ is the ground-state energy in the case of $t_2 = t_3 = 0$.

The good fit means that the change of the total energy is almost the same as the negative of the kinetic energy of the removed hole. This should only be possible if the spin background is hardly interrupted by the removed hole, i.e. the spin and charge degrees of freedom are almost separated dynamically. This separation would not be surprising if the spins were Néel ordered and if there were no quantum fluctuations. In the finite cluster, however, neither of these conditions are satisfied. We have thus increased the size of the transfer and investigated if the spin-charge separation persists. Figure 5 shows the results for the ground-state energies in the case of $t_2 = J$ and $t_3 = 0$. There is still a good agreement with the fit of Equation (5.2) at $(k_x, k_y) = (\pi, 0)$, but the dependence on momentum is much weaker than that of Equation (5.2) at other points in the Brillouin zone.

Thus with large transfer the separation of spin and charge does not happen in the present model. This is because the total energy of the system is lowered at the expense of spin excitations. The result shown in Figure 5 is interpreted in terms of a quasi-particle whose mass is enhanced to about twice of the bare value by interaction with the spin polarization cloud. It

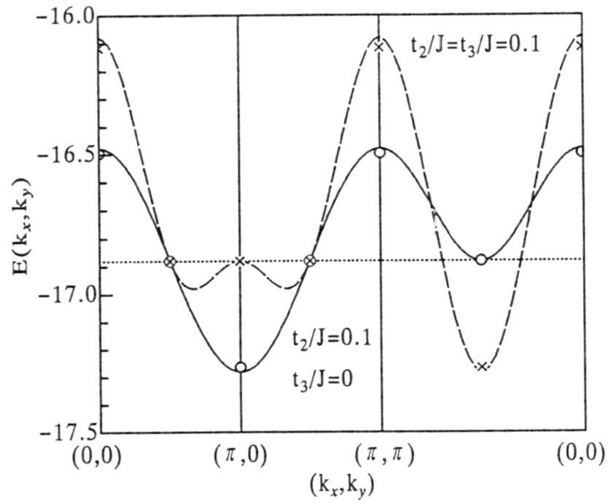

Fig. 4. Momentum dependence of the ground-state energy. The fit by Equation (5.2) is shown by the solid line for $t_2 = 0.1J$, $t_3 = 0$ and by the dashed line for $t_2 = t_3 = 0.1J$.

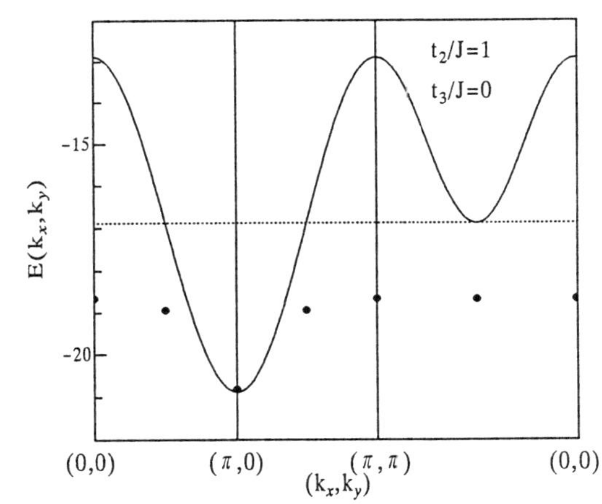

Fig. 5. Momentum dependence of the ground-state energy for the case of $t_2 = J$ and $t_3 = 0$. The black circle shows the numerical result and the line shows Equation (5.2).

should be interesting to study to what extent the gauge-theoretical formulation of the model [27] is useful to interpret our exact results on a small cluster.

6. Conclusions

Let us summarize the main points of the paper. With the ultimate aim of understanding the correlation effect in macroscopic systems we first take the multiband Hubbard-type model on a Cu_4O_{12} cluster and derived the low-energy levels by the exact diagonalization method. The structure of low-energy levels makes it possible to construct an effective Hamiltonian with fewer degrees of freedom. It is hoped that the effective Hamiltonian is more convenient to analyze the low-energy property of the system. In our exact numerical method the lower number of degrees of freedom permits a larger size of a cluster to be studied.

In the undoped case there is a significant correction to the Heisenberg model in the form of a 4-spin exchange interaction. We have derived the Raman spectrum and discussed how the 4-spin exchange influences the spectrum. However, for comparison with experimental results we have to analyze the size dependence of the result in more detail. It is expected that the 4-spin exchange should have important consequences not only in Raman scattering but in various static and dynamic properties of cuprates. For example it may be necessary to take the 4-spin exchange into account for an improved analysis of the dynamical spin susceptibility which has been measured by the neutron scattering [1].

In the doped case, we have discussed a particular situation where the transfer occurs only inside the same sublattice. It is found that the separation of spin and charge occurs if the transfer is much smaller than the nearest-neighbor exchange. If the transfer is about the same as the exchange there is a significant interaction between spin and charge degrees of freedom and, as a result, the effective mass of the quasi-particle is enhanced over the bare value.

Acknowledgements

It is a great pleasure of the author to dedicate this paper to Professor H. Kamimura in commemoration of his retirement from the University of Tokyo. The author is also grateful to his colleagues at Tohoku University, particularly Dr. H.J. Schmidt who is now at Cologne, Mr. Y. Honda and Professor T. Watanabe for stimulating cooperation. This work is supported by a Grant-In-Aid for Scientific Research from the Ministry of Education, Science and Culture.

References

1. For a recent review: *High Temperature Superconductivity*, K.S. Bedell, D. Coffey, D.E. Meltzer, D. Pines, and J.R. Schrieffer (eds.), Addison-Wesley, Redwood City (1990).
2. H.J. Schmidt and Y. Kuramoto: 1990, *Physica C* **167**, 263.
3. Y. Honda, Y. Kuramoto and T. Watanabe: 1991, to be published.
4. For example: N. Shima, K. Shiraishi, T. Nakayama, A. Oshiyama and H. Kamimura: 1989, *Proc. JSAP-MRS Int'l. Conf. on Elect. Mats.*, p. 51.
5. H. Kamimura and M. Eto: 1990, *J. Phys. Soc. Jpn.* **59**, 3053.
6. W. Weber: 1988, *Z. Phys. B* **70**, 323.
7. E.B. Stechel and D.R. Jennison: 1988, *Phys. Rev. B* **38**, 4632.
8. M.S. Hybertsen, M. Schlüter and N.E. Christensen: 1989, *Phys. Rev. B* **39**, 9028.
9. M.S. Hybertsen, E.B. Stechel, M. Schlüter and D.R. Jennison: 1990, *Phys. Rev. B* **42**, 11068.
10. E. Dagotto and A. Moreo: 1985, *Phys. Rev. D* **31**, 865.
11. Y. Kuramoto and H.J. Schmidt: 1989, *Springer-Series in Solid State Sciences*, Vol. 89, p. 88.
12. M. Takahashi: 1977, *J. Phys. C* **10**, 1289.
13. M. Imada: 1990, *Prog. Theor. Phys. Supplement* **101**, 391.
14. A. Yoshimori and S. Inagaki: 1978, *J. Phys. Soc. Jpn.* **44**, 101.
15. T. Moriya: 1967, *J. Phys. Soc. Jpn.* **23**, 490.
16. P.E. Sulewski, P.A. Fleury, K.B. Lyons, S.-W. Cheong and Z. Fisk: 1990, *Phys. Rev. B* **41**, 225.
17. S. Sugai, M. Sato, T. Kobayashi, J. Akimitsu, T. Ito, H. Takagi, S. Uchida, S. Hosoya, T. Kajitani and T. Fukuda: 1990, *Phys. Rev. B* **42**, 1045.
18. Y. Tokura, T. Arima, S. Koshihara, T. Ido, S. Ishibashi, H. Takagi and S. Uchida: 1990, *Phys. Rev. B* **41**, 11657.
19. E.R. Gagliano and C.A. Balseiro: 1987, *Phys. Rev. Lett.* **59**, 2999.
20. E. Dagotto and D. Poilblanc: 1990, *Phys. Rev. B* **42**, 7940.
21. E. Gagliano and S. Bacci: 1990, *Phys. Rev. B* **42**, 8772.
22. R.R.P. Singh, P.A. Fleury, K.B. Lyons and P.E. Sulewski: 1989, *Phys. Rev. Lett.* **62**, 2736.
23. H.J. Schmidt and Y. Kuramoto: 1991, to be published; 1990, *Phys. Rev. B* **42**, 2562.
24. F.C. Zhang and T.M. Rice: 1988, *Phys. Rev. B* **27**, 2359; H. Matsukawa and H. Fukuyama: 1990, *J. Phys. Soc. Jpn.* **58**, 2845.
25. T. Itoh, M. Arai and T. Fujiwara: 1990, *Phys. Rev. B* **42**, 4834 and references therein.
26. Y. Honda, Y. Kuramoto and T. Watanabe: 1991, to be published.
27. P.A. Lee: 1989, *Phys. Rev. Lett.* **63**, 680.

TWO-BAND MODELS FOR SUPERCONDUCTIVITY

HIDEO AOKI
Department of Physics, University of Tokyo, Hongo, Tokyo 113, Japan.

Abstract. Two-band models for superconductivity are reviewed. In particular, a recent two-band Hubbard model for strongly correlated electron systems is analysed. We consider various cases (signs and magnitudes) for the intraband and interband interactions, exhausting possible situations in the two-band system, to look into the off-diagonal order (superconductivity) as well as diagonal (magnetic and CDW) orders. Among the remarkable results are:

(i) there exists a mechanism for the superconductivity when all the intra- and interband interactions are repulsive,

(ii) superconductivity in band B can be enhanced by band A even when the interband and intra-A-band interactions are repulsive, and

(iii) the interband-pairing superconductivity due to the interband attraction exists only in the presence of the intraband repulsion.

Curious relevance and similarities to the fractional quantum Hall system, another prototype of strongly correlated system, is also predicted for both single-band and two-band cases.

1. Introduction

The theory of multi-band models for superconductivity has a long history dating back to the early 1960's [1–3], just after the BCS theory was established. Within the phonon mechanism BCS formalism, Kondo [3] has considered a pairing Hamiltonian as a model for an *sd* system, in which the β band has a positive Δ along with a negative Δ for α band and a negative interband Δ. T_C is shown to increase in the presence of interband Δ. There are also studies of superconductivity in multi-valley semiconductors by Cohen [4] and, more recently, by Rasolt [5]. For MOS inversion layers Takada has considered a two-subband system to propose an acoustic-plasmon mechanism for superconductivity [6].

After the high T_C copper oxides were discovered by Bednorz and Müller [7], high T_C superconductivity has inspired a new enthusiasm towards unconventional superconductivity [8] so far exemplified by heavy-fermion or organic superconductors. After a seminal suggestion by Anderson [9] of an electron mechanism of high T_C superconductivity, the Hubbard model, a prototype of the strongly correlated electron system, has been the subject of intensive studies [10–12]. This also gave an impetus to serious investigations on the superconductivity in multi-band systems. However, the fascination of two-band superconductivity is still some way from receiving wide recognition.

H. Aoki et al. (eds.), New Horizons in Low-Dimensional Electron Systems, 261–280.

The multi-band model has suffered a rather loose nomenclature in compound materials. Since the electronic structure of oxides involves both copper $3d$ orbitals and oxygen $2p$ orbitals, the so-called '$d-p$' models are often invoked and are sometimes called two-band models. The two-band model in its original sense, however, should have two kinds of orbitals for each site. These can be either (orthogonal) molecular orbitals of different symmetries, as in the case of oxides, or orbitals belonging to different layers, etc. Moreover, it has become increasingly clear that evidence for superconductivity does not seem to arise in $d-p$ models unless we make the model artificial, like in the '$t-J$' model. The importance of two kinds of orbitals, which are envisaged to be a_{1_g} and b_{1_g} orbitals in the case of cuprates, has been stressed by Kamimura, and the idea has been extensively developed into a theory for the spin-polaron mechanism of high T_C superconductivity and normal-state properties by Kamimura and coworkers [13].

Superconductivity is closely related to the magnetism in strongly correlated systems. Two-band models are interesting in this respect as well, which also applies to organic materials. For instance, the occurrence of ferromagnetism in organic materials as modelled by single- and two-band systems has been an important question. In addition, continuing advances in material design with epitaxy and other techniques will enable us to realize tailor-made two-band systems, in which intraband and interband interactions are controlled.

Theoretically, important questions for two-band systems are

(i) Are there mechanisms for superconductivity that are specific to two-band systems?

(ii) Can novel states with symmetries involving the band index space arise?

Specifically, we can paraphrase the first question – can superconductivity arise from purely repulsive interactions? Muttalib and Emery [14] have addressed this problem using a two-band Luttinger model for spinless fermions as a model for fluctuating-valence, organic and heavy-electron systems. They have concluded that superconductivity can exist in the Luttinger model with only repulsive interactions in a restricted condition, in which spin and charge degrees of freedom decouple. More recently, Kuroki and Aoki [15] have proposed a different mechanism for superconductivity in a two-band Hubbard model for spin 1/2 fermions, which involves correlated motion of electrons in the two bands.

The second question above is another key to understanding the ground states of multi-band systems. For instance, we can show from a rotational symmetry in the band-index space that an off-diagonal long-range order (superconductivity) and diagonal long-range orders (charge-density wave or spin-density wave) are degenerate under a special condition, and that only when this degeneracy is lifted (by doping, for instance) do we have superconductivity.

In the present article we explore these properties of two-band models that have recently been proposed. We also point out the curious relevance and similarities of the Hubbard system to the fractional quantum Hall system, another prototype of strongly correlated system, for both single-band and two-band cases.

2. Superconductivity in a Purely Repulsive Two-Band Hubbard Model

We first describe a mechanism for superconductivity due to Kuroki and Aoki [15], in which superconductivity specific to two-band systems for spin 1/2 fermions arises for purely repulsive interactions, from both intuitive considerations and quantum Monte Carlo results.

A general form for the two-band Hubbard Hamiltonian as conceived by Aoki and Kamimura [16] for cuprates is given in the standard notation by

$$\begin{aligned} H = &- \sum_{\nu=\alpha,\beta} \sum_{\langle i,j \rangle} \sum_{\sigma} t^{\nu}(c_{i\sigma}^{\nu\dagger} c_{j\sigma}^{\nu} + \text{H.c.}) + \sum_{i,\mu,\nu} U_{\mu\nu}\, n_{i\uparrow}^{\mu}\, n_{i\downarrow}^{\nu} \\ &+ \sum_{i\sigma} (U_{\alpha\beta} - J)\, n_{i\sigma}^{\alpha}\, n_{i\sigma}^{\beta} \\ &- J \sum_{i\sigma} c_{i\sigma}^{\alpha\dagger}\, c_{i-\sigma}^{\alpha}\, c_{i-\sigma}^{\beta\dagger}\, c_{i\sigma}^{\beta} \,, \end{aligned}$$

where $\langle i,j \rangle$ are nearest-neighbour chemical units, α, β label orbitals, U the Coulomb repulsion and J the intra-atomic exchange. If we neglect exchange processes we have

$$\begin{aligned} H = &- \sum_{\nu=\alpha,\beta} \sum_{\langle i,j \rangle} \sum_{\sigma} t^{\nu}(c_{i\sigma}^{\nu\dagger} c_{j\sigma}^{\nu} + \text{H.c.}) + \sum_{i} (U n_{i\uparrow}^{\alpha}\, n_{i\downarrow}^{\alpha} + F n_{i\uparrow}^{\beta}\, n_{i\downarrow}^{\beta}) \\ &+ \sum_{i,\sigma} (W n_{i\sigma}^{\alpha}\, n_{i-\sigma}^{\beta} + V n_{i\sigma}^{\alpha}\, n_{i\sigma}^{\beta}) \,, \end{aligned}$$

where $U(F)$ is the repulsion within the $\alpha(\beta)$ orbital and $V(W)$ is the interorbital interaction for the parallel (antiparallel) spins. Usually we have $V \leq W \leq U$.

An intuitive picture (Figure 1) indicates that an effective attraction between the β electrons arises when

$$V > \frac{1}{2}(U - F) \,,$$

since two β electrons experience the interaction energy of at least $2V$ when separate, but only $(U + F)$ when they are on the same site with the α electrons depleted from the site at the same time.

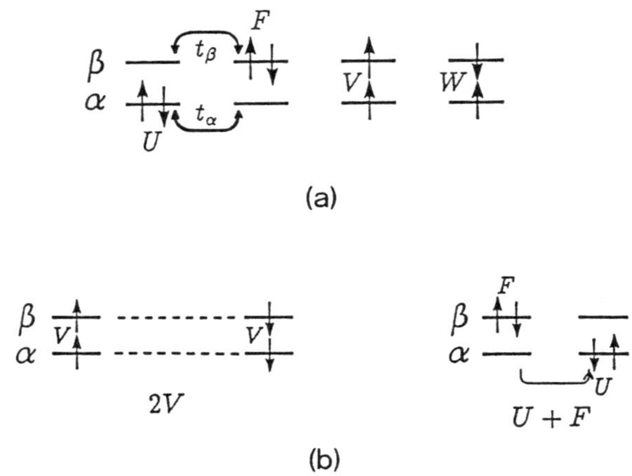

Fig. 1. (a) The interaction parameters in the two-band Hubbard model. (b) Two β electrons cost an interaction energy of at least $2V$ when separate, or $(U+F)$ when they are on the same site with the α electrons depleted from the site at the same time. (After Aoki and Kuroki [15].)

The emergence of superconductivity is indeed confirmed by an enhanced correlation of intraband pairing in the quantum Monte Carlo (QMC) result for finite 1D systems. To obtain the ground state, we have adopted the QMC method for ground states *à la* Sorella *et al.* [17,18] by extending the method described by Imada and Hatsugai [19] to the two-band model. The result for correlation functions in real space (Figure 2) shows the attraction in the density–density correlation and enhanced correlation of on-site pairing in the β band accompanied by suppressed antiferromagnetic correlation when the above criterion is met. The correlated motion of α and β electrons involved in the pairing mechanism is indeed seen in the interband density correlation, $\langle n_i^\alpha n_j^\beta \rangle$ (Figure 3). This confirms that the present mechanism is specific to two-band fermion systems.

If we include the effect of kinetic energies, the pairing condition must be relaxed to

$$2V - b > U + F - a$$

with $a, b \sim O(t)$. This is because the cost of energy to move an α electron off a site is not U but U^*, the energy required to dope the electron into the remaining $(N-1)$-electron band. In the Bethe-ansatz solution [20] for 1D single-band Hubbard model, $U - U^* = 1.2t$ for $U = 3$. The enhanced pairing when the relaxed condition is fulfilled is confirmed from the QMC results. However, it is too naïve to assume that the system can be precisely mimicked by the single-band *attractive* Hubbard model with an effective $U_{eff} = (2V -$

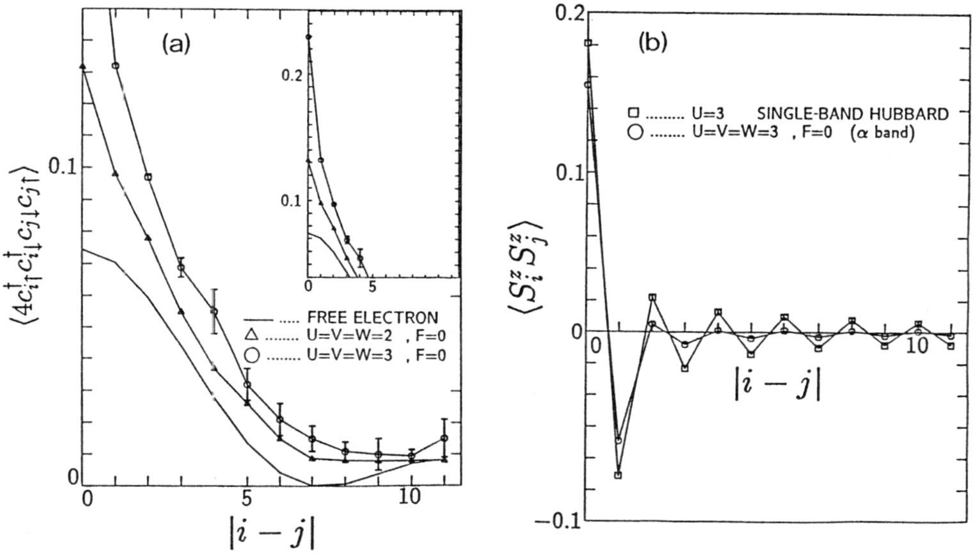

Fig. 2. Quantum Monte Carlo result for the correlation functions in real space for the intra-β-band pairing (a) and α-band spin (b) for repulsive intraband and interband interactions, $U = V = W = 2t(\triangle)$ or $3t(\bigcirc)$ after Aoki and Kuroki [15]. The system is a 22-site two-band Hubbard model in 1D with 22 electrons (half-filled) in the α band and 6 electrons in β in periodic boundary conditions. For comparison, the result for the free electron system is also shown by a solid curve.

$b)-(U+F-a)$. Numerical results indicate that the fit may be accomplished for a smaller $U_{eff}(= -0.5t$ for $U = V = W = 2t,\ F = 0)$. Thus the present mechanism of superconductivity involves a subtle balance between the interaction and kinetic energies inherent in the two-band system. The pairing criterion can be extended when a level offset between the α and β orbitals exists.

An important question is whether the above criterion for pairing is attainable in real materials. Eto [21] has estimated the value of interaction parameters in a CuO_6 cluster with the DVXα method. The result is $U = 10.6$ eV, $F = 10.3$ eV and $V = 10.4$ eV, where U is the repulsion of two electrons on the dz^2 orbital and F is the repulsion of two electrons on the $dx^2 - y^2$ orbital, and the method of transition states is employed in the estimation. Thus the values of $2V$ and $(U + F)$ are quite close, definitely in the same range when the kinetic energy corrections ($O(t) \sim 1$ eV) are included.

The T_C may be estimated from the Bethe-ansatz result for the energy gap, Δ, at $T = 0$ in the single-band attractive Hubbard model [22,23],

$$\Delta = 2[(t^2 + U_{eff}^2)^{1/2} - t]\ ,$$

if we pursue the analogy with the single-band model, although the estimate

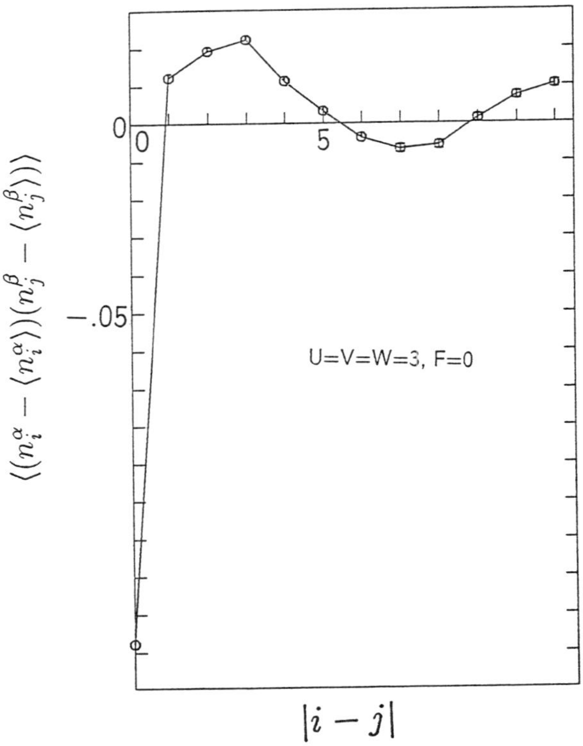

Fig. 3. Quantum Monte Carlo result for the interband density correlation function in real space for repulsive intra-band and inter-band interactions, $U = V = W = 3t$. The system is the same as in the previous figure.

of U_{eff} is subtle as mentioned above.

The momentum distribution of the β carriers in the superconducting phase is numerically shown (Figure 4) to be somewhat similar to the BCS formula,

$$|v_k|^2 = \frac{1}{2}\left(1 - \frac{\xi_k}{(\Delta^2 + \xi_k^2)^{1/2}}\right) ,$$

with $\xi_k = \varepsilon_k - \mu$. For $U = V = 3t$, we have $\Delta \sim 0.2t$.

As for the wavefunction of a pair, numerical results show that there is some enhancement in the nearest-neighbour pairing. Thus the pairing is of s symmetry, or may have an extended s symmetry, as has been evoked for heavy-fermion superconductivity [24].

We can also show that the superconductivity is destroyed as the carrier concentration, $n =$ (number of carriers)/(number of sites), is increased typically from 0.2 to 0.4. For a given band filling, narrower band gives more

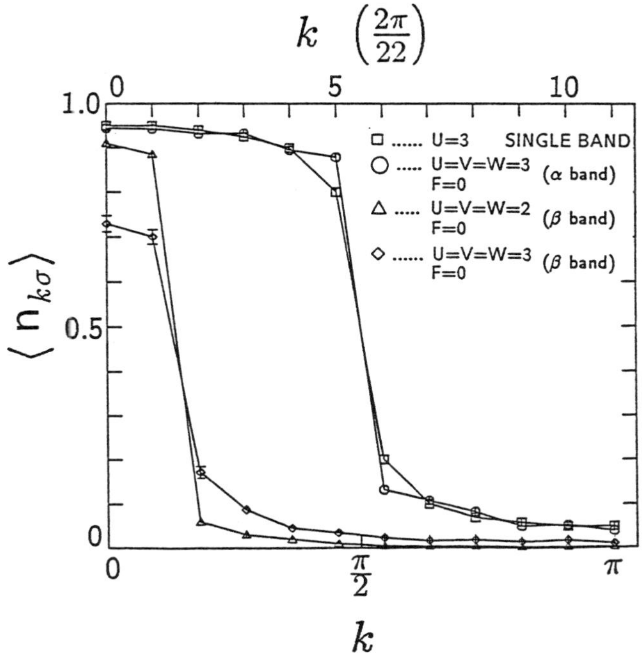

Fig. 4. Quantum Monte Carlo result for the momentum distribution for the α band (○) compared with the single-band case (□) and for the superconducting (β) band (△, ◇) for different strengths of interactions after Aoki and Kuroki [15]. The system is the same as in the previous figures.

enhanced superconductivity. There is also a dependence on the filling of the underlying α band, which is assumed to be around half-filled here.

The superconductivity considered here is irrelevant to the intraorbital exchange interactions that make the interactions between parallel spins and antiparallel spins different. QMC results in fact show that the introduction of the Hund's coupling with $V \neq W$ does not favour superconductivity.

A way to look at the present mechanism for the superconductivity is as follows. The Hubbard system has a dielectric function, $\varepsilon(q,\omega)$, and a spin susceptibility, $\chi(q,\omega)$, that are both strongly dependent on (q,ω). In other words, the Hubbard system has charge and spin excitation modes (in which charge and spin degrees of freedom are in general mixed unless $U=\infty$ [25]). Thus we can regard the accompanying band as a polarisable medium, and the superconductivity obtained here may be regarded as one mediated by a polarisable medium (α band) in an otherwise repulsive electron system (β band). The result that a superconductivity arises without the hybridisation between the two bands, does indeed allow this picture. Since $\varepsilon(q,\omega)$ and $\chi(q,\omega)$ become singular at the metal–insulator transition [35], we can expect

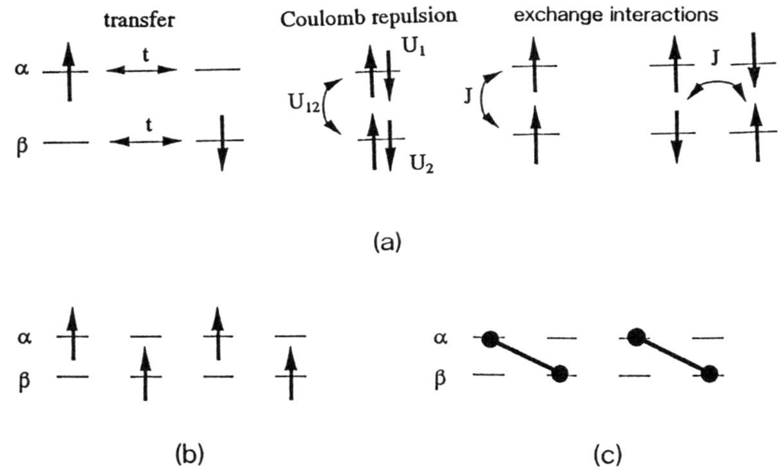

Fig. 5. (a) Interaction parameters in a two-band Hubbard model with exchange interactions. (b) A ferromagnetic phase conventionally assumed. (c) The lowest spin-unpolarised state, where a dumbbell represents an interorbital, intersite spin singlet. (After Kusakabe and Aoki [27].)

a peculiar effect when the accompanying band is near the Mott transition. That is one reason why we have set the α band around the Mott insulator.

On the other hand, if we include the hybridisation, the system becomes a periodic Anderson model, or Kondo lattice, which is a renowned model for the heavy-fermion systems, and the relation to this case is an interesting problem.

The presence of two bands is also expected to give peculiar normal-state properties such as the Hall effect. Schimizu and Kamimura [26] have shown theoretically that this is the case in the Hubbard I approximation.

Finally we mention the question of whether ferromagnetism or antiferromagnetism are realisable in the two-band Hubbard model. In the single-band Hubbard model, occurrence of high-spin states is related with a 'k-space Hund's coupling' of the electrons on the Fermi surface, and ferromagnetism does not arise except for special conditions [27]. In a two-band Hubbard model, we can promote the spin alignment by introducing interorbital exchange interactions (Figure 5a). In this case, a common belief is that ferromagnetic phase may be stabilised, in which a staggered ordering of orbital occupation is accompanied. A numerical result (Figure 5), however, shows that this state has to compete with the lowest spin-unpolarised state consisting of an array of interorbital, intersite spin singlets whose energy is only slightly higher and dependent on sample size [27]. Thus the competition of interaction and kinetic energies in a two-band system is at work also for magnetism.

3. Enhanced Superconductivity Due to Repulsive Interband Interactions

Apart from the question of superconductivity due to repulsive interactions, an equally important question is: can the superconductivity due to attractive interactions in band B be enhanced by the presence of band A, in which $A-A$ and $A-B$ interactions are all repulsive? We should recall that, even if the main source of attraction is from phonons in high T_C materials, strong electron–electron interactions cannot be ignored, so that it becomes an essential question whether the repulsive electron correlation helps the high T_C superconductivity.

We can show that the repulsive interband interactions indeed enhance the superconductivity under certain conditions [28]. Thus the enhanced superconductivity due to the presence of multi-bands is rather ubiquitous.

Figure 6 shows the QMC result for the enhanced pairing in the two-band Hubbard system with attractive $F < 0$ and repulsive $V, U > 0$ with $V \gtrsim U/2$. This is naturally understood along the lines described in the previous section, since the effective attraction, $2V - U - F$, derived there, is seen to be larger than the original $-F(> 0)$ when $V - U/2 > 0$. This result suggests in a way that a coexistence of electron–phonon and electron–electron interactions can favour a high T_C.

We mention in passing a point concerning phonons. Some authors stress the importance of anharmonic phonons in superconductivity [29]. In the case of oxides, anharmonicity may arise in the vicinity of a structural transition. We have shown from a non-emperical molecular-dynamics study for silicon oxide (silica) that we do indeed have a curious anharmonicity with large values of the non-Gaussian parameter [30].

4. Superconductivity Due to the Interband Attraction: A Competition of Diagonal and Off-Diagonal Long-Range Orders

The symmetry of the states in the band-index space for spin 1/2 fermion systems is an intriguing problem [5,31]. First, the two-band system highlights the subtlety of the occurrence of superconductivity, an off-diagonal long-range order (ODLRO), which has to compete with charge density wave (CDW) and spin density wave (SDW). Second, the pairing in the band-index space, rather than the spin space, is an interesting extension of the Cooper pairing.

To consider this problem, we study the two-band Hubbard model with attractive interband interactions ($V < 0$) along with repulsive intraband interactions ($U = F > 0$) [31]. Crudely speaking, negative V favours the interband pairing (of $(\alpha\sigma, \beta\sigma')$ electrons with spin σ, σ') or interband CDW, while positive U favours intra-band SDW.

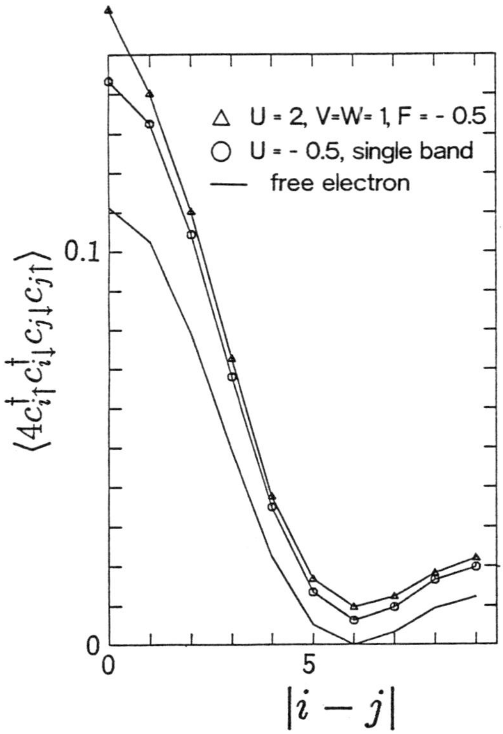

Fig. 6. QMC result (△) for the correlation function of the intra-β band pairing in real space for the attractive intra-β band interaction ($F = -0.5$) along with repulsive $\alpha-\beta$ ($V = 1$) and intra-α band ($U = 2$) interactions. For comparison, the result (○) for the single-band attractive Hubbard model with $U_{eff} = -0.5$ is also shown. The system is an 18-site two-band Hubbard model in 1D with 18 electrons (half-filled) in the α band and 6 electrons in β in periodic boundary conditions. The result for the free electron system is shown by a solid line. (After Aoki and Kuroki [28].)

The interband attraction considered here may be thought of as arising from a polarisable medium inserted between the two systems (chains or layers) as in Little's mechanism of superconductivity [32,33].

We can again start from an intuitive picture. When $U \sim -V$, we have relatively free motion of an interband pair, since a pair will have a cancelled net interaction with other single electrons.

From the rotation in the band-index (isospin) space, on the other hand, we can derive the following argument. In the special case in which both of the two bands are half-filled and U exactly equals V, an electron–hole transformation for the electrons in the α band changes the interband attraction into a repulsion, $-V(= U)$. Then all the interaction strengths are the same,

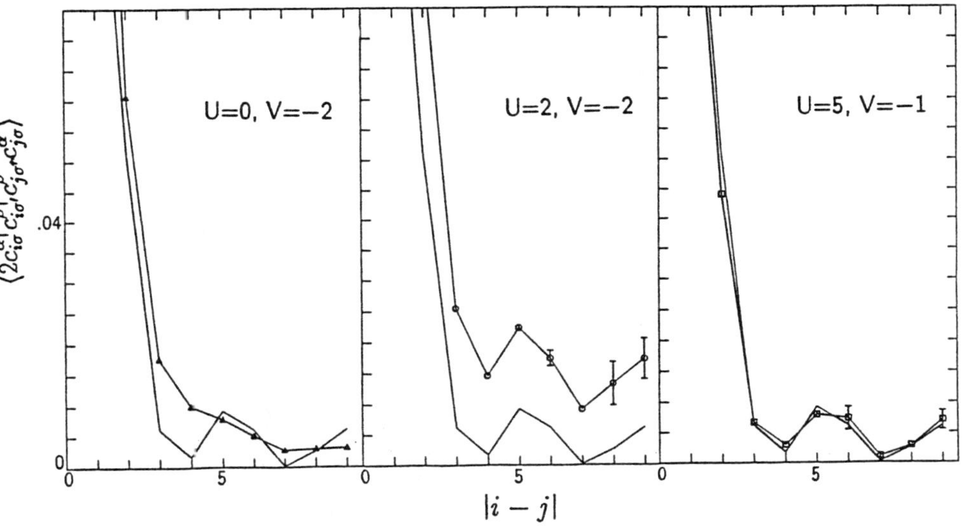

Fig. 7. QMC result for the correlation function in real space for the interband pairing in the two-band model for a 18-site (36-orbital) ring with 10 electrons (less-than-half-filled) in each band, after Kuroki, Aoki and Takada [31]. The interaction parameters are shifted for (△, □) from $U = -V$(○).

so that we have a rotational symmetry in the isospin space. This amounts, in the original model, that we interband pairing superconductivity is only degenerate with interband CDW and intraband SDW in the half-filled case despite the presence of interband attraction, so that a deviation from the half-filling is required to possibly stabilise the superconductivity.

QMC results indeed confirm the occurrence of an enhanced interband pairing for $U \sim -V$ in non-half-filled bands (Figure 7). From the results, we can then obtain a phase diagram (Figure 8) in the space of U, $|V|$ and n (band filling). There is a region for the off-diagonal long-range order (superconductivity) sandwiched between the phases with diagonal long-range orders (CDW or SDW), where the width of the superconducting region becomes finite as the band departs from the half-filling. The result of the mean-field theory, in which the pairing, CDW and SDW amplitudes are taken into account simultaneously, is also shown in the figure, and is consistent with the QMC result. Thus the interband pairing occurs only in the presence of intraband repulsion (U), which may be surprising at first, but true for spin 1/2 electron systems, since the CDW is stabilised if U is too small, while the SDW is stabilised if U becomes too large.

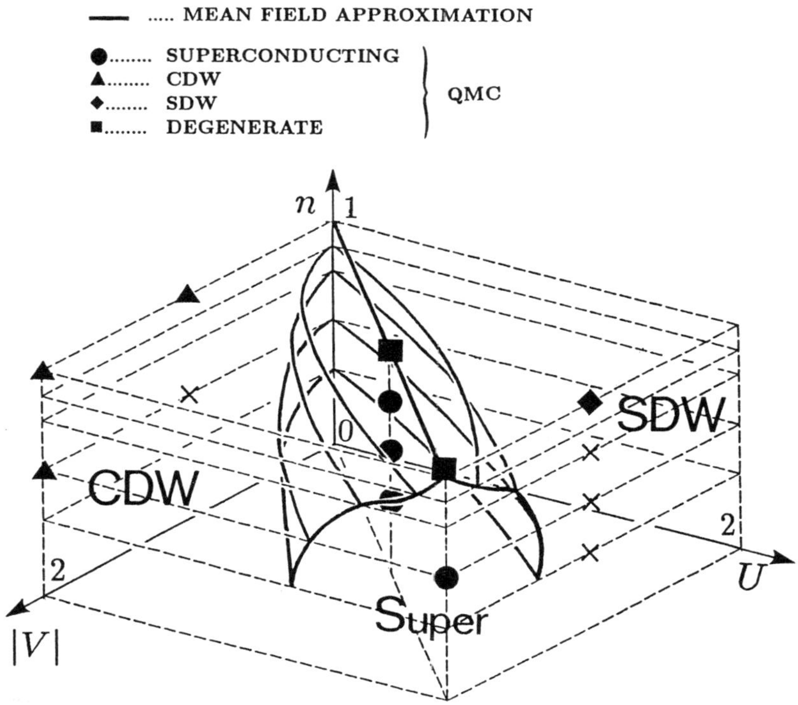

Fig. 8. The phase diagram in the space of inter-band attraction ($|V|$), intraband repulsion (U) and band filling (n) in 1D after Kuroki, Aoki and Takada [31]. The phase boundaries indicated by the mean-field approximation are shown by solid curves, and the phases determined by the quantum Monte Carlo result are superconducting (○), CDW (△), SDW (◇), all the three degenerated (□), indistinguishable (×).

If we combine all the results described in above sections, various cases (signs and magnitudes) of intra- and interband interactions in the two-band Hubbard model are covered. Emergence of interesting superconductivity, under favourable conditions in respective cases, may be tabulated as follows:

		Intraband	
		repulsion	attraction
Inter-band	repulsion	intraband pairing (*à la* Kuroki–Aoki)	enhanced intraband pairing
	attraction	interband pairing	mixed pairing

5. Relevance and Similarities to the Fractional Quantum Hall System

5.1. Single-band Hubbard and fractional quantum Hall systems

Now we are in position to point out curious similarities of the electronic structure of the Hubbard system to that of the fractional quantum Hall (FQH) system, which is a system of 2D electrons with repulsive interactions in strong magnetic fields in the perpendicular direction, another outstanding prototype of strongly correlated system [34,35].

Even within the usual single-band case, there exist interesting similarities between the Hubbard system and the FQH system. First let us look at the total spin, S, of finite systems as a function of the number of electrons (or the band filling) for the Hubbard system [27] and for the FQH system [36] in Figure 9. We can immediately notice a curious behaviour of S in both cases.

The singular dependence on the number of electrons is a reflection of the electron configurations in the two classes of systems. We can in fact look at the numerically obtained electron configurations in both the Hubbard [37] and FQH [38] models in Figure 10. In the Hubbard system, the occurrence of high-spin ground states for finite U is related with the double occupation of the sites, or the 'k-space Hund's coupling' of electrons on the Fermi surface in the weak-coupling ($U \ll t$) case [27]. Figure 9a is the perturbation result, which shows that S is controlled by the filling of the Fermi surface.

In the FQH system, the total spin is related with the double occupation of Landau orbits [36]. The parent state with the Landau-level filling of $\nu = 1/3$, for instance, has a fully spin-polarised ground state, in which dominant components have singly-occupied Landau orbits. The $\nu = 2/7$ daughter state, by contrast, has a spin-unpolarised ($S = 0$) ground state, in which dominant configurations include doubly-occupied orbits and spin singlets. Here the Zeeman energy of spins is neglected, so that the problem is the spin-configuration dependence of the exchange and direct Coulomb interactions. The kinetic energy is quenched in the Landau-quantised states, and the interactions are dominated by the peculiarity of the Landau orbits, in which the orbit centre has an uncertainty, $[X, Y] = i\,\ell^2$, where $X \propto p_y$ and ℓ is the cyclotron radius.

The picture emerging from the above discussions is as follows. Finite elementary lengths are essential in both cases: the lattice constant for the tight-binding electrons on a lattice and the cyclotron radius in the FQH system. Electron correlations impose constraints on the motion of electron on these length scales, in which exchange interactions are the key in both cases. These are in sharp contrast to usual interacting system in a continuous space such as an electron gas in a jellium.

In the Hubbard model, both Hubbard U and the band filling dominate the

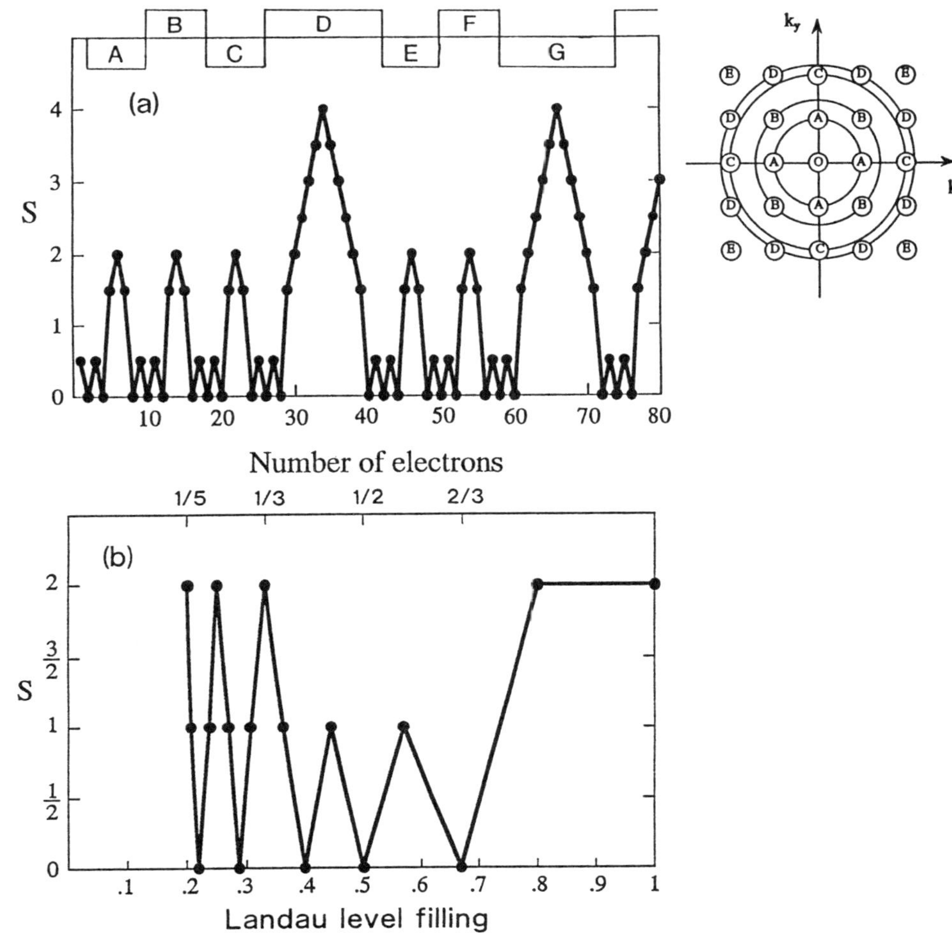

Fig. 9. (a) Dependence of the total spin, S, of the ground state on the number of electrons in the Hubbard model on a square lattice when the band filling is away from the half-filling so that Umklapp processes are absent. $A, B, \ldots$ represent the progressive Fermi surface. (After Kusakabe and Aoki [27].) (b) Dependence of the total spin, S, of the ground state on the Landau-level filling, ν, in the fractional quantum Hall system with four electrons after Maksym [36].

(direct and exchange) interaction of electrons in k space. In the FQH system, direct and exchange interactions among the Landau quantised electrons are relevant. These interactions are dominated by the Landau-level filling only, once the (usually coulombic) form of electron–electron interactions is given. One way to control the interaction is to consider a two-band or two-layer system as discussed in Section 5.2 below.

We can go one step further to discuss the ground-state phase diagram in the space of the Hubbard U and band filling n for the whole range including

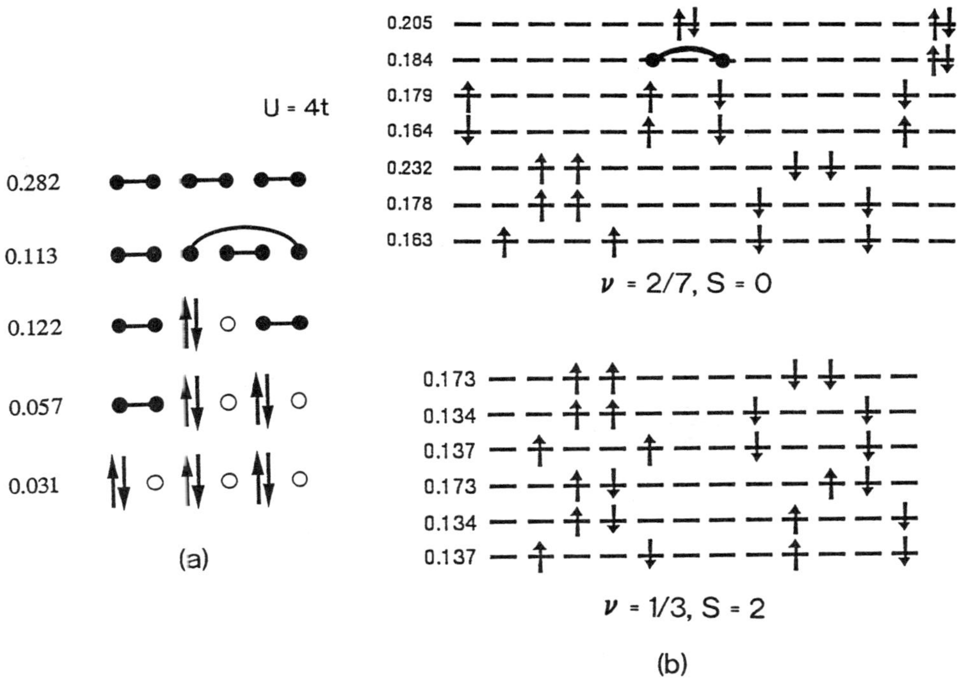

Fig. 10. (a) A typical electron configuration for the six-site Hubbard model [37]. A circle represents a site, while a dumbbell a spin singlet, $\frac{1}{\sqrt{2}}\left(|\uparrow\downarrow\rangle - |\downarrow\uparrow\rangle\right)$. The number attached to each of the representative configurations indicates the absolute value of the amplitude for $U = 4t$. (b) Typical electron configurations for the fractional quantum Hall system with four electrons for the fully spin-polarised ground state at the Landau-level filling $\nu = 1/3$ with $S = 2$ and its first daughter state at $\nu = 2/7$ with the spin-unpolarised ($S = 0$) ground state after Nakajima and Aoki [38]. A bar represents a Landau's basis function.

the strong U and half-filled band. A numerical result (Figure 11a) for the total spin of a finite 2D Hubbard system [27] exhibits a complex structure: a single-hole state undergoes a non-perturbative crossover to a fully spin-polarised state above a critical value of U, which hallmarks the onset of Nagaoka's state. In the vicinity of half-filling, there is another high-spin ground state with three holes, which emerges for even larger U. In the energy level scheme against U, these correspond to level crossings. One hole, three holes and their differences all correspond to an infinitesimal doping in the thermodynamic limit, and the phase diagram indicated here is thus pathological as a function of the band filling. This reminds us of the FQH state, which has quite a singular dependence on the Landau level filling as shown in Figure 11b [39].

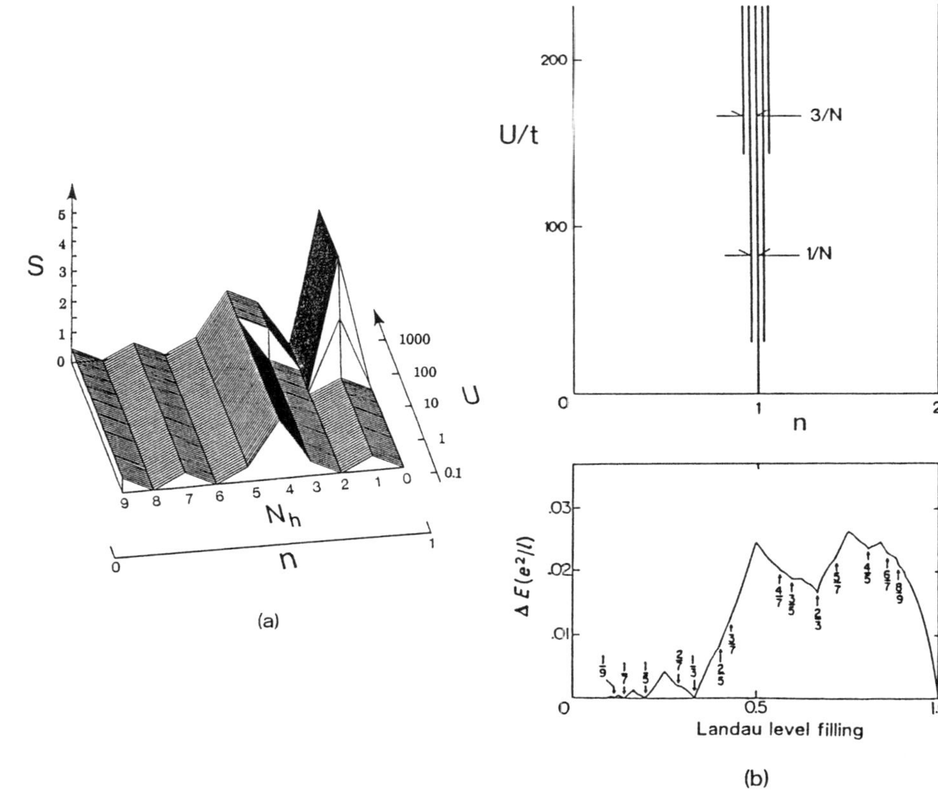

Fig. 11. (a) Numerical result for the total spin of the ground state in the 2D Hubbard model on a ten-site square lattice in periodic boundary conditions as a function of the number of holes, N_h (or the band filling, n) for varying Hubbard U, after Kusakabe and Aoki [27]. (b) Total energy of the fractional quantum Hall system against the Landau level filling, after Halperin [39]. Accompanying panel shows a schematic phase diagram for the Hubbard system.

5.2. Two-band Hubbard and fractional quantum Hall systems

In the light of the preceding sections, it is interesting to extend the analogy of the FQH system with the Hubbard model to two-band cases. The crossover of off-diagonal long-range orders and diagonal ones is of specific interest. Even for a single-band FQH system, there are extensive studies as to how we can regard the FQH state as a manifestation of an ODLRO [34]. This is related to whether we can give a description (with an appropriate gauge) for the FQH state as a condensation of anyons (fermions attached with fluxes). In field-theoretic approaches this may be realised by bosonising the fermion field [40].

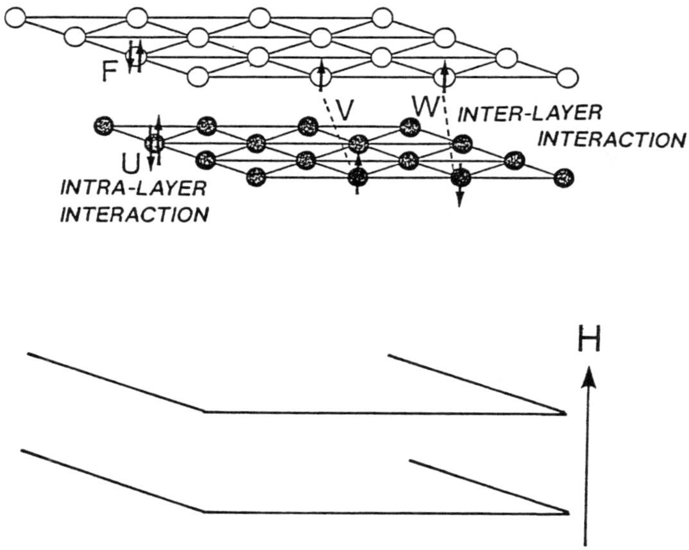

Fig. 12. Two-layer Hubbard and FQH systems.

By contrast, the problem of ODLRO becomes distinct in a two-band FQH system due to the coupling of the fermions in different bands. The two bands may again be thought of as arising either from different electronic bands or different layers, and the interband interactions may be tuned by, e.g., varying the interlayer spacing (Figure 12). There are a variety of interesting phases such as the double FQH and CDW states, or the excitonic phase for an electron-hole system [5,41]. There, a usual theoretical practice is to neglect the spin degrees of freedom to translate a two-band FQH system into a single-band FQH system with isospin up (band A) and down (band B) electrons. However, from the analogy with two-band Hubbard models, truly interesting effects should arise when one considers coupling of fermions in the two-band FQH systems for spin 1/2 electrons [38]. For instance, we can expect an interesting situation for the two-band FQH systems, in which interband interaction is attractive and intraband interaction is repulsive.

6. Concluding Remarks

In summary, we have clarified superconductivity and other orders in the ground state of the strongly correlated two-band electron system for various cases of intraband and interband interactions. A curious relevance to the fractional quantum Hall system, in both the single-band and two-band cases, is also worked out. To probe excited states as well as the ground states, the optical properties of these systems should be interesting. The present line

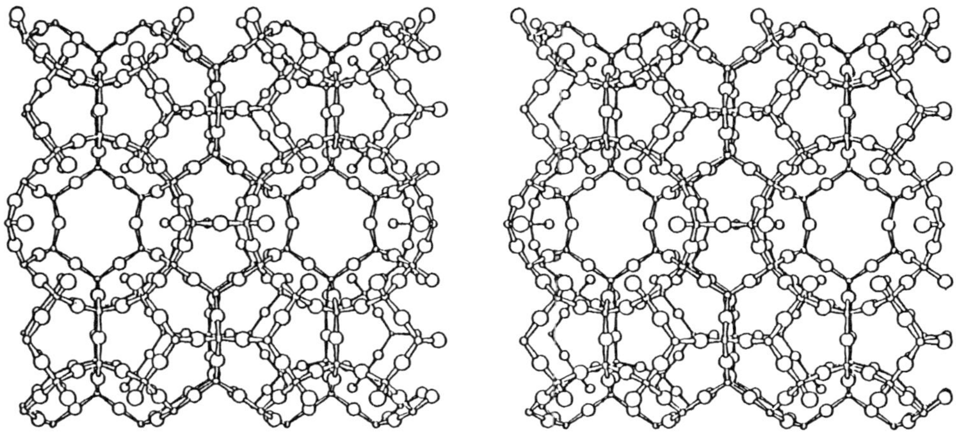

Fig. 13. Pure silica (SiO_2) with a hydrate-like structure predicted with a first-principles molecular dynamics by Tsuneyuki *et al.* [42]. The figure is a stereo pair with small spheres representing silicon atoms and large ones oxygen atoms.

of approach can be extended to superlattices with the 2D electron system stacked in the third direction, where we can expect new excitations and long-range orders. Realisation of these systems with controlled intraband and interband interactions in tailor-made materials is desired.

We can envisage the material design not only for layered structures, but also for three-dimensional structures such as zeolite-like open structures with possibly some molecules or microclusters incorporated, or quasi-one dimensional structures. We show an example of the former in Figure 13, which is a hydrate-structure pure silica (SiO_2), theoretically predicted with a non-empirical molecular dynamics [42]. Examples of quasi-1D systems include polyenes such as polysilene [43]. In both of these classes of systems we can explore the realization of multi-band structures.

Acknowledgements

It is an utmost pleasure to dedicate this paper to Professor Hiroshi Kamimura, who has introduced me to research in solid state physics and has inspired the enthusiasm that has accompanied me during all these years. Thanks are also due to Kazuhiko Kuroki, Yasutami Takada, Koichi Kusakabe, Tatsuya Nakajima and Shinji Tsuneyuki for fruitful collaboration. This work was in part supported by a Grant-in-aid (No. 02102002) from the Minstry of Education, Science and Culture of Japan.

References

1. H. Suhl, B.T. Matthias and L.R. Walker: 1959, *Phys. Rev. Lett.* **3**, 552.
2. J. Peretti: 1962, *Phys. Lett.* **2**, 275.
3. J. Kondo: 1963, *Prog. Theoret. Phys.* **29**, 1.
4. M.L. Cohen: 1964, *Phys. Rev.* **134**, A511; in *Superconductivity* Vol. 1, R.D. Parks (ed.), Marcel Dekker, New York, 1969, Ch. 12.
5. M. Rasolt in *Solid State Physics* **43**, H. Ehrenreich and D. Turnbull (eds.), Academic Press, 1990, p. 93.
6. Y. Takada: 1980, *J. Phys. Soc. Jpn.* **49**, 1713.
7. J.G. Bednorz and K.A. Müller: 1986, *Z. Phys. B* **64**, 189.
8. M. Sigrist and K. Ueda, *Rev. Mod. Phys.*, to be published.
9. P.W. Anderson: 1987, *Science* **235**; P.W. Anderson in *Frontiers and Borderlines in Many-Particle Physics*, North-Holland, Amsterdam, 1988, p. 1; P.W. Anderson in *Strong Correlation and Superconductivity*; H. Fukuyama, S. Maekawa and A.P. Malozemoff (eds.), Springer-Verlag, Berlin, 1989, p. 2.
10. H. Kamimura and A. Oshiyama (ed.), *Mechanisms of High Temperature Superconductivity*, Springer, Berlin, 1989, and references therein.
11. H. Fukuyama, S. Maekawa and A.P. Malozemoff (eds.), *Strong Correlation and Superconductivity*, Springer-Verlag, Berlin, 1989, and references therein.
12. G. Baskaran, A. Ruckenstein, E. Tosatti and Y. Lu (eds.), *Strongly Correlated Electron Systems*, World Scientific, Singapore, 1991, and references therein.
13. H. Kamimura: 1987, *Jap. J. Appl. Phys.* **26**, L627; H. Kamimura, S. Matsuno and R. Saitoh in *Mechanisms of High Temperature Superconductivity*; H. Kamimura and A. Oshiyama (eds.), Springer, Berlin, 1989, p. 8; S. Matsuno, K. Ishida and H. Kamimura: 1989, *Physica C* **162-164**, 781; M. Eto and H. Kamimura: 1990, *J. Phys. Soc. Jpn.* **59**, 3053.
14. K.A. Muttalib and V.J. Emery: 1986, *Phys. Rev. Lett.* **57**, 1370.
15. K. Kuroki and H. Aoki: 1990, *Solid State Commun.* **73**, 563; H. Aoki and K. Kuroki: 1990, *Phys. Rev. B* **42**, 2125; K. Kuroki and H. Aoki: 1990, *Physica B* **165 & 166**, 1011.
16. H. Aoki and H. Kamimura: 1987, *Solid State Commun.* **63**, 665.
17. S. Sorella, E. Tosatti, S. Baroni, R. Car and M. Parrinello: 1988, *Int. J. Mod. Phys. B* **1**, 993; S. Sorella, S. Baroni, R. Car and M. Parrinello: 1989, *Europhys. Lett.* **8**, 663.
18. S.R. White, D.J. Scalapino, R.L. Sugar, E.Y. Loh, J.E. Gubernatis and R.T. Scalettar: 1989, *Phys. Rev. B* **40**, 506.
19. M. Imada and Y. Hatsugai: 1989, *J. Phys. Soc. Jpn.* **58**, 3752.
20. E.H. Lieb and F.Y. Wu: 1968, *Phys. Rev. Lett.* **20**, 1445.
21. M. Eto, to be published.
22. N.M. Bogoliubov and V.E. Korepin: 1988, *Mod. Phys. Lett. B* **1**, 349; 1989, *Int. J. Mod. Phys.* **3**, 427.
23. R. Micnas, J. Ranninger and S. Robaszkiewicz: 1990, *Rev. Mod. Phys.* **62**, 113.
24. F.J. Ohkawa and H. Fukuyama: 1984, *J. Phys. Soc. Jpn.* **53**, 4344.
25. K. Kusakabe and H. Aoki: 1991, *Phys. Rev. B* **44**, 7863.
26. T. Schimizu and H. Kamimura: 1990, *J. Phys. Soc. Jpn.* **59**, 3691.
27. K. Kusakabe and H. Aoki, to be published.
28. H. Aoki and K. Kuroki, to be published.
29. J.R. Hardy and J.W. Flocken: 1988, *Phys. Rev. Lett.* **60**, 2191.

30. S. Tsuneyuki, H. Aoki, M. Tsukada and Y. Matsui: 1990, *Phys. Rev. Lett.* **64**, 776.
31. K. Kuroki, H. Aoki and Y. Takada, to be published.
32. W.A. Little: 1964, *Phys. Rev.* **134**, A1416; 1981, *Int. J. Quantum Chem.* **15**, 545.
33. J.E. Hirsch and D.J. Scalapino: 1985, *Phys. Rev. B* **32**, 117.
34. R.E. Prange and S.M. Girvin (eds.), *The Quantum Hall Effect*, 2nd Edn., Springer, New York, 1990.
35. H. Kamimura and H. Aoki, *The Physics of Interacting Electrons in Disordered Systems*, Oxford Univ. Press, 1989.
36. P. Maksym: 1989, *J. Phys.: Condensed Matter* **1**, 6299.
37. K. Kusakabe and H. Aoki, unpublished.
38. T. Nakajima and H. Aoki, to be published.
39. B.I. Halperin: 1984, *Phys. Rev. Lett.* **52**, 1583, 2390.
40. Z.F. Ezawa, M. Hotta and A. Iwazaki, *Phys. Rev. D*, to be published.
41. D. Yoshioka, A.H. MacDonald and S.M. Girvin: 1989, *Phys. Rev. B* **39**, 1932; D. Yoshioka and A.H. MacDonald: 1990, *J. Phys. Soc. Jpn.* **59**, 4211.
42. S. Tsuneyuki, H. Aoki and Y. Matsui, in *Proc. Int. Conf. on Computer Aided Innovation of New Materials*, Tokyo, 1990, p. 381.
43. K. Takeda and S. Kagoshima: 1988, *Phys. Rev. B* **37**, 6406; M. Springborg: 1989, *Phys. Rev. B* **40**, 7839.

AN ASPECT OF QUANTUM ANTIFERROMAGNETISM ON A SQUARE LATTICE

Y. NATSUME

Department of Physics, Faculty of Science, Chiba University, Yayoi-cho Chiba 260, Japan.

Abstract. We propose an accurate expression for the ground state of spin $-1/2$ Heisenberg antiferromagnetic system on a square lattice in order to understand the properties of the quantum ground state, on the basis of the extended RVB picture where a singlet pair with longer distance than nearest neighbor (NN) is taken into account as well as that for NN. Quite a few parameters introduced in consideration of geometrical properties for each RVB in a square lattice of spin number $N \leq 26$ is determined by the variational procedure. The agreement to numerical solution obtained by exact diagonalization is quite well. Properties of the ground state are discussed on the basis of such an agreement, in connection with the problem of Néel-like correlation containing the quantum fluctuation.

1. Introduction

A lot of the work on superconductivity in a CuO_2 plane in oxide compounds [1] is leading to the suggestion that such superconducting phenomena are essentially related to quantum antiferromagnetism [2] on the square lattice. Consequently, for the past four years, considerable interest has been manifest in the ground state on a square-lattice quantum Heisenberg antiferromagnet with $S = 1/2$ where the Hamiltonian is described as

$$\mathcal{H} = 2J \sum_{\langle i,j \rangle} \mathbf{S}_i \cdot \mathbf{S}_j \qquad (J > 0) , \tag{1}$$

though the problem of this ground state has been, needless to say, one of the most fundamental ones [3] to be investigated for the theory of spin systems in magnetic compounds. In Equation (1), sites i and j are nearest neighbors (NN) separated by a distance a_0. Each spin has two states expressed as α and β. Among much theoretical work on the ground state of Hamiltonian (1) on a square lattice from various points of view, we concentrate our attention on the numerical study, which has been extensive, especially in the recent few years by the great developments in the capacity of computers.

In fact, investigations on the basis of numerical calculation are divided into the following two approaches: one is the Monte Carlo study on the basis of some approximate function introduced by a physical picture. Another is the study of an exact diagonalization of (1). Monte Carlo studies [4,5] have provided fruitful information for the ground state energy [2], because the system with relatively large N can be treated. However, there is room for speculation on how the function used reflects the nature of the ground

H. Aoki et al. (eds.), New Horizons in Low-Dimensional Electron Systems, 281–294.

state itself, in consideration of the inevitable restriction of the Monte Carlo method. On the other hand, for cases of the exact diagonalization study [6–9], we essentially face the problem of how the physical picture is constructed in the numerical solution for the quantum ground state obtained as the list of the numerical values for an eigenvector, though the exact solution can be obtained for a system of finite N. As for the quantum state, the problem of how we understand the state is the same as the problem as what picture we use to describe it.

In the present paper, we discuss the nature of this quantum ground state in our physical picture [10–12] on the basis of a direct comparison between our proposed variational function and exact numerical solution for systems of $N = 10, 16, 18, 20, 26$.

2. The Ground State of a Heisenberg System

For a Heisenberg system, the square of the total spin $\mathbf{S}_t^2$, the z-component of the total spin S_t^z and the raising (lowering) operator for the total spin $S_t^x + iS_t^y$ $(S_t^x - iS_t^y)$ commute with the Hamiltonian (1). Therefore, the eigenstate can be assigned by $\mathbf{S}_t^2$ and S_t^z. Further, it can be easily proved that any energy level (including the ground state) is contained in the subspace of $S_t^z = 0$.

As for the ground state ψ_g of a Heisenberg system with two-sublattice structure (A sublattices and B sublattices), e.g. Hamiltonian (1) on a square lattice, Marshall has discussed [13–15] the conditions which should be satisfied. By the use of such Marshall's conditions, we can summarize the restriction as follows:

$MC1$: ψ_g is singlet, i.e. total spin $\mathbf{S}_t = 0$.

$MC2$: ψ_g is non-degenerate.

$MC3$: ψ_g is totally symmetric.

In detail, the irreducible representation is $\Gamma_1(\Gamma_2)$ according as $N/2 =$ even (odd) for a finite square lattice with periodic boundary condition.

$MC4$: Because ψ_g belongs to the subspace $S_t^z = 0$ as discussed above, ψ_g is expressed by the base where α spins (β spins) with the number of $N/2$ distributed over all spin sites. Further, it is proved that ψ_g should contain any base for such a distribution and sign for the coefficient is determined by the parity of the number of α spins on A sublattices.

3. RVB (Resonating Valence Bond)

3.1. Nearest neighbor RVB

As a physical picture for the expression of the ground state of quantum antiferromagnetic system, RVB (resonating valence bond) description by the products of a singlet pair (RVB)

$$[i,j] = \{\alpha(i)\ \beta(j) - \beta(i)\ \alpha(j)\}\ /\ \sqrt{2} \tag{2}$$

has attracted much attention through Anderson's proposal [16–18], though this is not the only approach; it should be noted that we can produce several basis sets expressing some quantum spin state. Though Anderson and his coworkers themselves have pointed out the possibility of the contribution of RVB with the long distance (whose length is more than a_0), much work as to RVB has been done in consideration of only the RVB with a nearest neighbor distance (called NNRVB) [19–22]. We would like to point out [10,11,23,24] that such an NNRVB picture fails even qualitatively for systems of $N \geq 16$, while it can reproduce the ground state to some extent for the system of $N \leq 10$ [21,22]. Thus, we introduce an RVB with the larger distance than a_0 as well as NNRVB, and we call this the extended RVB picture.

3.2. RVB with a long distance

In regard to the study of the comparison between trial function and numerical exact solution, we have found only a few papers [23–25] in which the contribution of RVB with long distances is taken into account. For example, Caspers and his coworkers [25] have carried out a numerical investigation of the system of $N \leq 16$ taking into account an effect of RVB with $\sqrt{5}\,a_0$. Nevertheless, the small size of the system treated has lead to such improper conclusion as the contribution to the ground state by RVB with the length more than a_0 is not appreciable.

Now, we would like to emphasize that we should adopt RVB with the longer distance than a_0 as well as NNRVB in order to express ψ_g by the complete set of bases. Here, we summarize the properties of such an extended RVB $[m,n]$ as follows, where sites m and n are never restricted to the nearest neighbor [10,11,23,26];

$$[n,m] = -[m,n] \tag{3}$$

$$[m,n]\ [k,l] + [m,k]\ [l,n] + [m,l]\ [n,k] = 0 \tag{4}$$

$$2(1/4 - \mathbf{S}_m \cdot \mathbf{S}_n)\ [m,n] = 2[m,n] \tag{5}$$

$$2(1/4 - \mathbf{S}_n \cdot \mathbf{S}_k)\ [m,n]\ [k,l] = [n,k]\ [l,m] \tag{6}$$

TABLE I
Number of bases for the expression of the ground state.

N	Number of patterns for extended RVB	Extended RVB in consideration of symmetric properties	Number of independent bases	Number of variational parameters
	$(N/2)!$	type		
10	120	7	3	1
16	40320	204	18	1
18	362880	2988	60	2
20	3628800	47124	269	2
26	6227020800		7401	3

We can prove that an extended RVB only with sites between A and B sublattices can make a complete set by adapting m, k (n, l) to $A(B)$ sublattices in Equation (4). Further, it is easily shown that the product of RVB belongs to the singlet subspace, i.e. $\mathbf{S}_t = 0$ by the use of these properties (3)–(6).

3.3. The number of bases for the expression of ground state

We now discuss the problem of how many bases can express ψ_g for the system of finite N, on the bases of the extended RVB picture introduced in the previous subsection.

The number of patterns where $N/2$ RVB for a pair between A and B sublattices are distributed on a square lattice N is $(N/2)!$ as listed in Table I. Here, we can reduce this enormous number in consideration of symmetric properties for the lattice, as shown in Table I [6]. In addition, the number of independent bases are also listed. Here, the discussion of this number is abbreviated [6,11,27]. In any case, we can obtain the exact expression of ψ_g for a system of finite N as coefficients for bases of a complete set.

3.4. Pattern analysis for extended RVB

If we adopt the RVB with a greater distance than a_0, we can make a complete set for bases and can express the ground state by investigating how such a pattern of distributed RVB changes each other by Hamiltonian (1). As a method of such an expression, we have proposed the pattern analysis [6,28]. This method is quite useful for the system of $N = 10$, as shown in Figure 1, because the analysis is systematic and easily performed in consideration of symmetry properties: Figure 1 shows types $A - G$ where the number of patterns belonging to these types is indicated. Further, we list the value of each coefficient for ψ_g. It is clear that RVB with a distance of $\sqrt{5}\ a_0$ contributes essentially to ψ_g. Although we have succeeded in adopting this

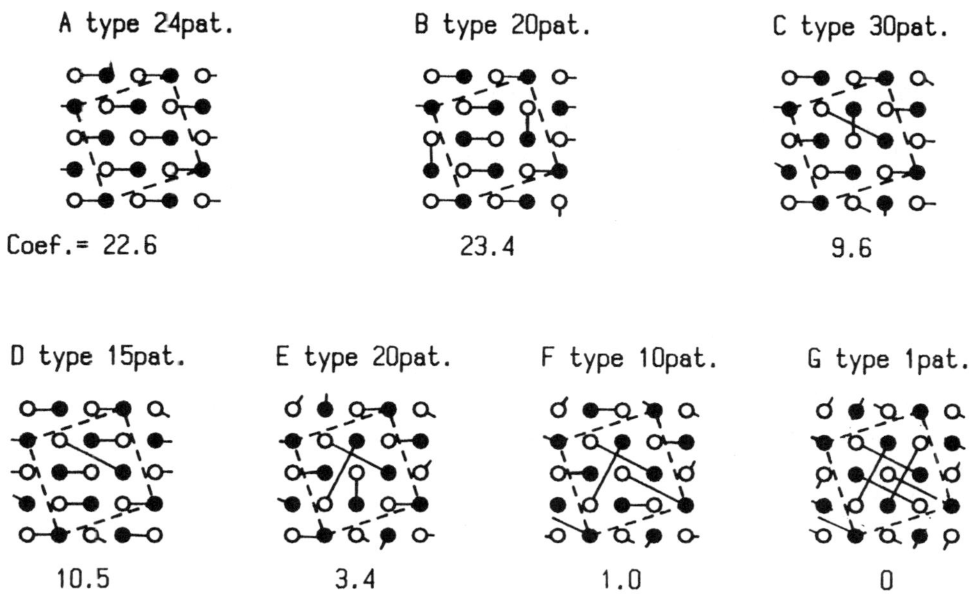

Fig. 1. Pattern analysis for the system of $N = 10$. Here, 120 patterns for distributed 5 RVB are classified into 7 types; $A-G$ types. The number of patterns belonging to each type is shown. The coefficient for each type in the ground state is also shown. Because the number of types is more than that for independent bases (i.e. over-complete) as shown in Table I, the expression is not unique.

method to the system $N = 16$ with the aid of a computer, we do not go any further into its description here.

4. Understanding the Quantum Ground State

For cases where N is more than 16, how should we investigate the ground state? In other words, what kind of picture should we use to describe it [29]? The problem of understanding a quantum state is very much the problem of finding of an expression for it. On the other hand, by the intensive use of a computer system, we can nowadays obtain the numerical solution of the ground state even for $N = 26$ [6,8] and $N = 32$ [9], using the exact diagonalization of the enormously large matrix. Nevertheless, such a solution is the eigenvector expressed numerically. Thus, we are faced with the question of what picture we use to understand it.

5. Proposition of a Trial Function

In this section we introduce a trial function [6,10,11] on the basis of the extended RVB picture. It is the purpose of this proposal to provide an understanding of the essential properties of the ground state comparing the trial function $\Phi_{\mathbf{var}}^N$ with the exact one $\Psi_{\mathbf{exa}}$. In brief, we perform the variational calculation for our trial function involving 1–3 parameters in a system of $N = 10 - 26$ in the following procedure: the square of the product

$$f = | < \Phi_{\mathbf{var}}^N \mid \Psi_{\mathbf{exa}} > |^2 \tag{7}$$

comes close to 1. Needless to say, the exact analytical solution is needed for any N to reproduce the perfect solution. However, we would like to emphasize that we can get quite a good agreement, as discussed in the following section, if we consider the geometrical properties in the variational procedure on the basis of Marshall's conditions.

We try to express the singlet ground state as follows:

$$\Phi_{\mathbf{var}}^N = \sum_{\substack{m\in A \\ n\in B}} w(m_1, n_1)\,[m_1, n_1]\,w(m_2, n_2)\,[m_2, n_2]\,\ldots$$

$$\ldots\,w(m_{(N/2)}, n_{(N/2)})\,[m_{(N/2)}, n_{(N/2)}] \tag{8}$$

where $[m(\gamma), n(\gamma)]$ represents an RVB between $\mathbf{r}_{m(\gamma)}$ and $\mathbf{r}_{n(\gamma)}$ sites. Here, the summation is taken over any pattern where $m(n)$ belongs to $A(B)$ sublattices. In consideration of MC1, MC2, MC4, any weight factor is expected to be positive. Up to this stage, the number of weight factors w is quite large. Now, we classify w into a few parameters x_ν considering the geometrical character of each RVB on a plane. Such a restriction to weight factor w is based on MC3. Each x_ν is shown in Figure 2 for systems of $N = 10 - 26$. One example for the determination of w is illustrated in Figure 3 for one pattern in $N = 26$. Here, it should be noted that variational parameters are ratios of x_1/x_0, x_2/x_0 and x_3/x_0. The number of these parameters is also listed in Table I.

6. Results of Calculation

6.1. The agreement of proposed function and exact one

We make the variational calculation for $f \to 1$ using the exact solution [6]. The parameters obtained and f are listed in Table II. The value for NNRVB is given by the use of only RVB for NN. While f for NNRVB departs from 1 with increased N, our proposed expression can keep quite close to 1. We would like to remark that RVB with distance than NN contributes more

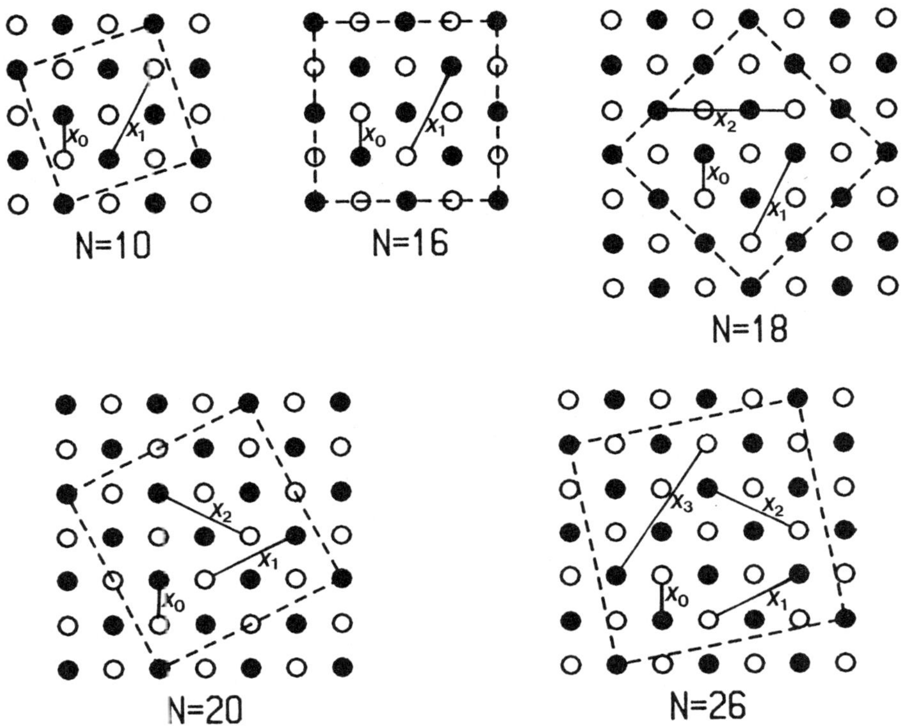

Fig. 2. The weight factor $w(m, n)$ for a singlet pair $[m, n]$ in the system of N is determined to be x_ν in consideration of geometric characteristics. Detailed discussion of this classification is given in the text.

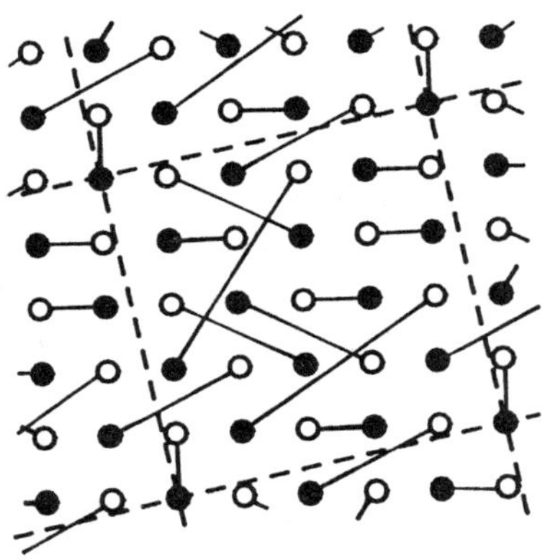

Fig. 3. An example for the determination of a product of weight factor $w(m, n)$ to a pattern of distributed RVB in the system of $N = 26$. For this pattern, it is given $x_0^6 \cdot x_1^2 \cdot x_2^3 \cdot x_3^2$ according to the rule discussed in the text.

TABLE II
Obtained values of f and ratios for x_1/x_0, x_2/x_0 and x_3/x_0 by the present calculation for $f \to 1$ using the procedure discussed in the text. Values for NNRVB represent f for the trial function including only NNRVB. Further, the ground state energy per spin, where unit is J, is listed for the exact solution and our variational function.

N	f	x_1/x_0	x_2/x_0	x_3/x_0	NNRVB	exact	variational
10	0.999998	0.3982			0.9747	−1.46001	−1.46001
16	0.999819	0.276			0.8567	−1.40356	−1.40342
18	0.999928	0.268	0.246		0.8559	−1.38797	−1.38792
20	0.999839	0.255	0.230		0.8152	−1.38162	−1.38151
26	0.999528	0.201	0.185	0.135	0.7158	−1.36890	−1.36867

substantially as N increases. We would like to emphasize that RVB for a pair with longer distance contributes essentially to ψ_g. This fact leads to the widely- believed suggestion that the ground state ψ_g has no energy gap, because the triplet excited state can close energetically to ψ_g by a singlet-triplet transition for a pair with a long distance [30].

Next we discuss properties for the ground state on the basis of the above close agreement between our variational function and the numerical solution.

The ground state energy per spin for each system of spin number N is also shown in Table II for the exact solution and our function. The agreement between the exact solution and our values is quite good even for large N. Here, we would like to mention the extrapolated value for infinite N : $-(1.358 \pm 0.010)\,J/N$.

Further, we investigate the correlation function expressed as

$$\mathrm{CR}(\mu) = 4 \cdot < S_0^z\; S_\mu^z > \tag{9}$$

for the μth nearest neighbor. Here, $|\mathrm{CR}(\mu)|$ is 1 for any μ in classical Néel state. As for the system of $N = 26$, the comparison between the exact solution and our variational function is shown in Table III. The agreement is quite good, because the wavefunction of Φ_{var}^N itself can reproduce quite well the exact solution by the present variational procedure. As for the staggered magnetization, we omit the calculated result [6,10,11,12] in this paper.

Moreover, we discuss the problem of how (and why) our variational solution resembles the exact one in more detail. Namely, we expand both solutions by common set of bases in order to compare expansion coefficients C_i [31]. We perform this with respect to the Ising bases i, where $\alpha(\beta)$ spins with the number of $N/2$ are distributed over all sites. Though the number of bases is enormous (i.e. 10 400 600 for $N = 26$), it can be reduced to a relatively small number (100 247 for $N = 26$) in consideration of symmetry properties [6,32]. Such reduced bases are called configurations. Figure 4

TABLE III
The correlation between spins for the system of $N = 26$. Values for the exact solution and our variational function are listed. The distances between A and B sublattices correspond to parameter x_ν discussed in the text.

distance	x_ν	exact	variational
a_0	x_0	−0.4563	−0.4562
$\sqrt{2}\,a_0$		+0.2806	+0.2804
$\sqrt{5}\,a_0$	x_1	−0.2548	−0.2548
	x_2	−0.2462	−0.2465
$2a_0$		+0.2567	+0.2569
$2\sqrt{2}\,a_0$		+0.2262	+0.2270
$\sqrt{13}\,a_0$	x_3	−0.2252	−0.2258

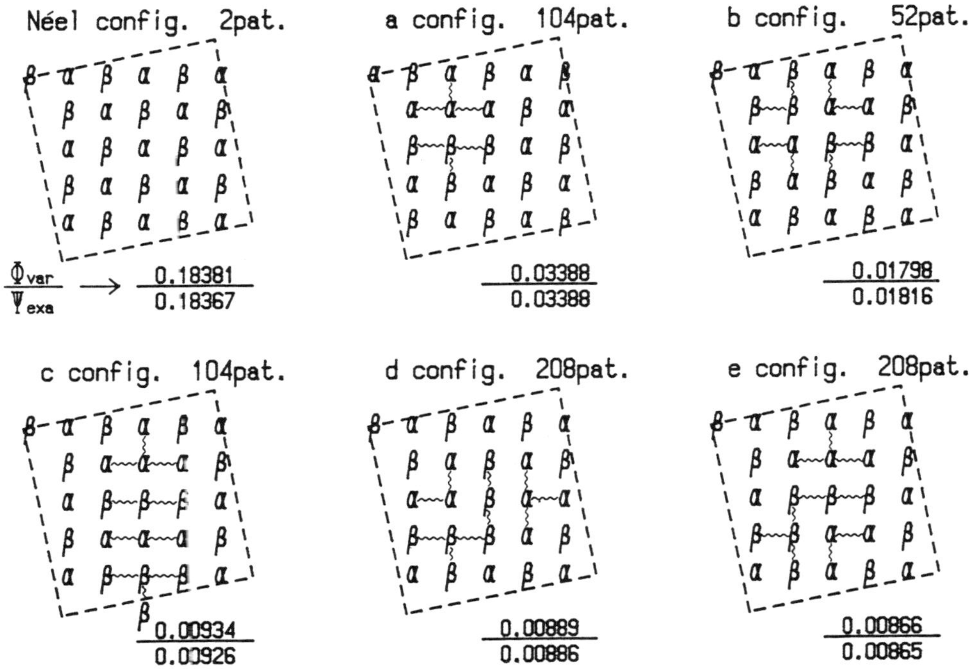

Fig. 4. Principal 6 configurations contributing essentially to the ground state among 100 247 for $N = 26$. The number of patterns belonging to each configuration is shown. The wavy line denotes a wrong bond, whose ends are $\alpha - \alpha$ or $\beta - \beta$. While the Néel configuration has no wrong bond, 6 wrong bonds appear in a configuration by one flip-flop from a Néel one. The b configuration has 8 wrong bonds, and $c - e$ configurations have 10 wrong bonds. It should be noted that configurations $b - e$ are formed by two flip-flops. The expansion coefficients C_i for a pattern belonging to each configuration for the exact solution Ψ_{exa} and the variational function Φ_{var} are listed for comparison.

shows several configurations which contribute substantially to the ground state. The expansion coefficients of both functions for a pattern belonging to these configurations are also listed. The agreement is quite good, not only for Néel configuration but also for other ones which construct the essential quantum fluctuation accompanying the Néel configuration.

6.2. A TRIAL OF STUDY BY WRONG BOND DENSITY

In this subsection we would like to briefly introduce a preliminary work using a new concept: 'wrong bond density' for the purpose of discussing properties of the ground state in connection with the relation between Néel state and the quantum fluctuation. Namely, as an approach to the analysis, paying attention to patterns in more detail, we try to adopt such a new concept as follows: for a case of the pattern where $N/2$ $\alpha(\beta)$ spin are distributed over sites, the bond whose ends are $\alpha - \alpha$ or $\beta - \beta$ (i.e. spins are parallel) is defined as a wrong bond. Further, the 'wrong bond density' is determined to be the ratio of the number of wrong bonds to that of all bonds $2N$ on a square lattice. In other words, it shows the energy fluctuation from Néel state by a quantum (off-diagonal flip-flop term) effect. Here, we calculate each distribution of probability amplitude $|C_i|^2$, where an Ising base i has a particular value of wrong bond density. Further, we sum over $|C_i|^2$, and express them as the histogram as shown in Figure 5 for $N = 26$. In this figure, the difference between our variational function and exact solution cannot be distinguish in the illustration because of the good agreement.

This behavior at $N = 26$ leads the suggestion that this ground state is essentially composed of quantum-fluctuating states, as well as the pure Néel state which has no wrong bond. However, we can suspect that such fluctuating states, especially 6-wrong bond states formed by one flip-flop from the pure Néel state, seem to assist the Néel ordering.

In order to confirm this suspicion, we are now proceeding with the calculation for systems of even larger N.

7. Concluding Remarks

In conclusion, we propose the variational function for the ground state of the quantum antiferromagnetic system, Equation (1), on a square lattice with the finite N. This function can get quite close to the numerical solution obtained by the exact diagonalization, with the use of only a few variational parameters which are introduced in consideration of geometric characteristics. It is certain that our function satisfies Marshall's conditions for the ground state. The present good agreement lead the suggestion of the possibility of expressing the present quantum state by the product of one-body RVB in consideration of the non-local quantum condition, such

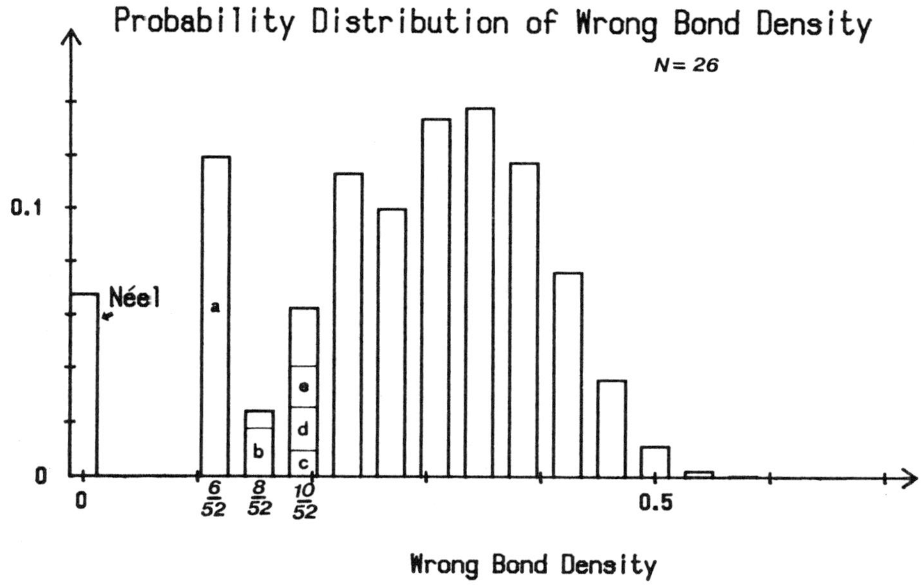

Fig. 5. The probability distribution of wrong bond density in the ground state for $N = 26$. The contributions of configurations refferred in Figure 4 are also indicated in the histogram. The difference between the exact solution and variational one cannot be found in this illustration. The relation between Néel state and quantum fluctuation is discussed in the text.

as Marshall's condition. This method of expression of a quantum state is quite similar to that of Hartree–Fock theory for a many-electron system. Though the problem of how to express the ground state cannot be solved perfectly until the exact analytical expression is obtained for any N, it can be expected that ground state properties are reflected essentially on our variational function. Further, we would like to hope that the investigation of our function provides a clue to obtaining the analytical expression.

The properties of the ground state are discussed on the basis of the good agreement between these solutions: we can presume the long range Néel-like ordering essentially contains the quantum fluctuation.

Acknowledgements

The author would like to express his thanks to T. Hamada of NEC Corporation, S. Nakagawa of Nippon Telegraph and Telephone Corporation, J. Kane of Fujitsu Laboratories Ltd., M. Katagiri and T. Toyoda of Chiba University for the numerical calculation and valuable discussions. Further, he would like to thank Dr. T. Nakayama of Chiba University and Dr. R. Saito of the University of Electro-Communications for helpful discussions.

Numerical calculation was performed by HITAC M680-D of Information Processing Center of Chiba University and HITAC M680-H/M682-H/S820 of the Computer Center of Tokyo University.

This work was supported by the Grant-in-Aid for Scientific Research from the Ministry of Education, Science and Culture. Further, the work was supported by the Grand-in-Aid for Research and Education from Nihon Suido Co. Ltd., Japan.

References

1. For example, *Proceedings of 19th International Conference on Low Temperature Physics*, Brighton (1990), *Physica B* **165 & 166** (1990).
2. Spin polaron mechanism for the superconductivity in oxide compounds has been proposed and investigated theoretically by H. Kamimura and his coworkers. H. Kamimura, *Proceedings of of the Adriatico Conference on Theoretical Understanding of High Temperature Superconductivity*, Trieste (1988), p. 163, World Scientific, Singapore (1989); H. Kamimura, S. Matsuno and R. Saito: 1988, *Solid State Commun.* **67**, 363; H. Kamimura, S. Matsuno and R. Saito, *Mechanisms of High Temperature Superconductivity*, Vol. 11 of Springer Series in Material Science, H. Kamimura and A. Oshiyama (eds.), Springer, Heidelberg (1989), p. 8.; H. Kamimura, S. Matsuno and K. Ishida: 1989, *Physica C* **162–164**, 781; M. Eto, R. Saito and H. Kamimura: 1989, *Solid State Commun.* **71**, 425; K. Ishida and H. Kamimura: 1990, *Phys. Rev. B* **42**, 8312; H. Kamimura and M. Eto: 1990, *J. Phys. Soc. Jpn.* **59**, 3053; T. Shimizu and H. Kamimura: 1990, *J. Phys. Soc. Jpn.* **59**, 3691; The problem of coexistence of magnetic correlation and superconductivity is also discussed in A. Ohta, M. Nojima, T. Suzuki and Y. Natsume: 1989, *J. Phys. Soc. Jpn.* **58**, 4147 and references therein.
3. For example, T. Oguchi, *The Statistical Theory of Magnetic Materials*, Shoukabou, Tokyo (1970).
4. One of the most noteworthy work is S. Liang, B. Doucot and P.W. Anderson: 1988, *Phys. Rev. Lett.* **61**, 365; we would like to point out that the criticism on this work is offered by D.A. Huse and V. Elser: 1988, *Phys. Rev. Lett* **60**, 2531.
5. For example, S. Miyashita: 1988, *J. Phys. Soc. Jpn.* **57**, 1934; Y. Okabe and M. Kikuchi: 1988, *J. Phys. Soc. Jpn.* **57**, 4351; M. Kikuchi and Y. Okabe: 1989, *J. Phys. Soc. Jpn.* **58**, 679; J.D. Reger and A.P. Young: 1988, *Phys. Rev. B* **37**, 5478; J.D. Reger, J.A. Riera and A.P. Young: 1989, *J. Phys.: Condens. Matter* **1**, 1855; D.A. Huse: 1988, *Phys. Rev. B* **37**, 2380; M. Gross, E. Sanches. Velasca and E. Siggia: 1989, *Phys. Rev. B* **39**, 2484; G. Gomez-Santos, J.D. Joannopoulos and J.W. Negele: 1989, *Phys. Rev. B* **39**, 4435.
6. S. Nakagawa, T. Hamada and J. Kane, M.Sc. Thesis, Chiba University, 1989. The algorithm for the calculation has been investigated by M. Matsushita, M.Sc. Thesis, Chiba University, 1987.
7. Systems for $N \leq 16$ are discussed by J. Oitmaa and D.D. Betts: 1978, *Can. J. Phys.* **56**, 897; The case of $N = 20$ is treated by S. Fujiki and D.D. Betts: 1986, *Can. J. Phys.* **64**, 876; The case of 2×2 2×12 is investigated by E. Dagotto and A. Moreo: 1988, *Phys. Rev. B* **38**, 5087.
8. As for the system of $N = 26$, results by exact diagonalization are reported as follows besides Reference [6]; S. Tang and J.E. Hirsch: 1989, *Phys. Rev. B* **39**, 4548.

9. Quite recently, the numerical study for $N = 32$ has been performed by H.Q. Lin.: 1990, *Phys. Rev. B* **42**, 6561.
10. S. Nakagawa, T. Hamada, J. Kane and Y. Natsume: 1990, *J. Phys. Soc. Jpn.* **49**, 1131.
11. Y. Natsume, T. Hamada, S. Nakagawa, J. Kane and M. Katagiri: 1990, *Physica B* **165 & 166**, 175.
12. The review article is Y. Natusme: 1990, *Butsuri* (the Physical Society of Japan) **45**, 241 [in Japanese].
13. W. Marshall: 1955, *Proc. Roy. Soc. A* **232**, 48.
14. E. Lieb and D. Mattis: 1962, *J. Math. Phys.* **4**, 749.
15. E. Lieb and D. Mattis: 1962, *Phys. Rev.* **125**, 164.
16. P.W. Anderson: 1987, *Science* **235**, 1196; P.W. Anderson: 1973, *Mat. Res. Bull.* **8**, 153; P.W. Anderson, S. Baskaran, Z. Zhon and T. Hsu: 1987, *Phys. Rev. Lett. B* **58**, 2790.
17. M. Ma: 1990, *Phys. Rev. B* **38**, 6813.
18. The approach using the mean-field theory is described by G. Baskaran, Z. Zou and P.W. Anderson: 1987, *Solid State Commun.* **63**, 973; D. Yoshioka: 1989, *J. Phys. Soc. Jpn.* **58**, 32.
19. The pioneer work has been performed by M. Inoue, T. Chikyu, X. Hu and M. Suzuki: 1988, *J. Phys. Soc. Jpn.* **57**, 3733.
20. M.D. Johnson and K.R. Subbaswamuy: 1988, *Phys. Rev. B* **37**, 9390; M. Kohmoto: 1988, *Phys. Rev. B* **37**, 3812; M. Kohmoto and Y. Shapir: 1989, *Phys. Rev. B* **37**, 9439; H. Tasaki: 1989, *Phys. Rev. B* **40**, 9183.
21. B. Sutherland: 1988, *Phys. Rev. B* **37**, 3786.
22. C. Dasgupta: 1989, *Phys. Rev. B* **39**, 386.
23. The role of RVB with long distance is also pointed out for various models by T. Oguchi and H. Kitatani: 1989, *J. Phys. Soc. Jpn.* **58**, 1403; the review article is by T. Oguchi: 1989, *Butsuri* (The Physical Society of Japan) **44**, 95 [in Japanese].
24. A Mathematical study for obtaining the complete set has been reported for the case of $N = 10$ in terms of permutation operations by R. Saito: 1989, *Solid State Commun.* **72**, 512. The completeness of the set he discusses has also been proved in R. Saito: 1990, *J. Phys. Soc. Jpn.* **59**, 482.
25. P.L. Iske and W.J. Caspers: 1987, *Physica* **142A**, 360; W.J. Caspers: 1989, *Spin Systems*, World Scientific.
26. As for a chain system, the essential role of RVB with long distance is pointed out on the basis of an exact solution by T. Hamada, J. Kane, S. Nakagawa and Y. Natsume: 1988, *J. Phys. Soc. Jpn.* **57**, 1891; Y. Natsume, T. Hamada, S. Nakagawa and J. Kane: 1989, *J. Phys. Soc. Jpn.* **58**, 3869.
27. As to the problem of how to determine this number, a comprehensive study on the basis of the group theory has been recently reported for $N \leq 20$ by T. Ishino, M.Sc. Thesis, Tokyo University, 1990; T. Ishino, R. Saito and H. Kamimura: 1990, *J. Phys. Soc. Jpn.* **59**, 3886. Further, the approach on the basis of Young's diagram is also noteworthy; R. Saito: 1990, *J. Phys. Soc. Jpn.* **59**, 482; R. Saito and K. Kusakabe: 1991, *J. Phys. Soc. Jpn.* **60**, 2388; R. Saito: 1989, *Solid State Commun.* **72**, 517.
28. Y. Natsume *et al.*, *Progress in High Temperature Superconductivity*, Vol. 12 — High Temperature Superconductivity and Other Topics, World Scientific, Singapore (1989), p. 300.

29. In the case of $N = 10$, the work treating this problem is reported by D. Poilblanc: 1989, *Phys. Rev. B* **39**, 140; however, his calculated system is too small to discuss this problem, and the consideration of symmetry properties is insufficient.
30. This suggestion is given by H. Fukuyama.
31. This method of comparison has been suggested by T. Oguchi.
32. Here, the symmetry of time inversion $\hat{K}$ is not taken into account. On the contrary, in Figure 4, the number of patterns belonging to each 'principal' configuration is determined in consideration of $\hat{K}$.
33. T.A. Kaplan, P. Horsch and W. von der Linden: 1989, *J. Phys. Soc. Jpn.* **58**, 3894; K. Kubo, T.A. Kaplan and J.R. Borysowicz: 1988, *Phys. Rev. B* **38**, 11550.

TWO-DIMENSIONAL QUANTUM MAGNETS – MEAN FIELD AND BEYOND

ULRICH SCHOLLWÖCK and DAVID YUK KEI KO
Department of Theoretical Physics, University of Oxford, Oxford, OX1 3NP, U.K.

Abstract. Following the recent interest in path integral approaches to studying the properties of low-dimensional quantum magnets we present an examination of the mean field properties of a two-dimensional quantum magnet, together with a detailed analysis of the validity of mean field theory in determining the phase diagram. We find that there are serious restrictions to the validity of the mean field approximation in these systems, the most important of which is an uncontrolled shifting of the zero of the free energy with respect to local spin rotation. We present a formalism which includes fluctuations beyond the mean field limit so that the problems may, in principle, be dealt with, and the effects of fluctuations on the mean field solutions may also be analysed.

1. Introduction

The phenomenon of magnetism may be approached from many different angles, but from the viewpoint of solid state physics it serves well to remember its origin in electron-electron correlations. It is within this context of strong electron correlation that the subject of two-dimensional magnetism and the Mott-Hubbard metal-insulator transitions was recently revived with the discovery of the high temperature superconducting copper oxide materials [1], where the interaction between magnetism and electron conduction is suspected to be important to the mechanism of the superconductivity. The exact role of this interplay between the two phenomena is still not fully understood and in the context of the high temperature superconductors the problem has an added dimension. In these materials the relevant aspect of the physical structure are the two-dimensional copper oxide layers in which the carriers are confined such that it is the *two-dimensional* antiferromagnetic ordering which is believed to be relevant. This reduced dimensionality has strong implications on the physics and is, in fact, crucial to some of the more exotic ideas proposed, such as non-Fermi liquids [2] and anyonic behaviour [3]. This is perhaps not surprising since a reduced dimensionality will strongly enhance quantum fluctuations in an antiferromagnet, and in the extreme quantum case of the two-dimensional spin-$\frac{1}{2}$ antiferromagnet one might almost expect there to be new physical phenomena associated.

Quantitatively, not much is known about the properties of the two-dimensional quantum magnets and questions on the fundamental nature of the ground state spin order itself are still unanswered. For example, it is not certain whether the classical Néel spin ordering exists even at zero temperature due to the disordering effects of strong quantum fluctuations.

H. Aoki et al. (eds.), New Horizons in Low-Dimensional Electron Systems, 295–315.

The fact that there are substantial quantum fluctuations is reflected in the breakdown of an assumed antiferromagnetic Néel order when a small static hole doping is introduced [4]. In some ways the presence of the strong fluctuations is also reflected by the variational mapping by Inui *et al.* [5] of the two-dimensional strong coupling Hubbard model with a small mobile hole concentration δ away from half filling to an effective frustrated Heisenberg system with second and third nearest neighbour antiferromagnetic interactions. Not surprisingly, there have been a large number of candidates proposed for the ground state [6]. These include spiral antiferromagnets, disordered dimer states, resonating valence bond states and also the more exotic chiral states. The theoretical support for these proposals vary, but most are obtained within some sort of a mean field limit. What is known is that within spin wave approximation Chandra and Douçot [7] showed that the classical limit for the second nearest neighbour frustrated Heisenberg model is unstable to quantum fluctuations. This is true especially for $J_1/J_2 \approx \frac{1}{2}$ where the classical solutions are unstable even for large S. Schwinger boson functional integral formalisms have also been introduced to this problem, notably by Arovas and Auerbach [8] who considered the unfrustrated system. Within this path integral context a variety of work has been carried out. Read and Sachdev [9] found in the large N limit, where N is the number of Schwinger boson flavours, that a spin-Peierls dimer state or resonating valence bond solid may exist. Mean-field theory by Kane *et al.* [10] has further suggested that in the doped two-dimensional antiferromagnet doping favours a spiral spin phase. This has been extended by Chakraborty *et al.* [11] to show that for larger dopings a time-reversal symmetry breaking flux phase may also occur. More recently, Mila *et al.* [12] analysed the stability of the Néel state within a Schwinger boson framework with the conclusion that its stability is enhanced beyond the classical boundary by quantum fluctuations, with a gap opening in the spin fluctuation spectrum. This is a quantum version of order from disorder first discovered by Villain [13]. The possibilities of a twisted spin ordering has also been raised by Dagotto and Moreo [14], though more recent numerical calculations by Singh and Narayanan [15] have questioned this, at least in the limit of a small system.

With the obvious lack of consensus over even the fundamental nature of the ground state path integral approaches have emerged with a certain theoretical appeal. They inherently sample over all possible configurations, and therefore do not rely on any preassumed ground states (compare, for example, with the Néel state assumed in the usual spin wave approaches [16, 7]). These path integral approaches often carry with them, however, inherent and often quite subtle problems. The result is that it is difficult to judge sensibly the regime of their validity and this has inadvertently contributed to the confusion in the subject. What we aim at in this paper is first to present aspects of the physics of the quantum magnet in the mean field limit of these path integral approaches, and then to examine the more important of the problems inherent in these otherwise appealing methods so that, hopefully, some assessment of the extent to which their results may be applicable may be reached. We shall finish by presenting a formalism which analyses the problem beyond mean field theory so that the free energies for different phases may be compared more rigorously.

Before we begin the analysis proper we should point out specifically what the inherent problems with the path integral approaches are and what their consequences are. This is important to give the rest of the paper some perspective, and also to avoid the possibility of confusion due to the technical details which follow. The most serious drawback of path integral approaches is the well-known fact that they are really only tractable within a saddle point approximation. As with most field theories such an approximation is basically uncontrolled and can lead to unphysical breakings of inherent symmetries. In the case of the quantum magnet it is the $SU(2)$ symmetry of local spin rotations which is broken. The implication is that the free energy is no longer $SU(2)$ gauge invariant. Consequently, the zero of the free energy for a given ground state configuration is gauge dependent, and in a way which is not accessible to analysis at the mean field level. Thus we cannot determine the degree to which an optimal $SU(2)$ gauge rotation may be chosen for a given spin configuration such that mean field theory is at its most accurate, nor can we compare results obtained under different $SU(2)$ gauges.

Another important problem concerns the role of fluctuations in the system. This again is associated with the fact that the formalism is only tractable in the saddle point limit so that only mean field results, where all fluctuation effects are absent, may be obtained. Clearly, if one believes that there are strong fluctuations present it is not clear to what extent mean field results can be trusted. In other words it is not obvious that the various order parameters which have been discussed in these systems in the mean field context can survive in the real physical system.

In terms of both of the problems mentioned what is needed is a tractable means of going beyond the saddle point approximation. This should recover the inherent symmetry of the system, and allow us to assess the importance of fluctuation effects. Up till now there has really been very little that has been carried out with respect to this beyond numerical simulations on small systems. In the second half of this paper we give the formalism for developing a diagrammatic method for such an expansion beyond the mean field limit. The result is suggestive that development beyond mean field theory is feasible, and that the major problems associated with the mean field approximation may be avoided. Before we go on to that, however, we shall begin by giving the background behind the mean field limit.

2. Mean-Field Theory

The starting point is the quantum magnet in two-dimensions on a bipartite lattice which, in general, need not be restricted to nearest neighbour interactions. The Hamiltonian for the case up to second nearest neighbours is

$$H = J_1 \sum_{1^{st}} \mathbf{S}_i\mathbf{S}_j + J_2 \sum_{2^{nd}} \mathbf{S}_i\mathbf{S}_j. \tag{1}$$

Within the path integral formalisms, there are essentially two approaches one usually takes, both of which replace the non-commuting spin operators by simple bosons or fermions. This simplifies the problems of the spin algebra but, in some ways, is responsible for the subsequent problems. For the bosonic representation Schwinger bosons [17] are usually chosen. For spin-$\frac{1}{2}$ these replace each spin raising and lowering operator by two boson operators such that $S_i^+ = b_{i\uparrow}^\dagger b_{i\downarrow}$, $S_i^- = b_{i\uparrow} b_{i\downarrow}^\dagger$ and $S_i^z = \frac{1}{2}(b_{i\uparrow}^\dagger b_{i\uparrow} - b_{i\downarrow}^\dagger b_{i\downarrow})$, or more compactly $\mathbf{S}_i = \frac{1}{2} b_{i\sigma}^\dagger \boldsymbol{\sigma}_{\sigma\sigma'} b_{i\sigma'}$ where $\boldsymbol{\sigma} = (\sigma_1, \sigma_2, \sigma_3)$ are the Pauli spin matrices. A simple interpretation of these boson operators may be obtained by imagining each S^z eigenvalue as corresponding to a different bosonic excitation. For a fermionic representation, the idea is again similar in that one replaces the spin operators by combinations of fermion operators obeying anticommutation relations. In this case $S_i^+ = c_{i\uparrow}^\dagger c_{i\downarrow}$, $S_i^- = c_{i\downarrow}^\dagger c_{i\uparrow}$ and $S_i^z = \frac{1}{2}(c_{i\uparrow}^\dagger c_{i\uparrow} - c_{i\downarrow}^\dagger c_{i\downarrow})$, or again $\mathbf{S}_i = \frac{1}{2} c_{i\sigma}^\dagger \boldsymbol{\sigma}_{\sigma\sigma'} c_{i\sigma'}$.

In terms of these new operators the Hamiltonian (1) may be rewritten to give

$$H = \sum_{ij} J_{ij}\mathbf{S}_i\mathbf{S}_j = \sum_{ij} J_{ij}[\frac{1}{2} : D_{ij}^\dagger D_{ij} : -S^2] \tag{2}$$

where $D_{ij}^\dagger = b_{i\uparrow}^\dagger b_{j\uparrow} + b_{i\downarrow}^\dagger b_{j\downarrow}$ for the bosonic representation, or

$$H = \sum_{ij} J_{ij}\mathbf{S}_i\mathbf{S}_j = \sum_{ij} J_{ij}[S^2 - \frac{1}{2} : C_{ij}^\dagger C_{ij} :] \tag{3}$$

with $C_{ij} = c_{i\uparrow}^\dagger c_{j\uparrow} + c_{i\downarrow}^\dagger c_{j\downarrow}$ for the fermionic representation. The partition function for the system is obtained in a functional integral following standard procedures [19]. This involves expressing $Z = \mathrm{Tr}\, e^{-\beta\mathcal{H}}$ in the form

$$Z = \int_{c-i\infty}^{c+i\infty} \mathrm{d}\lambda_j \int \mathcal{D}[b^\dagger, b] \exp - \int_0^\beta \mathrm{d}\tau \{ \sum_{j\sigma} (\frac{1}{2} b_{j\sigma}^\dagger \overleftrightarrow{\partial}_\tau b_{j\sigma} - \lambda_j (b_{j\sigma}^\dagger b_{j\sigma} - S)) + H \}, \tag{4}$$

for the bosonic representation, with the operators expressed by complex fields. From this the mean field free energy is calculated. Note that the second term in (4) is a constraint over the number of Schwinger bosons, and arises in this form as a result of expressing the usual delta function constraints in an inverse Laplace transformation. The λ_j is a Lagrangian multiplier field which in actual calculations is customarily treated as uniform over the lattice. This means that the contraint is satisfied globally rather than locally on each site. The effect of this will be discussed later. The corresponding expression for the partition function for the fermionic representation differs only in that the operators are Grassmann fields.

It is perhaps appropriate at this point to discuss the local $SU(2)$ symmetry of the system. This arises due to an invariance of the physics of the quantum magnet to *local* spin rotations. Let us first recall that one of the conventional approaches to calculating the statistical properties of classical bipartite Heisenberg antiferromagnets is first to map it to a ferromagnet by means of a *global* rotation of one sublattice. The essential point here is that with a quantum (frustrated) system it too may also be mapped in a similar way as long as the relevant rotations are now *locally* determined. The result is the introduction of a local $SU(2)$ gauge field which represents the local rotations of the spins at their respective sites, and an associated vector potential which acts as a book-keeping for the various changes in the axes of rotations which have occurred. Why is this important? After all, the cost of the mapping is to produce a much more complicated system. One has now not only the original spins to consider, but also an additional gauge field and the associated vector potential. The answer lies in the fundamental physical assumption behind mean field theory.

It is the definition of mean field theory that the field felt by any spin may be taken as the simple averaged contribution of the field from all other spins. Consequently, it will be most accurate when the physical situation corresponds most closely to this assumed situation. That is, when all the spins are identically aligned. It follows then that the mean field approximation, on which as we have mentioned the path integral formulation is fundamentally based, is best for ferromagnetic situations where the spins do align identically. With this in mind, it is therefore desirable first to transform whatever spin configurations to a ferromagnetic arrangement before the mean field limit is taken. In a non-frustrated bipartite antiferromagnet, this is the well-known global rotation of one sublattice. For a general quantum magnet with frustration this rotation to a ferromagnetic configuration may be obtained only by a local spin rotation of the type we have just discussed. Essentially, one exploits the $SU(2)$ invariance of the quantum spin-$\frac{1}{2}$ magnet and the corresponding gauge freedom to perform suitable local spin rotations so that the resulting configuration is optimal for the accuracy of a mean field analysis.

This, as shown by Chandra, Coleman and Larkin [20], corresponds to expressing the local $SU(2)$ spin rotations as rotations in our operators $b^{\dagger}_{i\sigma} = b'^{\dagger}_{i\sigma} \exp(\mathrm{i}\frac{1}{2}\boldsymbol{\theta}_i.\boldsymbol{\sigma})$ where $\boldsymbol{\theta}_i$ gives the rotation at the site i with respect to the rotation axis. The transformed spin then reads as $\mathbf{S}'_j = \frac{1}{2} b'^{\dagger}_{j\sigma}\boldsymbol{\sigma}_{\sigma\sigma'} b'_{j\sigma'}$. The situation for the fermionic case is essentially identical. The result, if we further restrict the local rotation between spins at sites i and j to the form $\boldsymbol{\theta}_{ij} = \mathbf{q}.(\mathbf{r}_i - \mathbf{r}_j)$ about a fixed axis $\hat{z}$ is that we arrive at a gauge invariant Hamiltonian [20]

$$H = \frac{1}{2}\sum_{ij} J_{ij}\{\gamma^+[\mathbf{S_i}\cdot\mathbf{S_j}] - \gamma^-[\mathbf{S_i}\cdot\mathbf{S_j} - 2S^z_i S^z_j]\} \tag{5}$$

where $\gamma^{\pm} = \frac{1}{2}(1 \pm \cos \mathbf{q}(\mathbf{r_j} - \mathbf{r_i}))$ and $\mathbf{q}$ is an $SU(2)$ gauge vector such that $\mathbf{q} = (0,0)$ implies that no rotation of the spins is necessary so that the system starts off as a ferromagnet. This thus corresponds to the "ferromagnetic" gauge. At the other extreme $\mathbf{q} = (\pi,\pi)$ reverses the orientation of every alternate spin such that the original configuration is that of the classical Néel magnet. This is therefore the "antiferromagnetic" gauge.

Within such a gauge invariant Hamiltonian, we are further free to define composite operators which reflect the two ferromagnetic and antiferromagnetic limits of the possible spin configurations. Thus, in the bosonic representation by defining operators $D^{\dagger}_{ij} = b^{\dagger}_{i\uparrow}b_{j\uparrow} + b^{\dagger}_{i\downarrow}b_{j\downarrow}$ for a ferromagnetic pairing and similarly $B^{\dagger}_{ij} = b^{\dagger}_{i\uparrow}b^{\dagger}_{j\downarrow} + b^{\dagger}_{i\downarrow}b^{\dagger}_{j\uparrow}$ for the sublattice rotated situation appropriate for an antiferromagnet, the $SU(2)$ gauge invariant Hamiltonian (5) may be rewritten as

$$H = \sum_{bonds} \frac{1}{2} J_{ij}\{\gamma^+ : D^{\dagger}_{ij}D_{ij} : -\gamma^- B^{\dagger}_{ij}B_{ij}\} - \sum_{ij}\frac{1}{2}J_{ij}S^2 \cos \mathbf{q}(\mathbf{r_j} - \mathbf{r_i}). \tag{6}$$

The corresponding expression for the fermionic representation is

$$H = \sum_{bonds} \frac{1}{2} J_{ij}\{-\gamma^- : C^{\dagger}_{ij}C_{ij} : +\gamma^+ E^{\dagger}_{ij}E_{ij}\} - \sum_{ij}\frac{1}{2}J_{ij}S^2 \cos \mathbf{q}(\mathbf{r_j} - \mathbf{r_i}) \tag{7}$$

with $E^{\dagger}_{ij} = c^{\dagger}_{i\uparrow}c^{\dagger}_{j\downarrow} + c^{\dagger}_{i\downarrow}c^{\dagger}_{j\uparrow}$ in addition to the $C^{\dagger}_{ij}$ defined previously. The summations over the bonds signify a summation over pairs of sites connected by an exchange interaction J_{ij}. Note the interchange of $\gamma^{\pm}$ in the fermionic Hamiltonian $\mathbf{q}$ has been appropriately redefined.

We thus arrive at an explicitly $SU(2)$ gauge invariant form of the partition function, from which the thermodynamical properties of the two-dimensional quantum magnet may be obtained. To go on further we first have to decouple the terms which are quadratic in the composite boson operators D_{ij}, B_{ij} or the fermion operators C_{ij} and E_{ij}. This may be done using a Gaussian identity known as the Hubbard-Stratonovich transformation [18, 19]

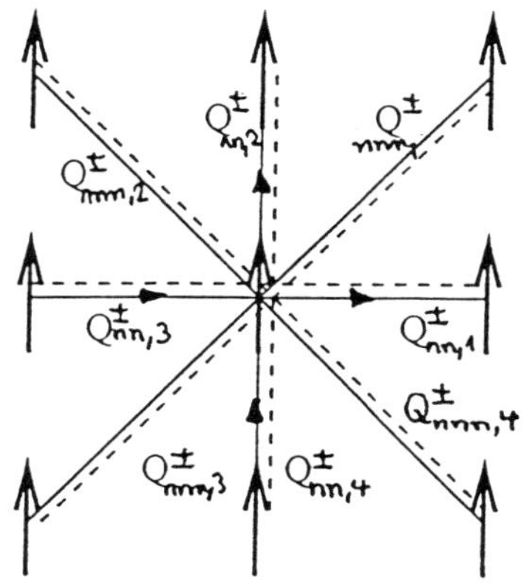

Fig. 1. Hubbard–Stratonovich auxiliary fields for a uniform configuration without symmetry breaking. The dashed and the solid lines show the ferromagnetic and the antiferromagnetic bonds respectively

$$\int \mathcal{D}[X^\dagger, X] e^{\int \mathrm{d}\tau X^\dagger V X} = \frac{\int \mathcal{D}[Q^\dagger, Q, X^\dagger, X] e^{\int \mathrm{d}\tau - Q^\dagger V^{-1} Q + X^\dagger Q + X Q^\dagger}}{\int \mathcal{D}[Q^\dagger, Q] e^{-\int \mathrm{d}\tau Q^\dagger V^{-1} Q}}, \tag{8}$$

where X represents a composite operator. This introduces, see figure 1, two auxiliary field $Q^{\pm}_{ij}$ for each bond between sites i and j, one corresponding to the ferromagnetic gauge and the other corresponding to the antiferromagnetic gauge term of the Hamiltonian (5). The denominator is a normalisation factor which is $SU(2)$ gauge dependent. The partition function now becomes

$$\begin{aligned} Z = N^{-1} \int_{c-\mathrm{i}\infty}^{c+\mathrm{i}\infty} \mathrm{d}\lambda_j \int \mathcal{D}[b^\dagger, b] \exp - \int_0^\beta \mathrm{d}\tau \sum_{j\sigma} (\frac{1}{2} b^\dagger_{j\sigma} \overleftrightarrow{\partial}_\tau b_{j\sigma} \\ - \lambda_j (b^\dagger_{j\sigma} b_{j\sigma} - S)) + \mathcal{H} \end{aligned} \tag{9}$$

with the Hamiltonian $\mathcal{H}$ for the boson representation modified to

$$\begin{aligned} \mathcal{H} = \sum_{bonds} - \frac{2}{J_{ij}\gamma^+} |Q^+_{ij}|^2 + \frac{2}{J_{ij}\gamma^-} |Q^-_{ij}|^2 + D^\dagger_{ij} Q^+_{ij} + D_{ij} Q^{+*}_{ij} \\ + B^\dagger_{ij} Q^-_{ij} + B_{ij} Q^{-*}_{ij} - \sum_{ij} \frac{1}{2} J_{ij} S^2 \cos \mathbf{q}.(\mathbf{r}_j - \mathbf{r}_i). \end{aligned} \tag{10}$$

The expression is again similar for the fermionic representation.

The next steps towards obtaining the mean-field solution are now relatively straightforward. One first makes the connection that the lattice symmetry of the ground state spin configuration is reflected in the symmetry of the Hubbard-Stratonovich auxiliary fields. That is, these auxiliary fields act as an effective order parameter arising from the interaction between sites such that, for example, the flux phase may be expressed through variations in the phases of the auxiliary fields $Q_{ij}^{\pm}$ [21], or the dimer state may be chosen by setting to zero appropriate $Q_{ij}^{\pm}$'s. Next the terms in the exponential of the partition function may be diagonalised by a Fourier transform of the remaining operators $b_{i\sigma}$, $b_{i\sigma}^{\dagger}$or $c_{i\sigma}$, $c_{i\sigma}^{\dagger}$into momentum space and followed by a Bogoliubov transformation. This gives for the partition function an expression

$$Z = \frac{\int \mathrm{d}\lambda \int \mathcal{D}[b^{\dagger}, b] \int \mathcal{D}Q_i \exp\{-\int_0^{\beta} \mathrm{d}\tau\ Q_i V_{ij}^{-1} \overline{Q_j} + \mathcal{H}_R(b^{\dagger}, b, \lambda, Q_i)\}}{\int \mathrm{d}\lambda \int \mathcal{D}[b^{\dagger}, b] \int \mathcal{D}Q_i \exp\{-\int_0^{\beta} \mathrm{d}\tau\ Q_i V_{ij}^{-1} \overline{Q_j}\}}. \quad (11)$$

in which the term $\mathcal{H}_R$ is diagonal in the transformed boson (fermion) operators. The matrix V_{ij}^{-1} is diagonal and contains the coefficients to the quadratic contributions in the Hubbard-Stratonovich auxiliary fields in the Hamiltonian (10). After integrating out the boson or fermion operators the saddle point partition function Z_{SP} may be calculated by a saddle point approximation and the corresponding free energy $F_{SP} = -(1/\beta) \log Z_{SP}$ obtained. Note that since the normalisation factor in the denominator is neither a function of the auxiliary fields nor of the Lagrangian constraint it is usually ignored.

As an example let us consider a situation with no lattice symmetry breaking in a frustrated system, as illustrated in Figure 1. For this, there are four inequivalent auxiliary fields, two for each of the ferromagnetic and antiferromagnetic gauge terms corresponding to the nearest-neighbour and the next-nearest-neighbour spin pairs. The effective Hamiltonian in the partition function becomes

$$\begin{aligned}\mathcal{H} &= -\frac{4N}{J_1(1+\cos q_{x,y})}|Q_1^+|^2 - \frac{4N}{J_2(1+\cos(q_x \pm q_y))}|Q_2^+|^2 \\ &\quad + \frac{4N}{J_1(1-\cos q_{x,y})}|Q_1^-|^2 + \frac{4N}{J_2(1-\cos(q_x \pm q_y))}|Q_2^-|^2 + \mathcal{H}_{\mathcal{R}} \qquad (12)\end{aligned}$$

with

$$\mathcal{H}_{\mathcal{R}} = \sum_{\mathbf{k},\sigma} \omega_{\mathbf{k}}(b_{\mathbf{k}\sigma}^{\dagger} b_{\mathbf{k}\sigma} + \frac{1}{2}) - \frac{S^2}{2} \sum_{ij} J_{ij} \cos \mathbf{q}(\mathbf{r_j} - \mathbf{r_i}) - N\lambda(2S+1). \quad (13)$$

The auxiliary fields for the ferromagnetic gauge have been chosen to be real for convenience. This Hamiltonian corresponds to a system with excitation energies $\omega_{\mathbf{k}} = \sqrt{\eta_{\mathbf{k}}^2 - |\xi_{\mathbf{k}}|^2}$ where $\eta_{\mathbf{k}} = \lambda + 2Q_1^+(\cos k_x + \cos k_y) + 4Q_2^+ \cos k_x \cos k_y$ and $\xi_{\mathbf{k}} = 2Q_1^-(\cos k_x + \cos k_y) + 4Q_2^- \cos k_x \cos k_y$. The resulting saddle-point free energy F_{SP} is given by

$$F_{SP} = \frac{2}{\beta}\int_{B.Z.}\frac{d^2\mathbf{k}}{(2\pi)^2}\ln(2\sinh\frac{\beta\omega_{\mathbf{k}}}{2}) \\ -S^2(J_1(\cos q_x + \cos q_y) + 2J_2\cos q_x\cos q_y) - (2S+1)\lambda + E_{HS}. \quad (14)$$

E_{HS} is a term quadratic in the Hubbard–Stratonovich auxiliary fields. This form for the free energy obtained at the saddle point limit is generic and as we can clearly see, it is explicitly dependent on the $SU(2)$ gauge vector $\mathbf{q}$.

To obtain the mean-field free energy F_{MF} for a given gauge vector $\mathbf{q}$ we need of course to minimise F_{SP} with respect to the auxiliary fields and the Lagrangian constraint.

$$\left.\frac{\partial F_{SP}}{\partial\lambda}\right|_{F_{SP}=F_{MF}} = \left.\frac{\partial F_{SP}}{\partial Q_1^{\pm}}\right|_{F_{SP}=F_{MF}} = \left.\frac{\partial F_{SP}}{\partial Q_2^{\pm}}\right|_{F_{SP}=F_{MF}} = 0. \quad (15)$$

In Figure 2 we compare the free energies at different temperatures for different configurations of the auxiliary fields in the unfrustrated system, obtained using a pure antiferromagnetic gauge vector $\mathbf{q} = (\pi, \pi)$ in the bosonic representation throughout. The corresponding situation in the fermionic representation, again with $\mathbf{q} = (\pi, \pi)$, is shown in Figure 3. Note that for the bosonic representation there is no mean field solution corresponding to a dimer state. The solid lines show the free energies and the dashed lines show the corresponding auxiliary field's magnitude. For configurations we have chosen there is only one auxiliary field strength. The results (Figure 2) show that in the bosonic representation the uniform phase, where the auxiliary fields on all the bonds are identical, has a lower free energy than the flux phase at all temperatures. The flux phase we consider has a plaquette flux of π distributed over the four Hubbard–Stratonovich bonds in the form 0, $\frac{\pi}{2}$, 0, $\frac{\pi}{2}$. There is a transition in both the uniform and the flux phase to the paramagnetic phase, where all the auxiliary fields are trivially zero, at a temperature of $0.9J_1$. In the fermionic situation the results are again similar, except this time the phase with the lowest free energy is the dimer phase. Again, there is a transition to a paramagnet, this time at a lower temperature of $0.25J_1$.

For a given $SU(2)$ gauge then, what our results seem to suggest is that the phase diagram for the two-dimensional quantum magnet may be obtained by comparing the free energies of the different possible ground states. However, as the fact that the saddle point approximation gives positive free energies for the fermionic representation is indicating, this is not the whole story. Since the accuracy of the mean-field approximation depends on the appropriateness of the gauge vector used with respect to the assumed ground state, we cannot simply compare the free energies of different ground states for an arbitrary gauge vector, but should compare the free energies at their

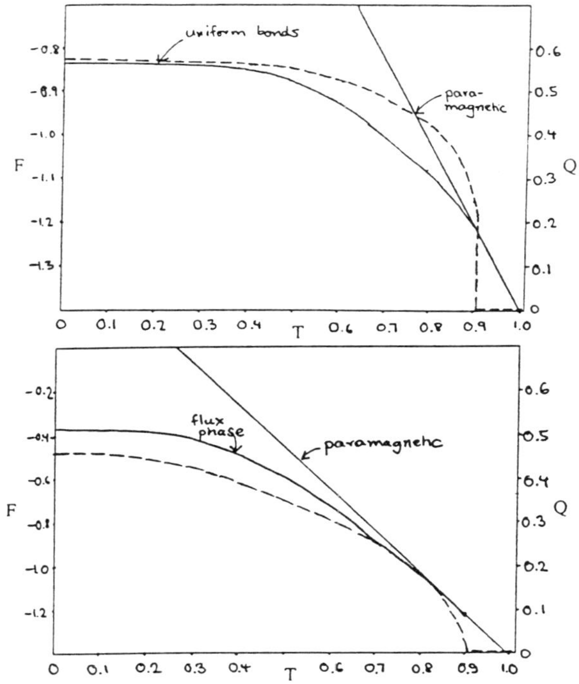

Fig. 2. Temperature dependence of the free energy in the bosonic representation for different auxiliary field configurations. The dashed lines show the corresponding magnitude of the auxiliary field

corresponding optimal gauge vectors. This brings on the problem that although the normalisation factor in the Hubbard–Stratonovich transformation (8) neglected in the saddle point approximation does not depend on the auxiliary field configuration, it is actually $SU(2)$ gauge *dependent* and contributes to the free energy. It is partly due to neglecting this contribution in the saddle point approximation that the mean field free energy becomes explicitly $SU(2)$ gauge dependent. Thus, when we are going to a different gauge vector the zero of the mean field free energy shifts in an unknown manner, governed essentially by the normalisation factor, such that the free energies obtained using different gauge vectors may not be compared. We cannot easily evaluate the normalisation factor itself, since the integral form expressed in (8) is not always well defined when it taken out of the discrete form in which it is defined into a continuum limit.

One thing that we may do is to find a gauge appropriate for all ground state symmetries of the system. This condition implies is that the gauge vector $\mathbf{q}$ should be independent of the Hubbard–Stratonovich auxiliary fields

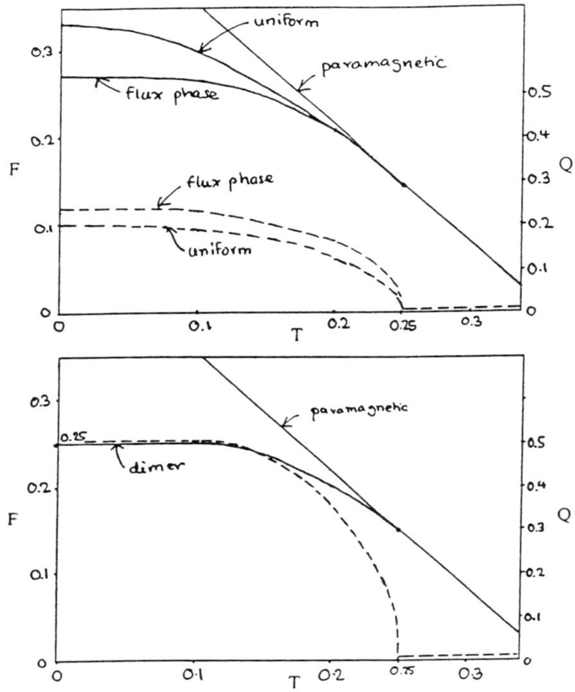

Fig. 3. Temperature dependence of the free energy in the fermionic representation for different auxiliary field configurations. The dashed lines show the corresponding magnitude of the auxiliary fields

$Q^{\pm}_{ij}$ since the symmetry properties of the ground state are reflected in the values of these auxiliary fields. Carrying this out then leads to the gauge fixing condition

$$\nabla_{\mathbf{q}}[J_1(\cos q_x + \cos q_y) + 2J_2 \cos q_x \cos q_y] = 0 \tag{16}$$

from which we find that for all values of J_i both $\mathbf{q} = (0,0)$ and $\mathbf{q} = (\pi,\pi)$ lead to gauge-stationary solutions. They correspond to purely ferromagnetic and purely antiferromagnetic systems, and the first (second) solution may be dismissed if considering antiferromagnets (ferromagnets). For $J_2/J_1 \geq 0.5$, a further solution with

$$\mathbf{q} = (\arccos -J_1/2J_2, \arccos -J_1/2J_2)$$

branches from the antiferromagnetic solution. This reflects the physically sensible results derived from spin-wave theory and expected from the classical limit. The branching off of a second solution corresponds with the spin-wave prediction of an ordered Néel state for $\kappa = J_2/J_1 < 0.5$ and a

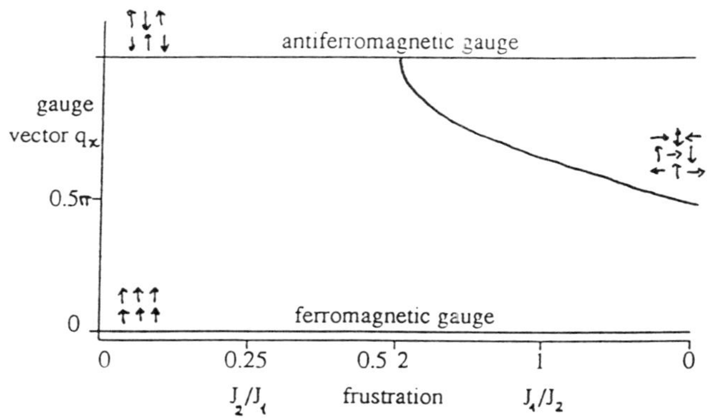

Fig. 4. Dependence of the approximated gauge vector component q_x on the exchange energies J_1 and J_2. The inset shows the corresponding classical spin order

phase transition when $\kappa = 0.5$. Taking the physical situation to the limit of $J_2 \gg J_1$ one would expect the emergence of sublattices with opposite magnetization which exhibit a lattice constant of $2a$ instead of $\sqrt{2}a$ as in the ordinary Néel case (Figure 4). This is partially reflected by the limiting value of $\mathbf{q} = (\pi/2, \pi/2)$. Rigorously, however, such a gauge vector universally appropriate to all symmetry breakings does not exist, and the free energies obtained under different gauges will have to be compared. Physically, at the saddle point level, this is meaningless as the gauge invariance of the free energy has been violated by the approximations used.

Partially to compensate for this, we may demand that the mean-field free energy F_{MF} is at least locally gauge invariant for the gauge chosen. That is, in addition to (15) we require that

$$\left.\frac{\partial F}{\partial q_x}\right|_{F_{SP}} = \left.\frac{\partial F}{\partial q_y}\right|_{F_{SP}} = 0. \tag{17}$$

In Figure 5 the dependence of the mean-field free energy on the gauge vector is shown for a temperature of $T = 0.1J_1$ in an unfrustrated system. We can see that this condition does not fully eliminate the problem with the dependence on the gauge vector as there are in both the bosonic (Figure 5.a) and the fermionic (Figure 5.b) representations more than one stationary point in the free energy with respect to the gauge vector. The appropriate choice can only be made by guessing which choice of the gauge vector will lead to the most physically intuitive result.

All this is clearly not particularly satisfactory, though it may, in many cases, be all that one can hope for. There is, however, a further point which we have not discussed so far. The actual real ground state must be one which is stable with respect to the quantum and thermal fluctuations present in the

(5.a)

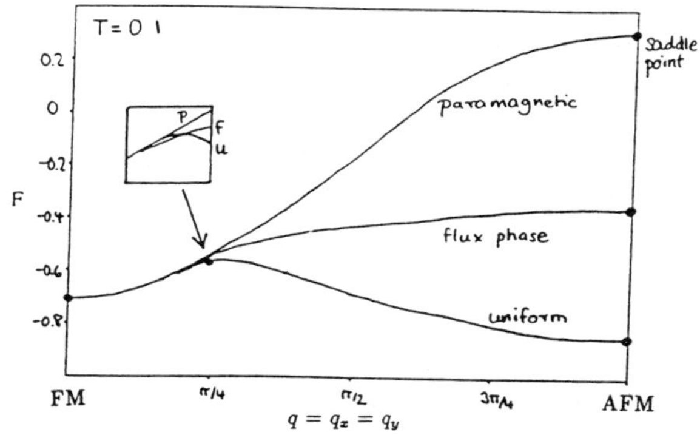

(5.b)

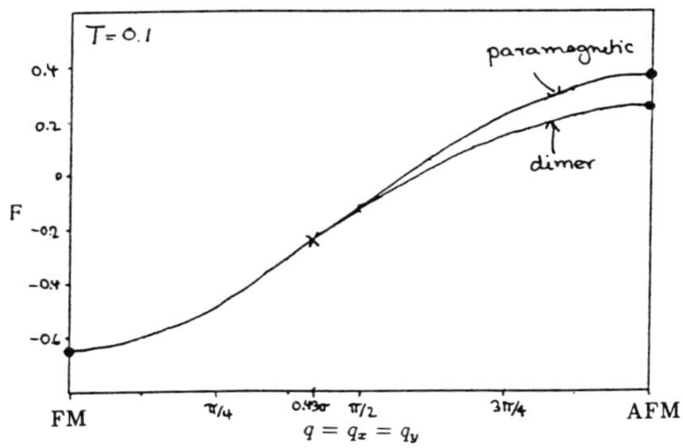

Fig. 5. Gauge dependence of the free energy on the $SU(2)$ gauge vector in the bosonic representation (Fig. 5a) and the fermionic representation (Fig. 5b) for different auxiliary field configurations

real system. In two-dimensions these fluctuations are believed to be strong so that before a mean field phase diagram, however obtained, may be judged to be realistic the order parameters must be shown to be stable. Thus, in order to develop a controlled assessment of the free energy and to be able to examine the validity of the phases predicted by the mean field results, it is necessary to go to an expansion beyond the saddle-point approximation.

3. Beyond Mean-Field

The approach we take to develop a formalism beyond mean field theory is to ignore initially the idea of expanding about a small parameter, but to expand about the remainder in our action that is left after the saddle point approximation and perturb in the auxiliary fields. This is physically attractive since if the mean field result is good the remainder would naturally be small. Furthermore, it is also equivalent to a $1/N$ expansion in the particular case of the quantum magnet. Before we proceed we should, however, ask what is the relevance of a finite temperature expansion when it is known through the Mermin–Wagner theorem [22] that in two-dimensions there is no long range order at finite temperatures. Consequently, any expansion might be expected to diverge such that whether the zero-temperature limit can be taken is doubtful. However, one expects that there is still short range order remaining in the system from which the zero temperature state may be deduced. In our expansion then, we shall explicitly exclude long wavelength infrared fluctuations which disorder the system by a cutoff which is chosen to correspond to the correlation length of the short range order, with the believe that the convergence behaviour of the free energy and the temperature behaviour of the cutoff will give us an indication of the stability of the different candidate phases. One would expect this simple procedure described to be sufficient for certain ground state configurations, such as the dimer state, which are in some sense already disordered at the mean field level.

The idea of a pertubation series in terms of the auxiliary fields was first developed by Blaizot and Orland [23] in nuclear physics. We develop here a somewhat different formalism more appropriate to the quantum magnet problem. The starting point is again the partition function

$$Z = \frac{\int \mathcal{D}B \int \mathcal{D}Q_i \exp^{\int_0^\beta \mathrm{d}\tau \ \frac{1}{2}\sum Q_i V_{ij}^{-1} Q_j - H(B,Q_i)}}{\int \mathcal{D}Q_i \exp^{\int_0^\beta \mathrm{d}\tau \ \frac{1}{2}\sum Q_i V_{ij}^{-1} Q_j}}. \tag{18}$$

where we have for the moment ignored the integration over the Lagrangian constraint by selecting its mean field value as a fixed constant. We have also absorbed a minus sign into the V_{ij}^{-1}. Since the technical details become somewhat involved for the general case, we assume for simplicity the auxiliary field configuration which corresponds to a lattice symmetry broken antiferromagnet in the fermionic representation. We note that the mean field results for the fermionic representation all give zero for the Lagrangian constraint λ. The Fourier transformed Hamiltonian becomes

$$H = -\frac{1}{2}\sum_i Q_i^2(\tau)V_i^{-1} + \sum_{\mathbf{k}\sigma} \xi_{\mathbf{k}}(\tau) c_{\mathbf{k}\sigma}^{A\dagger} c_{\mathbf{k}\sigma}^{B} + \xi_{\mathbf{k}}^{\dagger}(\tau) c_{\mathbf{k}\sigma}^{B\dagger} c_{\mathbf{k}\sigma}^{A} \tag{19}$$

with $\xi_{\mathbf{k}} = Q_1^{\dagger} e^{ik_x} + Q_2^{\dagger} e^{ik_y} + Q_3 e^{-ik_x} + Q_4 e^{-ik_y}$ and $c_{\mathbf{k}\sigma}^{A}$ and $c_{\mathbf{k}\sigma}^{B}$ are the Fourier transforms of the fermion operators on sublattice A and B respectively. Note that ξ is linear in the auxiliary fields Q_i. We now split the Hamiltonian into the mean field part H_0 and a remainder H_R

$$H = [-\frac{1}{2}\sum_i (Q_i^0)^2 V_i^{-1} + \sum_{\mathbf{k}\sigma} \xi_{\mathbf{k}}^0 c_{\mathbf{k}\sigma}^{A\dagger} c_{\mathbf{k}\sigma}^{B} + \xi_{\mathbf{k}}^{0\dagger} c_{\mathbf{k}\sigma}^{B\dagger} c_{\mathbf{k}\sigma}^{A}]$$

$$
\begin{aligned}
&+[-\frac{1}{2}\sum_i (Q_i^2 - (Q_i^0)^2)V_i^{-1} + \sum_{\mathbf{k}\sigma} \xi'_{\mathbf{k}} c^{A\dagger}_{\mathbf{k}\sigma} c^B_{\mathbf{k}\sigma} + \xi'^{\dagger}_{\mathbf{k}} c^{B\dagger}_{\mathbf{k}\sigma} c^A_{\mathbf{k}\sigma}] \\
&= H_0 + H_R
\end{aligned} \tag{20}
$$

and use the definition of a thermal average $\langle A\rangle$ in the interaction picture for an operator A, $\langle A\rangle = \frac{\mathrm{Tr}A\ e^{-\beta H_0}}{\mathrm{Tr}\ e^{-\beta H_0}}$, to rewrite the partition function

$$
Z = N^{-1}\int \mathcal{D}Q_i \exp(\frac{1}{2}\int \mathrm{d}\tau \sum_i (Q_i)^2 V_i^{-1}) \tag{21}
$$

$$
\times \int \mathcal{D}[c^{\dagger}_{\mathbf{k}\sigma}, c_{\mathbf{k}\sigma}] \exp(-\{\int \mathrm{d}\tau\ H_0 + \sum_{\mathbf{k}\sigma} \xi'_{\mathbf{k}}(Q'_i) c^{A\dagger}_{\mathbf{k}\sigma} c^B_{\mathbf{k}\sigma} + \xi'^{\dagger}_{\mathbf{k}}(Q'_i) c^{B\dagger}_{\mathbf{k}\sigma} c^A_{\mathbf{k}\sigma}\})
$$

as

$$
\begin{aligned}
\frac{Z}{Z_0} &= N^{-1}\int \mathcal{D}Q'_i\ e^{(\frac{1}{2}\int \mathrm{d}\tau \sum_i (Q'_i)^2 V_i^{-1})} \\
&\left\langle e^{-\int \mathrm{d}\tau \left(\sum_{\mathbf{k}\sigma} \xi'_{\mathbf{k}}(Q'_i) c^{A\dagger}_{\mathbf{k}\sigma} c^B_{\mathbf{k}\sigma} + \xi'^{\dagger}_{\mathbf{k}}(Q'_i) c^{B\dagger}_{\mathbf{k}\sigma} c^A_{\mathbf{k}\sigma}\right)}\right\rangle .
\end{aligned} \tag{22}
$$

$N = \int \mathcal{D}Q'_i \exp(\frac{1}{2}\int \mathrm{d}\tau \sum_i (Q'_i)^2 V_i^{-1})$ is the normalization factor in the Hubbard–Stratonovich transformation (8). Note that Q_i has been rewritten as $Q_i = Q_i^0 + Q'_i$, with the mean field value denoted by the superfix 0[1]. The general technique is now to express the operators $c^A_{\mathbf{k}\sigma}$, $c^B_{\mathbf{k}\sigma}$ in terms of their transformed operators $b^A_{\mathbf{k}\sigma}$ and $b^B_{\mathbf{k}\sigma}$

$$
\begin{pmatrix} c^A_{\mathbf{k}\sigma} \\ c^B_{\mathbf{k}\sigma} \end{pmatrix} = \frac{1}{\sqrt{2}} \begin{pmatrix} e^{-i\phi_k/2} & -e^{-i\phi_k/2} \\ e^{i\phi_k/2} & e^{i\phi_k/2} \end{pmatrix} \begin{pmatrix} b^A_{\mathbf{k}\sigma} \\ b^B_{\mathbf{k}\sigma} \end{pmatrix} \tag{23}
$$

where $e^{i\phi_{\mathbf{k}}} = \xi^0_{\mathbf{k}}/|\xi^0_{\mathbf{k}}|$ and A, B now denote operators corresponding respectively to the positive and negative eigenvalues $\omega^{\pm}_{\mathbf{k}} = \pm|\xi_{\mathbf{k}}|$ of the *diagonalised* mean field Hamiltonian. Note that $\phi_{\mathbf{k}}$ does not depend on Q'_i. Using this representation one arrives after straightforward but lengthy algebra at

$$
\begin{aligned}
\frac{Z}{Z_0} &= N^{-1}\int \mathcal{D}Q'_i \exp(\tfrac{1}{2}\int \mathrm{d}\tau \sum_i (Q'_i)^2 V_i^{-1}) \\
&\times \left\langle \exp\left(-\left\{\int \mathrm{d}\tau \sum_i Q'_i \sum_{\mathbf{k}\sigma} f^i_{\mathbf{k}}[b^{A\dagger}_{\mathbf{k}\sigma} b^A_{\mathbf{k}\sigma} - b^{B\dagger}_{\mathbf{k}\sigma} b^B_{\mathbf{k}\sigma}] + g^i_{\mathbf{k}}[b^{A\dagger}_{\mathbf{k}\sigma} b^B_{\mathbf{k}\sigma} - b^{B\dagger}_{\mathbf{k}\sigma} b^A_{\mathbf{k}\sigma}]\right\}\right)\right\rangle .
\end{aligned} \tag{24}
$$

$f^i_{\mathbf{k}}$ and $g^i_{\mathbf{k}}$ are geometrical factors arising from the transformation, and depend on i, $\mathbf{k}$ and Q^0_i only. These effectively provide the coupling between the different sublattices due to the fluctuations. For a general ground state configuration there will be other operator products of the form $b^{\dagger}_{\mathbf{k}\sigma} b^{\dagger}_{-\mathbf{k}\sigma}$,

[1] We are dropping a linear term here for compactness of notation, which does not change the basic idea.

$b_{\mathbf{k}\sigma}b_{-\mathbf{k}\sigma}$present. These will not change the approach we take, but will make the situation technically much more involved.

To evaluate the thermal average in (25) we first carry out the sum over the sites before evaluating the integral over τ, and expand the exponential in the different τ integrals, each containing one of the variables Q'_i, around 0 in a multivariable Taylor series. We then take the two summations over n and α out of the thermal average to obtain

$$\frac{Z}{Z_0} = N^{-1}\int \mathcal{D}Q'_i \exp(\tfrac{1}{2}\int \mathrm{d}\tau \sum_i (Q'_i)^2 V_i^{-1}) \times \sum_{n=0}^{\infty}\sum_{|\alpha|=n}\left\langle \frac{(-1)^n}{\alpha^n}\{\int \mathrm{d}\tau Q' \sum_{\mathbf{k}\sigma} f_{\mathbf{k}}[b^{A\dagger}_{\mathbf{k}\sigma}b^{A}_{\mathbf{k}\sigma} - b^{B\dagger}_{\mathbf{k}\sigma}b^{B}_{\mathbf{k}\sigma}] + g_{\mathbf{k}}[b^{A\dagger}_{\mathbf{k}\sigma}b^{B}_{\mathbf{k}\sigma} - b^{B\dagger}_{\mathbf{k}\sigma}b^{A}_{\mathbf{k}\sigma}]\}^{\alpha}\right\rangle \tag{25}$$

where $\alpha = (\alpha_1, \ldots, \alpha_m)$, $\alpha_i \geq 0$, $|\alpha| = \sum_{i=1}^{m}\alpha_i$, $\alpha! = \prod_{i=1}^{m}\alpha_i!$ and $Q^{\alpha} = \prod_{i=1}^{m} Q_i^{\alpha_i}$. We now perform a twofold "integration", one over the $Q'_i(\tau)$, the other one over $\mathrm{d}\tau$. The expansion series produces multiple time integrations. As usual, the $Q'_i(\tau)$ integrand resembles derivatives of Gaussian integrals so that we may introduce a source term $\sum_i j_i(\tau)Q'_i(\tau)$ in the exponential and evaluate its derivatives at $\mathbf{j} = 0$ in the standard way to generate correlation functions. Expanding the path integral back into its defining discrete imaginary time slice integrals to make explicit the various dependences on the different times, and noticing that the Gaussian integral does not depend on n or α, and that the multiple time integrals in the series also do not affect the Gaussian integral, we may pull the corresponding sums over n and α in front to give

$$\frac{Z}{Z_0} = \sum_{n=0}^{\infty}\sum_{|\alpha|=n}\frac{(-1)^n}{\alpha!}\prod_{l=1}^{n}(\sum_{r_l=0}^{M-1}\Delta) \times (\coprod_n Q')\frac{\int \prod_{i=0}^{m}\prod_{s=0}^{M-1}\frac{\mathrm{d}Q'_{i,s}}{\sqrt{2\pi i}}\exp(\sum_{i=0}^{m}\sum_{s=0}^{M-1}\frac{1}{2}\Delta(Q'_{i,s})^2V_i^{-1} + \Delta j_{i,s}Q'_{i,s})}{\int \prod_{i=0}^{m}\prod_{s=0}^{M-1}\frac{\mathrm{d}Q'_{i,s}}{\sqrt{2\pi i}}\exp(\sum_{i=0}^{m}\sum_{s=0}^{M-1}\frac{1}{2}\Delta(Q'_{i,s})^2V_i^{-1})} \times \left\langle\left(\sum_{\mathbf{k}\sigma} f_{\mathbf{k}}[b^{A\dagger}_{\mathbf{k}\sigma}b^{A}_{\mathbf{k}\sigma} - b^{B\dagger}_{\mathbf{k}\sigma}b^{B}_{\mathbf{k}\sigma}](r_l) + g_{\mathbf{k}}[b^{A\dagger}_{\mathbf{k}\sigma}b^{B}_{\mathbf{k}\sigma} - b^{B\dagger}_{\mathbf{k}\sigma}b^{A}_{\mathbf{k}\sigma}](r_l)\right)^{\alpha}\right\rangle \tag{26}$$

Here we have introduced a new notation $\coprod_n Q'$ which signifies a product over n $Q_{i,r}$'s, with i and r determined as follows. For a multiindex $\alpha = (\alpha_1, \alpha_2, \ldots)$ the first index is 1 for the first α_1 factors, 2 for the next α_2 factors and so on. r is taken to be the current value of r_l for the lth factor. The r_l in the thermal average signifies that one of the time summations is assigned to each operator in each factor of the product with the corresponding time τ.

The whole expression can now be written as a derivative:

$$\frac{Z}{Z_0} = \sum_{n=0}^{\infty} \sum_{|\alpha|=n} \frac{(-1)^n}{\alpha!} \prod_{l=1}^{n} (\sum_{r_l=0}^{M-1} \Delta)$$
$$\times (\coprod_n \frac{\delta}{\Delta \delta j})\Big|_{j=0} \frac{\int \prod_{i=0}^{m} \prod_{s=0}^{M-1} \frac{\mathrm{d}Q'_{i,s}}{\sqrt{2\pi i}} \exp(\sum_{i=0}^{m} \sum_{s=0}^{M-1} \frac{1}{2}\Delta (Q'_{i,s})^2 V_i^{-1} + \Delta j_{i,s} Q'_{i,s})}{\int \prod_{i=0}^{m} \prod_{s=0}^{M-1} \frac{\mathrm{d}Q'_{i,s}}{\sqrt{2\pi i}} \exp(\sum_{i=0}^{m} \sum_{s=0}^{M-1} \frac{1}{2}\Delta (Q'_{i,s})^2 V_i^{-1})}$$
$$\times \left\langle \left(\sum_{\mathbf{k}\sigma} f_{\mathbf{k}} [b^{A\dagger}_{\mathbf{k}\sigma} b^{A}_{\mathbf{k}\sigma} - b^{B\dagger}_{\mathbf{k}\sigma} b^{B}_{\mathbf{k}\sigma}](r_l) + g_{\mathbf{k}} [b^{A\dagger}_{\mathbf{k}\sigma} b^{B}_{\mathbf{k}\sigma} - b^{B\dagger}_{\mathbf{k}\sigma} b^{A}_{\mathbf{k}\sigma}](r_l)\right)^{\alpha} \right\rangle \tag{27}$$

Reversing the order of differentiation and integration, and evaluating first the Gaussian integral, we arrive at

$$\frac{Z}{Z_0} = \sum_{n=0}^{\infty} \sum_{|\alpha|=n} \frac{(-1)^n}{\alpha!} \prod_{l=1}^{n} (\sum_{r_l=0}^{M-1} \Delta) (\coprod_n \frac{\delta}{\Delta \delta j})\Big|_{j=0} \exp\left(\sum_{i=0}^{m} \sum_{s=0}^{M-1} -\frac{1}{2} \Delta V_i\, j_{i,s}^2\right)$$
$$\times \left\langle \left(\sum_{\mathbf{k}\sigma} f_{\mathbf{k}} [b^{A\dagger}_{\mathbf{k}\sigma} b^{A}_{\mathbf{k}\sigma} - b^{B\dagger}_{\mathbf{k}\sigma} b^{B}_{\mathbf{k}\sigma}](r_l) + g_{\mathbf{k}} [b^{A\dagger}_{\mathbf{k}\sigma} b^{B}_{\mathbf{k}\sigma} - b^{B\dagger}_{\mathbf{k}\sigma} b^{A}_{\mathbf{k}\sigma}](r_l)\right)^{\alpha} \right\rangle. \tag{28}$$

so that reintroducing the integral notation, we obtain

$$\frac{Z}{Z_0} = \sum_{n=0}^{\infty} \sum_{|\alpha|=n} \frac{(-1)^n}{(\alpha/2)!} \int_0^{\beta} \prod_{l=1}^{n/2} \mathrm{d}\tau_l (\coprod_n \frac{\delta}{\delta j})|_{j=0} \exp(-\frac{1}{2} \int_0^{\beta} \mathrm{d}\tau \sum_{i=0}^{m} V_i\, j_i^2(\tau))$$
$$\times \langle (\sum_{\mathbf{k}\sigma} f_{\mathbf{k}} [b^{A\dagger}_{\mathbf{k}\sigma} b^{A}_{\mathbf{k}\sigma} - b^{B\dagger}_{\mathbf{k}\sigma} b^{B}_{\mathbf{k}\sigma}](\tau_l) + g_{\mathbf{k}} [b^{A\dagger}_{\mathbf{k}\sigma} b^{B}_{\mathbf{k}\sigma} - b^{B\dagger}_{\mathbf{k}\sigma} b^{A}_{\mathbf{k}\sigma}](\tau_l))^{\alpha} \rangle. \tag{29}$$

Note that the $\coprod$ notation has been kept, but now assigns the value of the integration variable of one time integration to each of the derivatives. Note that the number of time integrations has been halved. This is the perturbation series we require. The Hubbard-Stratonovich transformation normalization N^{-1} has been treated *exactly* when we integrated over the auxiliary fields in (27). Essentially, the integration has carried out the reverse of the Hubbard-Stratonovich transformation so that there is now no longer the problem of the $SU(2)$ gauge dependent normalisation. One is thus free to compare free energies obtained using different gauge vectors.

Although the above expression (29) is complicated, it may be treated by a diagrammatic method. In the remainder of this paper we will show how this may be carried out.

4. Diagrammatic Expansions

The expansion series (29) may be divided into two parts. The first is the functional differentiation of the source term, which gives the interaction energies in the expansion, and the second is thermal average of the operators which can be dealt with using a Wick's expansion. Since the exponential is quadratic in j, successive functional derivatives with respect to j corresponds

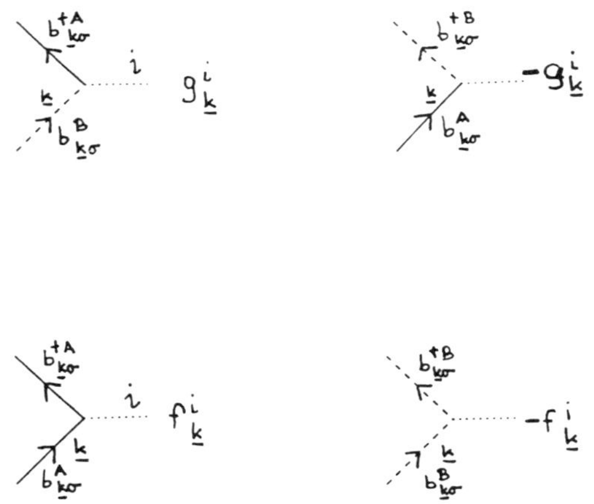

Fig. 6. Diagrams corresponding to the first order vertices with a dotted line showing the interaction V_i, and the Green's functions representated by the solid and the dashed lines. The geometric factors are also given for the corresponding vertices

$(\coprod_n \frac{\delta}{\delta j})\Big|_{j=0} = (-1)^n(\Delta V/2)^{\alpha/2}\prod_{i=1}^n \prod_{j=1,n_i} H_{n(i,j)}(0)$, where $H_n(0)$ are the Hermite polynomials of order n evaluated at 0. Since $H_n(0) = 0$ for all odd n it may be shown that the only contributions come from products of polynomials of degree two, where an interaction energy V_i is brought down from the exponential. With this result our expansion may be understood by observing that terms of order $2n$ essentially correspond to the nth order of a two-particle vertex, with the pairing in the functional derivatives effectively reconstructing an interaction potential via the V_i's. The Wick expansion of the thermal average gives the corresponding Green's functions.

For the particular example we considered above the single particle vertices are shown in Figure 6, where the dotted line represents an interaction line originating from the site i, carrying a factor of $\sqrt{V_i}$ such that the factor of V_i is obtained by joining two interaction lines. In any diagram the interaction line should close over the same site so that zero momentum pairing of the excitations is ensured. At each vertex, the wavevector $\mathbf{k}$ is to be integrated together and the spins summed over. The diagrams here are different from the usual diagrams encountered in many-body theory in that the vertices are in the space of the auxiliary field Q_i, but over each vertex there is an internal momentum summation. The solid and the dashed lines are Green's function lines corresponding to $\langle b^{A\dagger}_{\mathbf{k}\sigma} b^A_{\mathbf{k}\sigma}\rangle$ and $\langle b^{B\dagger}_{\mathbf{k}\sigma} b^B_{\mathbf{k}\sigma}\rangle$ respectively, and to each vertex there is a factor of $\pm f^i_{\mathbf{k}}$ or $\pm g^i_{\mathbf{k}}$ depending on the nature of the lines entering and leaving it (Figure 6). Note that these factors are $\mathbf{k}$ and i dependent and must be included in the momentum summation. By combining these vertices with the appropriate Green's functions all the terms

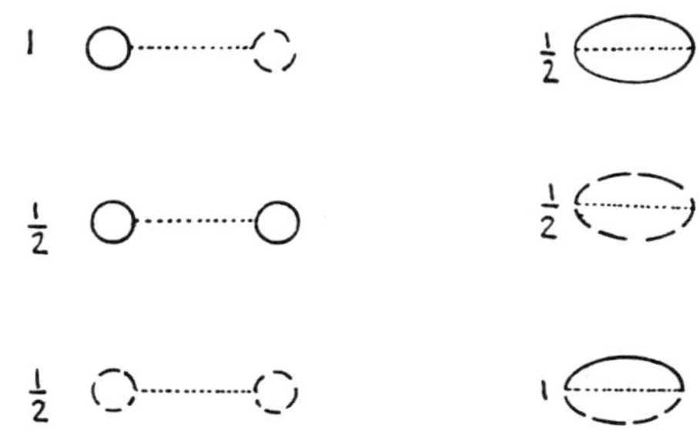

Fig. 7. Second order diagrams which contribute to the free energy. The symmetry factor of the diagrams are also given. The solid and the dashed lines represent Green's functions, the dotted line represents an interaction V_i

of the perturbation expansion may be recovered. An example for the 2nd order diagrams are shown in Figure 7. Finally the rules for the diagrams corresponding to the perturbation series may be summarised as follows.

1. For every multiindex α at the order $2n$ draw n vertices and label them according to the multiindex in any one particular way, then draw all topological distinct diagrams of order n.
2. For every one of these generating diagrams consider all distinct variations including the different Green's functions allowed. Include a factor of $\frac{2^n}{S}$, where S is the symmetry factor of the generated diagram.
3. For every vertex include a factor of $(-V)^{\alpha/2}$.
4. For every diagram include the geometric factors $\pm f^i_{\mathbf{k}}$, $\pm g^i_{\mathbf{k}}$ according to the Green's functions which enter and leave as shown in Figure 6.
5. At every vertex include *one* $\mathbf{k}\sigma$-summation. This recovers the zero-momentum pairing inherent to the path integral formalism chosen. For every vertex also include one time-integration over the interval $[0, \beta]$. For every closed loop include a factor of (-1). (1 for bosons. This is the only explicit difference to a bosonic formalism.)

We have thus shown that a perturbation series beyond mean field theory which recovers the $SU(2)$ gauge symmetry of the problem is possible. For certain ground state configurations the number of different vertex types is restricted and it should be possible to evaluate corrections at least to RPA order. However, depending on the mean field configuration of Hubbard Stratonovich auxiliary fields the calculation may become very involved, and it is not yet clear whether there are further ways of simplifying the calculation. In view of the great uncertainty over the mean field results, and of the likely richness in the physics of the two-dimensional quantum magnet some work beyond mean field theory does need to be carried out.

We have so far avoided any mention of the Lagrangian constraint. It was introduced as a field to restrict the number of excitations allowed on each site. In taking the field to a uniform limit, that is, by going from λ_i to λ we have changed the constraint from a local requirement to a global requirement. In the bosonic representation where the statistics of the particles do not restrict the occupation of any site, the result is that the model changes from a Heisenberg model to a spherical model where the spin on any site may take any value up to $2NS$, where N is the number of sites. Thus, a significant number of states allowed are unphysical in that their spin quantum number greatly exceeds the physical value of $\frac{1}{2}$. In the mean field limit one can argue that the nature of the mean field approximation requires all sites to be equivalent so that the local constraint will be maintained. Once we consider fluctuations, however, intuition and preliminary results from the expansion show that no configuration is stable, with the disordering effects coming largely from fluctuations due to the unphysical states where $S \neq \frac{1}{2}$. In the fermionic representation the inherent statistics of the particles automatically restricts the occupation to no more than one excitation of each type. This is probably the reason that mean field theory consistently showed that the constraint λ is unnecessary ($\lambda = 0$ for all mean field solutions), and suggests that the fermionic theory would have better convergence properties in going beyond the mean field limit.

5. Summary

We have examined the two-dimensional quantum magnet in a path integral formalism and have shown that at least in the fermionic representation within a selection of possible ground states the dimer state may be a good candidate for the ground state at low temperatures. We have, however, shown explicitly that the process of assigning ground states from mean field results alone is highly dubious. In particular, the accuracy of the mean field results depends on the choice of an appropriate local rotation of the spins. This together with the unphysical breaking of the $SU(2)$ invariance due to the saddle point approximation means that one is unable to say with certainty what the state with the lowest free energy is. In response to these problems we have developed and presented a formalism which recovers the gauge invariance by going explicitly beyond mean field theory. Although the calculations are likely to be quite involved in general, it should be sufficiently tractable to provide also a more realistic assessment of the strength of the quantum fluctuation effects.

Acknowledgements

One of us (David Ko) would like to thank Prof. Hiroshi Kamimura for all the help and support he gave during my stay in his group in Japan. Without it the stay could not have been half as enjoyable or profitable.

References

1. J. G. Bednorz and K. A. Müller, *Z. Phys.* **64**, 189 (1986); Y. Tokura, H. Takagi and S. Uchida, *Nature* **337**, 345 (1989).
2. C. M. Varma, P. B. Littlewood, S. Schmitt-Rink, E. Abrahams and A. E. Ruckenstein, *Phys. Rev. Lett.* **63**, 1996 (1989); P. W. Anderson, *Phys. Rev. Lett.* **64**, 1839 (1990); P. W. Anderson, *Phys. Rev. Lett.* **65**, 2306 (1990); G. T. Zimanyi and K. S. Bedell, *Phys. Rev. Lett.* **66**, 228 (1991).
3. R. B. Laughlin, *Science* **242**, 525 (1988); see also *Fractional Statistics and Anyon Superconductivity*, F. Wilczek, ed. (World Scientific 1990).
4. D. Y. K. Ko *Phys. Rev. Lett.* **65**, 116 (1990).
5. M. Inui, S. Doniach and M. Gabay, *Phys. Rev. B* **38**, 6631 (1988).
6. See, for example, D. Y. K. Ko, *Mod. Phys. Lett. B* **4**, 1325 (1990).
7. P. Chandra and B. Douçot, *Phys. Rev. B* **38**, 9335 (1988).
8. D. P. Arovas and A. Auerbach, *Phys. Rev. Lett.* **61**, 2376 (1988); *Phys. Rev. B* **38**, 316 (1988); also in *Field Theories in Condensed Matter Physics*, Ed. Z. Tešanović (Addison-Wesley 1990).
9. N. Read and S. Sachdev, *Phys. Rev. Lett.* **62**, 1694 (1989).
10. C. L. Kane, P. A. Lee, T. K. Ng, B. Chakraborty and N. Read, *Phys. Rev. B* **41**, 2653 (1990).
11. B. Chakraborty, N. Read, C. Kane and P. A. Lee, *Phys. Rev. B* **42**, 4819 (1990).
12. F. Mila, D. Poilblanc and C. Bruder, *to be published* (1991).
13. J. Villain, *J. Phys. (Paris)* **38**, 26 (1977); J. Villain, R. Bidaux, J. P. Carton, R. Conte, *J. Phys. (Paris)* **41**, 1263 (1980).
14. E. Dagotto and A. Moreo, *Phys. Rev. Lett.* **63**, 2148 (1989).
15. R. R. P. Singh and R. Narayanan, *Phys. Rev. Lett.* **65**, 1072 (1990).
16. P. W. Anderson, *Phys. Rev.* **86**, 694 (1952); M. Takahashi, *Phys. Rev. B* **40**, 2494 (1989).
17. J. Schwinger, *On Angular Momentum*, U. S. Atomic Energy Commision preprint NYO-3071 (1952); reprinted in *Quantum Theory of Angular Momentum*, L. Biedenharm and H. Van Dam, eds. (Academic Press, 1965).
18. R. L. Stratonovich, *Sov. Phys. Dokl.* **2**, 416 (1958); J. Hubbard, *Phys. Rev. Lett.* **3**, 77 (1959).
19. For an introduction into the functional integral field theoretical approach to condensed matter, see *Quantum Many-particle Systems, Frontiers in Physics Series* **68**, by J. W. Negele and H Orland (Addison-Wesley 1988).
20. P. Chandra, P. Coleman and A. I. Larkin, *J. Phys.: Condens. Matter* **2**, 7933 (1990).
21. I. Affleck and J. B. Marston, *Phys. Rev. B* **37**, 3774 (1988).
22. D. Mermin and H. Wagner, *Phys. Rev. Lett.* **17**, 1133 (1966)
23. J. P. Blaizot and H. Orland, *Phys. Rev. C* **24**, 474 (1981); see also chapter 12 of *Quantum Theory of Finite Systems* by J. P. Blaizot and G. Ripka (M.I.T. Press 1985).

PART IV

Surfaces, Mesoscopic Systems and One- and Zero-Dimensional Systems

ELECTRONIC STRUCTURE AND RECONSTRUCTION OF Si(100) SURFACES

M. TSUKADA, N. SHIMA[1], Z. ZHU[2], H. KAGESHIMA[3] and J. YAMAUCHI
Department of Physics, Faculty of Science, University of Tokyo,
Hongo 7-3-1, Bunkyo-ku, Tokyo 113, Japan

Abstract. The first-principles LDA calculation is applied to the electronic states of Si(100) reconstructed surfaces. Among various higher order reconstruction structures of asymmetric dimers, $c(4 \times 2)$ structure is found to be the most stable structure. The first-principles molecular dynamics calculation provided a supporting evidence for the stability of asymmetric dimer structure. The energy band of the $c(4 \times 2)$ surface reveals the semiconducting surface band dispersion, which gives an excellent agreement with the ARUPS result for the single domain surface. The stabilization mechanism of the $c(4 \times 2)$ surface is ascribed to the hybridization interaction between the orbitals of the up dimer atoms in the adjacent row, and not to the ionic interaction as $p(2 \times 2)$ structure. The theoretical simulation of STM/STS on the Si(100) surfaces explained fairly well the experimental feature. The 'seemingly symmetric' dimer in the STM image does not necessarily correspond to the real symmetric dimer.

1. Introduction

The atomic and electronic structures of semiconductor surfaces have always provided important issues of modern surface physics. The elucidation of properties of well characterized semiconductor surfaces is the basis of the microfabrication of new materials or new artificial systems, which can be applied to future ultrafine electronic devices. In particular a clean surface of a covalent semiconductor shows a variety of reconstructed structures, depending on the crystallographic geometry, preparation condition, temperature and so on. The elucidation of a mechanism causing the reconstruction is impossible without a clue from the surface electronic structure. In particular the Si(100) surface has long been studied as a prototype of surface reconstruction, because the structure is apparently simple and the mechanism is thought to be textbook-like. Moreover, this surface is quite important from the viewpoint of the nanostructure technology. For example, an ultrafine quantum wire is proposed with use of the step edge of the Si(100) surface. In spite of a rather long history of research and the relative simplicity of the structure, many problems on this surface are yet to be solved and new, challenging mysteries have arisen as the research has developed.

[1] Present address: Dept. of Material Science, Fac. of Science, Himeji Inst. of Technol., Kanaji, Akohgun, Hyogo-ken, Japan.
[2] Present address: Dept. of Physics and Astronomy, The State University of New Yersey, Rutgers, U.S.A.
[3] Present address: NTT LSI Laboratories, 3-1 Morinosato-Wakamiya, Atsugi-si, Kanagawa-ken, Japan.

H. Aoki et al. (eds.), New Horizons in Low-Dimensional Electron Systems, 319–335.

As for the Si(100) 2×1 structure the dimer model had been proposed as early as 1952 by Schiler and Farnsworth [1], but its direct confirmation had to await the invention of the scanning tunneling microscopy (STM) in 1982 [2–4]. Trom *et al.* [5] and Hamers *et al.* [6] observed very clear STM image of the dimer rows over the Si(100) 2×1 surface. In a defect-free region of the surface, they observed the symmetric dimer rows, although near the defects, such as missing dimers or at the step edge, the zigzag rows of the asymmetric dimers have been observed. The predominance of the symmetric dimers in the defect free region is rather strange, because most of the experiments supported the asymmetric dimer structure as the most stable structure. For example LEED (low energy electron diffraction) observation of a carefully prepared surface revealed the transition from the 2×1 to the $c(4 \times 2)$ structure below about 200 K [7]. As will be described below, the ARUPS (angle revolved ultra-violet photoemission spectroscopy) observed for the single domain Si(100) $c(4\times2)$ surface revealed the surface band dispersion showing an excellent agreement with the theoretical calculation [9]. Most of the first-principles calculations predicted that the asymmetric dimer structure has lower energy than the symmetric dimer structure [9–12], though some authors claimed the opposite based on the spin polarized calculation using cluster models [13,14].

So, in the present article, we discuss the stabilization mechanism for various reconstructed structures of Si(100) surface based on careful LDA (local density functional) calculations of the electronic state. We first investigate the feature of the surface bands peculiar to each reconstructed structure and its relation to the stabilization mechanism. It will be shown that the surface band dispersion of the $c(4 \times 2)$ structure gives a remarkable agreement with the experimental one observed recently by the ARUPS [8] for a single domain surface. We also use the total energy minimization procedure for searching the optimum relaxed structures around the three different models $c(4 \times 2)$, $p(2 \times 2)$ and uniformly buckled 2×1, and compare the total energies thus obtained with each other. A first-principles molecular dynamics approach is also applied to the search of the stable structure. To resolve the contradiction between the STM observation of the symmetric dimer and other experiments for the asymmetric dimer structures, we also investigate the STM/STS data based on a theoretical simulation with the LDA calculations.

2. Models of the Dimerized Si(100) Surfaces and Theoretical Methods

We have studied the symmetric dimer structures and the three asymmetric dimer structures of the uniformly buckled (2×1), alternately buckled $p(2\times2)$ and $c(4\times2)$ surfaces. These three structures are shown in Figure 1. Though the uniformly buckled (2×1) surface has not been observed experimentally,

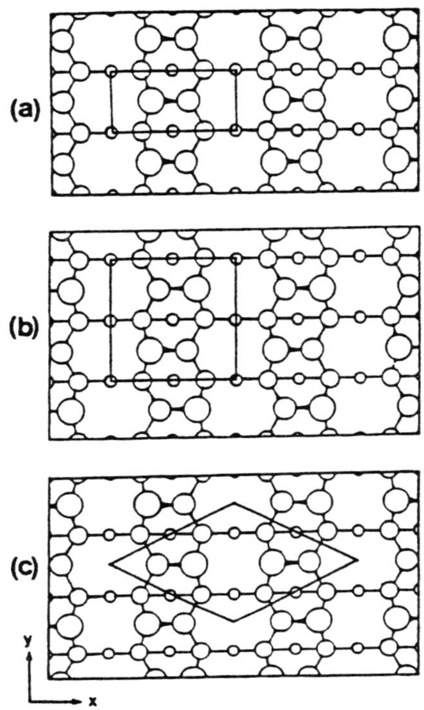

Fig. 1. Top view and the unit cell of (2×1), $p(2 \times 2)$, $c(4 \times 2)$ structure of Si(100) surface.

we calculated the electronic states of this surface to provide a reference for the comparison of the stabilization mechanism of the other models. For the LDA energy minimization and for the calculation of the surface energy band dispersion we used the repeated slab model, each slab of which is composed of ten atomic layers. Both sides of the surface have dimers in this model. On the other hand, for the application of the first-principles molecular dynamic method described in Section 5.2, a five-layer slab model is used in order to save computation time. In this case the dangling bonds of the backside, outermost layer are saturated by hydrogen atoms. The norm-conserving pseudopotential of Bachelet, Schlüter and Hamann [15] is used throughout the whole calculation. The cutoff energy of the plane waves is 5.8 Ry unless otherwise described. We also investigate the STM image of an STS spectra of the Si(100) surfaces with the use of the five layer slab model.

For the numerical simulation of STM/STS, the DV-LCAO method is used as described later in Section 6.

3. Band Dispersion of the Surface States

As is well known, there are two energy bands per dimer in the surface unit cell, one corresponding to the bonding state and the other to the antibonding state of the dangling bond. Early theoretical calculations by Chadi [10], and by Ihm, Cohen and Chadi [16] concluded that the asymmetric dimer surfaces become semiconducting, due to the repelling of the two surface bands π and π^*. This feature corresponds well with the photoemission experiments. However, as far as theoretical calculation is concerned, the situation is more subtle. Our more accurate calculations with much larger cutoff energy and improved (norm conserving and nonlocal) pseudopotential concluded a small but finite overlap of the two bands for the asymmetric dimer structures of the uniform buckled (2×1), and $p(2 \times 2)$ surfaces.
The same conclusion is obtained by Kruger and Pollman [17] with use of the Green's function scattering formalism. In our calculation the semiconducting surface electronic structure is obtained only for the $c(4 \times 2)$ structure. However the difference of the energy dispersions of the occupied π band is found rather small between the two structures $c(4 \times 2)$ and $p(2 \times 2)$, and both roughly agree with that obtained by the ARUPS for the Si(100) 2×1 surface [18,19]. This fact would indicate that even the 2×1 surface is formed by the asymmetric dimer structures, though they lose the long range order. In fact Monte Carlo simulations of the surface structure reproduce the temperature dependence of the LEED intensity, especially the behavior of the streak pattern [20]. These simulations revealed a reminicsense of the zigzag array of the asymmetric dimers especially along the dimer row direction.

Figure 2 shows the energy dispersion of the surface band of the π dangling bond for the single domain $c(4 \times 2)$ surface compared with the experiment observed by ARUPS [8]. The agreement is excellent. It should be remarked that our report on the theoretical dispersion curve appeared first [9] and later it was confirmed by the experiment by Kono's group [8]. For the calculation of the dispersion relation of Figure 2, we used a modified structure of the optimized geometry provided by Yin and Cohen [11]. Further completely optimized structure of the $c(4 \times 2)$, especially in the deeper atomic layer, causes lowering of the energy around the Γ ponit of the surface Brillouin zone (Figure 3), so the agreement with the experiment becomes worse. One of the reasons for this would be ascribed to the fact that the actual surface structure, because of the presence of defects, is different in the deeper layers from the calculated one, which has an ideal condition.

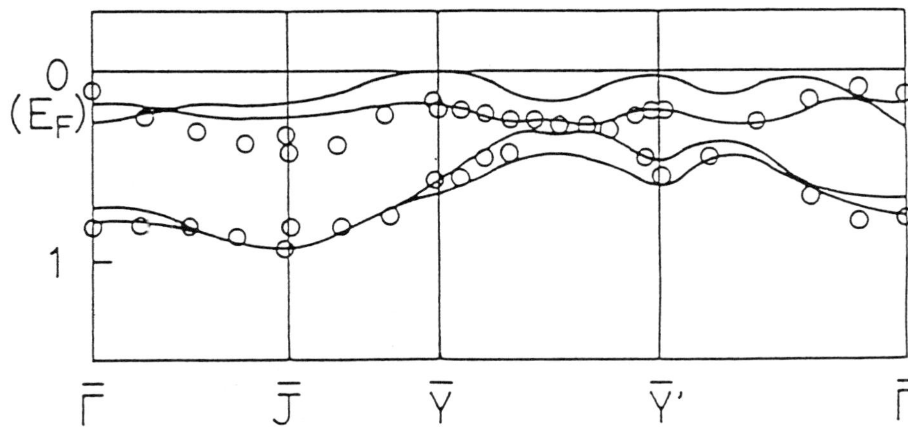

Fig. 2. The calculated surface band dispersion of $c(4 \times 2)$ surface compared with the ARUPS observation [8].

4. Stabilization Mechanism of the $p(2 \times 2)$ and $c(4 \times 2)$ Structures

The reason why the asymmetric dimer is more stabilized than the symmetric dimer is understood by a mechanism similar to Jahn–Teller effect [21]. There are two equivalent dangling bonds (DBs) per symmetric dimer associated with the respective atoms in the dimer. One electron is accomodated in each DB. If one of the dimer is raised while the other is pushed down to the bulk, the DB of the raised atom adds the more s-like orbital character and that of the lowered atom loses the s-orbital character. This results in the lowering and raising of the DB energy for the raised and the lowered atom in the dimer, respectively. In the calculated surface energy bands, this is reflected as the repelling of the π and π^* band. Therefore transferring some part of the electron density from the latter to the former DB, the lowering of the total energy is realized as compared to the symmetric dimer structure.

There are various ways for the arrangement of the asymmetric dimer directions in the surface layer. The relative stability among the different asymmetric dimer arrangements is determined by a somewhat subtle dimer–dimer interaction. Comparison of the surface energy bands among these structures provides a basis for the discussion. Figure 3 shows the surface band structure of $p(2 \times 2)$ and $c(4 \times 2)$ as compared with the uniformly buckled (2×1) surface [9]. The geometry used here is the modified Yin–Cohen model [11]. The latter band is shown in the folded Brillouin zones for the ease of comparison. As will be expected by the comparison of Figure 3a and Figure 3c, the interaction between the electron pocket and the hole pocket is found to be strongest for the $c(4 \times 2)$ surface. This causes the semiconducting character only for this surface, as mentioned before, as well

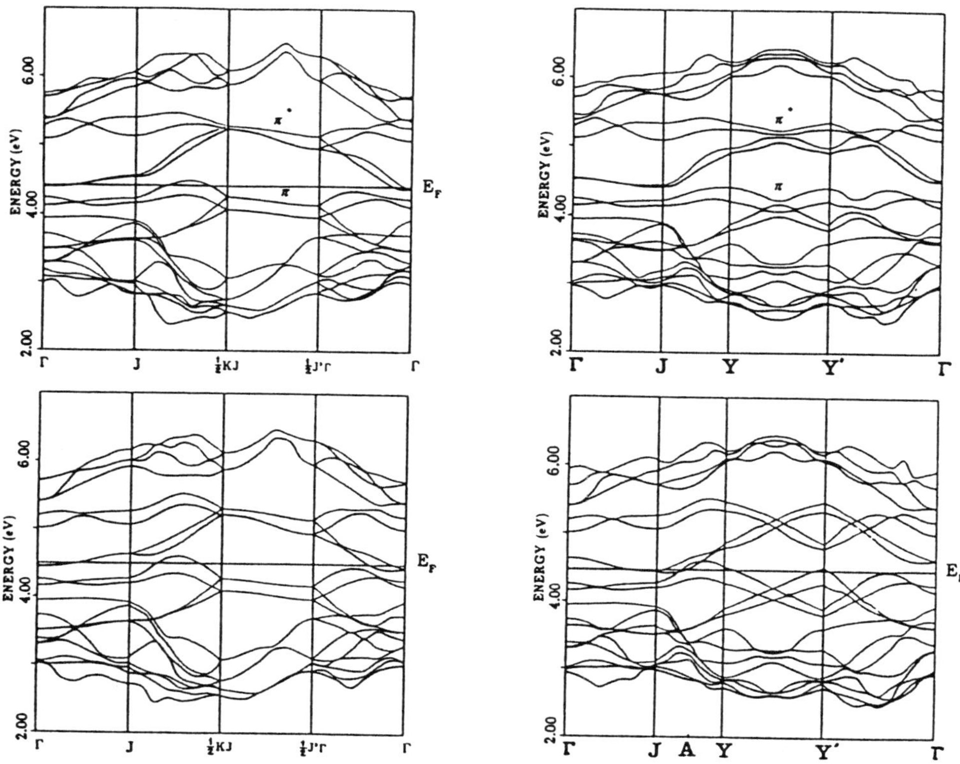

Fig. 3. Surface band structure of (a) $p(2 \times 2)$ surface compared with (b) the folded one for (2×1), (c) is the surface band structure of $c(4 \times 2)$ compared with (d) folded one for (2×1).

as its unique stabilization mechanism.

On the other hand, from the view-point of the electrostatic interaction among dimers, the $p(2\times 2)$ structure is expected to be the most stable. This is because this structure keeps the greatest distances among the up dimer atoms or those among down dimer atoms. Table I shows the comparison of the decomposed energies for the three structures. As seen in the table the core–core interaction energy, E_{c-c} and the electron–electron interaction energy, E_{e-e} are most stabilized for the $p(2\times 2)$ structure, while these energies are the highest for the $c(4 \times 2)$ structure. However the lowering of the

TABLE I
Differences of energies between asymmetric (2 × 1), $p(2\times2)$, and $c(4\times2)$ systems.

Energy component (eV/dimer)	$E_{p(2\times2)} - E_{(2\times1)}$	$E_{c(4\times2)} - E_{(2\times1)}$	$E_{c(4\times2)} - E_{p(2\times2)}$
E'_{c-c}	−0.616	0.204	0.820
E'_{e-e}	−0.933	0.398	1.331
E'_{e-c}	1.545	−0.671	−2.216
E_{kin}	0.013	0.064	0.051
Exchange	−0.022	−0.018	0.003
Total energy	−0.013	−0.023	−0.011

total energy is realized most effectively for the $c(4\times2)$ structure, and this is caused by the decomposed energy term E_{c-e}.

As is partly known from the negative sign of E_{c-e}, the mechanism of the energy lowering of the $c(4\times2)$ structure is due to the bonding hybridization interaction of the orbitals of the nearest neighboring up dimer atoms between the adjacent dimer rows. This effect is most efficient around the intermediate region along the $J-Y$ line of the Brillouin zone, where the electron and the hole pocket overlaps (see Figure 4). Figure 5 shows how the wavefunction at the point A shown in Figure 4 changes by the structure, i.e. uniformly buckled (2 × 1) and $c(4\times2)$. As compared to the case of the uniformly buckled 2×1 asymmetric dimer structure, the amplitude of the upper band tends to be more localized on the down dimer atom and that of the lower band concentrates to the up dimer atom. Such change of the orbital shape is induced by the hybridization effect of the orbitals of the up dimer atoms between the adjacent rows. That this overcomes the raised Coulomb energy among the dimers is seen by Table I. Thus the stabilization mechanism is quite dissimilar between the $p(2\times1)$ and $c(4\times2)$ surface.

5. Optimization of the Surface Structure

5.1. Energy minimization method

The stable structure of each reconstructed surface, i.e. uniformly buckled 2 × 1, alternately buckled $p(2\times2)$ and $c(4\times2)$ is searched by the energy minimization method described in Section 2 relaxing all ion configurations, including deeper substrate ions. If we compare it with the respective fully relaxed structures, the total energy difference between the $c(4\times2)$ and the uniform buckled (2×1) surface is −0.067 eV/dimer, while it is −0.025 eV/dimer between the $p(2\times2)$ and the uniform buckled (2×1) surface. Thus the $c(4\times2)$ surface is the most stable structure among the three. This seems consistent

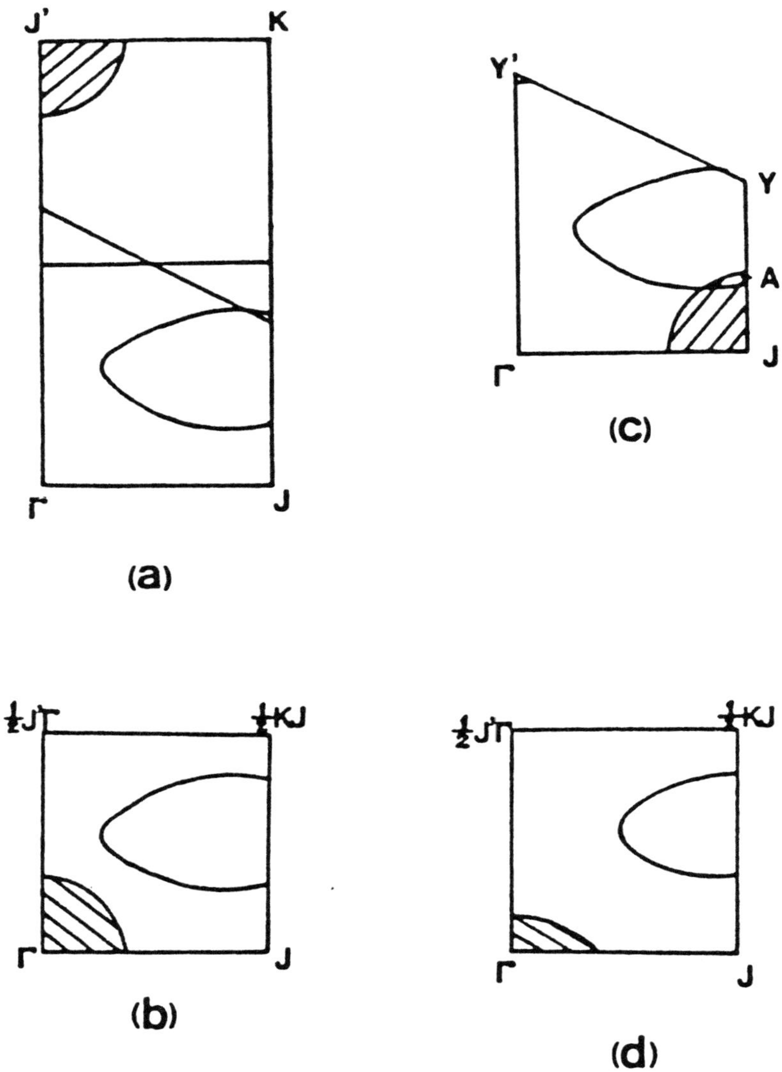

Fig. 4. (a) Fermi surface of (2×1) surface, (b) the same as (a) but folded in $p(2 \times 2)$ Brillouin zone, (c) the same as (a) but folded in $c(4 \times 2)$ Brillouin zone, (d) Fermi surface of actual $p(2 \times 2)$ structure.

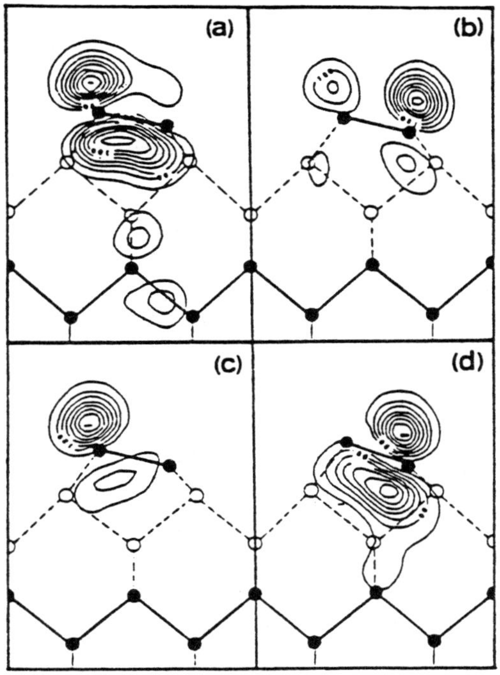

Fig. 5. Contour maps of the square of wavefunctions at point A in $c(4\times2)$ Brillouin zone. (a) surface bonding and (b) antibonding states of the (2×1) surface, and (c) surface bonding and (d) antibonding states of the $c(4\times2)$ surface.

with experimental facts, i.e. the well prepared single domain Si(100) surface shows the $c(4\times2)$ LEED spot at lower temperatures [7]. The atomic positions of the optimized structure of the Si(100) $c(4\times2)$ surface is listed in Table II compared with experiment [22]. These results show rather good agreement with each other.

The bond length of the surface dimer of the $c(4\times2)$ surface is calculated to be 2.27 Å, which is shorter than the bulk bond length of 2.35 Å. On the other hand this is slightly longer than the bond length, 2.22 Å, of the uniformly buckled 2×1 asymmetric dimer. The reason for this could be ascribed to the depletion of the DB electron density around the dimer bond region for the case of the $c(4\times2)$ surface compared with the uniformly buckled case. The tilting angle of the dimer for the $c(4\times2)$ surface is 14°, which is larger than that of the uniformly buckled surface (8.5°), and shows a good agreement with the observed value, 15.5° [22]. For the $c(4\times2)$ surface the length of the back bond connecting the surface dimer is 2.34 Å and 2.33 Å for the up dimer atom and 2.33 and 2.30 Å for the down dimer atoms, which are nearly equal to the bulk values.

TABLE II
The atomic displacements (in Å) of the $c(4 \times 2)$ surface obtained by the present calculation and experiment.

	Atoms (k,l,m)	Expt. Δx	Expt. Δy	Expt. Δz	Present calc. Δx	Present calc. Δy	Present calc. Δz
Layer 1	(0,0,0)	0.593	0.000	0.170	0.667	0.012	−0.034
	(2,0,0)	−0.954	0.000	−0.460	−0.957	0.010	−0.577
	(6,0,0)	−0.593	0.000	0.170	−0.667	−0.012	−0.034
	(4,0,0)	0.954	0.000	−0.460	0.957	−0.010	−0.577
Layer 2	(0,1,−1)	0.092	−0.113	0.071	0.122	−0.077	−0.030
	(2,1,−1)	−0.092	0.113	0.071	−0.122	0.077	−0.030
	(4,1,−1)	0.092	0.113	0.071	0.139	0.091	−0.020
	(6,1,−1)	−0.092	−0.113	0.071	−0.139	−0.091	−0.020
Layer 3	(1,1,−2)	0.000	0.000	−0.057	0.000	0.000	−0.241
	(3,1,−2)	0.000	0.004	0.166	0.021	0.005	0.153
	(5,1,−2)	0.000	0.000	−0.057	0.000	0.000	−0.241
	(7,1,−2)	0.000	−0.004	0.166	−0.021	−0.005	0.153
Layer 4	(−1,2,−3)	0.000	0.000	0.137	0.000	0.000	0.123
	(1,2,−3)	−0.033	0.000	−0.029	−0.012	0.005	−0.193
	(3,2,−3)	0.000	0.000	0.127	0.000	0.000	0.123
	(5,2,−3)	0.033	0.000	−0.029	0.012	−0.005	−0.193
Layer 5	(0,2,−4)	−0.056	0.000	0.035	−0.096	−0.002	−0.014
	(2,2,−4)	0.027	0.000	0.056	0.082	0.010	−0.008
	(4,2,−4)	−0.027	0.000	0.056	−0.082	−0.010	−0.008
	(6,2,−4)	0.056	0.000	0.035	0.096	0.002	−0.014

5.2. The first-princples MD method

We tried also the Car–Parrinello method [23] in order to search of the optimized structure of the uniformly buckled (2×1) surface. The five atomic layer models are used, in which the bottom layer is formed by hydrogen atoms to eliminate the dangling bonds. The ion positions in the lowest two atomic layers are fixed in the molecular dynamics calculation. The same, Bachelet, Schlüter, Hamann pseudopotential [15] is used with a cutoff energy of the basis being 6.25 Ryd. Some characteristic values of the dimer geometry calculated by this method are presented in Table III and compared with the results by the usual energy minimization method mentioned in Section 5.1. The values by the two methods are slightly different from each other, but the difference comes mainly from the difference of models, and not from the theoretical method.

Yamauchi [25] also investigated the absorbed C monolayer structure on the Si(100) surface by the model obtained by the substitution of C atoms

[4] ZST; the result reported in Reference [9].

TABLE III
Bond lengths of Si(100) surface.

Bond	Si(Å) asymmetric	Si(Å) symmetric	Si(ZST)[4] (Å) asymmetric
No. 9–11	2.311	2.309	2.29
No. 10–12	2.367	2.309	2.33
No. 11–12	2.268	2.350	2.22

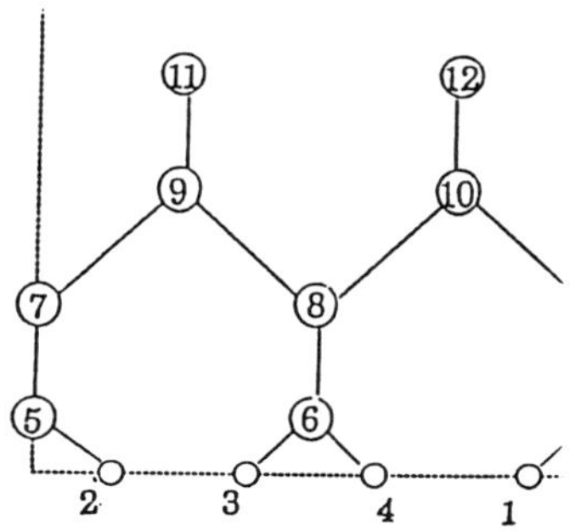

for all the Si dimer atoms on the uniformly buckled (2 × 1) model. The C–Si back bond of the dimer is rather shorter than the corresponding back bond of the pure Si surface, while the C–C dimer bond length becomes elongated as compared to the bond length in diamond crystal. But the result is not conclusive, because the calculation is performed with the cutoff energy 9.6 Ryd, which is not enough for C including system.

6. Simulation of the STM Image and STS Spectrum of Si(100) Surfaces

6.1. Features of the observed STM images

Though the clear STM images of Si(100) surfaces so far observed established the dimer model as an essential ingredient of the reconstructed surfaces, they introduced a new mystery [5,6]. Namely, in the large defect-free part of the surface, the 'seemingly symmetric' dimers are predominantly observed. The zigzag rows of the asymmetric dimers are observed in rather limited parts of the surface such as near the missing dimer or around the step. Whether the 'seemingly symmetric' dimers are actually the symmetric dimers or not is the issue of heated controversies, because the total energy calculations used by several theoretical groups, including us [9–12], predicted that the asymmetric dimer structure is more stable than the symmetric one. And this corresponds well with much experimental evidence such as LEED, ARUPS and ISS. As supporting evidence the STM image of a Ge(100) 2 × 1 surface clearly shows the asymmetric dimer rows [26].

To resolve the contradiction, two ideas are postulated. One assumes the flip-flop type oscillation of the asymmetric dimer with much shorter time period than the residence time of the tip on a particular surface position during scanning. Thus the time averaged STM image would appear symmetric. Another idea assumes that the interaction between the tip and the surface causes the distortion of the surface atoms with the movement of the tip. For example, suppose that the single apex atom of the tip attracts the Si dimer atom. Then the closer atom in the dimer is always raised toward the tip apex and no direct information of the original dimer geometry is obtained by the tip.

6.2. Method and model of theoretical simulation

To clarify the situation actually happens in the STM observation, theoretical simulation of the STM/STS based on the first-principles electronic state calculation is quite helpful. In the research group of the present authors, a theoretical method of STM/STS simulation has been successfully developed [27–30]. The theoretical procedure has been reported elsewhere, so we summarize the essence of the method briefly. Bardeen's perturbation formula [30] is used as the basis of the calculation for the tunnel current. The electronic states of the tip and that of the sample surface are calculated as separate systems, with the use of LDA [27–29]. As for the model of the tip, clusters formed by 10–20 atoms are used and, for the surface, an isolated or repeated slab model is used. The wavefunctions of the tip and the surface are usually calculated by the LCAO method, since this is convenient for the evaluation of the tunnel matrix elements. However, to reproduce rather accurately behavior of the surface wavefunction in the very tail part of the surface, the 'connected vacuum tail method' is used [31], which utilizes the wave solved by the Laue representation connected to the wavefunctions inside the surface.

6.3. Results

Figure 6 shows the simulated STM images of the symmetric dimer Si(100) 2×1 surface by the connected vacuum tail method [31]. For the model of the tip, the W10 [111] cluster with the axis directed to the [111] direction is adopted. The distance between the tip and the surface is 5.2 Å and the surface bias voltage is -1.0 eV and 1.0 eV for Figures 6a and 6b, respectively. Thus Figure 6a represents the occupied states and Figure 6b the unoccupied states of the surface dimer. The simulated image of the occupied state seemingly looks like the observed STM image of the 'seemingly symmetric' dimers in the defect-free region. However, the simulated image of the unoc-

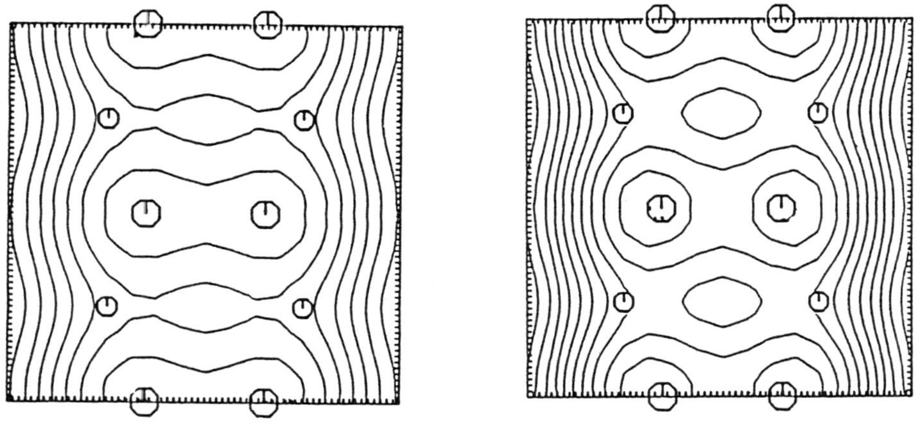

(a) (b)

Fig. 6. Simulated STM image of (2 × 1) symmetric dimer surface at the tip height 5.3 Å for (a) the surface bias −1.0 V and +1.0 V.

cupied states is rather different from the experimental observation. Namely, for this bias condition, a rather deep trough is running through middle of the dimer bonds. This trough, according to the experimental condition, is deeper than that between the adjacent dimer rows. On the other hand, this feature does not appear in the theoretically calculated image, which looks rather similar to the image of the occupied states (see Figure 6). The deep trough might be caused by the nodal structure of the antibonding π^* state. But this is not necessarily the case, because for a symmetric dimer surface both π and π^* bands contribute with an almost similar weight even for the positive surface bias because of the overlaps of the two energy bands. The other reason comes from the effect of the tip; the tip orbitals, which are not axial symmetric with respect to the tip axis, can pass a stronger tunnel current at the nodal line of the surface wavefunction, which causes significant distortion of the STM image as compared to the contour map of the LDOS.

The distinct contradiction appears in the STS spectrum. Figures 7a and b show the calculated STS spectra for the symmetric dimer surface and $c(4 \times 2)$ asymmetric dimer surface, respectively. Figure 7c illustrates the experimental results by Hamers *et al.* [32]. For the symmetric dimer surface, a significant differential conductance of the tunnel current appears at the zero bias region, which is not observed in the experiments. This is because the symmetric dimer surface is metallic due to the band overlap of the π and π^* bands.

Figures 8a and b illustrate the theoretically calculated image of the $c(4 \times 2)$ surface by the same tip model and the tip height as Figure 6.

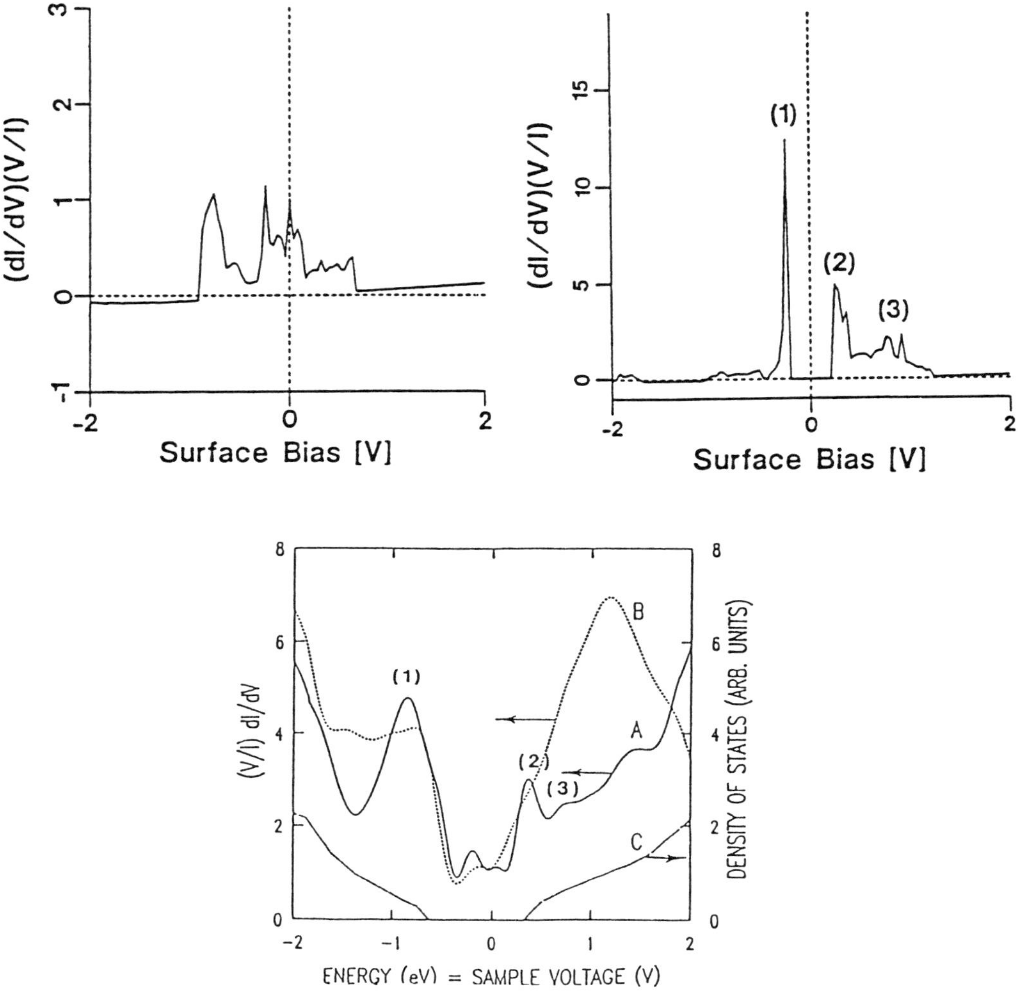

Fig. 7. Simulated STS spectra for (a) the symmetric dimer surface and (b) for the $c(4 \times 2)$ surface compared with the experimental observation (curve A) [33].

For the image of the occupied state at the surface bias -1.0 eV, the zigzag pattern connecting the up dimer atoms appears, which resembles the experimental observation. The zigzag pattern also appears for the unoccupied states image (see Figure 8b), but a stronger current region now corresponds to the down dimer atoms. The corrugation of the image is much reduced as compared to the zigzag pattern for the occupied states.

As seen in Figure 7, the calculated STS spectra for the $c(4 \times 2)$ surface correspond remarkably well with experiments [32] which are obtained at the region of the symmetric dimers. The peak labelled as 1, 2 and 3 correspond,

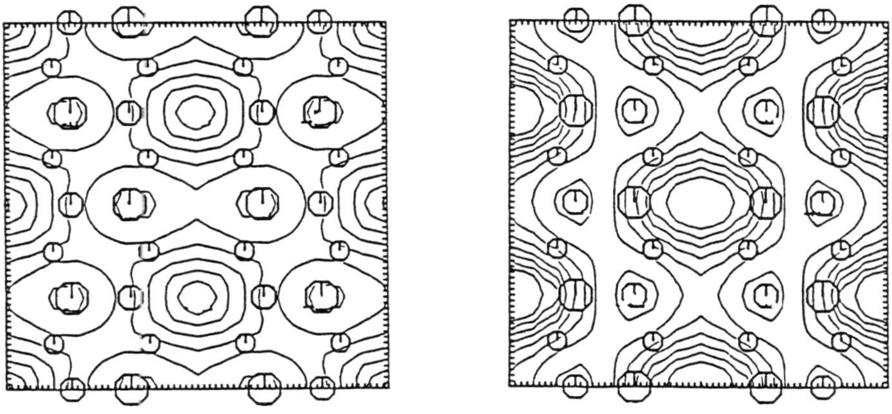

Fig. 8. Logarithm contour maps of LDOS of $c(4 \times 2)$ surface at the tip height 5.3 Å. (a) Occupied surface band at $E = -2.66$ eV, (b) unoccupied surface band at $E = -1.66$ eV [32].

respectively, to the upper edge of the π band, the lower and the upper edges of the π^* band.

Based on the similarity between the simulated image and the experimental image of the zigzag pattern, one may conclude that the latter surely corresponds to the asymmetric dimer arrays, where the direction of the dimer inclination alternates along the dimer row. The fact that the feature of the experimental STS at the 'seemingly symmetric' dimer region resembles well with that of the $c(4 \times 2)$ surface and quite dissimilar to that of the uniformly buckled (2×1), surface may mean that the observed 'seemingly symmetric' dimers are essentially of the same character as the asymmetric dimer. It might be the time averaged image of the oscillating asymmetric dimer, or some effect of the tip-surface interaction that induces the change of the tilting angle during the relative tip motion. The very deep dip along the center of the dimer bond may be explained by the latter effect.

7. Summary and Discussion

Based on the first-principles LDA calculation, the electronic structure of various reconstructed structures of Si(100) dimer surfaces has been investigated. It was found that the most stable higher order reconstruction structure among symmetric dimer 2×1, asymmetric dimer $p(2 \times 2)$, $c(4 \times 2)$ and uniformly buckled 2×1 structures is the $c(4 \times 2)$ structure. This is consistent with the LEED observation for the single domain surface in lower temperatures. Moreover only this surface gives the semiconducting electronic states.

The stabilization mechanism of this structure is caused by the bonding hybridization interaction between the up dimer atoms in the adjacent rows. The dispersion of the surface band of $c(4 \times 2)$ structure showed an excellent agreement with the ARUPS result. The first-principles molecular dynamics study also supported the stability of asymmetric dimer structure.

Careful theoretical simulation of the STM image and STS spectra have been performed based on the LDA theory. The simulated STS spectra of the $c(4 \times 2)$ structure model showed a good correspondence with the observed STS in the seemingly 2×1 surface region. The simulated STM image of this model also reproduces well the zigzag pattern observed around the defect or step edge regions. Though the simulated results of the STM image of the 2×1 symmetric dimer surface represented the experimental feature of the 'seemingly symmetric' dimer regions, the image of the unoccupied states has not been so well reproduced. This might indicate, as well as a large difference between the simulated and the observed STS spectra, that the 'seemingly symmetric' dimer does not actually correspond to the real symmetric dimer, but is caused by some artifact due to the tip–surface interaction.

As discussed in the present article, considerably detailed features of the Si(100) surface have been clarified. However, with the increase of our knowledge of this surface more and more new problems are revealed. For example the tip–surface interaction, as mentioned above, might be a very interesting issue and would be the starting point for the study of atomic control of the surface. Some authors might claim that the flip-flop type oscillation would induce the 'seemingly symmetric' dimer image. At the present stage, this remains only speculatin, but the confirmation of this, if it is a fact, also provides a challenging target of surface physics. The diffusion of Si and other atoms on this surface are also very important subjects. Recently both experimental and theoretical works have made considerable progress. For example, the difference in the morphology between the two types of the step has been observed by the STM and clearly explained by the theoretical works. The topic related to alkali chemisorption on this surface is another exciting problem with various controversies. This we will discuss elsewhere, as well as other recent developments.

Acknowledgements

One of the authors (M.T.) and his research group enjoyed a good cooperation with Professor H. Kamimura for a long period. We are most grateful for this, and wish him hearty congratulations for his sixtieth birthday. Numerical calculations are performed at the Computer Center of the University of Tokyo, and the Computer Center of the Institute for Molecular Science. This work is supported by the Grand-in-Aid, from Ministry of Education, Science and Culture, Japan.

References

1. R.E. Schlier and H.F. Farnsworth: 1959, *J. Chem. Phys.* **30**, 917.
2. G. Binnig, H. Rohrer, Ch. Gerber and E. Weibel: 1982, *Phys. Rev. Lett.* **49**, 57.
3. P. Hansma and J. Tersoff: 1987, *J. Appl. Phys.* **61**, R1.
4. M. Tsukada, H. Kageshima, N. Isshiki and K. Kobayashi: *Surface Sci.*, submitted.
5. R.M. Tromp, R.J. Hamers and J.E. Demuth: 1985, *Phys. Rev. Lett.* **55**, 1303.
6. R.J. Hamers, R.M. Tromp and J.E. Demuth: 1986, *Phys. Rev. B* **34**, 1303.
7. T. Tabata, T. Aruga and Y. Murata: 1986, *Surface Sci.* **179**, L63.
8. T. Enta, S. Suzuki and S. Kono: 1990, *Phys. Rev. Lett.* **63**, 2704.
9. Z. Zhu, N. Shima and M. Tsukada: 1989, *Phys. Rev. B* **40**, 11868.
10. D.J. Chadi: 1979, *Phys. Rev. Lett.* **43**, 43.
11. M.T. Yin and M.L. Cohen: 1981, *Phys. Rev. B* **24**, 2303.
13. E. Artacho and F. Yndrain: 1989, *Phys. Rev. Lett.* **62**, 2491.
14. I.P. Batra: 1990, *Phys. Rev. B* **41**, 5048.
15. G.B. Bachelet, D.R. Hamann and M. Schlüter: 1982, *Phys. Rev. B* **26**, 4199.
16. J. Ihm, M.L. Cohen and D.J. Chadi: 1980, *Phys. Rev. B* **21**, 4592.
17. P. Kruger and J. Pollmann: 1988, *Phys. Rev. B* **38**, 10578.
18. R.I.G. Uhrberg, G.V. Hansson, J.M. Nicholls and S.A. Flogstrom: 1981, *Phys. Rev. B* **24**, 4684.
19. Y. Enta, S. Suzuki, S. Kono and T. Sakamoto: 1990, *J. Phys. Soc. Jpn.* **59**, 657.
20. A. Ypshimori, *Ordering at Surfaces and Interfaces*, Springer, 1991.
21. W.A. Harrison, *Electronic Structure and Properties of Solids, The Physics and Chemical Bond*, Freeman, San Francisco, 1980.
22. R.M. Tromp, R.G. Smeenk, F.W. Saris and D.J. Chadi: 1983, *Surface Sci.* **133**, 137.
23. R. Car and M. Parrinello: 1985, *Phys. Rev. Lett.* **55**, 2471.
24. J. Yamauchi, Master's Thesis, University of Tokyo, 1991.
25. J.A. Kubby, J.E. Griffith, R.S. Becker and J.S. Vickers: 1987, *Phys. Rev.* **36**, 6079.
26. M. Tsukada and N. Shima: 1987, *J. Phys. Soc. Jpn.* **56**, 2875.
27. M. Tsukada, K. Kobayashi and S. Ohnishi: 1990, *J. Vac. Sci. and Techn. A* **8** (1), 155.
28. M. Tsukada *et al.*, *Surface Sci. Report*, in press.
29. N. Isshiki, K. Kobayashi and M. Tsukada: 1990, *Surface Sci.* **238**, L439.
30. J. Bardeen: 1961, *Phys. Rev. Lett.* **15**, 57.
31. H. Kageshima, Doctor's Thesis, University of Tokyo, 1991.
32. R.J. Hamers, Ph. Avouris and F. Bozso: 1987, *Phys. Rev. Lett.* **59**, 2071.

INITIAL STAGE OF OXIDATION OF SILICON: A STATE-OF-THE-ART CALCULATION

ATSUSHI OSHIYAMA and YOSHIYUKI MIYAMOTO
Fundamental Research Laboratories, NEC Corporation Miyukigaoka, Tsukuba 305, Japan

Abstract. We present our microscopic calculations for mechanisms of oxidation of silicon. The results provide us with characteristic features in the oxidation of covalent materials: metastability in oxygen adsorption on the surface, oxygen penetration and bond formation in the silicon substrate, the resulting expansion in volume, and significant relaxation at the highly mismatched Si/SiO_2 interface.

1. Introduction

Oxidation phenomena are well-known from ancient times and have been investigated for many years. The first problem is, how the oxygen and the oxidized atoms diffuse through the oxides, and what determines the rate of oxidation. Extensive works, both theoretical and experimental have revealed the apparent features of the oxidation phenomena, but an understanding of their microscopic origins still remains poor [1]. The study of the oxidation of silicon has the origin in its importance in silicon technology. The inferiority of Si in respect of its electronic properties as compared to Group III–V compound semiconductors is overcome by the gift of a high-quality insulator, silicon dioxide, which is rather easily formed by the oxidation of Si substrate. It has been said that were it not for this gift, Si would lose its premier status in semiconductor technology. With decreasing size of current electronic devices, however, the situation becomes stringent: in current dynamic random access memories (DRAM) the thickness of silicon dioxides decreases to the order of ten Ångstroms so that it is imperative to understand the Si oxidation from the microscopic viewpoint. Microscopic understanding of the Si oxidation, in turn, provides us with important knowledge about the exciting issues in condensed matter physics: e.g. mechanisms of dissociation of an oxygen molecule on the surface or in the Si substrate and the subsequent formation of a new bond network with Si atoms, diffusion of oxygen in silicon dioxides which are usually of vitreous forms, and the atomic relaxation in the vicinity of the resulting Si/SiO_2 interface which is highly mismatched.

In this paper we present our microscopic calculations for energetics in the initial stage of oxidation of Si [2,3] and for atomic reconstruction at the highly-mismatched Si/SiO_2 interface [4]. Limitation of conventional analysis in X-ray photoemission spectroscopy (XPS) and a possibility of a new analysis in the determination of atomic structures from the measured core-level shifts is also discussed.

H. Aoki et al. (eds.), New Horizons in Low-Dimensional Electron Systems, 337–350.

2. Calculation

The total-energy and force calculations have been performed within the local-density functional formalism [5] using norm conserving pseudopotentials [6]. The localized-orbitals basis-set, consisting of Gaussian orbitals located at atomic sites, are utilized in the calculation.

First we construct the norm conserving non-local pseudopotentials following the procedure described in Reference [6]. We then place the pseudopotentials at nucleus sites in a given atomic geometry. Distribution of the valence electrons is obtained by the local density approximation with the Ceperley–Alder form [7] for the exchange-correlation energy. In that self-consistent field calculation the Gaussian orbitals basis set is used to express the wavefunctions or the valence electron charge density. The advantage of the Gaussian orbitals basis set is its compactness in size as compared with the well-converged plane-wave basis set. The disadvantage is its incompleteness, or the difficulty in the systematic improvement of the completeness. Yet, if we carefully determine the exponents of the Gaussian orbitals, the results from the well-converged plane-wave basis set are satisfactorily reproduced by the compact Gaussian orbitals basis set. In this calculation we use three exponents for each of Si s, O s and p orbitals, and two exponents for the Si p orbital. The values of the exponents are determined by fitting the pseudo-orbitals obtained numerically and then by minimizing the total energy of the solids. The experimental values for the structural properties of Si and α quartz are well reproduced by the present basis set: e.g. the deviation in the lattice constant from the experimental value is 0.3% for Si and 2.6% for α quartz, respectively.

We next calculate the force (Helmann–Feynman force plus the correction) acting on each atom in the geometry. In the localized-orbitals basis set the correction term contributes considerably to the total force [8]. According to the calculated forces we move the atoms toward a new atomic geometry with the lower total energy. Repeating this procedure, i.e. separate total energy optimization of electron degrees of freedom and the nucleus degrees of freedom, we obtain the total-energy minimized (optimized) atomic geometry. The path toward the optimized geometry is thus the intrinsic reaction coordinates. In the optimized geometry the magnitude of the forces are less than 10^{-2} Ryd per atomic unit.

3. Oxygen Dissociation and Adsorption on the Si(100) Surface

In the first place we present the results for dissociation of an oxygen molecule and the subsequent adsorption on the Si(100) surface [2,3]. The problem is, whether an O_2 molecule dissociates on the surface, and what geometry is realized for the oxygen adsorption.

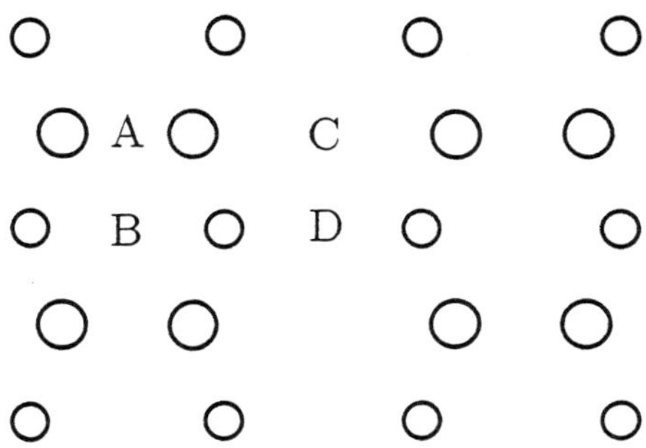

Fig. 1. Top view of Si(100) surface. The top layer Si (large circles, forming the dimers) and the second-layer Si (small circles) atoms are shown.

Figure 1 shows a top view of the Si(100) surface. Top-layer Si atoms form dimers and there are several sites for oxygen approach: the site on the dimer (A), between the two dimers (B), and the sites between the dimer rows (C and D). These four points are representatives of the sites on the Si(100) surface. We place an O_2 molecule above these sites and move the atoms according to the calculated forces. One of our findings is that an O_2 molecule dissociates at any site on the Si(100) surface.

Figure 2 shows an example. First an O_2 molecule is placed above the site B with its axis 45° rotated both for [100] and [011] directions (top panel of Figure 2). Then the molecule approaches to the surface and rotates its axis being parallel to the surface and perpendicular to the Si dimer axis (middle panels of Figure 2). When the molecule is located between the Si dimers in the vicinity of the surface (bottom panel of Figure 2), the forces still act on the two O atoms so as to make the molecule dissociate. After the dissociation, each O atom indeed settles down at one of the two adsorption sites near the Si dimer that are (meta)stable for the O atom: one is the site on the dimer (Figure 4) and the other between the two Si atoms of the dimer (the bridge site, Figure 5). This reaction of dissociation and the subsequent atomic adsorption is exothermic. The calculated heat of formation per O atom in the reaction is 2.5 eV for the adsorption site on the dimer and 3.0 eV for the bridge site, respectively. The obtained local vibration energy and the electron density of states for the adsorption sites are consistent with the experimental data [2]. The mechanism of the dissociation is visible in the valence electron density shown in Figure 3: more than one electron is transferred from Si to the antibonding π orbital of the O_2 molecule so that the bond between the O atoms diminishes.

Similar behaviors are observed for other sites A, C, and D. The example for the site C is shown in Figure 6. An O_2 molecule approaches the surface, rotating its axis. The charge transfer from Si to O occurs so that

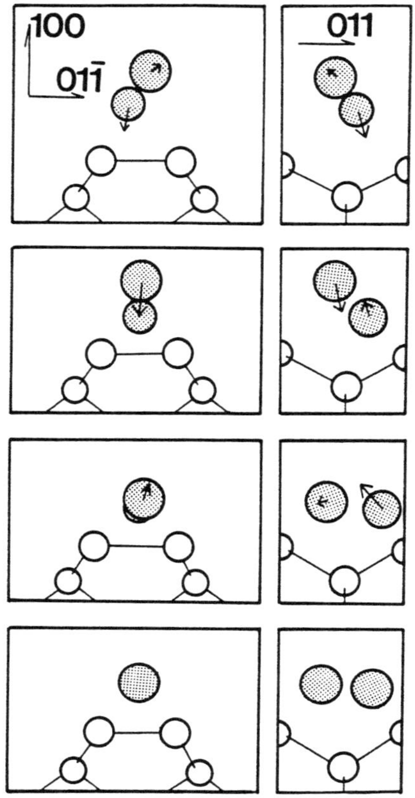

Fig. 2. Dissociation of an O_2 molecule (dotted circles) approaching above the site B to the Si(100) surface (from the top panel to the bottom panel). At first (top panel), the center of the molecule is placed by 2.2 Å above the surface, and the forces (the arrows) of 0.05 Ryd/a.u. and 0.09 Ryd/a.u. act on the 2 O atoms of the molecule. In the geometry shown in the bottom panel, the forces of 0.09 Ryd/a.u. and 0.07 Ryd/a.u. act on the O atoms which then reach one of the two (meta)stable adsorption sites: the site 1.5 Å above the Si dimer or the bridge site (see text and Figure 4 and 5).

the molecule dissociates and the resulting O atoms are adsorbed. The calculated heat of formation of the reaction shown in Figure 6 is 1.4 eV per O atom. Moreover we have found that the adsorption site for the resulting O atom is determined by where the preceding O_2 molecule dissociates: when the molecule approaches to the surface above the site A or B, the resulting O atom is adsorbed after dissociation at the site on the Si dimer or at the bridge site, as is stated above. On the other hand, when the O_2 molecule approaches to the site between the dimer rows (i.e. the site C or D), the molecule dissociates and each of the resulting two O atoms is adsorbed at

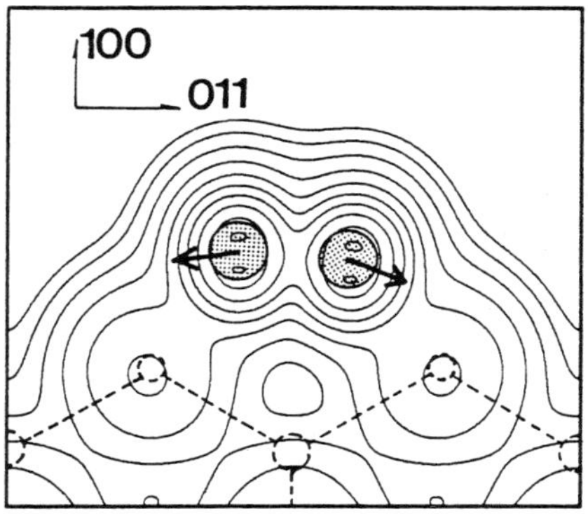

Fig. 3. The contour plot of the valence electron charge density in the geometry shown in the bottom panel of Figure 2. The maximum value of the contour is 1.1 electrons/(a.u.)^3 and the subsequent lines differ by a factor of 2.

the site aside from the Si dimer. It is of interest that the atomic adsorption is accompanied with the peculiar relaxation of the top-layer Si atoms: the Si dimers are twisted in Figure 6, decomposed in Figure 5, and preserved in Figure 4.

The effect of spin polarization has been examined by the local-spin-density calculation. It is found that, as the distance between the O_2 molecule and the Si surface decreases, the spin state most energetically favorable changes from the triplet to the doublet, and that at the distance at which the dissociation of the O_2 molecule occurs no spin-polarized state can be obtained. It is thus concluded that the O_2 molecule loses its spin due to enough electron transfer from Si bonds and then dissociates at any site on the Si(100) surface.

4. Oxygen Penetration and Formation of New Bonds in Si Substrate

We next present the results on penetration of the O_2 molecule through the oxygen-covered surface and successive bond formation between the O and Si atoms. In the case of one-monolayer coverage of oxygen, the most energetically favorable site for the oxygen adsorption is found to be the bridge site: Figure 7 is the total-energy minimized geometry in which the Si dimer is decomposed by the intervening O atom. We then place an oxygen molecule above the O-covered surface and seek for the path of the penetration of the O_2 molecule.

We have found that the calculated forces acting on the O_2 molecule are repulsive from the surface unless the O atoms already adsorbed and the top-layer Si atoms are dislodged: i.e. there is an energy barrier for the O_2

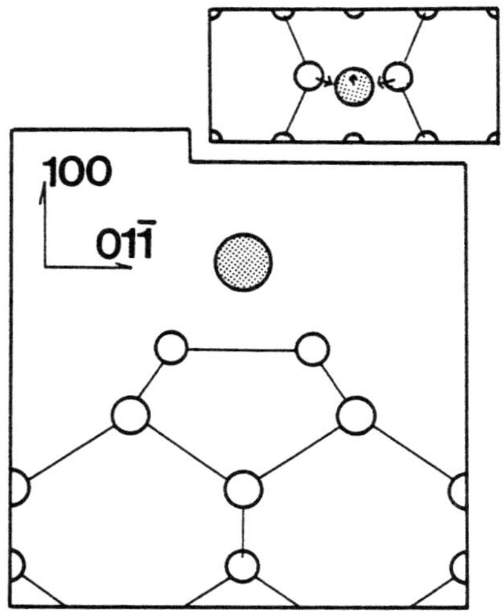

Fig. 4. A metastable atomic geometry for oxygen adsorption shown on the (011) plane. Dotted and open circles denote O and Si atoms, respectively. When the oxygen atom is displaced from the (011) plane (upper panel), restoring force of 0.054 Ryd/(a.u.) acts on it.

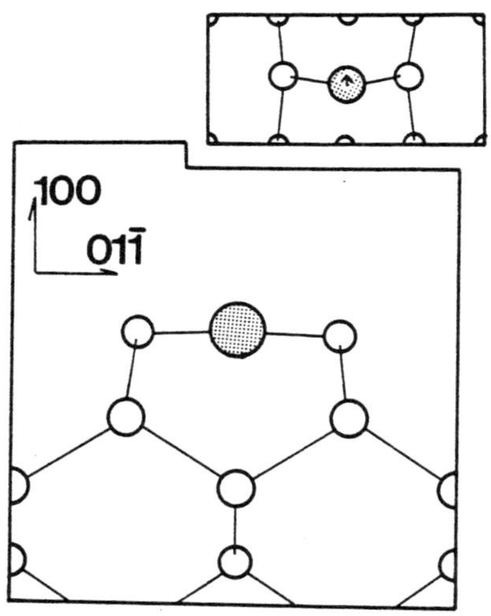

Fig. 5. Stable atomic geometry for oxygen adsorption shown on the (011) plane. Dotted and open circles denote O and Si atoms, respectively. When the oxygen atom is displaced from the (011) plane (upper panel), restoring force of 0.026 Ryd/(a.u.) acts on it.

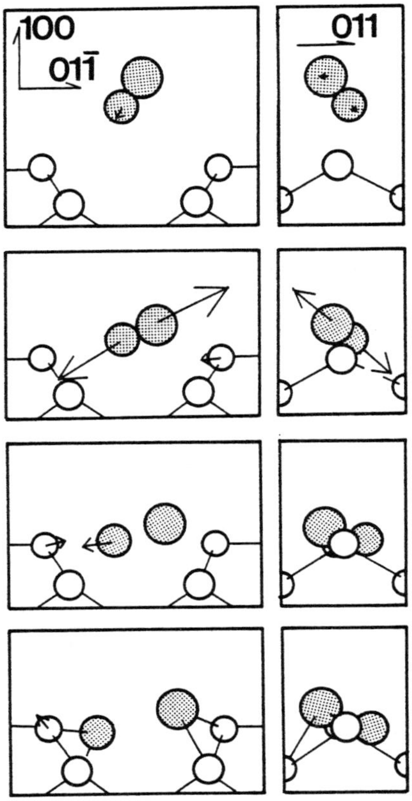

Fig. 6. Dissociation of an O_2 molecule (dotted circles) approaching above the site C to the Si(100) surface (from the top panel to the bottom panel).

penetration in case of no relaxation of neighbor atoms. When the top-layer O and the Si atoms are moved along [011] direction by half angestroms, however, the O_2 molecule is found to penetrate through the oxygen-covered surface. After penetration, the molecule dissociates in the vicinity of the bond center sites of the top- and the second-layer Si atoms. The dissociation is again due to the electron transfer from the Si bonds to the π orbital of the O_2 molecule. This suggests that in the oxidation reaction the dissociation of the O_2 molecule occurs not on the O-covered surface or in the SiO_2 film but at the interface where the electrons triggering the dissociation can be provided from the Si–Si bonds.

The final geometry we have found is shown in Figure 8 along with the valence electron charge density. We can see the peculiar bond formation of Si and O after the penetration and the dissociation. This reaction of the penetration, the dissociation and the new bond formation is exothermic: the calculated heat of formation is 1.2 eV per O atom. The remarkable feature of

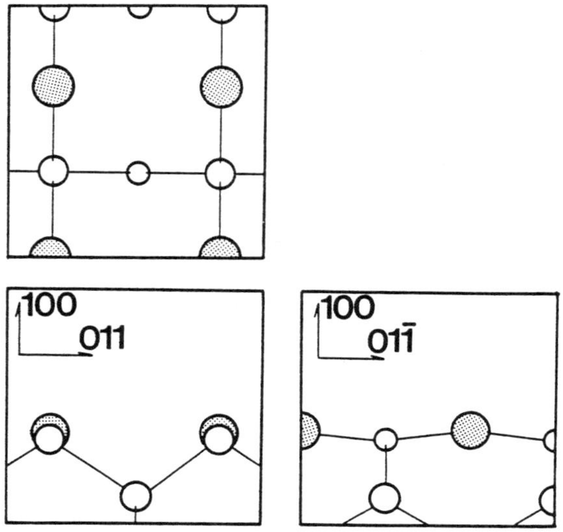

Fig. 7. Total-energy minimized geometry in the case of one-monolayer coverage of oxygen on the Si(100) surface (upper panel: top view, lower panels: horizontal views). Dotted and open circles denote O and Si atoms, respectively.

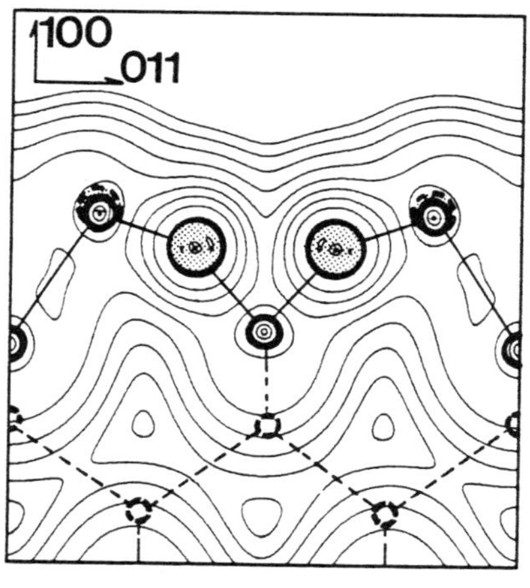

Fig. 8. The stable atomic geometry after penetration and dissociation of the O_2 molecule near the Si(100) surface. The O_2 molecule penetrates through the 1 monolayer oxygen covered surface (Figure 7), dissociates and forms the Si–O bonds between the top and second layer Si atoms. We notice the sizable relaxation of the 1st and the 2nd layer Si atoms. The maximum value of the contour plot of the valence electron charge density is 1.1 electrons/(a.u.)^3, and the subsequent lines differ by a factor of 2.

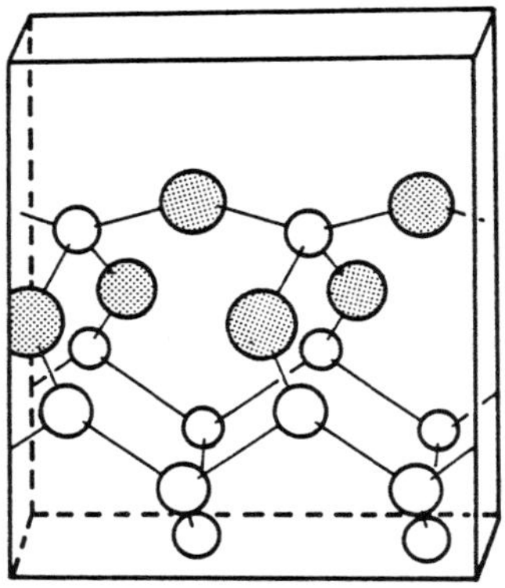

Fig. 9. A metastable atomic geometry in the oxidized Si. Swelling of the Si–O–Si local bond network from the Si(100) surface is observed. The O atoms (dotted circles) intervene into all the bond center sites between the 1st and the 2nd layer Si atoms. Because of the limited room the O atoms are squeezed so that the adjacent Si–O–Si bonds are rotated in the opposite directions.

the stable geometry is in its local structure. The obtained Si–O bond length (1.6 Å) and the Si–O–Si bond angle (144°) in Figure 8 are surprisingly close to the values observed in α quartz, the stablest crystalline form of SiO_2. The Si and the O atoms strongly favor this local bond configuration in many cases (e.g. the isolated O impurity in bulk Si [8]). Yet the O–Si–O bond angle (81°) is quite a lot smaller than the corresponding value (about 110°) in the crystalline SiO_2 under low pressure (i.e. quartz, cristobalite or coesite [9]). Under high pressure, however, the SiO_2 has another crystalline form, stishovite. The value of the O–Si–O angle in stishovite ranges from 81° to 98°, and is comparable with the value obtained in the geometry in Figure 8. This coexistence of the low and the high pressure crystalline forms is a consequence of the local strain due to the sizable difference in Si–Si distance between SiO_2 and Si substrate; lattice mismatch.

The top-layer Si atoms in Figure 8 are surrounded by three O atoms and one Si atom. To attain the bond configuration topologically the same as in SiO_2 (i.e. four oxygen atoms around a Si atom), additional O atoms are required to intervene into the remaining bond center sites. In fact, due to the large upward displacement of the top-layer Si atoms (0.5 Å from the original site) the remaining bonds are weakened considerably, as is clearly seen in the contour plot in Figure 8. Yet the resulting Si–Si distances are still tight to accommodate the additional O atoms. A natural consequence is therefore a swelling of the Si–O–Si unit from the original Si surface. Figure 9 shows the obtained metastable geometry after penetration of additional O atoms, in which the top-layer Si atoms are surrounded by four O atoms and the Si–O–Si units swell from the original Si surface. This reaction of oxygen uptake is again exothermic: the calculated heat of formation in the reaction of adding a new O atom is 1.5 eV. The structure in Figure 9 is again a local mixture of the low and the high pressured crystalline forms of SiO_2. The swelling of the Si–O–Si unit obtained in Figure 9 corresponds to the volume expansion of the region containing the O and the top- and the second-layer Si atoms by 83%. This value can be compared with the volume difference between Si and silica (e.g. the ratio in volume per Si atom of α quartz to Si is 1.88). It is of interest to notice that the expansion has begun from the very early stage of the oxidation.

Another important finding shown in Figures 8 and 9 is the existence of different oxidation states of Si atoms. In Figure 8, the top layer and the second layer Si atoms are surrounded by three and two O neighbors, respectively, and in Figure 9 the top layer Si atom is surrounded by four O neighbors. The difference in the coordination numbers that we have found indeed results in the different amount of the core-level shift of the Si atom, which is indeed observed in the X-ray photoemission spectroscopy (XPS) measurements for Si $2p$ core levels at the SiO_2/Si interface [10,11]. In the usual analysis of the XPS, however, the spin-orbit splitting is treated independently of the local-environment effect. This treatment is not applicable to the case in which the splitting due to anisotropic Coulomb field is comparable to the shift due to the charge transfer, and thus the amount of spin-orbit splitting is sensitive to the local environments [12].

Moreover, in case of the Si surrounded by three O atoms, there is the unbalance in length between the two different bonds: the Si–Si and the Si–O bonds. The consequence is the significant stretching of the Si–Si bond (see the contour plot in Figure 8). The stretching in this stable geometry causes further oxidation, as is demonstrated above and, on the other hand, reduces the formation energy of the Si self- interstitial atom at the interface. This is the microscopic origin of the observed injection of the self-interstitial atoms to the Si substrate upon oxidation [13].

The inherent features in the initial stage of the oxidation are the followings: the thermal penetration of the O_2 molecule, the dissociation of the molecule due to electron transfer from Si, the characteristic bond formation of the Si and O atoms accompanied by the large stretching of the Si–Si bond, and the resulting expansion in volume of the local atomic configurations. These features are not limited to the initial stage. In the case where the SiO_2 film has been already formed, although the thermal diffusion of the O_2 molecule plays a role, the essential features of the oxidation stated above are expected at the frontier of the oxidation: the SiO_2/Si interface. It is further noticed that the local bond lengths and angles of the Si and O atoms formed during the oxidation are similar to the values in the most stable form of silica, α quartz. In the stable geometries in Figures 8 and 9, however, the Si atoms near the O atom are significantly dislodged, and the

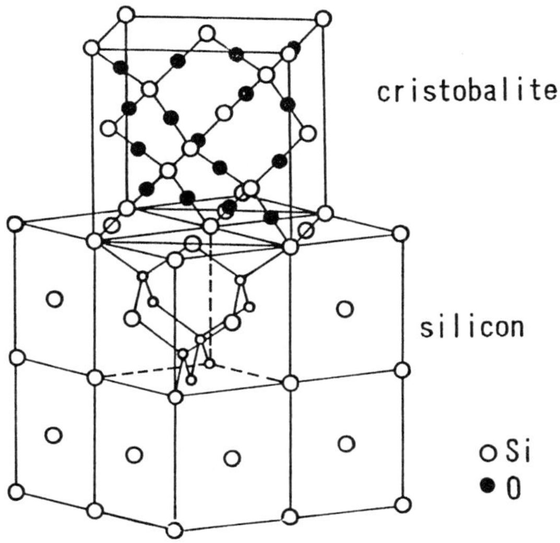

Fig. 10. Schematic view of ideal interface for cristobalite on Si(100).

high pressure form of SiO_2 appears. Further, the network in the Si crystal is topologically different from the Si network in the crystalline forms of silica. Thus the vitreous form of SiO_2 is realized.

5. Crystalline SiO_2 at the Amorphous SiO_2/Si Interface

We now turn to the atomic structure of highly mismatched SiO_2/Si interface. The structure of the SiO_2 film formed by the oxidation is vitreous. Yet at the interface the crystalline SiO_2 could exist [4,14] if the strain energy is released by the relaxation of the interface atoms. Among the variety of crystalline forms of SiO_2, the cristobalite and the tridymite are in the topological matching with the Si(100) surface [9]. There remains an enormous mismatch in atomic distance, however, at the interface: about 5% and 8% mismatch for cristobalite and tridymite, respectively. We here examine a possibility that the cristobalite structure is preserved on the Si(100) surface after the relaxation of the interface atoms [4].

Figure 10 shows a view of the interface in which Si[100] || cristobalite [100] and Si[011] || cristobalite [010]. In the calculation we use the slab which consists of three SiO_2 layers and three Si layers plus vacuum layers, and perform the total-energy and force calculation to reach the stable atomic geometry. The calculated results are examined by repeating the calculation with the thicker slab consisting of five SiO_2 layers and four Si layers. The artificial surfaces of the slabs are terminated by H atoms.

The calculated forces in the unrelaxed and the total-energy optimized

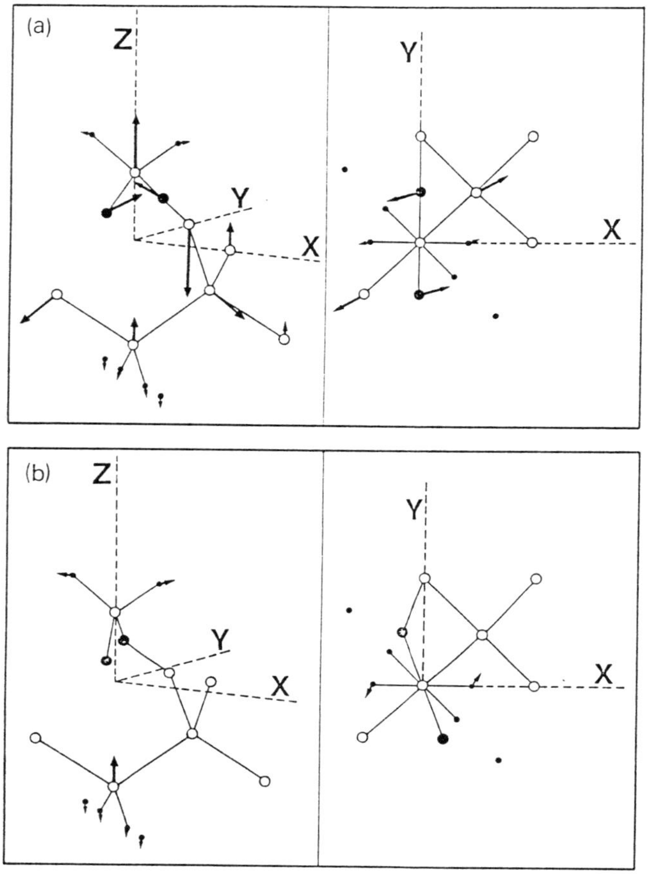

Fig. 11. Atomic geometries (upper panel: unrelaxed, lower panel: relaxed) and calculated force acting on each atom in the geometries. Open and solid circles denote Si and O atoms, respectively. Small dots are H atoms which are introduced to eliminate the artificial surface effect in the slab model [4].

geometries are shown in Figure 11. In the unrelaxed geometry the large forces act on the interface atoms. As the interface atoms are relaxed, the Si–O–Si bond angles decrease from 180°. In the final optimized geometry the Si–O–Si angle becomes 144°, which is the value commonly observed in α quartz. As is shown in Figure 11, the forces on the interface atoms virtually disappear in the optimized geometry. There remains the force acting on the third layer Si atom. Yet the amount of the force is small, 0.047 Ryd/a.u., and could be diminished through the long range relaxation. We thus conclude that the interface of cristobalite and Si(100) is stable after relaxation of the interface Si and O atoms. More elaborate calculations to compare the free energies of the cristobalite/Si and the tridymite/Si interfaces are now underway.

Another finding is the splitting of the two dangling bond states. Half of the interface Si atoms (Figure 10) have two dangling bonds each since they are not connected to the O atoms. The dangling bonds are expected to give partially occupied degenerate states in the energy gap. What we have found, however, is the contrary. The *s* and *p* orbitals of the Si atom re-hybridize and the two dangling bond states are split. The calculated gap of the two states is 1.3 eV. The interface is therefore of a semiconducting character. This splitting is the consequence of the electrostatic field due to the electron transfer from the Si bonds to the oxygen atoms through the interface [4].

6. Conclusion

In conclusion, we have presented a microscopic picture of the oxidation of Si on the basis of the *ab initio* total-energy electronic-structure calculations. The results provide a firm theoretical framework for understanding the oxidation phenomena and the atomic structures of the resulting SiO_2 films. The state-of-the-art calculations utilized in the present work would be the promising approach toward the clarification of phenomena in the cross-disciplinary field between science and technology.

Acknowledgements

We wish to thank M. Hane and A. Ishitani for fruitful collaboration in some aspects of the present work. Also, on the occasion of Professor Hiroshi Kamimura's retirement from University of Tokyo, we wish to express our true respect for his activity for many years in research and education, and hope that he sets out again on a new journey full of adventure and knowledge.

References

1. See, for a review, the special issue of *Phil. Mag.* **55**, No. 2 and No. 5 (1987). Also, useful information is avaiable in the pioneer experiments: M. Tabe, T.T. Chiang, I. Lindau and W.E. Spicer: 1986, *Phys. Rev. B* **34**, 2706 and references therein.
2. Y. Miyamoto and A. Oshiyama: 1990, *Phys. Rev. B* **41**, 12680.
3. Y. Miyamoto and A. Oshiyama: 1991, *Phys. Rev. B* **43**, 9287. A. Oshiyama and Y. Miyamoto, *Proc. 20th Int. Conf. Physics on Semiconductors*, Thessaloniki 1990, World Scientific, p. 87.
4. M. Hane, Y. Miyamoto and A. Oshiyama: 1990, *Phys. Rev. B* **41**, 12637.
5. For a review see *The Inhomogeneous Electron Gas*, N.H. March and S. Lundqvist (eds.), Plenum, New York, 1984.
6. D.R. Hamann, M. Schlüter and C. Chiang: 1979, *Phys. Rev. Lett.* **43**, 1494.
7. D.M. Ceperley and B.J. Alder: 1980, *Phys. Rev. Lett.* **45**, 566.

8. M. Saito and A. Oshiyama: 1988, *Phys. Rev. B* **38**, 10711.
9. R.W.G. Wyckoff, *Crystal Structures*, Interscience, NY 1963, Vol. 1, pj. 312.
10. F.J. Grunthaner, P.J. Grunthaner, R.P. Vasquez, B.F. Lewis, J. Maserjian and A. Madhukar: 1979, *Phys. Rev. Lett* **43**, 1683.
11. G. Hollinger and F.J. Himpsel: 1983, *Phys. Rev. B* **28**, 3651; F.J. Himpsel, F.R. McFeely, A. Taleb-Ibrahimi, J.A. Yarmoff and G. Hollinger: 1988, *Phys. Rev. B* **38**, 6084.
12. Details will be found in Y. Miyamoto and A. Oshiyama, to be published in *Phys. Rev. B*, September, 1991.
13. See, e.g., P.M. Fahey, P.B. Griffin and J.D. Plummer: 1989, *Rev. Mod. Phys.* **61**, 289.
14. There is a controversy on the existence of the crystalline SiO_2 at the interface: e.g., A. Ourmazd, D.W. Taylor, J.A. Rentschler and J. Bevk: 1987, *Phys. Rev. Lett.* **59**, 213; J.M. Gibson and M.Y. Lanzerotti: 1989, *Nature* **340**, 128.

MESOSCOPIC PHENOMENA IN SEMICONDUCTOR QUANTUM STRUCTURES

T. ANDO, H. TAMURA and H. AKERA
Institute for Solid State Physics, University of Tokyo, 7–22–1 Roppongi, Minato-ku, Tokyo 106, Japan

Abstract. A review is given of recent theoretical studies of transport properties of semiconductor quantum structures in magnetic fields. The topics covers conductance fluctuations and localization in quantum wires, their crossover from one to two dimensions, effects of boundary roughness scattering on the Hall effect and the magnetoresistance in quantum wires, and the ballistic conduction at quantum point contacts.

1. Introduction

Recent developments in the technology of ultrafine lithography and crystal growth have enabled the fabrication of quantum point contacts and ultranarrow wires at semiconductor heterostructures. The size of such quantum structures is much smaller than the electron mean free path and comparable to the Fermi wavelength, giving only a few occupied subbands. These systems have revealed quite new features of transport, such as the conductance quantization [1,2], the Aharonov–Bohm effect [3,4], fluctuations [5,6], anomalies in the low-field Hall effect [7–9], a bend resistance [10], and a negative resistance [11]. The purpose of the present paper is to give a brief review of some of recent theoretical studies on the transport properties in magnetic fields.

In Section 2 conductance fluctuations in quantum wires are studied. The fluctuations are *not universal* in narrow quantum wires. A crossover between one (1D) and two dimensions (2D) will also be discussed. One of the key ingredients in the transport in high magnetic fields is edge states which are localized in the vicinity of wire boundaries. Section 3 deals with the localization of such edge states due to mixing with bulk Landau levels and its dependence on the wire width. Effects of inherent boundary roughness scattering are discussed in Section 4. The scattering gives rise to a singular current distribution among different subbands, which manifests itself as a suppression of the weak-field Hall effect and also a positive magnetoresistance. In Section 5, results of quantum-mechanical calculations of the ballistic transport at quantum point contacts are presented. It will be shown in Section 6 that the channel wavefunction can be observed directly through the magnetic-field dependence of the conductance at a series of two point contacts.

H. Aoki et al. (eds.), New Horizons in Low-Dimensional Electron Systems, 351–368.

2. Conductance Fluctuations

A periodic oscillations of the conductance measured in metallic wires as a function of a magnetic field have been known to be a manifestation of universal conductance fluctuations [12]. The universality has been shown theoretically in the diffusive regime where the system size is much larger than the mean free path [13,14]. The fluctuation has been calculated as 0.729 for 1D wires, 0.862 for 2D squares, and 1.088 for 3D cubes in units of e^2/h. We can expect a different behavior in quantum wires where only a few well-defined 1D subbands are occupied.

The conductance in quantum wires with a finite length can easily be calculated by a numerical technique [15]. We consider a continuum system confined within the finite width W and containing short-range scatterers. The S matrix for a single scatterer, consisting of transmission and reflection coefficients for traveling and evanescent channels, is calculated and the total S matrix of a long wire is constructed from those of individual scatterers. The conductance is given in terms of the transmission probability by the multi-channel version of Landauer's formula [16].

Figure 1 shows the fluctuation as a function of wire length for several wires with different width W [17]. Note that the number of occupied 1D subbands is given by $2W/\lambda_F$, where λ_F is the Fermi wavelength. The fluctuation first increases with increasing length, attains a maximum, and then starts to decrease gradually. The initial increase corresponds to the simple fact that the conductance is quantized into the integer multiple of $e^2/\pi\hbar$ in the ballistic regime. The gradual decrease is due to the localization effect that is very important in narrow wires. A half of the localization length calculated from the length dependence of the conductance is denoted by vertical arrows in the figure. For sufficiently wide wires, the fluctuation exhibits a universal behavior, i.e. independent of the length, in the range much larger than the mean free path Λ in 2D and smaller than the localization length.

Figure 2 gives the results for the case that the broadening Γ is larger than the subband energy separation. In such 2D cases, the fluctuation becomes larger and shows the length dependence close to the curve obtained by perturbation calculations.

3. Localization – Mixings between Edge and Bulk States

The states are all localized in 1D however small the randomness may be. The situation changes drastically in a strong magnetic field. When $W \gg l$ with $l = (c\hbar/eH)^{1/2}$ and H the magnetic-field strength, the subbands are well separated into edge states and bulk Landau levels. The edge states become more and more extended with increasing magnetic field as long as their energy is away from those of bulk Landau levels. When their energy

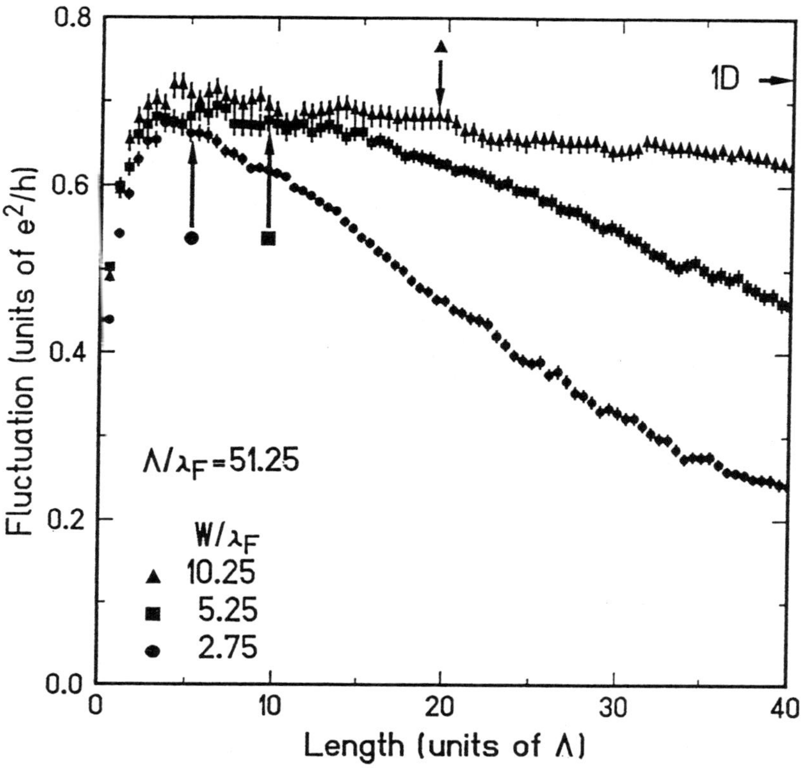

Fig. 1. Calculated conductance fluctuations as a function of wire length measured in units of Λ of different widths. The vertical arrows indicate a half of the localization length and the horizontal arrow the universal value in 1D wires calculated perturbationally.

lies in the region of broadened bulk Landau levels, they are mixed strongly.

A numerical study of the localization of the edge states has been performed in high magnetic fields [18]. Use is made of a lattice model which simulates a continuum system and randomness is introduced through fluctuactions of site-diagonal energies. Figure 3 gives the calculated subband-energy dispersion for $\hbar\omega_c/E_W = 10$ $(W/l = 7.02)$ with $E_W = (\hbar^2/2m)$ $(\pi/W)^2$ and $\omega_c = eH/mc$.

Examples of calculated density of states and inverse localization length $\alpha(E)$ in strong magnetic fields are given in Figure 4. Effects of scattering are characterized by $\Gamma_W/\hbar$, representing the probability of backscattering for electrons with wavevector $k = \pi/W$ in the lowest subband or the corresponding mean free path Λ_W. With increasing magnetic field, the density of states becomes closer to that in 2D at the energies of the bulk Landau

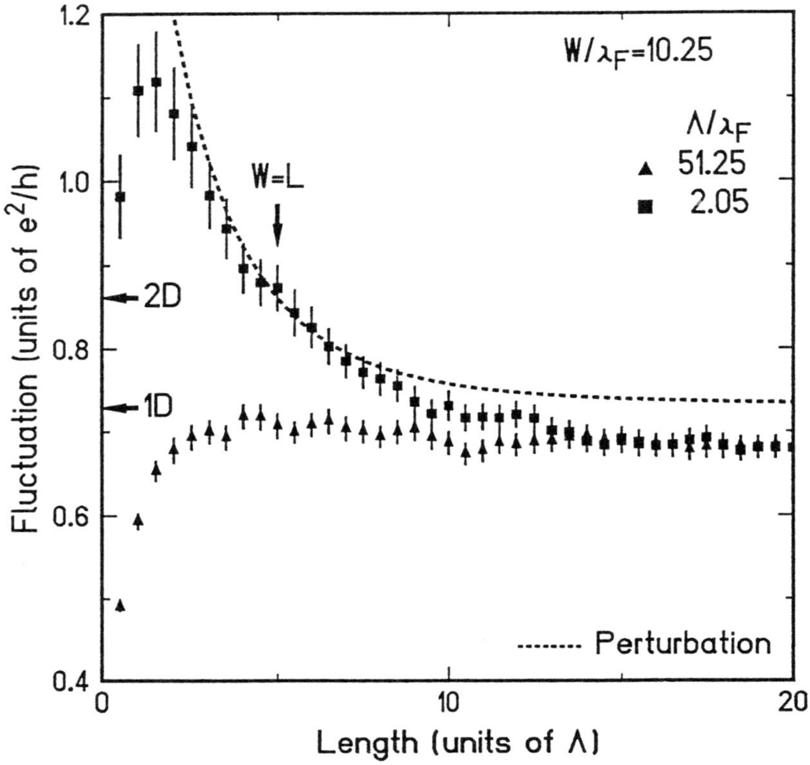

Fig. 2. Calculated conductance fluctuations for wide wires. The squares represent the results in the case that the broadening is larger than the subband energy separation and the vertical arrow shows the length corresponding to a 2D square. The triangles are the same as one of those given in Figure 1.

levels, while it is given well by that of the unperturbed edge states for energies between the Landau levels. The broadening of the bulk Landau levels has been calculated in the self-consistent Born approximation [19–21] as $\Gamma_H = [(2/\pi)\,\hbar\omega_c\,\Gamma]^{1/2}$, where Γ is the broadening in 2D in the absence of a magnetic field and related to Γ_W through $\Gamma = (2\pi/3)\,\Gamma_W$.

The inverse localization length becomes extremely small because of the formation of the edge states when the energy is away from the density of states of the bulk Landau levels. There is a remarkable difference in the behavior between the lowest and first excited Landau levels, which is the reflection of the obvious fact that the edge states are present only above the lowest Landau level. In the energy region corresponding to the first excited Landau level, $\alpha(E)$ becomes large due to strong mixings with bulk states.

The dependence of $\alpha(E)$ on W reflects how the amplitude of the edge

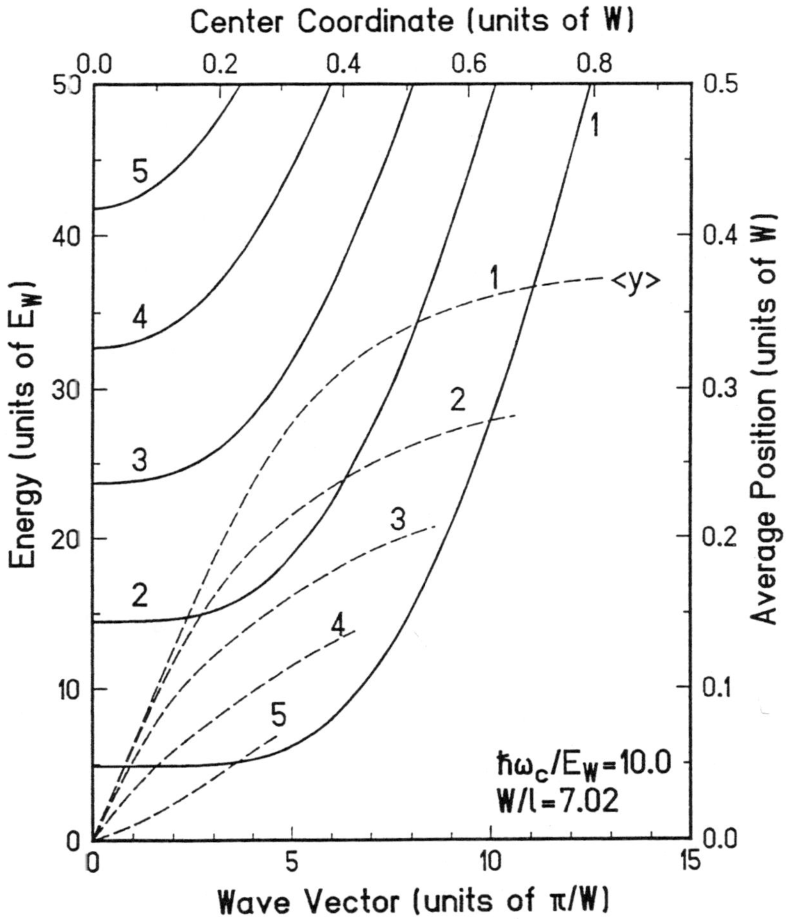

Fig. 3. Examples of the energy dispersion calculated in a lattice model for $\hbar\omega_c/E_W = 10$. The solid lines represent the subband energy and the dashed lines the average value of y.

state wavefunction decays inside the wire away from the boundary. The results can be fitted to $\alpha(E) \propto \exp(-\beta W)$. This suggests that the wavefunction decays exponentially away from the boundary. The extent of the edge states in the vicinity of the center of the first excited Landau level is roughly $\beta^{-1} \sim 15l$. The energy dependence of β within the broadened first excited subband is not symmetric and the edge states lying higher in energy are more strongly confined in the vicinity of the boundary. This suggests that mixing depends strongly on the detailed form of the confining potential at the boundary.

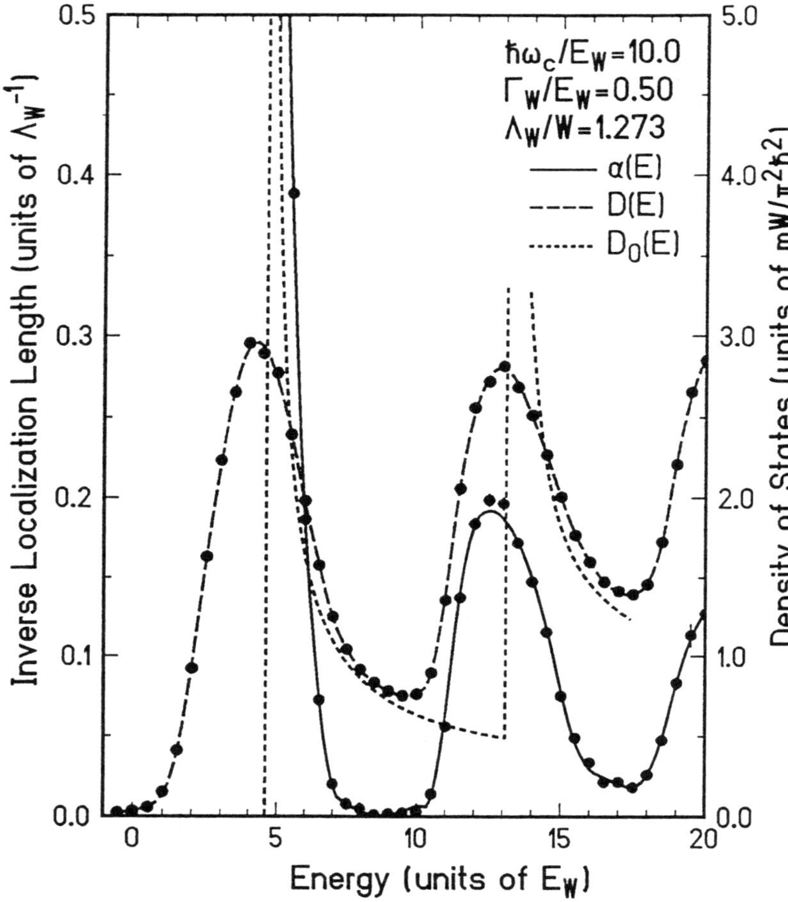

Fig. 4. Calculated inverse localization length (solid line) in units of Λ_W^{-1} and density of states (dashed line). The dotted line represents the density of states in the absence of scattering. $\hbar\omega_c/E_W = 10$ and $\Gamma_W/E_W = 0.5$.

4. Boundary Roughness Scattering

Boundary roughness is considered to be inherent to artificially confined quantum wires and can constitute a major cause of scattering. Classically, effects of boundary roughness are treated in terms of a single parameter p, representing the probability of specular scattering [22]. It is clear that such a treatment is inadequate and a quantum mechanical calculation is necessary in quantum wires.

We consider a wire infinitely long in the x direction and confined in the y direction by the potential

$$V(y) = \begin{cases} \frac{1}{2} E_F \, y_e^{-2} (y + \frac{1}{2} W_0)^2 \ , & \text{if} \quad y < -\frac{1}{2} W_0 \ ; \\ 0 \ , & \text{if} \quad -\frac{1}{2} W_0 < y < \frac{1}{2} W_0 \ ; \\ \frac{1}{2} E_F \, y_e^{-2} (y - \frac{1}{2} W_0)^2 \ , & \text{if} \quad y > \frac{1}{2} W_0 \ , \end{cases} \tag{4.1}$$

where y_e is the width of the edge region, the effective width W is given by $W = W_0 + 2y_e$, and E_F is the Fermi energy in 2D. Roughness is introduced through random fluctuations of the edge position, i.e. $\pm W_0 \to \pm W_0 + \Delta_\pm(x)$. The roughness is conventionally characterized by its root mean square deviation Δ and correlation length d, i.e.

$$\begin{aligned} \langle \Delta_+(x) \rangle = \langle \Delta_-(x) \rangle = \langle \Delta_+(x) \, \Delta_-(x') \rangle = 0 \ , \\ \langle \Delta_+(x) \, \Delta_+(x') \rangle = \langle \Delta_-(x) \, \Delta_-(x') \rangle = \Delta^2 \exp[-(x - x')^2 / d^2] \ . \end{aligned} \tag{4.2}$$

It is straightforward to calculate the conductivity if we use the Bolzmann transport equation. The results show that the boundary roughness causes a very singular current distribution among subbands in the absence of a magnetic field as long as $d < \lambda_F$ [23]. In the case of a hard-wall boundary ($y_e = 0$), for example, the current carried by subband n is proportional to n^{-2}, which means that most of the current is carried by the lowest few subbands.

This peculiar current distribution among subbands causes various interesting phenomena. A typical example is a positive magnetoresistance [24]. This occurs due to the fact that the lowest subbands become equally affected by the boundary roughness in magnetic fields. Figure 5 gives an example of calculated resistivity determined by boundary roughness. The resistivity increases first and then decreases with magnetic field. As the confining potential becomes softer, the peak in the positive magnetoresistance becomes sharper and at the same time the resistance starts to have a longer taili at higher magnetic fields. A positive magnetoresistance has been observed in wires formed on a GaAs/AlGaAs heterointerface [25]. The experimental results can be best fitted by the results for $y_e \sim 0.3 \, W$.

Another example is the weak-field Hall effect. Figure 6 shows an example of calculated Hall resistance in a crossed wire geometry in the presence of strong boundary roughness [23]. The Hall resistance is strongly suppressed from the classical linear dependence in the weak-field region. The lowest subband corresponds classically to a state moving predominantly in the direction of the wire and has a smaller probability of being transmitted into the wire corresponding to the Hall probes, leading to a reduction of the Hall voltage. Note that rounding of corners at the crossing of two wires is not taken into account in the calculation. Therefore, the present roughness effects is completely different from the so-called nozzle effect [26] or the

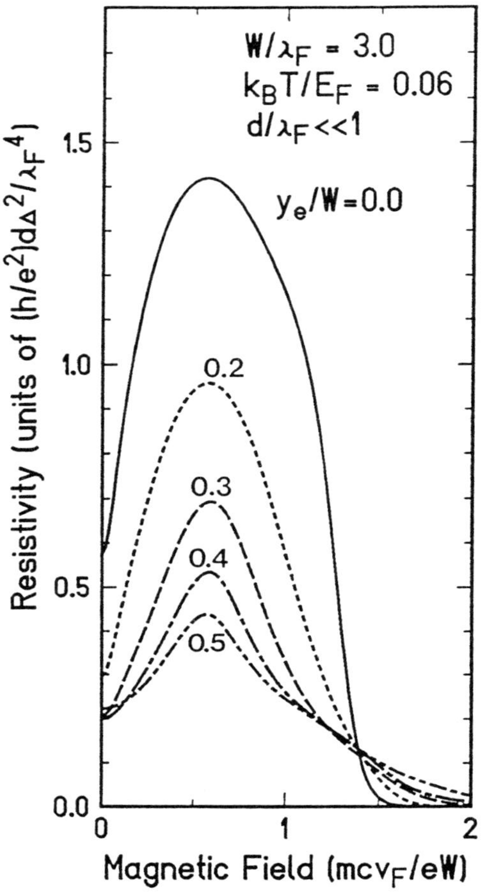

Fig. 5. Calculated magnetoresistance as a function of magnetic field for a wire with width $W/\lambda_F = 3$ in the limit that the correlation length is much smaller than the Fermi wavelength for varying softness of the confining potential. $k_BT/E_F = 0.06$.

scrambling effect [27] in which the large rounding of the junction corners is essential.

5. Quantum Point Contact – Conductance Quantization

The quantum point contact is an ideal system where quantum effects on the transport can be directly studied. The typical example is the quantization of the conductance into integer multiples of $e^2/\pi\hbar$ [1,2]. Following the interesting experimental finding, there have appeared various calculations of conductances across narrow constrictions [28–37]. The conductance of vari-

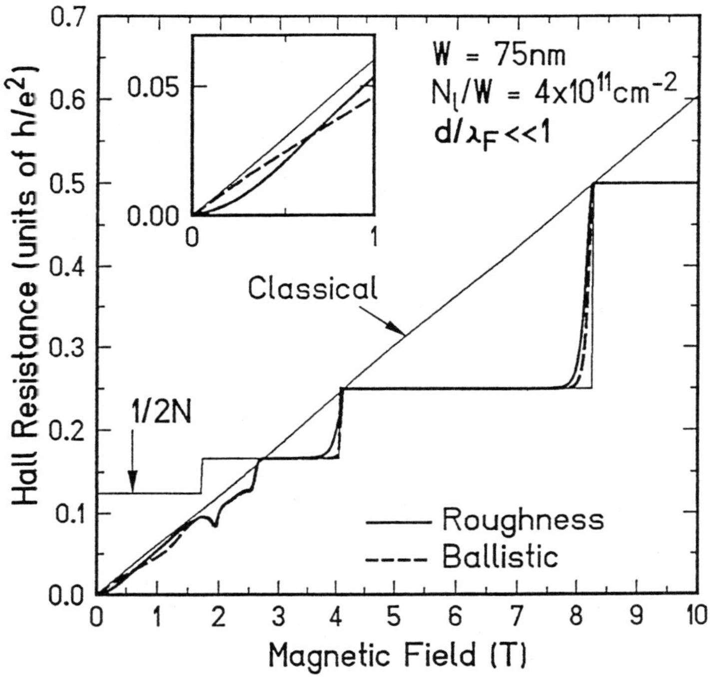

Fig. 6. Calculated Hall resistance for scattering from boundary roughness. $W = 750$ Å. Inset: blowup of the region of low magnetic fields.

ous structures consisting of several point contacts has been measured [38,39] and analyzed in terms of a collimation effect [40]. There are various ways of calculating conductance in ballistic structures. The simplest way is to use a lattice model, i.e. a discretization of space. It is particularly convenient in the presence of a magnetic field which is incorporated simply by a Peierls' phase factor of the transfer integral.

We consider a wire with length L_x and width $L_y(-L_x/2 < x < L_x/2$ and $-L_y/2 < y < L_y/2)$ and introduce the following potential which confines electron motion in the y direction in the vicinity of the center of the wire.

$$V(x,y) = \frac{V}{2}\left[1+\cos\left(\frac{2\pi x}{L_x}\right)\right] + E_F \sum_{\pm} \left(\frac{y-y_\pm(x)}{\Delta}\right)^2 \theta[y^2 - y_\pm(x)^2] \,, \tag{5.1}$$

where $\theta(t)$ is a step function, defined by $\theta = 1$ for $t > 0$ and $\theta = 0$ otherwise, and

$$y_\pm(x) = \pm\frac{L_y}{4}\left[1-\cos\left(\frac{2\pi x}{L_x}\right)\right] . \tag{5.2}$$

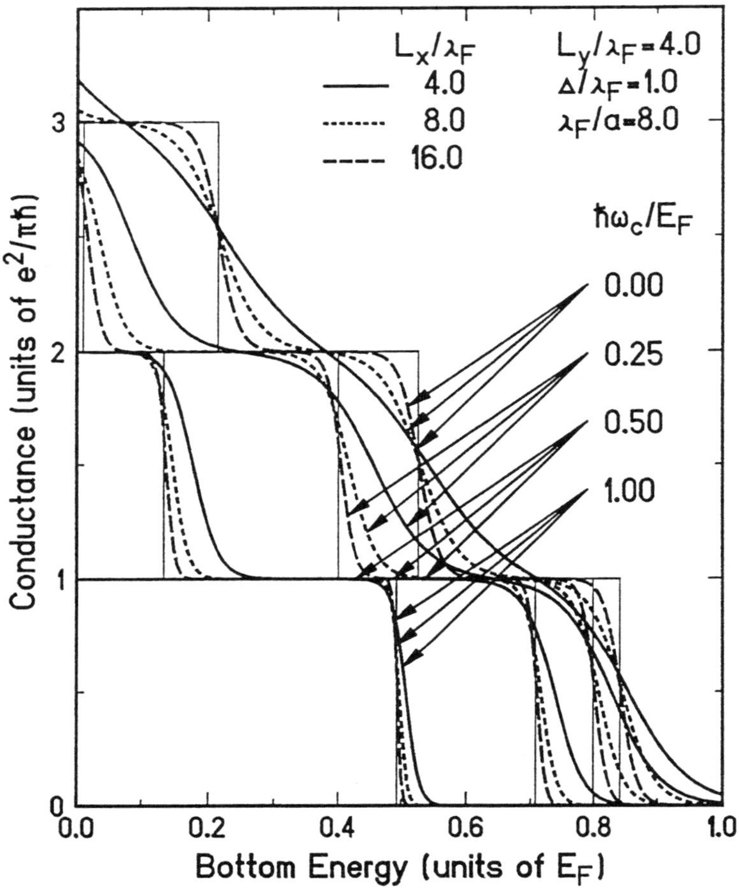

Fig. 7. An example of calculated conductance at a quantum point contact as a function of energy V of the point-contact bottom. The dotted lines represent the quantized conductance in the adiabatic limit, obtained by the number of the allowed channels.

The parameter Δ/λ_F characterizes the width of the channel of the point contact and V represents the maximum bottom energy at $x = y = 0$. For a fixed width L_y, the wire length L_x determines the length scale over which the potential varies. This model is expected to simulate the potential variation in the vicinity of actual quantum point contacts for appropriate values of the parameters. The adiabatic limit corresponds to $L_x/\lambda_F \gg 1$.

Figure 7 gives an example of the conductance as a function of V for different magnetic fields. The degree of the quantization depends sensitively on L_x/λ_F. In the absence of a magnetic field, the conductance varies al-

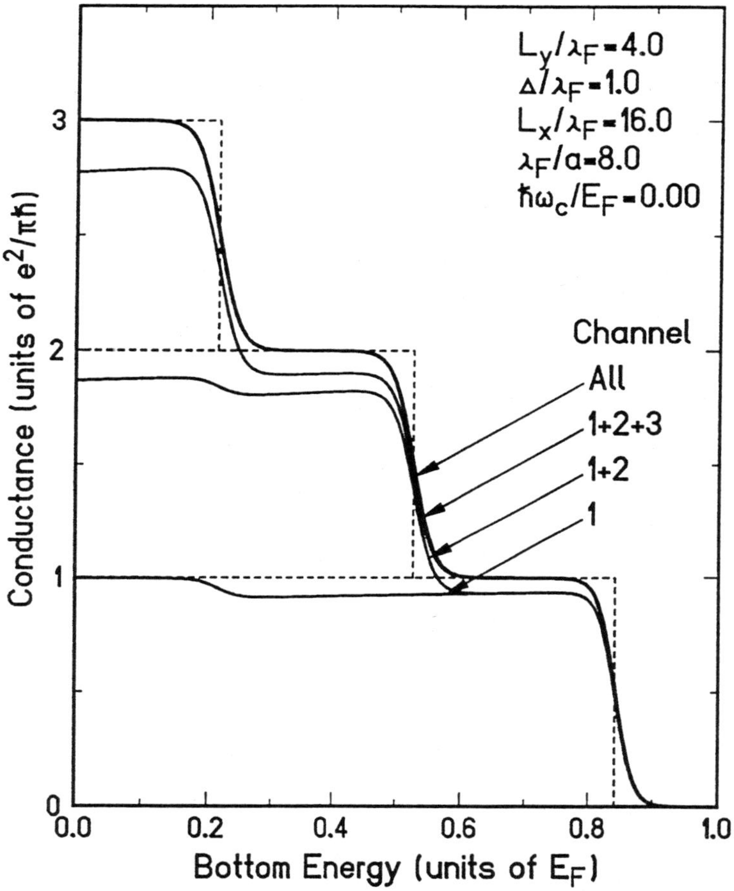

Fig. 8. Contribution of different channels to the conductance across a point contact as a function of energy V of the point-contact bottom. The dotted lines represent the quantized conductance in the adiabatic limit, obtained by the number of the allowed channels.

most linearly with V for $L_x/\lambda_F = 4$, while it becomes well quantized for $L_x/\lambda_F = 16$. It can clearly be seen that the conductance quantization becomes better in higher magnetic fields as has been observed in experiments [41] and discussed theoretically [42,43]. This is closely related to the appearance of edge channels discussed in the previous sections, which reduces backscattering probabilities at the point contact and the electron motion becomes more adiabatic.

Figure 8 gives the current carried by each out-going channel for the case where the conductance is well quantized [$L_x/\lambda_F = 16$ at $\hbar\omega_c/E_F = 0$]. An appreciable part of the current is carried by channels different from those at the point contact although the total conductance is well quantized. Inter as well as intra-channel scattering due to nonadiabaticity of the confining potential has been shown to be exponentially small in the slowly-varying case [44–47]. Therefore, the present result shows that the conductance is quantized even when the potential is far from being in the adiabatic limit.

6. Series of Two Point Contacts – Channel Wavefunction

Combinations of several point contacts are particularly interesting, because they provide means to measure the subband wavefunctions directly. Quite recently, Okada *et al.* [48] measured the conductance across a series of two point contacts as a function of magnetic field and found an interesting dependence on the channel number of the source contact. They have observed a single peak as a function of the magnetic field for the case of a single channel and splitting into two peaks when the channel number is two.

Figure 9 schematically shows the configuration of a series of two point contacts. The confining potential at both emitter (left, $-$) and collector (right, $+$) contacts is parabolic, i.e. given by $E_F(y/\Delta_\pm)^2 + V_\pm$. The boundary between the region of flat and parabolic potential changes according to $(x/L_\pm)^p + (2y/L_y)^p = 1$ with $p = p_\pm$ in the vicinity of the contacts. Infinitely long wires with width L_x are attached at both top and bottom sides, i.e. only outgoing electron waves are allowed at $y = \pm L_y/2$. The distance between the point contacts is $D = L_x + L_+ + L_-$. For typical $\lambda_F \sim 500$ Å, we have $D \sim 1$ μm, which is close to the distance of the two point contacts used by Okada *et al.* [48].

Figure 10 gives the calculated conductance as a function of magnetic field or the distance of the two point contacts divided by the classical cyclotron radius. The channel number at the collector contact is fixed as three and that of the emitter is varied. The conductance depends essentially on the channel number of the emitter only except for the presence of oscillations due to quantum interference. The interference fluctuations are almost chaotic in the sense that it depends sensitively on the detailed form of the confinement potential as well as the distance between the emitter and collector contacts.

At nonzero temperatures, such interference fluctuations disappear because of averaging due to the broadening of the electron energy distribution. An example of the results of such energy averaging at $k_BT/E_F = 0.02$ is shows in Figure 11, which demonstrates the disappearance of the fluctuations and shows clearly the features corresponding to the channel wavefunctions. The conductance depends also on the channel number of the collector point contact. An example is given in Figure 12 for the channel number $N_+ = 2$. The difference of the conductance for $N_- = 1$ and 2 is still appreciable but the conductance remains essentially the same for $N_- = 2$ and 3. The

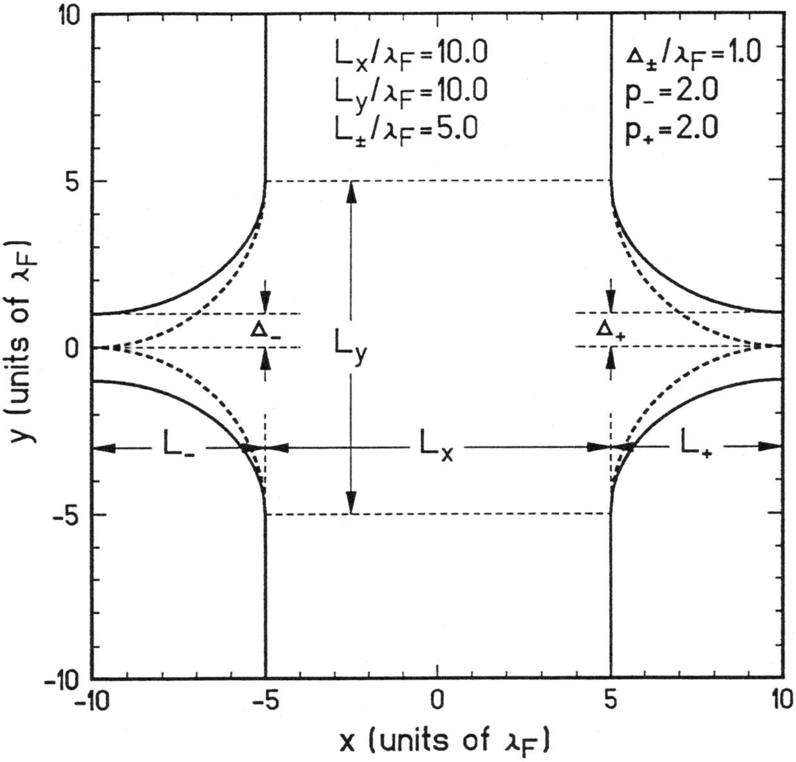

Fig. 9. A schematic illustration of a series of two point contacts. Electrons are injected from the left hand side and collected at the right hand side. Only outgoing electron waves are allowed at the top and bottom.

same feature can be seen when $N_+ = 1$, where the conductance for different N_+ is very close to each other. This interesting dependence on the collector channel has not been reported yet.

7. Summary and Conclusion

A review has been given on recent theoretical studies of transport properties in quantum wires and point contacts. It has been shown that the conductance fluctuations in quantum wires are not universal and quite different from those in metallic wires. Edge states, which play a dominant role in transport in strong magnetic fields, mix strongly with bulk Landau states when their energies are degenerate but decay exponentially away from the wire boundary. Boundary roughness inherent in artificially confined wires causes a peculiar current distribution among one dimensional subbands,

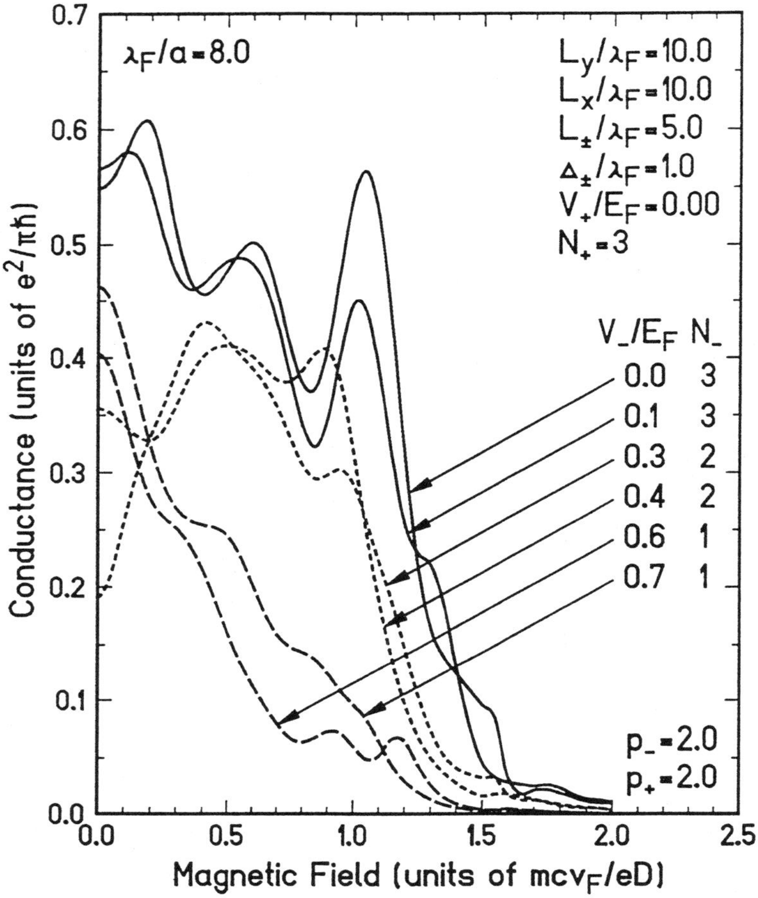

Fig. 10. An example of calculated conductance across the two point contacts as a function of a magnetic field for different numbers of the channels at the emitter point contact. The channel number at the collector contact is fixed as three.

which manifests itself as a strong positive magnetoresistance and a suppression of the weak-field Hall effect. The conductance across quantum point contacts has also been calculated and several new findings have been obtained. The condition for the quantization does not require the complete adiabaticity of the confining potential. The channel wavefunction is directly observable in magnetic field dependence of the conductance across series of two point contacts.

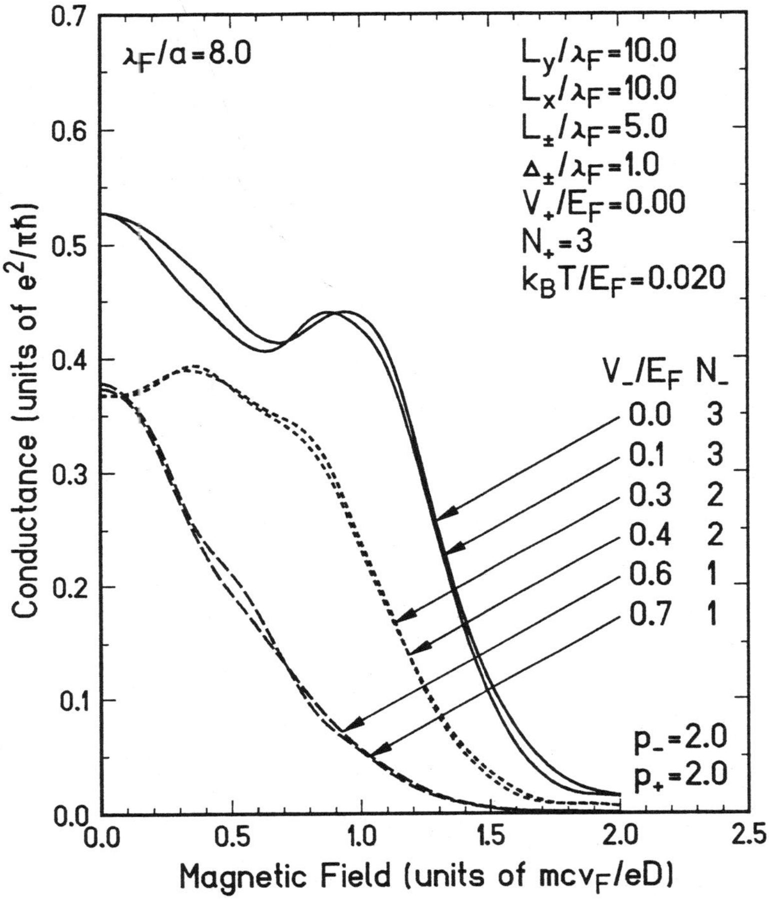

Fig. 11. Examples of calculated conductance across a series of two quantum point contacts as a function of magnetic field at a nonzero temperature. The parameters are $L_x/\lambda_F = L_y/\lambda_F = 10$, $L_\pm/\lambda_F = 5$, $p_\pm = 2$, $\Delta_\pm/\lambda_F = 1$, $N_+ = 3$, and $k_BT/E_F = 0.02$.

Acknowledgements

This work is supported in part by the Industry-University Joint Research Program 'Mesoscopic Electronics' and by the Grant-in-Aid for Scientific Research on Priority Area 'Electron Wave Interference Effects in Mesoscopic Structures' from the Ministry of Education, Science and Culture, Japan. One of the author (T.A.) would like to thank M. Okada, M. Saito and N. Yokoyama of Fujitsu Laboratories Ltd. for showing their experimental results prior to publication.

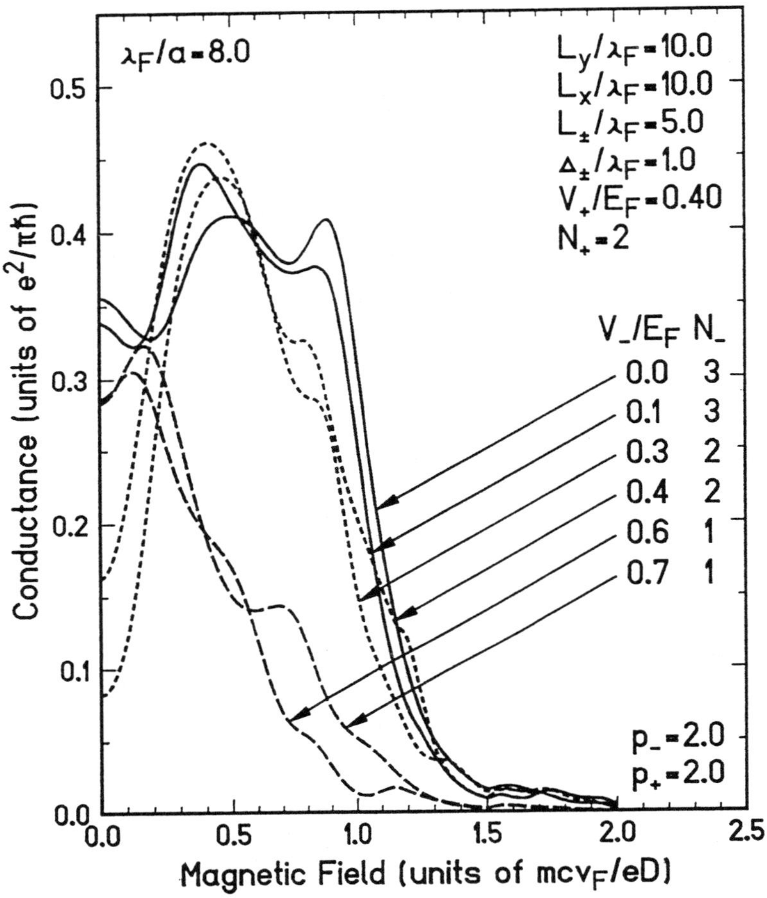

Fig. 12. Examples of calculated conductance across a series of two quantum point contacts as a function of magnetic field for $L_x/\lambda_F = L_y/\lambda_F = 10$, $L_\pm/\lambda_F = 5$, $p_\pm = 2$, $\Delta_\pm/\lambda = 1$, and $N_+ = 2$.

References

1. B.J. van Wees, H. van Houten, C.W.J. Beenakker, J.G. Williamson, L.P. Kouwenhoven, D. van der Marel and C.T. Foxon: 1988, *Phys. Rev. Lett.* **60**, 848.
2. D.A. Wharam, T.J. Thornton, R. Newbury, M. Pepper, H. Ahmed, J.E.F. Frost, D.G. Hasko, D.C. Peacock, D.A. Ritchie and G.A.C. Jones: 1988, *JPC* **21**, L209.
3. K. Ishibashi, K. Nagata, K. Gamo, S. Namba, S. Ishida, K. Murase, M. Kawabe, and Y. Aoyagi: 1987, *Solid State Commun.* **61**, 385.
4. M. Mizuno, K. Ishibashi, S.-K. Noh, Y. Ochiai, Y. Aoyagi, K. Gamo, M. Kawabe and S. Namba: 1989, *Jpn. J. Appl. Phys.* **28**, L1025.
5. K. Ishibashi, Y. Takagaki, K. Gamo, S. Namba, S. Ishida, K. Murase, Y. Aoyagi and M. Kawabe: 1987, *Solid State Commun.* **64**, 573.

6. G. Timp, P.M. Mankiewich, P. de Vegvar, R. Behringer, J.E. Cunningham, R.E. Howard, H.U. Baranger and J.K. Jain: 1989, *Phys. Rev. B* **39**, 6227.
7. G. Timp, A.M. Chang, P.M. Mankiewich, R.E. Behringer, J.E. Cunningham, T.Y. Chang and R.E. Howard: 1987, *Phys. Rev. Lett.* **59**, 732.
8. M.L. Roukes, A. Scherer, S.J. Allen, Jr., H.G. Craighead, R.M. Ruthen, E.D. Beebe and J.P. Harbison: 1987, *Phys. Rev. Lett.* **59**, 3011.
9. C.J.B. Ford, S. Washburn, M. Büttiker, C.M. Knoedler and J.M. Hong: 1989, *Phys. Rev. Lett.* **62**, 2724.
10. G. Timp, H.U. Baranger, P. de Vegvar, J.E. Cunningham, R.E. Howard, R.E. Behringer and P.M. Mankiewich: 1988, *Phys. Rev. Lett.* **60**, 2081.
11. Y. Takagaki, K. Gamo, S. Namba, S. Takaoka, K. Murase, S. Ishida, K. Ishibashi and Y. Aoyagi: 1988, *Solid State Commun.* **68**, 1051; 1989, **68**, 811.
12. S. Washburn and R.A. Webb: 1986, *Adv. Phys.* **35**, 375; See also: 1988, *IBM J. Res. Develop.* **32**, 303–437.
13. B.L. Altshuler: 1985, *JETP Lett.* **41**, 648.
14. P.A. Lee and A.D. Stone: 1985, *Phys. Rev. Lett.* **55**, 1622.
15. S. Datta, M. Cahay and M. McLennan: 1987, *Phys. Rev. B* **36**, 5655; M. Cahay, M. McLennan and S. Datta: 1988, *ibid.* **37**, 10125.
16. R. Landauer: 1957, *IBM J. Res. Dev.* **1**, 223; 1970, *Philos. Mag.* **21**, 863.
17. H. Tamura and T. Ando: 1991, *Phys. Rev. B* **44** (in press).
18. T. Ando: 1990, *Phys. Rev. B* **42**, 5626.
19. T. Ando and Y. Uemura: 1974, *J. Phys. Soc. Jpn.* **36**, 959; T. Ando: 1974, *ibid.* **36**, 1521; 1974, **37**, 622; 1974, **37**, 1233.
20. T. Ando: 1975, *J. Phys. Soc. Jpn.* **38**, 989.
21. T. Ando, Y. Matsumoto and Y. Uemura: 1975, *J. Phys. Soc. Jpn.* **39**, 279.
22. K. Fuchs: 1938, *Proc. Camb. Philos. Soc.* **34**, 100.
23. H. Akera and T. Ando: 1990, *Phys. Rev. B* **41**, 11967.
24. H. Akera and T. Ando: 1991, *Phys. Rev. B* **43**, 11676.
25. T.J. Thornton, M.L. Roukes, A. Scherer and B.P. Van de Gaag: 1989, *Phys. Rev. Lett.* **63**, 2128.
26. H.U. Baranger and A.D. Stone: 1989, *Phys. Rev. Lett.* **63**, 414.
27. C.W.J. Beenakker and H. van Houten: 1989, *Phys Rev. Lett.* **63**, 1857.
28. L. Escapa and N. Garcia: 1989, *J. Phys. Condens. Matter* **1**, 2125; N. Garcia and L. Esacapa: 1989, *Appl. Phys. Lett.* **54**, 1418.
29. E.G. Haanappel and D. van der Marel: 1989, *Phys. Rev. B* **39**, 5484; D. van der Marel and E.G. Haanappel: 1989, *ibid.* **39**, 7811.
30. G. Kirczenow: 1988, *Solid State Commun.* **68**, 715; 1989, *J. Phys. Condens. Matt.* **1**, 305; E. Castano and G. Kirczenow: 1989, *Solid State Commun.* **70**, 801.
31. A. Szafer and A.D. Stone: 1989, *Phys. Rev. Lett.* **62**, 300.
32. E. Tekman and S. Ciraci: 1989, *Phys. Rev. B* **39**, 8772; 1989, **40**, 8559.
33. Song He and S. Das Sarma: 1989, *Phys. Rev. B* **40**, 3379.
34. Y. Avishai and Y.B. Band: 1989, *Phys. Rev. B* **40**, 12535.
35. A. Kawabata: 1989, *J. Phys. Soc. Jpn.* **58**, 372.
36. I.B. Levinson: 1988, *JETP Lett.* **48**, 301.
37. A. Matulis and D. Segzeda: 1989, *J. Phys. Condens. Matt.* **1**, 2289.
38. D.A. Wharam, M. Pepper, H. Ahmed, J.E.F. Frost, D.G. Hasko, D.C. Peacock, D.A. Ritchie and G.A.C. Jones: 1988, *J. Phys. C* **21**, L887.

39. L.W. Molenkamp, A.A.M. Staring, C.W.J. Beenakker, R. Eppenga, C.E. Timmering, J.G. Williamson, C.J.P.M. Harmans and C.T. Foxon: 1990, *Phys. Rev. B* **41**, 1274.
40. C.W.J. Beenakker and H. van Houten: 1989, *Phys. Rev. B* **39**, 10445.
41. B.J. van Wees, L.P. Kouwenhoven, H. van Houte, C.W.J. Beenakker, J.E. Mooij, C.T. Foxon and J.J. Harris: 1988, *Phys. Rev. B* **38**, 3625.
42. L.I. Glazman and A.V. Khaetskii: 1989, *J. Phys. Condens. Matt.* **1**, 5005.
43. K.B. Efetov: 1989, *J. Phys. Condens. Matt.* **1**, 5535.
44. L.I. Glazman, G.B. Lesovik, D.E. Khmelnitskii and R.I. Shekter: 1988, *JETP Lett.* **48**, 239.
45. A. Yacoby and Y. Imry: 1990, *Phys. Rev. B* **41**, 5341.
46. L.I. Glazman and M. Jonson: 1989, *J. Phys. Condens. Matt.* **1**, 5547.
47. M.C. Payne: 1989, *J. Phys. Condens. Matt.* **1**, 4939.
48. M. Okada, M. Saito, K. Kosemura, T. Nagata, H. Ishiwari and N. Yokoyama, to be published in *Proc. 5th Int. Conf. Superlattices and Microstructures*, Berlin, 1990.

MESOSCOPICS AND SUPERCONDUCTIVITY

HIDETOSHI FUKUYAMA and HIDEO YOSHIOKA
Institute for Solid State Physics, University of Tokyo
7-22-1 Roppongi, Minato-ku, Tokyo 106, Japan

Abstract. The essential feature of mesoscopic systems is the observability of the phase coherence of electronic states in the current flow of electrons. Another and well-known example of the persistence of the phase coherence is superconductivity; a macroscopic quantum phenomenon. An interesting interplay between these two is discussed.

1. Introduction

The essential feature of mesoscopic systems is the manifestation of the phase coherence of wavefunctions in such electronic motions as transport processes and in the persistent current [1]. Above all in systems where electrons are frequently scattered by normal impurities and their motions are diffusive the characteristic length, L_ε, where the phase coherence remains is determined by the inelastic scattering and is given by $L_\varepsilon = \sqrt{D\tau_\varepsilon}$, where D and τ_ε are the diffusion constant and the life time due to inelastic scattering, respectively. Noteworthy is the fact that this L_ε is generally far longer than the mean free path and moreover it can be divergent as the temperature, T, tends to absolute zero since the inelastic scattering is to be quenched as $T \to 0$. This is the reason why the phase coherence can be observed even in systems of submicron sizes, which is much longer than the microscopic scale.

This is understood as follows; before electrons are scattered inelastically and their energies are changed the electronic states are governed by the eigenfunctions $\varphi_\alpha(\mathbf{r})$ of the Schrödinger equation for one-particle

$$\left[\frac{\mathbf{p}^2}{2m} + U(\mathbf{r})\right] \varphi_\alpha(\mathbf{r}) = \varepsilon_\alpha \varphi_\alpha(\mathbf{r})$$

with the eigenenergy, ε_α, where $U(\mathbf{r})$ is the random potential. The relative phases of the wavefunction, $\varphi_\alpha(\mathbf{r})$, at any two different positions $\mathbf{r}_1$ and $\mathbf{r}_2$ are well defined since $\varphi_\alpha(\mathbf{r})$ is the eigenfunction and the phase coherence persists as far as the energy of electronics is not changed. But once an inelastic scattering sets in an electron is scattered into a different state without any coherence of the phases of the wavefunction. Hence $L_\varepsilon = \sqrt{D\tau_\varepsilon}$ follows.

Such a persistence of the phase coherence of the one-particle state is clearly manifested in the Aharonov–Bohm effect; as shown in Figure 1 an incoming electron wave with a fixed energy is split into two paths 1 and 2 at P enclosing a magnetic flux, Φ, and is gathered at Q resulting in the interference of two waves

H. Aoki et al. (eds.), New Horizons in Low-Dimensional Electron Systems, 369–375.

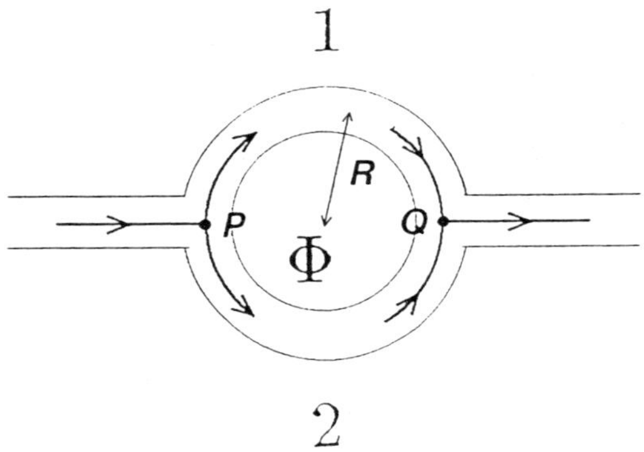

Fig. 1. The Aharonov–Bohm (AB) ring.

$$|\varphi|^2 = |\varphi_1 + \varphi_2|^2 = |\varphi_1|^2 + |\varphi_2|^2 + 2 \text{ Re } \varphi_1^* \varphi_2 \tag{1.1}$$

$$\text{Re } \varphi_1^* \varphi_2 = |\phi_1^0 \varphi_2^0| \cos(\gamma + 2\pi\Phi/\Phi_0) \tag{1.2}$$

where $\Phi_0 = hc/e$ is the flux quantum and $|\varphi_i^0|$ $(i = 1, 2)$ is the amplitude of the wavefunction along the ith path and γ is the difference of the phases of two wavefunctions φ_1^0 and φ_2^0 at Q in the absence of the magnetic flux. The oscillation in Equation (1.2) with the period Φ_0 is the Aharonov–Bohm effect (AB oscillation) and γ is a random variable dependent on each energy state. Hence, in a case where many states (say N) with essentially the same energy are contributing to this, the relative amplitude of the AB oscillation to that of incoherent contributions are reduced as $N^{-1/2}$, since γ in Equation (1.2) now takes random values γ_α and the summation over γ_α result in

$$\sum_\alpha \cos(\gamma_\alpha + 2\pi\Phi/\Phi_0) = \text{ Re } e^{i2\pi\Phi/\Phi_0} \sum_\alpha e^{i\gamma_\alpha} \simeq \sqrt{N} \cos(\bar{\gamma} + 2\pi\Phi/\Phi_0) \tag{1.3}$$

where $\bar{\gamma}$ is defined in Figure 2. We noted that the problem here is the random walk in a complex plane with the width of each step being one. Equation (1.3) is to be compared with incoherent contributions of the order N.

This fact indicates that the AB oscillation is due to fluctuations around the mean value and that it is then to be observed only in mesoscopic systems as is clarified by Stone and Imry [2]. Moreover if effects due to finite T are taken into account the relative amplitude of AB oscillation is given as follows:

$$\frac{1}{\sqrt{N}} \exp(-\pi R/L_\varepsilon) \cos(\bar{\gamma} + 2\pi\Phi/\Phi_0) \tag{1.4}$$

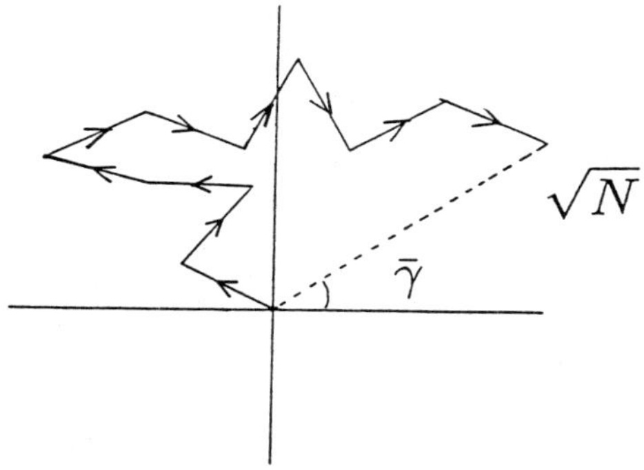

Fig. 2. The random walk in the plane of the complex variable with a fixed step width.

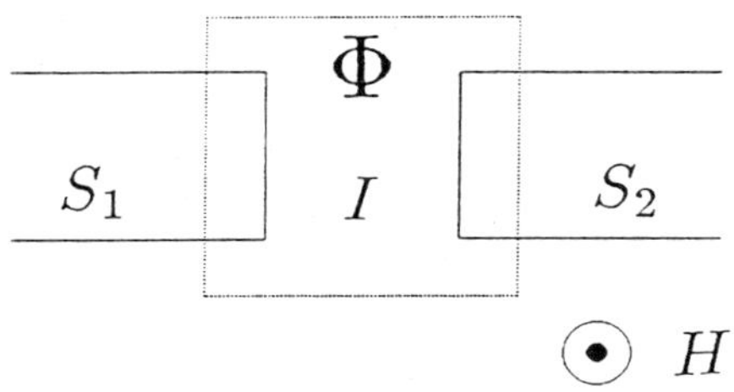

Fig. 3. Superconductor–Insulator–Superconductor Josephson junction with the applied magnetic field, H.

where R is the characteristic radius of the ring. These theoretical considerations have been verified in various beautiful experiments in rings of normal metals [3,4].

2. A Josephson Junction in a Magnetic Field

An interference effect similar to AB oscillation has been observed in Josephson junctions in the presence of the magnetic field [5]. In this case of superconductivity, however, the interference is due to the coherence of Cooper pairs as is well known. For a case where the junction consists of an insulator, I, sandwiched by two superconductors, S_1 and S_2, as shown in Figure 3 the maximum Josephson current, I_M, exhibits the Fraunhofer pattern as a function of Φ as in Figure 4. Here the period of the damped oscillation is $\Phi_0/2$.

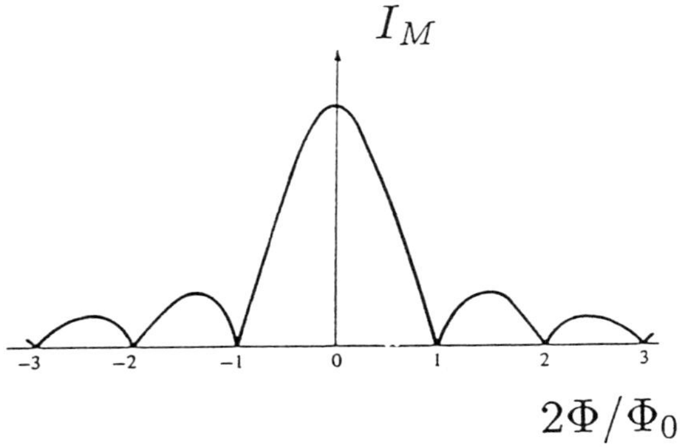

Fig. 4. Fraunhofer pattern of the maximum Josephson current.

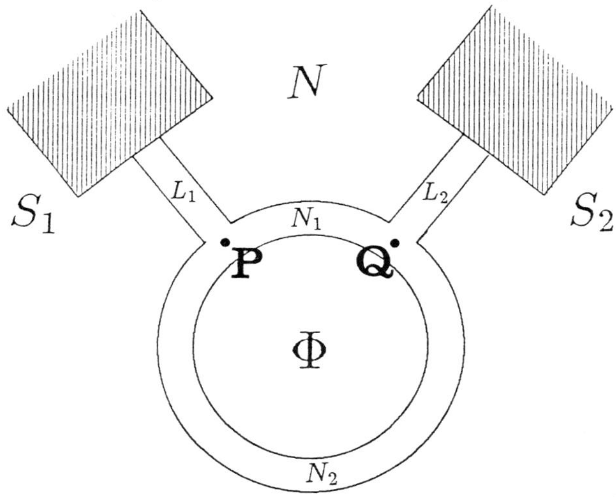

Fig. 5. Superconductor–Normal metal–Superconductor junction with a AB ring in the N region.

3. Proximity Effect in a Normal Metal Ring

An interesting interplay between AB oscillation and the Fraunhofer pattern in a Josephson junction can be studied in a SNS (Superconductor–Normal metal–Superconductor) system, where the N region consists of a ring as shown in Figure 5 [6]. Due to the proximity effect the amplitude of Cooper pair wavefunction is finite in the N region and the characteristic length of this penetration of the amplitude, ξ, is dependent on temperature, T; $\xi = \sqrt{D/2\pi T}$ in the dirty limit and $\xi = v_F/2\pi T$ in the clean limit [7].

In the structure of Figure 5 there exist different types of interference. One is the case where two electrons of a Cooper pair move along either path N_1 or N_2 and these two waves of a Cooper pair can interfere (process A). Another is the case where two electrons of a Cooper pair get split at P and

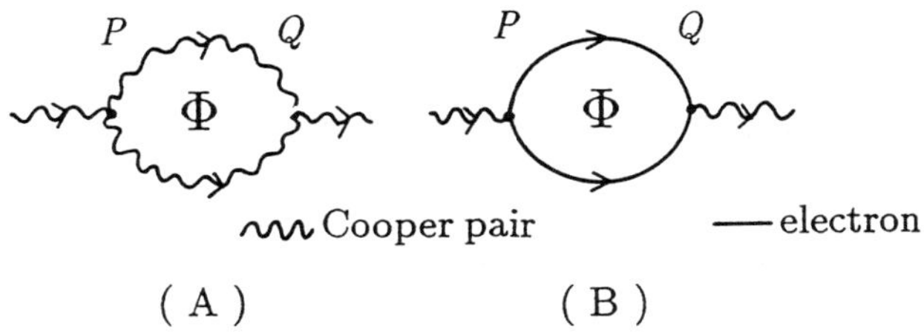

Fig. 6. The schematic representation of two different processes for the proximity effect in the structure of Figure 5.

one of them propagates along N_1, while the other goes along N_2, and merge at Q (process B). The latter is somewhat similar to the case of Figure 1, but the energy and momentum of two electrons are different here. The total contributions of these processes to the Josephson current have recently been investigated for a case where the electronic state in the N region is purely one-dimensional. The result in the absence of impurity scattering is given as follows in the lowest order in the order parameters of two superconductors and in the spatial decay

$$j = \frac{8eT}{v_F^2} \exp\left[-\frac{L_1 + L_2}{\xi}\right] |\Delta^l(\pi T)||\Delta^r(\pi T)| \tag{2.1}$$

$$\times\Big[A_1 \sin(\Psi - 4\pi\Phi/\Phi_0) + A_2 \sin \Psi$$

$$+2\sqrt{A_1 A_2} \sin(\Psi - 2\pi\Phi/\Phi_0) \cos 2k_F R(\pi - \theta)\Big]$$

where v_F and k_F are the Fermi velocity and the Fermi wave number, respectively, and $\Delta^{(l,r)}(\pi T)$ is proportional to the energy gap of S_l or S_r at SN boundary. In Equation (2.1) $A_1 = \exp[-(2\pi - \theta) R/\xi]$, $A_2 = \exp[-\theta R/\xi]$ with $\xi = v_F/2\pi T \equiv \xi_c$ and $\Psi = \Theta + 2\theta\, \Phi/\Phi_0$ is gauge invariant, where Θ is the difference of the phases of Δ^l and Δ^r. The first two terms in r.h.s. of Equation (2.1) are due to the process A and the last term due to process B, where the factor $\cos[2k_F R(\pi - \theta)]$ is due to the difference of the phases of two plane wave merging at Q which is the same for all pairs of electrons contributing to the process B as far as the system is strictly clean. In the presence of the impurity scattering, however, even in the clean limit $\tau T > 1$, each pair of waves in the process B will have a random phase due to impurity scattering and by use of the discussion of Section 1 Equation (2.1) will be modified as

$$j = j_0 \exp\left[-\frac{L_1 + L_2}{\xi}\right] |\Delta^l(\pi T)||\Delta^r(\pi T)| \times \tag{2.2}$$

$$\times\Big[A_1 \sin(\Psi - 4\pi\Phi/\Phi_0) + A_2 \sin \Psi +$$

$$+2(\sqrt{A_1 A_2}/\sqrt{N_B}) \sin(\Psi - 2\pi\Phi/\Phi_0) \cos \bar{\zeta}\Big] .$$

Here N_B is the effective number of pairs of waves contributing to the process B and j_0 is some constant weakly dependent on T. The formal expression of Equation (2.2) is valid even in the dirty limit if ξ is understood as $\xi_d \equiv \sqrt{D/2\pi T}$ and the actual number of N_B is redefined. The same is true if the wire consists of normal metals with the Fermi wavelength being much shorter than the width of the wire, W, and then the one-particle electron state can be considered as three-dimensional. Even in the case the diffusion of the Cooper pair can be considered as one-dimensional and then contributions from process A can be given as in the purely one-dimensional case, where the contributions from the process B consist of various pair of electrons split into those moving along N_1 or N_2.

Then the maximum current, I_M, is given by

$$I_M = j_0 \exp\left[-\frac{L_1 + L_2}{\xi}\right] F(2\pi\Phi/\Phi_0) \tag{2.3}$$

$$F(x) = \Big[A_1^2 + A_2^2 + 4A_1 A_2 \cos^2 \bar{\zeta}/N_B + 2A_1 A_2 \cos 2x + \tag{2.4}$$

$$+4(A_1 + A_2) \sqrt{A_1 A_2/N_B} \cos \bar{\zeta} \cos x\Big]^{1/2}$$

It is to be noted that there exists an interesting competition of two different periodicities in Equation (2.4) with regard to Φ, i.e. $\Phi_0/2$ and Φ_0; the former is due to the Fraunhofer interference pattern of the Cooper pair and the latter is due to the AB effect resulting from the process B. The relative intensity of the latter to the former, f, is given by

$$f = 2 \left(\sqrt{\frac{A_1}{A_2}} + \sqrt{\frac{A_2}{A_1}}\right) / \sqrt{N_B} \tag{2.5}$$

besides a random factor $\cos \bar{\zeta}$ representing the fact that this is the fluctuation effect. It is seen from Equation (2.5) that this fluctuation effect can be appreciable if $A_1/A_2 \gg 1$ or $A_1/A_2 \ll 1$, which is possible in the case where the leads L_1 and L_2 are attached to the ring in an asymmetric way. The effective number, N_B, of the pairs electrons split at point P depends on the actual structure of the N region and will have various possibilities [8], e.g. $N_B \simeq (L_1 + L_2 + \pi R)/\xi_c$ in the clean limit and $N_B \simeq (L_1 + L_2 + \pi R)/\ell$ in the dirty limit in the case of one-dimensional systems, where ℓ is the mean free path.

The present competition of Φ_0 and $\Phi_0/2$ oscillations in the maximum Josephson current will also have an intriguing consequence on the flux quantization in the ring [8].

Acknowledgements

This article is based on the research financially supported by a Ministry–Industry Joint Research Program 'Mesoscopic Electronics'.

References

1. See reviews by S. Washburn and R.A. Webb: 1986, *Adv. Physics.* **35**, 375; Y. Imry, *Directions in Condensed Matter Physics*, G. Grinstein and E. Mazenko (eds.), World Scientific, Singapore, 1986, p. 101.
2. A.D. Stone and Y. Imry: 1986, *Phys. Rev. Lett.* **56**, 189.
3. R.A. Webb, S. Washburn, C.P. Umbach and R.B. Laibowitz: 1985, *Phys. Rev. Lett.* **54**, 2696.
4. C.P. Umbach, C. van Haesendonck, R.B. Laibowitz, S. Washburn and R.A. Webb: 1986, *Phys. Rev. Lett.* **56**, 386.
5. See, e.g., M. Tinkham, *Introduction to Superconductivity*, McGraw-Hill, 1975, Chap. 6.2.
6. H. Yoshioka and H. Fukuyama: 1990, *J. Phys. Soc. Jpn.* **59**, 3065.
7. See, e.g., K.K. Likharev: 1979, *Rev. Mod. Phys.* **51**, 101.
8. H. Yoshioka and H. Fukuyama, unpublished.

FIELD-INDUCED CHARGES IN POLYACETYLENE MIS AND MISFET DEVICES

J.H. BURROUGHES and R.H. FRIEND
Cavendish Laboratory, Madingley Road, Cambridge CB3 0HE, U.K.

Abstract. Conjugated polymers which can be processed to form thin, coherent films can be used as the active layers in semiconductor device structures. We have used the Durham precursor route to polyacetylene to fabricate MIS diodes and MISFETs. These allow injection of charge to form accumulation of inversion layers at the interface between the polyacetylene and insulator layer. The fundamental excitation of the *trans*-polyacetylene chain is the self-localised soliton-like kink defect in the bond alternation pattern along the chain; the soliton has associated with it an energy state at mid-gap of non-bonding p_z character. We have been particularly concerned to demonstrate the formation of solitons from charges injected into the polyacetylene layer in these device structures, and have measured the changes in the optical properties that accompany soliton formation. For the MIS structures formed on an insulator layer of SiO_2 working in accumulation mode we find the 'mid-gap' transition from soliton level to the band edge at energies at 0.8 eV. We present here measurements of the electrical characteristics of MISFETs formed also with SiO_2 as the insulator layer. These devices can operate well, but carrier mobilities are low, being limited by thermally-activated transport between chains. We discuss mechanisms for the charge transport processes under the conditions of operation of these devices.

1. Introduction

Conjugated polymers have provided an important class of organic semiconductor for both experimental and theoretical work over the past decade. They offer, in principle, the advantages of polymer processibility together with interesting electronic properties, and the feasibility of controlling materials to give structures with well-defined properties has provided an important impetus for the experimental work. These polymers, in which there is a p_z orbital on each site along the polymer backbone, form a model class of organic semiconductors, with a range of energy gaps between the π valence and π^* conduction bands between 1.5 and 3 eV.

Polyacetylene has provided the prototypical polymer for this field. The *transtransoidal* isomer (referred to here as *trans*) is the thermodynamically stable form at room temperature, and all the work discussed here will be based in this isomer. This is a rigid rod polymer, which shows a strong tendency to organise in a crystalline fashion, but the chain rigidity prevents solubility in common solvents and pushes the melting temperature above the decomposition temperature of the polymer (typically 200°C). It is therefore a difficult material to process, since it is effectively intractable one formed.

Interest in the electronic properties dates back to the work of the group of Shirakawa who managed to synthesise polyacetylene in the form of a fibrillar mat by polymerisation of acetylene on a surface treated with a Ziegler-Natta

H. Aoki et al. (eds.), New Horizons in Low-Dimensional Electron Systems, 377–400.

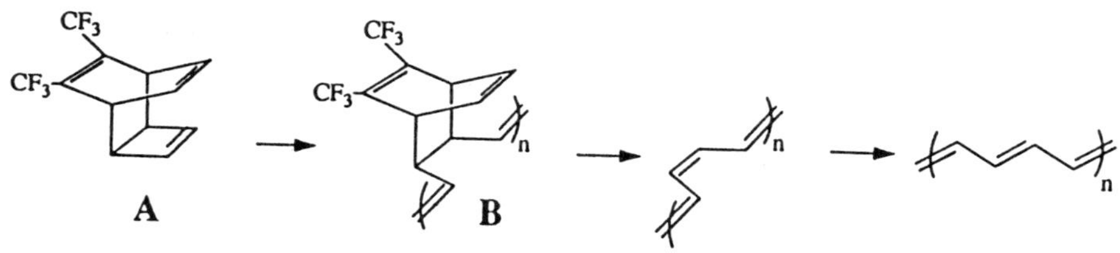

Fig. 1. The Durham route to polyacetylene. Monomer (A) is polymerised to the 'precursor' polymer (B) which is processed from solution and is converted to polyacetylene by thermal elimination of hexafluoro-oxylene, initially to *cis*-rich and then to all-*trans* isomer.

catalyst [2]. Chemical oxidation or reduction of this material produces a material with metallic conductivity, initially reported at about 10^3 S/cm at room temperature [3], but more recently, there have been several reports of conductivities in excess of 10^5 S/cm for iodine-doped polyacetylene prepared by a modified synthesis [4]. The processes of oxidation and reduction are similar to the intercalation chemistry of graphite, with the formation of chains or sheets of the charged dopant species sitting in well-defined sites alongside the polymer chain.

The metallic properties of the heavily-doped polymers provided the initial focus of interest in the electronic properties, but the theoretical framework for the modelling of the electronic excitations due to Rice [5] and Su *et al.* [6], which for the case of *trans*-polyacetylene take the form of solitary-wave excitations of the chain, has broadened the area of interest to cover in particular the optical and nonlinear optical properties of this class of polymers.

Experimental work on polyacetylene has been primarily carried out with polymer prepared via the Shirakawa route or modifications of it. This material has an open, fibrillar morphology (with up to 2/3 voids) and is very well suited for rapid doping, achieved either chemically or electrochemically. It is, however, much less suited for many of the measurements of the semiconductor properties, and is not readily used within device structures. An alternative route for the synthesis of polyacetylene, developed by Edwards and Feast [7–10], does allow very much improved processing, and it is material prepared by this route which has been used for the present work. This route, which we call here the Durham route, is shown in Figure 1. Polymerisation of the monomer (A) is achieved by ring-opening metathesis, to give the polymer (B) which is readily soluble in for example acetone and 2-butanone, and it is this 'precursor' polymer which we process, usually to form thin films by spin-coating from solution. Heating, typically to 100°C, then gives *in-situ* elimination of the side-group which is evolved as volatile hexafluoro-oxylene, to give initially *cis*-rich and then all-*trans*-polyacetylene, as shown in Figure 1.

We have found that the polymer prepared in this way is very well suited for use in semiconductor device structures in which semiconductor of one carrier type only is required (unipolar devices). The polymer as prepared is extrinsically doped with p-type carriers, to a concentration in the range 10^{16} to 3×10^{17} cm^{-3}, and these dopants are not readily mobile under the applied electric fields within these structures. We have made and measured Schottky-barrier diodes, MIS (metal insulator semiconductor) diodes and MISFETs (MIS field effect transistors), and we have been developing these structures to allow characterisation of the mode of operation of these devices and elucidation of the microscopic electronic processes that determine the device operation [11–17].

We present here some of the electro-optical characterisations of the MIS device which demonstrate that the charge injected into these devices is stored in the nonlinear excitations of the chain (solitons) and present also the electrical properties of MISFETs which give information about the charge transport processes in the polymer. The interest in developing such device structures lies both in the possible applications that may be found for them, and also in the possibilities for carrying out detailed and controlled experiments that allow characterisation of the semiconductor electronic excitations of this polymer.

2. Electronic Excitations in Conjugated Polymers

Polyacetylene is a semconductor with an energy gap of about 1.7 eV between the filled π band and empty π^* band. Su *et al.* [6] reduce the electronic structure to decoupled $sp^2\,\sigma$ bonds and π bands, and use a Hückel model for the π electrons, with the electron-lattice interaction introduced through a bond-length modulation of the nearest-neighbour transfer integral, and the σ electrons parametrised through the force constant for the bonds. The π-electron system of the *trans* isomer of polyacetylene can be considered as an example of a one-dimensional, half-filled electron band, and the bond dimerisation that is found for the ground state of the chain can thus be considered as the result of a Peierls distortion of the regular chain. Within the Peierls electron-phonon coupling model for the chain dimerisation and $\pi - \pi^*$ band gap, the gap $(2\Delta_0)$ is determined by the dimensionless electron phonon coupling constant, λ and a cutoff energy, E_c, comparable to the bandwidth, $4t$, through the relation [18]

$$\Delta_0 = 2E_c \, \exp\{-1/2\lambda\} \tag{1}$$

where λ is the dimensionless electron-phonon coupling constant. Fixing to experimental values for the gap, of 1.7 eV and the bandwidth, of 12 eV, the value of λ is required to be of order 0.2, rather larger than can be attributed to electron-phonon coupling alone [19,20], and it is widely held that the enhanced bond dimerisation is due to the effects of the on-site Coulomb repulsion energy, U_0. This is shown for example in the work on Baeriswyl and Maki [21], in which, for U_0 small ($< 4t$), bond dimerisation is enhanced. It is possible to choose realistic values of $\lambda(= 0.1)$ and $U_0/4t(= 0.6 - 0.7)$ and to obtain a satisfactory value for the energy gap.

In spite of the clear evidence for the importance of Coulomb interactions, and their role in determining the properties of conjugated molecules, the SSH model and its continuum version [18] are widely used, particularly for the modelling of polyacetylene in its excited states with the various parameters (bandwidth, electron-phonon coupling etc.) chosen empirically. The fundamental excitations of the half-filled band Peierls- distorted chain are known to be phase kinks, or solitons, in the pattern of the bond alternation. This was shown for polyacetylene [5,6] to take the form of the bond alternation defects shown in Figure 2. An important insight into the nature of these excitations from the work of Su *et al.* [6] is that the bond alternation defect is not localised at a single carbon site, as indicated for convenience in the schematic representation in Figure 2, but is spread over some 10 to 15 carbon sites. This delocalisation is crucial to the energetics of the stabilisation of the soliton, and is clearly demonstrated experimentally. For this situation, it is possible to use a continuum model for the polyacetylene chain, within which various simple analytic results are found. Thus, the gap parameter, Δ varies through the soliton as

$$\Delta(x) = \Delta_0 \tanh(x/\xi) \tag{2}$$

with the soliton half-width, $\xi = v_F/\Delta_0$ where v_F is the Fermi velocity [18]. The extra lattice energy associated with the formation of the soliton is $2\Delta/\pi$. The soliton has associated with it an energy level which within one- electron theory lies at mid-gap and is of p_z non-bonding character. This is as shown in Figure 2, where for the injection of negative charge, this state is doubly occupied. The corresponding positively charged state is unoccupied.

Within the SSH model and its continuum version [18], storage of added charge on the mid-gap state created with the soliton is energetically favoured over storage in the band states; the energy of creation of the soliton is $2\Delta/\pi$, but the availability of the mid-gap state allows storage of the charge in a state of energy Δ lower than the appropriate band edge state. Thus the energy of stabilisation of the soliton over charge storage in the band edges is $\Delta(1 - 2/\pi)$; this is about 0.3 eV with the experimentally measured value of $\Delta \approx 0.8$ eV. We note here that this has direct consequences for the operation of devices such as the MIS structure; charges injected onto the chains are stored in soliton states, and the threshold for injection into these states is at energies of $\pm\Delta(1 - 2/\pi)$ relative to the gap centre, rather than at the conduction and valence band edges, at $\pm\Delta$.

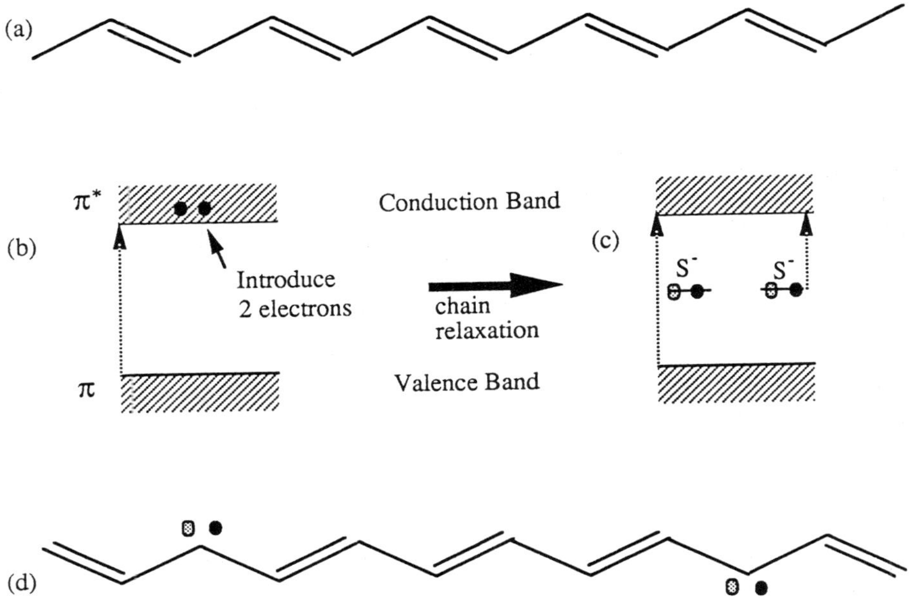

Fig. 2. Schematic representation of the formation of a pair of solitons on a polyacetylene chain. (a) shows the undistorted chain and (b) shows the accociated band scheme. The interband $\pi - \pi^*$ optical transition is shown as a dotted arrow. If two electrons are introduced onto the chain, initially into the conduction band as shown in (b), the chain relaxes to the form shown in (d), with a reversed sense of bond alternation in the centre of the chain separated by two solitons. Associated with the two solitons are nonbonding π states in the gap, created from one (doubly occupied) valence and one (empty) conduction band state. These two states are doubly occupied, as shown in (c), and each carries a negative charge. New optical transitions from the soliton level to the conduction band are indicated with a dotted arrow; the oscillator strength for the interband transition is weakened through the loss of the band states. A complementary picture holds for positive charge, which is accommodated as unoccupied solitons levels.

The formation of solitons through the addition of charge to the polymer chain modifies the optical properties, with new optical transitions allowed between the mid-gap soliton levels and the band edges, as indicated for the negative soliton to conduction band transition in Figure 2. The dipole matrix element for this transition is large, and is enhanced above that for the $\pi - \pi^*$ interband transition by a factor of $\pi^2\xi/a$, where a is the average carbon–carbon spacing along the chain [23,24]. The effect of Coulomb interactions, characterised through the on-site Coulomb repulsion, U_0, is to modify the energies of the electronic transitions between band states and the localised soliton states [25,26]. Transitions involving charged solitons are reduced in energy, and modelling the effects of U as a perturbation in the one-electron model [9],

$$\Delta E = \Delta_0 - \frac{aU_0}{3\xi} = \Delta_0 \left(1 - \frac{aU_0}{3\hbar v_F}\right) . \tag{3}$$

A value of $aU_0/3\xi$ of about 0.4 eV is consistent with the measured values of ΔE obtained from photo-induced absorption measurements [25]. Associated with the formation of the 'mid-gap' soliton absorption band is the loss of π and π^* band states and the bleaching of the $\pi - \pi^*$ interband transitions.

3. Device Fabrication

The Durham route is shown in Figure 1, and as mentioned in Section 1 we have made use of the solution processibility of the precursor polymer (B), poly (B), poly((5,6-bis(trifluoromethy)-bicyclo[2,2,2]octa-5,7-diene-2,3-diyl)-1,2 -ethenediyl), for the fabrication of the devices investigated here. The standard procedure for the fabrication of a thin-film devices is to spin-coat the precursor polymer in solution onto the required substrate, and then heat treat to convert to the polyacetylene, by elimination of hexafluoro-o-xylene. The thermodynamics and kinetics of the transformation and isomerisation reactions have been thoroughly investigated, and heat treatment at 100°C achieves conversion from precursor to polyacetylene and full isomerisation from the *cis* to *trans* isomers [27]. The hexafluoro-o-xylene leaving group is found to be absent in thin films of the converted polymer (as used in the present device structures), though for thicker films (> 1 μm), some of this is trapped inside the polymer film. The precursor polymer has very good film-forming properties, and it is straightforward to control the thickness of the resultant polyacetylene film over the range 100 Å to several μm [11–17].

Polyacetylene produced by the Durham route can be obtained in a variety of morphologies. If prepared in thin-film form by spin-coating the precursor onto a substrate and thermally converting to polyacetylene *in situ* the polymer retains much of the disorder that was transferred from the precursor polymer in solution. Although polyacetylene, as a rigid-rod polymer, shows a strong tendency to crystallise, these films show coherence lengths as determined from X-ray measurements of typically of no more than 40 Å [28]. In contrast, if a free-standing film of the precursor polymer is stretched during the thermal conversion, it is possible to obtain a very high degree of orientation and an increase in the crystallinity [29–31]. We have found that the electronic properties of the polyacetylene are very sensitive to the structure of the polymer, and we find that the unoriented films obtained by spin-coating onto substrates and conversion *in situ* show considerably increased band-gaps over that for the stretch-oriented free-standing films.

Device fabrication requires the successive deposition of a series of layers of metal, polymer and insulators, with appropriate patterning in the case of the

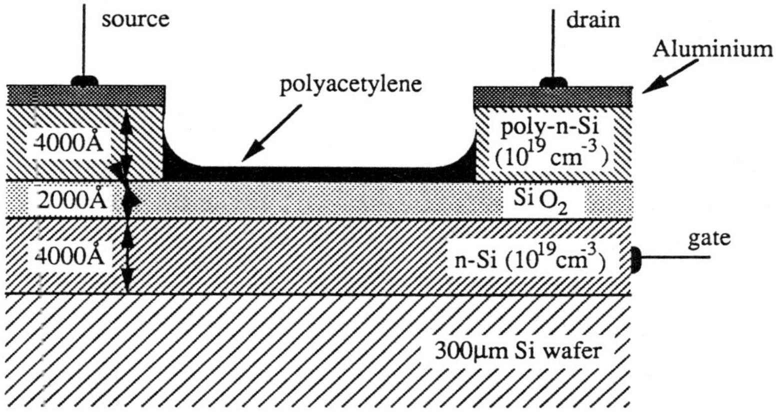

Fig. 3. Schematic diagram for a polyacetylene MISFET structure. Dimensions shown are to scale, except the channel width (20 μm) and length (1.5 m).

MISFET for the formation of the source and drain areas. The requirement for an insulator suitable for use in a MIS device is that it form films free of defects, especially pin-holes, and that the product of the relative dielectric constant, ε_r, and the dielectric breakdown field, F_b, be high. Silicon dioxide has excellent insulating properties and we have therefore made extensive use of silicon processing techniques to provide both gate, as a heavily n-doped layer, and insulator, as silicon dioxide grown on the silicon substrate. This structure provides an MIS device when coated with polyacetylene and capped with an evaporated thin gold top contact. This can be readily extended to give the MISFET structure; it is convenient to put the source and drain contacts onto the insulator layer, and put the polyacetylene layer down on top as the last stage in the construction. In the example of the MISFET structure is shown in Figure 3, the source and drain contacts are of poly-n-silicon; we have also made structures with gold.

As mentioned in Section 1, Durham route polyacetylene as prepared by thermal conversion from the precursor is p-doped, to a concentration in the range 10^{16} to 10^{18} cm^{-3}, and we have been able to use the polymer without further doping, as this range gives convenient values for the thickness of the charge depletion layer in our devices. In addition to the convenient concentration of the dopants responsible for these carriers (in the range 1 to 1000 parts per million carbon atoms), the majority of the dopants are not found to diffuse under the applied fields. This is an essential requirement for useful device operation, and is not in general likely to be met in chemically-doped conjugated polymers. In particular, post-doping of the polymer, either chemically or electrochemically, requires that dopant diffusion be sufficiently rapid to allow the doping reaction to proceed, and these diffusion rates will then also govern movement of the dopants under the action of the applied electric fields in the devices. Evidence for dopant diffusion can be obtained through hysteresis in the C(V) characteristics for MIS devices, and has been readily obtained, for example, with poly(3-hexylthienylene) [32,33]. The origin of the dopants in Durham polyacetylene is not well established, though we consider that the may be due to the presence of metathesis catalyst residues at the chain ends (the degree of polymerisation of the precursor does give appropriate concentrations of chain ends [34]).

4. The MIS Field Effect Device

The function of the insulator layer in the MIS device is to allow the application of a strong electric field to the semiconductor at the interface with the insulator by control of the gate voltage. This field can then raise or lower the energy levels in the semiconductor with negative or positive gate voltages respectively. In the conventional case of a 'rigid-band' semiconductor such as crystalline silicon, it is possible to force the band states to pass through the Fermi energy in the semiconductor, and this results in the formation of a surface charge layer at the interface with the semiconductor. For a semiconductor extrinsically doped to be *p*-type, negative bias lifts the valence band through the Fermi energy, resulting in the formation of a surface charge layer of holes (by flow of electrons through the bulk of the semiconductor to the back ohmic contact), the accumulation layer. As the gate voltage is taken to positive values, the device passes through a regime with a depletion layer present at the interface with the semiconductor, and at large positive gate voltages, the bottom of the conduction band will be lowered sufficiently close to the Fermi energy to give population of these states, and the formation of a negative surface charge layer. For a *p*-type semiconductor, these carriers are of the opposite sign to those present in the bulk of the semiconductor, the inversion layer.

The description of the formation of the charge inversion and accumulation layers in the case where the semiconductor can adjust its structure to the local charge on the chain must be very different to that for the 'rigid-band' semiconductor. Though we can still expect to see the band states for the polyacetylene bent by the applied field as the interface with the insulator is approached as these approach the Fermi energy we expect to see rearrangement of the structure and the formation of mid-gap soliton levels which will then store the charge in the accumulation or inversion layers. The effect of this is to change the position of the Fermi energy, E_F, relative to the band edges in comparison to the conventional description, in which E_F lies close to E_v for *p*-type majority carriers and to E_c for *n*-type majority carriers. Let us consider first the condition for charge accumulation, as shown in Figure 4(a). Injection of a hole results in formation of a soliton which stores

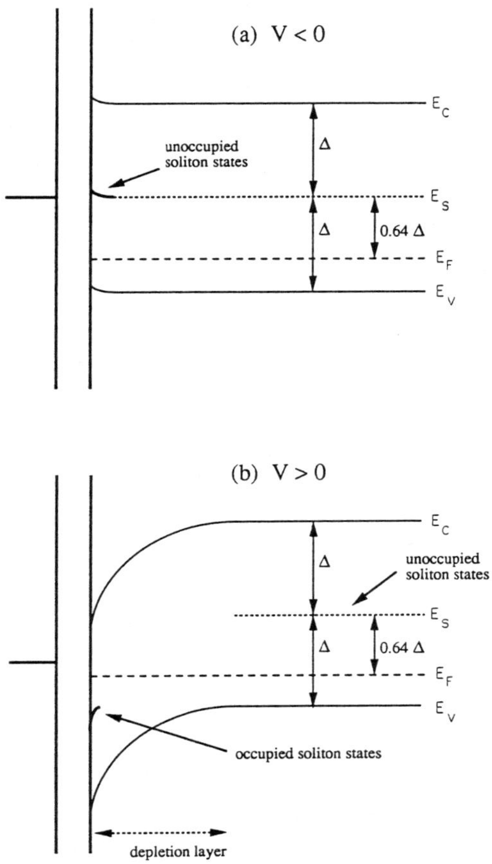

Fig. 4. MIS band scheme appropriate for p-type polyacetylene. (A) shows the energy level scheme present for charge accumulation. As discussed in the text, the threshold for hole injection occurs when E_F is $\Delta(1 - 2/\pi) = 0.36\Delta$ above E_v, and the charge injected is accommodated in the mid-gap states associated with the solitons created to store this charge. (B) shows the scheme that describes the onset of strong inversion; there is a charge depletion region, in which the soliton states associated with the extrinsic carriers are absent, and inversion occurs as E_c is brought down to $\Delta(1 - 2/\pi)$ above E_F. The negative charges in the inversion layer are stored in the mid-gap soliton levels created accommodate this charge (these are now doubly occupied).

the injected charge at $E_v + \Delta$, but at a cost of the energy of formation of the soliton, $2\Delta/\pi$. Thus the threshold for injection of holes occurs when the Fermi energy is $\Delta(1 - 2\Delta/\pi)$ above the top of the valence band; relative to E_F, the injected hole is stored at an energy of $-(\Delta - \Delta(1 - 2/\pi)) = 2\Delta/\pi$ and this is balanced by the energy of formation of the soliton, $2\Delta/\pi$. Thus, accumulation occurs well before the valence band is brought up to the Fermi energy, and we see that E_F cannot move closer to E_v than this. Similarly, injection of electrons for formation of the inversion layer occurs when E_c is brought down to $\Delta(1 - 2\Delta/\pi)$ above E_F, as shown in Figure 4(b), and we see that E_F can never move closer to the band edges than as set by these limits, i.e.

$$|E_F - E_{v,c}| \geq \Delta(1 - 2\Delta/\pi) . \tag{4}$$

The position of E_F in the extrinsically-doped p-type region must lie close to the energy threshold for accumulation, as shown in Figure 4(a). Finally, in the charge depletion region, present in Figure 4(b), there are no mid-gap soliton levels present.

The formation of the inversion layer is inherently different from that for the accumulation layer; for the two-terminal device considered here it requires thermal generation of an electron-hole pair and the subsequent diffusion of the hole back through the depletion layer into the bulk of the semiconductor. Even for the case of the MIS device built with crystalline silicon, this process can be extremely slow, and differential capacitance measurements at frequencies above 10 Hz at room temperature are at too high a frequency to show up the inversion layer [35]. For the case of polyacetylene we can expect a very much slower rate. Firstly, the energy gap is higher, so that the Boltmann factors are smaller. Secondly, the one-dimensional character of the charge carrier dynamics will impede the separation of an electron-hole (or positive soliton/negative soliton) pair, as the diffusion process is ineffective in one dimension. Thus, we can expect to see a high proportion of charge pairs recombine before separation. We do not therefore expect to see charge inversion in the polyacetylene MIS structure under the experimental conditions used here. In contrast, the addition of an extra terminal to give the FET structure of the type shows in Figure 3, allow injection of electrons directly into the inversion layer, and as with silicon devices, we can expect rapid formation of the inversion layer in this case.

The formation of accumulation and depletion may be demonstrated through the behaviour of the device capacitance with respect to the bias voltage. The measured capacitance, C, is that of the series combination of the insulator capacitance, C_i, and the capacitance of the active region of the semiconductor, C_d, and is given by $C = C_i C_d/(C_i + C_d)$. Since C_d is large for the accumulation and inversion layers, C is equal to the geometric capacitance of the insulating layer, but C falls to a lower value for movement of

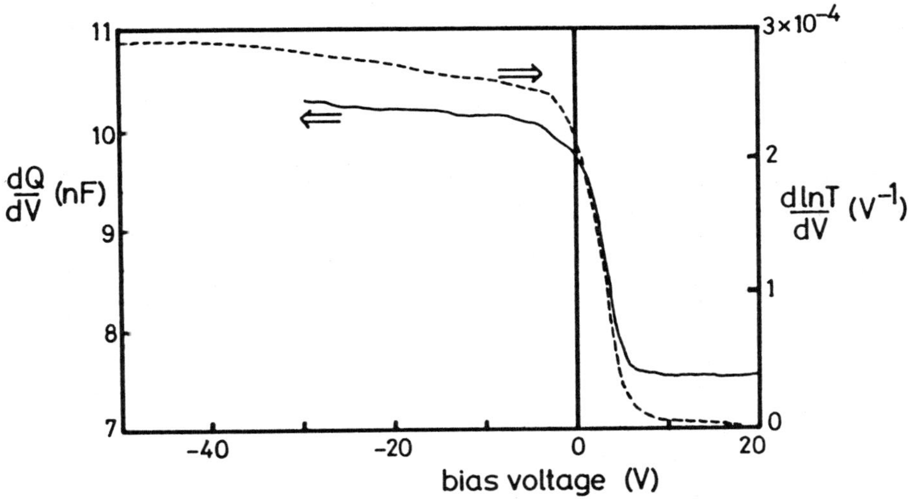

Fig. 5. $\partial \ln(T)/\partial V$, where T is the optical transmission at 0.8 eV, and the differential capacitance, $\partial Q/\partial V$, versus bias voltage for the MIS device shown in Figure 5. Both quantities are measured with an a.c. modulation of 0.25 V at 500 Hz.

the depletion layer. The capacitance versus bias voltage response for a polyacetylene MIS structure is shown in Figure 5. As expected, the capacitance for negative gate voltages flattens out to the geometric capacitance of the insulator, indicating the formation of an accumulation layer. The decrease in the measured capacitance for positive voltages displays the formation of the depletion layer. Saturation of the capacitance for large positive biases sets in when the depletion layer extends across the polyacetylene film (estimated to be about 1200 Å for this device). Analysis of the variation of the capacitance in the depletion regime allows determination of the extrinsic charge carrier concentration [36], and this is found to be in the range 10^{16} to 3×10^{17} cm^{-3}.

5. Electro-Optical Properties of the MIS Structure

The movement of charge to form the accumulation layer in the MIS device is expected to result in the formation of new mid-gap soliton states at the expense of the band states, and thus to produce changes in the optical properties of the device. We have measured the optical transmission through these structures as a function of the bias voltage on the gat. We expect a decrease in the device transmission below the band-edge as the device is driven towards accumulation, and new charged solitons are introduced onto the polyacetylene chains at the interface with the insulator. The voltage-modulated optical transmission (T) spectrum between 0.4 and 1.2 eV in

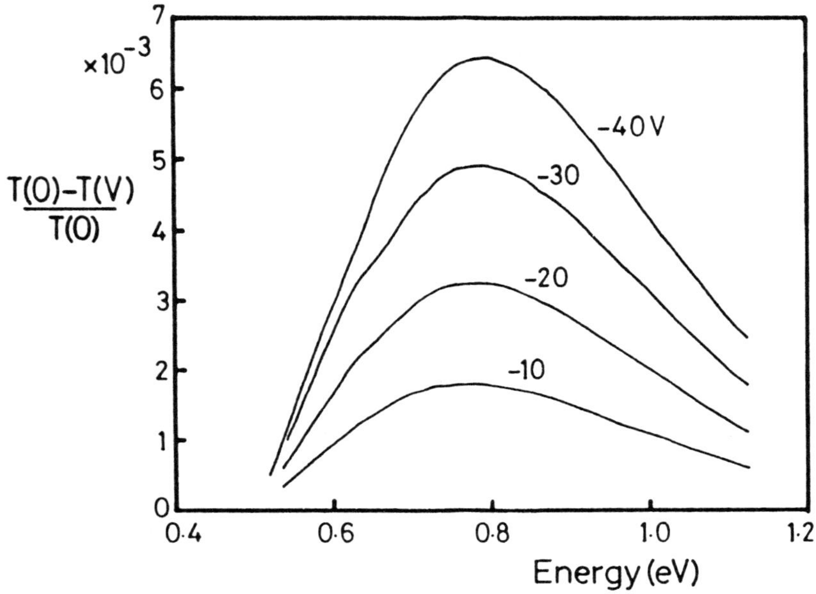

Fig. 6. Fractional change in the optical transmission, T, for a n-Si/SiO_2/polyacetylene MIS device (SiO_2 thickness 200 nm, polyacetylene thickness 60 nm), obtained between gate voltages of 0 and V, with $V = -10, -20, -30$ and -40 V. Modulation frequency is 330 Hz.

Figure 6 shows that there is a decrease in the transmission through the device for negative bias. The peak value of $\Delta T/T$, at 0.8 eV, is about 0.64%. If all the charge at the polymer-insulator interface is stored in soliton-like states then the electro-optical response, $\partial \ln T/\partial V$ should scale with $\partial Q/\partial V$, the differential capacitance, as the bias voltage is varied. This is shown to be the case in Figure 5, in which $\partial \ln T/\partial V$ and the differential capacitance are both plotted against bias voltage.

For the low areal soliton concentrations, N^a, found in the device structures, the modulation in transmission, $\Delta T/T$ can be expressed as σN^a, where σ is the optical cross-section for the dopant-induced mid-gap absorption. The determination of N^a is particularly simple for the MIS device driven into accumulation, with $N^a = \varepsilon_r \varepsilon_0 V/qd$, where V is the bias voltage and d the insulator thickness. The value for $\partial \ln T/\partial V$ seen in Figure 5 is constant with negative bias voltages, as expected in this simple model. Typical results, such as those shown in Figure 5, give the magnitude of the optical cross-section at the peak in the 'mid-gap' absorption as 1.2×10^{-15} cm^2, in very close agreement with the value measured from dopant-induced absorption [37]. Note that the maximum depth of optical modulation is fixed by the maximum value for N^a, which is determined for the MIS structure by the product $\varepsilon_r E_{\max}$ for the insulator, where $E_{\max}$ is the dielectric breakdown field. For silicon oxide, the maximum value for N^a is about 2×10^{13} cm^{-2}, and the maximum value for $\Delta T/T$ is thus about 2.4%.

Besides the electronic transitions associated with the soliton state, there are extra vibrational modes associated with the charged soliton. We have observed both IR activity is due to new vibrational modes which couple to motion of the charged soliton along the polymer chain and Raman activity associated with the amplitude mode of the soliton [11,14]. The position of the electronic 'mid-gap' transition is found here for devices built with SiO_2 as insulator to be near 0.8 eV, considerably higher than the 0.55 eV found for photo-generated carriers [38]. We consider that this is due to an increased level of disorder in the polyacetylene at the SiO_2/polyacetylene interface, which produces a locally-increased bandgap; this is supported by evidence also from the strength and frequency of the induced IR modes [17]. When we use a polymer as the insulator layer, such as poly(methylmethacrylate) or poly(imide), the energy of the mid-gap transition is indeed at 0.55 eV, and we take this to indicate that the structure of the polyacetylene at the interface with these insulator layers is similar to that in the bulk. This does provide evidence that devices fabricated with SiO_2 as insulator are not ideal, but we have used these for the work presented in the following section on MISFETs since we can readily obtain the necessary pattering that we require for these structures.

6. MISFET Devices

The MIS field effect transistor, or MISFET, operates by modulation of the conductance of the semiconductor in the depletion, accumulation or inversion modes, with control through the potential on the gate. We have made a variety of structures, with the general form shown in Figure 3. These devices are built on silicon substrates as 'buried gate' devices, with the gate formed as a doped layer of silicon underneath the thermally-grown oxide which functions as the insulator layer. Interdigitated source and drain contacts are next deposited, and the polyacetylene layer is applied as the final stage. We have worked with two types of source/drain contacts, gold for which results have been published elsewhere [13], and poly-n-silicon capped with aluminium which we discuss here. The reason for working with these source/drain contacts was to promote the formation of the charge inversion layer for positive gat voltages.

The variation of the channel conductance with gate voltage for constant V_{ds} defines the transconductance, $g_m = \Delta I_{ds}/\Delta V_{gs}$. For low values of V_{ds}, well below the level for pinch-off, g_m gives a direct measure of the carrier mobility, through the relation

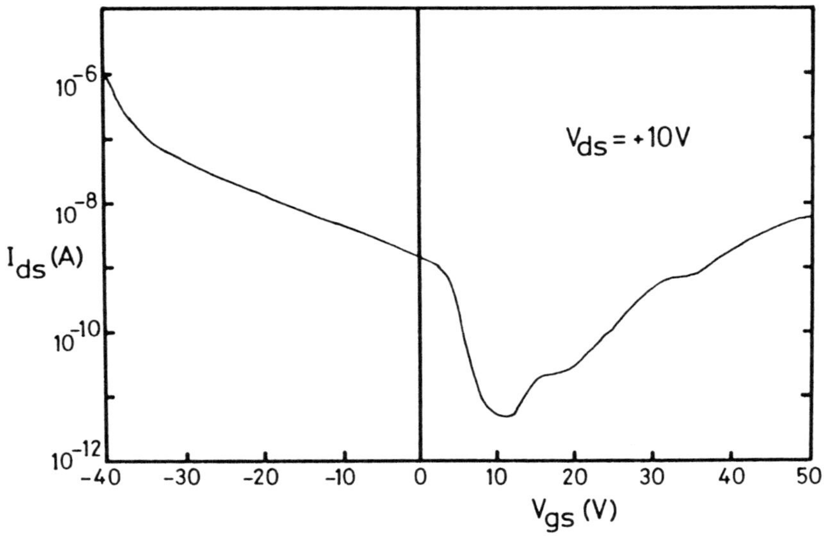

Fig. 7. MISFET structure with poly-*n*-silicon source and drain contacts with polyacetylene layer thickness of 20 nm. Shown here is the variation of I_{ds} with V_{gs} for $V_{ds} = +10$ V.

$$g_m = \frac{W}{L}\,\mu C_i\,V_{ds}\ . \tag{5}$$

The saturation current, I_{sat}, for the channel at high values of V_{ds} is conventionally modelled by the relation

$$I_{\text{sat}} = \frac{\mu C_i\,W}{2L}\,(V_{gs} - V_{\text{th}})^2 \tag{6}$$

where V_{th} is the threshold voltage at which the accumulation channel forms, C_i is the gate capacitance of the insulator layer, and μ is the charge carrier mobility [35].

The structure of these devices is shown in Figure 3. The polyacetylene layer for these devices can be kept thin, typically to no more than 20 nm; this allows full depletion at the appropriate bias, and thus gives very good 'on' to 'off' g_m ratios. Figure 7 shows the variation of I_{ds} with V_{gs} for fixed $V_{ds} = +10$ V. I_{ds} has a minimum at $V_{gs} = +10$ V, corresponding to full depletion across the polyacetylene layer. For voltages above this the conductivity rises rapidly, and is identified with the formation of a charge inversion layer at the polyacetylene/SiO_2 interface. For voltages lower than +10 V I_{ds} rises very sharply over one or two decades. This sharp rise is most likely due to the decrease in width of the depletion layer, so that the 'knee' in the curve, seen at around +5 V, corresponds to the flat-band voltage at which the measured conductivity is just that of the bulk polymer. Further

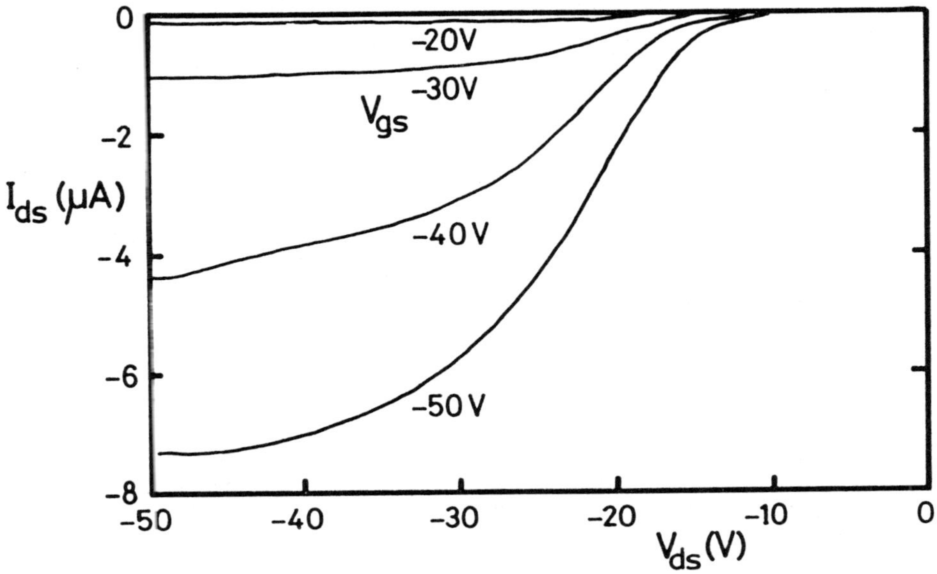

Fig. 8. MISFET structure shown in Figure 7. Shown here is the variation of I_{ds} with V_{ds} for various values of V_{gs}.

reduction in the gate voltage produces a further increase in I_{ds} and we associate this with the formation of a charge accumulation layer, even though we had expected that the poly-n-silicon source and drain would give blocking contacts to the majority carriers in the accumulation layer. The change in conductivity for these structures is rather higher than achieved with the gold source/drain contacts [13], with a ratio of 10^5 between $V_{gs} = +10$ and -40 V. This MISFET structure shows characteristic FET behaviour in accumulation, as seen in the I_{ds} versus V_{gs} curves shown in Figure 8. Saturation due to pinch-off is seen at high negative values of V_{ds}.

Even though the source/drain contacts are able to promote minority charge injection into the polymer film, characteristic MISFET behaviour for the structure in inversion was very difficult to observe. The formation of a conducting inversion layer would be inhibited if the negatively charged dopant ions are able to drift towards the insulator surface when the gate was positively biased, and we do find that only those devices fabricated with thin polyacetylene layers, such as that shown in Figures 7 and 8, give good inversion characteristics. These are shown for the same FET in Figure 9.

A plot of I_{ds} versus V_{gs} for $V_{ds} = -10$ V for a MISFET is shown in Figure 10. There is a very rapid increase in I_{ds} with V_{gs}, as is evident from the almost linear variation of $\log(I_{ds})$ with V_{gs}. In the non-saturated regime we can use Equation 5 to model the transconductance, and we find from this

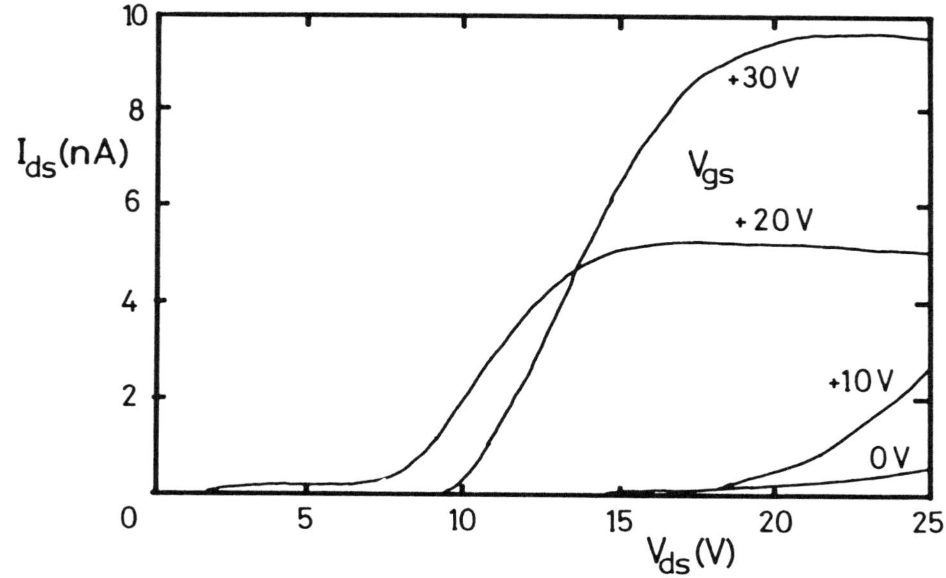

Fig. 9. MISFET structure shown in Figures 7 and 8. Shown here is the variation of I_{ds} with V_{ds} for various values of V_{gs} in the inversion regime. The device was held at 315 K for these experiments.

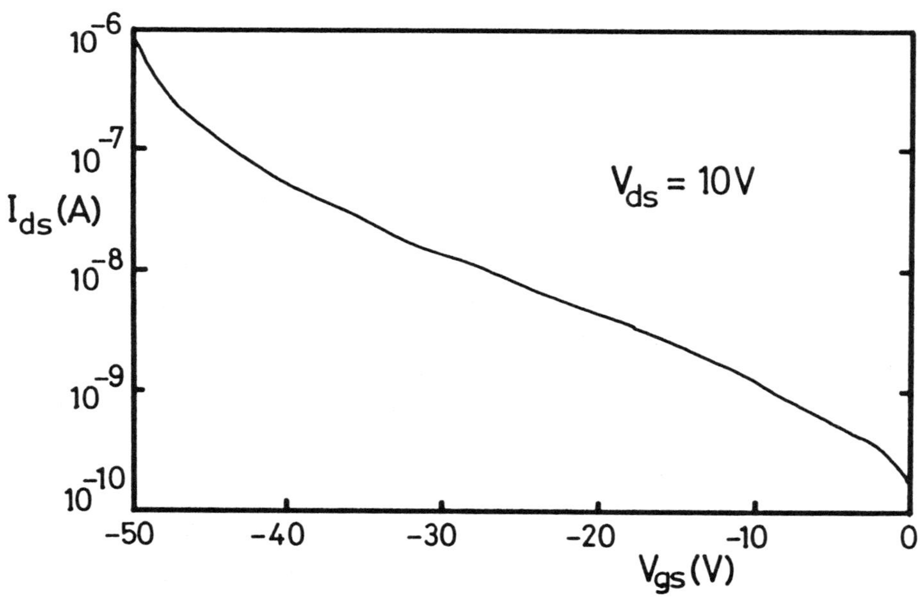

Fig. 10. I_{ds} versus V_{gs} with $V_{ds} = -10$ V for a MISFET with poly-n-silicon source/drain, and a thin polyacetylene layer (estimated to be about 60 nm in thickness). The thickness of the SiO_2 layer for this device is 120 nm.

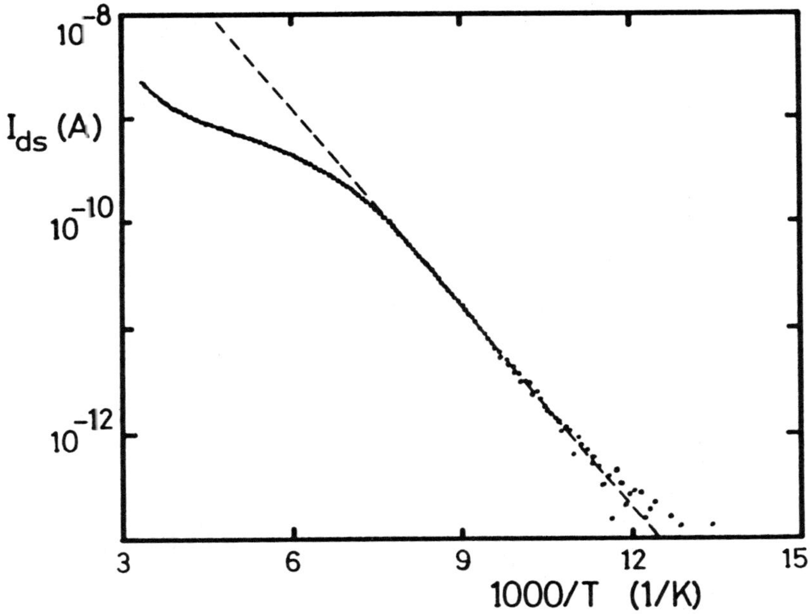

Fig. 11. I_{ds} versus reciprocal temperature for the MISFET shown in Figure 10, with $V_{ds} = +10$ V and $V_{gs} = -40$ V. The broken line fitted to the linear region of the response has a slope corresponding to an activation energy of 130 meV.

that the carrier mobility rises very quickly with the size of the gate voltage. Thus, the transconductance for $V_{gs} = -40, \ -45$ and -50 V is estimated to be 9, 29 and 840 sec, and the carrier mobilities derived from Equation 5 for these gate voltages are 0.07, 0.23 and 6.6×10^{-4} cm^2/Vsec, respectively. We estimate that the surface charge concentration in the accumulation layer at the highest gate voltage (-40 V) and mobility (6.6×10^{-4} cm^2/Vsec) is 7×10^{12} cm^{-2}.

The strong variation of the apparent carrier mobility with gate voltage for these polyacetylene MISFETs is in marked contrast to the behaviour of silicon devices, for which the apparent mobility falls at high gate voltages. We consider that this is the expected behaviour in polyacetylene, and comes about because the carrier mobility is strongly dependent on the charge carrier density. We discuss this further in Section 7.

We have obtained a small amount of data on the temperature dependence of the channel currents. Figure 11 shows the variation of I_{ds} with reciprocal temperature for the MISFET shown in Figure 10, with a small value of V_{ds} and a high value of V_{gs} so that the device is in the non-saturated regime. The device was cooled from room temperature so that the accumulation surface charge layer was unaltered throughout the experiment (the time response

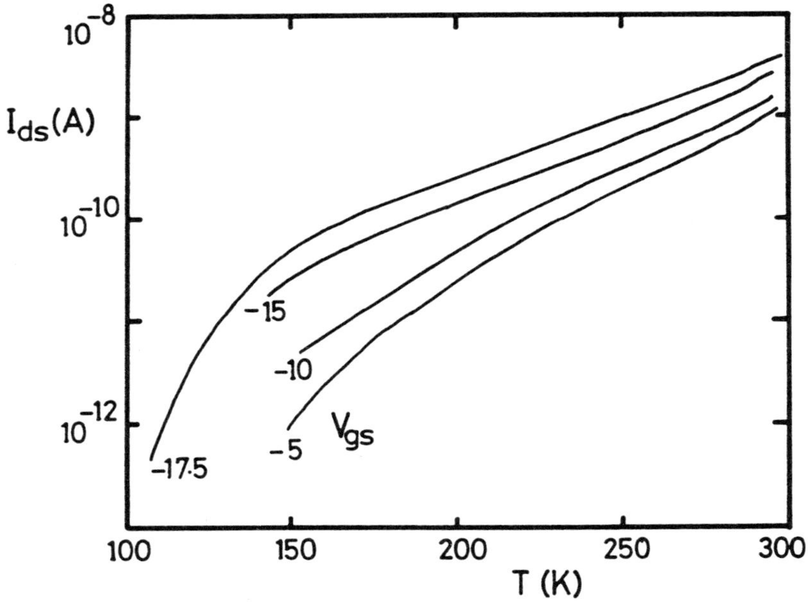

Fig. 12. I_{ds} versus temperature for a MISFET structure with polyacetylene thickness estimated to be 100 nm, and SiO_2 thickness 120 nm. Data are shown with $V_{ds} = +5$ V and with various values of V_{gs}.

for the formation of the accumulation layer would have been very slow at the lowest temperatures reached, 80 K). I_{ds} decreases with falling temperature, showing a weak variation with temperature down to about 140 K, and a more rapid variation below this. In the range 140 – 80 K, the mobility shows activated behaviour, with an energy of activation of about 130 meV. Figure 12 shows similar data obtained with another MISFET device; I_{ds} is plotted here against temperature, and is measured with a fixed value of $V_{ds}(= -5$ V) and at various values of V_{gs}. The weak variation of I_{ds} at temperatures above 140 K is more clearly seen, empirically, $\log(I_{ds})$ is found to vary linearly with temperature in this regime. There is a large variation in I_{ds} with V_{gs}, not just at room temperature, as in Figure 10, but also at lower temperatures. At the highest gate voltage used here (−17.5 V), the low temperature behaviour (below 104 K) is modelled with an activation energy of 170 meV; note that this value for the activation energy, which might be associated with the interchain hopping energy, is higher than that obtained for the device shown in Figure 11, but is obtained with a significantly lower value of gate voltage and hence accumulation layer charge density.

7. Charge Transport in Polyacetylene

The mechanism for charge transport in the disordered films of polyacetylene used in the present work is clearly one that involves motion of charge between chains. The strong self-localisation of charge onto a given chain with formation of the soliton, must reduce the rate at which charge can move between chains. First, there is the topological problem that solitons cannot move singly, and secondly, interchain charge motion must overcome the self-trapping energy and will therefore require thermal activation. Townsend and Friend [38] have measured mobilities of dark carriers and of photocarriers in the form ofpolyacetylene used here. In both cases the mobility is low (10^{-5} $cm^2/Vsec$ at room temperature) and thermally-activated with an energy of activation in the range 0.3–0.4 eV. They apply a model for 'bipolaron hopping' of like-charged solitons between chains, and for three-dimensional hopping, the mobility, μ, is given as

$$\mu = \frac{ea^2\omega}{6k_B\Theta} \exp\left(\frac{-W_h}{k_B\Theta}\right) \tag{7}$$

where a is the distance between hops, W_h is the hop energy, ω is the attempt-to-hop frequency. If W_h is identified as the experimentally measured mobility for the photocarriers and ω is taken to be the average of the frequencies of the optic phonons that modulate the dimerisation amplitude (3.9×10^{13} Hz), then a value for a of 3 nm is obtained. This is considerably larger than the interchain separation (0.42 nm), and can be taken as a measure of the average distance travelled between interchain hops. This value is close to the dimension of the crystallites in the unoriented Durham material [28]. For the case of the oriented polyacetylene for which the carrier mobility along the chain direction is a factor of 1000 higher than as found for the unoriented samples, we note that modelling the mobility through Equation 7 requires that there is a very much larger value for the average distance between interchain, thermally activated hops, a.

It is well known that the carrier mobilities in chemically doped polyacetylene can reach very high values; in the case of material produced by the BASF modified Shirakawa route, values of up to 1000 $cm^2/Vsec$ are found in the metallic regime [4]. There is, thus, a transition from the low mobility that is characteristic of the 'bipolaron' hopping process, which we expect at low charge carrier concentrations, to high mobility in the 'metallic' regime, where the cooperative effects of the high concentration of charges present reduces the interchain hopping energy. There is a considerable literature on the control of conductivity through control of the concentration of dopants introduced through chemical or electrochemical doping [39], to give $[(CH)A_y]_x$ where A is the dopant species. There is a large increase in mobility in the range of dopant concentration, y, of 0.001 to 0.01, and this is loosely described as the transition from semiconducting to metallic behaviour. A detailed model for the electronic structure at these dopant concentrations is not well established. There is ample spectroscopic evidence for mid-gap states with associated electronic and vibrational escitations, and it generally accepted that charge storage is in soliton-like wavefunctions. The description of the electronic structure at higher dopant concentrations ($y > 0.06$) is also far from complete; it is clear that the density of states at E_F must be high in order to give the very high levels of conductivity in this regime, but the spectroscopic results still show the behaviour expected for charge storage in solitons. One proposal for the electronic structure is that of a 'polaron lattice' [40,41].

These results obtained from chemical doping provide a point of comparison for the behaviour of the MISFETs. However, it is important to distinguish between the chemically doped material where there must be structural changes to accommodate the dopants, and the case of the charge inversion and accumulation layers in the MIS devices in which the charges are injected without any accompanying dopants. One of the issues that has dogged discussion and interpretation of the chemically-doped samples at low dopant concentrations is that of the homogeneity of the doping. It is generally expected that the formation of what is, in effect, an intercalation complex, should give inhomogeneous doping; in the simplest model we expect two phases present, one 'doped' and the other 'undoped'. This is driven by considerations af lattice strain energy; if a channel has been opened to receive one dopant species, it is then easier to accomodate subsequent dopants within the same channel rather than forming new channels for each of these dopants. If this is the case, then the variation of the apparent carrier mobility with dopant concentration will be determined by the percolation paths between conducting (doped) islands. It is not the place here to reopen this issue, but we caution against direct comparison of the mobilities in chemically-doped polyacetylene and those obtained in the MIS structures.

The surface charge concentration in the accumulation and inversion layers is limited by the dielectric strenght of the insulator layer in the MIS device, and with SiO_2 as insulator, this limit is set at 2×10^{13} cm^{-2}. We have not given much consideration to the variation in charge density in the direction normal to the plane of the device. For the case of devices fabricated with crystalline silicon the form of the potential confining the charge to the interface is well studied [42]. The electron wavefunctions in the inversion layer, for example, are particle-like in the two in-plane directions, but are solutions to the Schrodinger equation for the electron confined in 'triangular' potential set up by the barrier presented by the insulator layer and the spatially-varying potential in the depletion layer in the semiconductor, giving a series of particle-in-box solutions. With occupation of the lowest of these, the charge in the accumulation layer in confined to within typically less than 4 nm of the interface.

Turning to the polyacetylene MIS devices, the description of the electron states must be very different. There are several considerations to be made. First, since the charge is stored in soliton-like states formed by the relaxation of the chain to accommodate the injected charge, we must consider the electrons to be localised to individual chains. Second, we can expect that the chains at the interface with the insulator lie flat in the plane of the interface (there is some experimental evidence for this), so that we can defined the potential on a chain just by its distance from the interface. Third, whereas for the Si device, the injected carriers can form a degenerate gas of Fermi particles in the plane of the device, the particles here (solitons) have integer spin and are bosons, and we note that Brazovskii and Kirova have speculated that is may be possible to obtain a Bose condensation of these to a superconducting ground state [43]. Fourth, there is ample evidence for structural modifications to the polymer at the surface, with increased disorder and an increase in the $\pi - \pi^*$ gap, and such disorder must control the distribution of the charge. There are general considerations that are applicable here; the dielectric constant for the semiconductor will determine the screening of the charge set up in the surface charge layer, and we do expect to see stronger confinement of charge in the polyacetylene layer than in the Si devices, and consider that an appropriate figure for this is 2 nm.

We can use this figure for the confinement of the surface charge layer to the interface with the insulator layer to obtain the volume charge densities. With a surface charge density of 2×10^{13} cm^{-2}, we thus obtain a value of 10^{20} cm^{-3}, which is to be compared with the volume density of (CH) repeat units which is about 5×10^{22} cm^{-3}. Thus the upper limit to the charge concentration that we can obtain in the MIS structure is about 0.2%. This is very much in the middle of the range over which the chemically-doped material shows a rapid variation in mobility with charge concentration, and we consider that the mobilities in the surface charge layers will similarly be increasing rapidly with charge concentration. We consider that this accounts for the behaviour of the I_{ds} versus V_{gs} plots in the non-saturated regime shown in Figures 11 and 12.

How much do the results we have obtained here tell us about the fundamental limits to the mobilities of carriers in devices fabricated with polymer that is very much better ordered than the polyacetylene that we have used here? There are recent reports of very much improved mobilities for devices on sexithiophene (the six repeat unit oligomer of polythiophene), with a value of 0.4 cm^2/Vsec now reported [44], and there is now considerable interest in the development of polymer FETs as large areas thin film transistors, with interest in polythiophene derivatives [45] and in poly(arylenevinylenes) such as poly(2,5-thienylenevinylene) [46]. We can see from the optical characterisations of the MIS devices that the surface layer of polyacetylene formed on SiO_2 is very much more disordered than the bulk material, but we have not made FET devices with the polymer inslator layers which give better ordered structures as characterised optically. We must presume that the extra disorder in the SiO_2 devices reduces the carrier mobility. If we use Equation 7 for the mobility limited by thermally-activated bipolaron hopping, but use the experimentally measured values for the activation energy in the MISFETs (Figures 11 and 12), we obtain room temperature mobilities of around 0.1 to 1 $cm^2/Vsec$, and we consider that the discrepancy between this and the measured mobilities is indeed due to the disorder present in the surface layer of the polymer. It is clearly of interest to develop improved methods for the fabrication of polymer devices in order to realise such improvements in device performance.

The development of working FET devices put renewed interest into the understanding of the charge transport processes in polyacetylene in the regime in which the charge concentration is low. The transition from the low mobilities found at very low charge concentrations, which is modelled by the bipolaron hopping process discussed in Section 7, to the higher mobilities above 0.1% charge concentration is not well understood. We see experimentally that there is a reduction in the activation energy for conductivity as the charge concentration is brought into this regime (Figures 11 and 12). The activation energy is associated within the bipolaron hopping model with the bipolaron binding energy, i.e. the energy of relaxation associated with formation of the soliton/anti-soliton pair. This is easy to identify for the case of well-separated charges, but as the concentration of charge is increased, interactions between them may become important. If the presence of some charge on a chain reduces the binding energy for formation of solitons to accommodate further charges placed on that chain, then the activation energy for hopping of charge from an adjacent chain will be reduced. We consider that this is a model which can account for the fall in the activation energy for the conductivity from the value of 0.3 – 0.4 eV in the low charge regime to the values found in the MISFETs under accumulation in the range 0.13 – 0.17 eV.

Acknowledgements

This work has been dependent on excellent support for the chemistry and materials science of the Durham route polyacetylene, and we thank Professor W.J. Feast and his group at Durham, and the British Petroleum Research Centre at Sunbury-on-Thames for this support. It is a pleasure to acknowledge the wide ranging contributions that Professor H. Kamimura has made to this field of low dimensional conductors.

References

1. *Proceedings of the International Conference on Synthetic Metals*, Santa Fe, New Mexico, June 1988, *Synthetic Metals*, **27–29** (1988,1989).
2. T. Ito, H. Shirakawa and S. Ikeda: 1974, *J. Polym. Sci. Poly. Chem. Ed.* **12**, 11.
3. C.K. Chiang, C.R. Fincher, Y.W. Park, A.J. Heeger, H. Shirakawa, E.J. Louis, S.C. Gau and A.G. MacDiarmid: 1977, *Phys. Rev. Lett.* **39**, 1098.
4. N. Basescu, Z.X. Liu, D. Moses, A.J. Heeger, H. Naarman and N. Theophilou: 1987, *Nature* **327**, 403.
5. M.J. Rice: 1979, *Phys. Rev. A* **71**, 152.
6. W.P. Su, J.R. Schrieffer and A.J. Heeger: 1979, *Phys. Rev. Lett.* **42**, 1698; 1980, *Phys. Rev. B* **22**, 2099; 1983, *erratum B* **28**, 1138.
7. J.H. Edwards and W.J. Feast: 1980, *Polym. Commun.* **21**, 595.
8. J.H. Edwards and W.J. Feast: 1980, *Polymer* **21**, 595.
9. J.H. Edwards, W.J. Feast and D.C. Bott: 1984, *Polymer* **25**, 395.
10. W.J. Feast and J.N. Winter: 1985, *J. Chem. Soc. Chem. Commun.*, 202.
11. J.H. Burroughes, C.A. Jones and R.H. Friend: 1988, *Nature* **335**, 137.
12. J.H. Burroughes, C.A. Jones and R.H. Friend: 1989, *Synthetic Metals* **28**, C735.
13. J.H. Burroughes, R.H. Friend and P.C. Allen: 1989, *J. Phys. D* **22**, 956.
14. R.A. Lawrence, J.H. Burroughes and R.H. Friend: 1989, *Springer Series on Solid State Sciences* **91**, 127.
15. R.H. Friend and J.H. Burroughes: 1989, *Faraday Discussion Chem. Soc.* **88**, 213.
16. J.H. Burroughes, C.A. Jones, R.A. Lawrence and R.H. Friend, NATO ARW, Conjugated Polymeric Materials: Opportunities in Electronics, Optoelectronics and Molecular Electronics, Mons, Belgium, *NATO-ASI Series E: Applied Sciences* **182**, 221, Kluwer, Dordrecht, 1990.
17. J.H. Burroughes and R.H. Friend, Materials Research Research Society Fall Meeting, Boston, November 1989, *MRS Symposium Proceedings* **173**, 425 (1990).
18. H. Takayama, Y.R. Lin-Liu and K. Maki: 1980, *Phys. Rev. B* **21**, 2388.
19. D. Baeriswyl, in *Electronic Properties of Polymers*, J. Mort (ed.), Wiley, 1982.
20. D. Baeriswyl, D.K. Campbell and S. Mazumdar (preprint).
21. D. Baeriswyl and K. Maki: 1985, *Phys. Rev. B* **31**, 6633.
22. B. Horovitz: 1982, *Solid State Commun.* **41**, 729.
23. N. Suzuki, M. Ozaki, S. Etemad, A.J. Heeger and A.J. MacDiarmid: 1980, *Phys. Rev. Lett.* **45**, 1209; *erratum* **45**, 1463.
24. S. Kivelson, T-K. Lee, Y.R. Lin-Liu, I. Peschel and L. Yu: 1982, *Phys. Rev. B* **25**, 4173.
25. Z. Vardeny and J. Tauc: 1985, *Phys. Rev. Lett.* **54**, 1844; 1986, *Phys. Rev. Lett.* **56**, 1510.
26. D. Baeriswyl, D.K. Campbell and S. Mazumdar: 1986, *Phys. Rev. Lett.* **56**, 1509.
27. D.C. Bott, C.S. Brown, C.K. Chai, N.S. Walker, W.J. Feast, P.J.S. Foot, P.D. Calvert, N.C. Billinham and R.H. Friend: 1986, *Synthetic Metals* **14**, 245.
28. C.S. Brown, M.E. Vickers, P.J.S. Foot, N.C. Billingham and P.D. Calvert: 1986, *Polymer* **27**, 1719.
29. D. White and D.C. Bott: 1984, *Polymer Commun.* **25**, 98.
30. H. Kahlert and G. Leising: 1985, *Mol. Cryst. Liq. Cryst.* **117**, 1.
31. M. Sokolowski, E.A. Marseglia and R.H. Friend: 1986, *Polymer* **27**, 1714.

32. K.E. Ziemelis, A.T. Hussain, D.D.C. Bradley, R.H. Friend, J. Rühe and G. Wegner: 1991, *Phys. Rev. Lett.* **66**, 2231.
33. G. Gustafsson, M. Sundberg, O. Inganäs and C. Svensson, *J. Mol. Electron* (in press).
34. K. Harper and P.G. James: 1985, *Mol. Cryst. Liq. Cryst.* **117**, 55.
35. S.M. Sze, *Physics of Semiconductor Devices*, Wiley, New York, 1969.
36. A. Goetzberger and E.H. Nicollian: 1966, *Appl. Phys. Lett.* **9**, 12.
37. J. Orenstein, in *Handbook of Conducting Polymers*, T.J. Skotheim (ed.), Marcel Dekker, New York, 1986.
38. P.D. Townsend and R.H. Friend: 1989, *Phys. Rev. B* **40**, 3112.
39. Y.W. Park, A.J. Heeger, M.A. Druy and A.G. MacDiarmid: 1980, *J. Chem. Phys.* **73**, 946.
40. S. Kivelson and A.J. Heeger: 1985, *Phys. Rev. Lett.* **55**, 308.
41. S. Stafström and J.L. Brédas: 1988, *Phys. Rev. B* **38**, 4180.
42. T. Ando, A.B. Fowler and F. Stern: 1982, *Rev. Mod. Phys.* **54**, 437.
43. S. Brazovskii and N. Kirova: 1985, *Solid State Commun.* **55**, 187.
44. F. Garnier, G. Horovitz, X. Peng and D. Fichou: 1990, *Advanced Materials* **2**, 592.
45. H. Akimichi, K. Waragai, S. Hotta, H. Kano and H. Sakaki: 1991, *Appl. Phys. Lett.* **58**, 1500.
46. A. Tsumara, H. Fuchigami and H. Koezuka: 1991, *Synthetic Metals* **41–43**, 1181.

DYNAMICS OF SPIN-DENSITY WAVE IN ORGANIC CONDUCTORS $(TMTSF)_2X$

TAKASHI SAMBONGI and KAZUSHIGE NOMURA
Department of Physics, Hokkaido University, Sapporo, Japan

Abstract. The dynamic properties of the spin-density wave in the organic conductors $(TMTSF)_2X$ are discussed. Recent experimental results on the nonlinear conductivity and narrow noise are presented. Frequency dependent conductivity and narrowing of NMR signal at high d.c. bias current are also briefly reviewed.

1. Introduction and Historical Background

In 1960 Overhauser [1] proposed that the ground state of an electron gas is the spin-density wave (denoted as SDW hereafter), in which the number density of conduction electrons is uniform but the density of their spins is modulated in space with the wavevector equal to the diameter of its Fermi sphere. The SDW was observed first in metallic chromium. Its antiferromagnetism below $T_N = 312$ K was assigned as the SDW by neutron diffraction [2]; the wavevector is incommensurate with the lattice period and is temperature dependent. Overhauser's original proposal was modified by Lomer [3]. The Fermi surface of Cr is composed mainly of an electron and a hole pocket of similar size and significant numbers of states on these pockets are connected to each other by a vector approximately equal to $\Gamma - H$ in the Brillouin zone. The SDW wavevector determined by neutron diffraction is in fair agreement with the nesting vector estimated from a band structure calculation. It has been expected that the SDW manifests itself more frequently in a one-dimensional electron system, because it can be stabilized in planar Fermi sheets by the Overhauser mechanism.

At the end of the 1970s, the SDW was found in some low-dimensional organic conductors of formula $(TMTSF)_2X$, where (TMTSF) is the abbreviated notation for (2,3,6,7-tetramethyl-1,4,5,8-tetraselenafulvalenium) and X is one of the single valence anions PF_6, ClO_4, NO_3 etc., as the low temperature non-metallic phase below the transition temperature $T_c \sim 10$ K [4].

The charge-density wave (CDW), in which the electron density is modulated periodically in space without spin polarization, is another class of ground states of low-dimensional electron systems. The CDW was found in two-dimensional layered conductors like TaS_2 and $NbSe_2$ in the middle of the 1970s and then in various quasi-one dimensional compounds like TTF–TCNQ, $NbSe_3$, TaS_3 and $K_{0.3}MoO_3$. One of the most prominent properties of the CDW system originates in the fact that the phase of the periodic

H. Aoki et al. (eds.), New Horizons in Low-Dimensional Electron Systems, 401–414.

modulation of condensed electrons is arbitrary in perfect lattices, as long as its wavevector is incommensurate with the lattice period. In actual materials which contain a finite concentration of impurities and other types of defects, however, the CDW is stabilized by deforming itself around these lattice imperfections. In high applied electric fields, the CDW is displaced relative to the host lattice. As a result, various phenomena are observed in one-dimensional CDW systems, such as an increase of d.c. conductivity above a threshold electric field, a strongly frequency-dependent a.c. conductivity and a voltage/current oscillation under high d.c. current/voltage. Low field conductivity shows time-dependent variation and hysteresis against electric field sweep, because there are a variety of spatial configurations of the CDW pinned by imperfections and they are almost degenerate in energy. Comprehensive reviews on the CDW systems and their properties, both static and dynamic, have been published in the papers listed in Reference [5].

Similar phenomena are expected in the SDW as well; the low-energy excitation, the so-called sliding mode, is common to these two types of condensation of low-dimensional conduction electrons. Intuitively, electrons condensed into the SDW as a whole are regarded as a charged macroscopic body, on which the applied field acts as an external force. The SDW would be able to couple to the impurity potential if internal deformation is allowed. Nevertheless, experimental research on the SDW dynamics has started to progress only recently, because the SDW has been found in a limited members of materials, and that was rather recently.

Walsh *et al.* [6] found that the conductivity of $(TMTSF)_2PF_6$ in the SDW phase increases with increasing electric current, and suggested that this is indicative of the SDW sliding in an electric field higher than the threshold in order to overcome the pinning force by impurities. On the other hand, Chaikin *et al.* [7] were negative to their idea. The reason was that no threshold field was found; the conductivity is non-ohmic even in the lowest field. Another reason was the absence of narrow band voltage 'noise' under d.c. current, which would be strong evidence for the SDW sliding. Absence of frequency dependence of the conductivity up to 80 MHz was regarded as further negative evidence for the SDW sliding.[1]

In the CDW systems, the correspondence between low threshold electric field for depinning and strong dependence of the a.c. conductivity on frequency has been found; the so-called rigid body model [9] has been successful to some extent, in which the CDW in an a.c. field behaves as a particle in the harmonic impurity potential. At that time, it was generally accepted that the threshold field for the SDW depinning would be much lower than that for the CDW because the charge density, which is uniform in the undistorted

[1] Positive results of later measurements [25] in higher frequencies are reviewed briefly in Section 4.

SDW, would couple to non-magnetic impurities only weakly; the pinning force by the latter would be considerably weaker than that on the CDW, in which the modulation of charge density is directly coupled to the charged impurities.

An alternative interpetation, given by Chaikin *et al.* [7] and subsequently refined by Conwell and Banik [8], is that the increased conductivity is associated with the enhanced mobility of normal carriers near the SDW gap. Crystals of $(TMTSF)_2X$ grown by the standard electro-chemical method is self-purified during the growth process. Because of the small SDW gap the band mass near the gap is so small that a high mobility of normal carriers is expected. Conwell *et al.* proposed that normal carriers are warmed even in such a weak field; the density of electronic states decreases with excitation energy, and that effect is manifested in one-dimensional energy band as the increase of the mobility μ with electric field E as $\Delta\mu/\mu \propto E^2$. Because such a large increase of conductivity as is observed is not expected from the estimated mobility in $(TMTSF)_2PF_6$ as large as $\sim 10^{+6}$ $cm^2/Vsec$ at 4 K, it was suggested that the strain field around dislocations makes the SDW energy gap inhomogeneous and the warming of normal carriers is enhanced in the compressed region near edge dislocations [8].

It has been noted that the nonlinearity of conductivity in the SDW phase of $(TMTSF)_2X$ is correlated with the discontinuous change in the ohmic conductivity, which can happen during cooling the sample from room temperature. This 'resistance jump' indicates the generation of micro-cracks[2] by, e.g., thermal stress, within the sample, but no detailed analysis has been presented as to how these micro-cracks are correlated with the nonlinearity of conductivity in the SDW phase.

2. SDW in $(TMTSF)_2X$

$(TMTSF)_2X$ materials were first synthesized by Bechgaard *et al.* [4], therefore they are sometimes called Bechgaard salts. The molecule (TMTSF), shown in Figure 1, works as the electron donor in the crystal, while anions X (PF_6 etc.) are single valence strong acceptors. The crystal structure of $(TMTSF)_2X$ is triclinic and the unit cell contains one chemical unit, as illustrated in Figure 2. (TMTSF) molecules stack along the a-axis and the stack is repeated along the b-axis. The two-dimensional stacks of (TMTSF) separated by X-sheets carry the conduction band. There is one hole in the unit cell, but the conduction band is conventionally regarded as 3/4-filled with the pseudo-unit cell of $(TMTSF)X_{1/2}$, because (TMTSF) molecules are approximately equidistant within the stack. The conduction band on the (TMTSF) sheets is two-dimensional in the sense that the transfer integral,

[2] Local degradation of current contacts also should be examined as another source of 'resistance jump'.

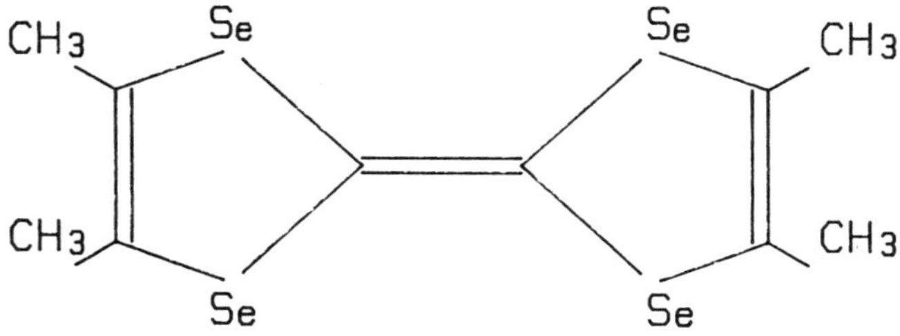

Fig. 1. (TMTSF) molecule

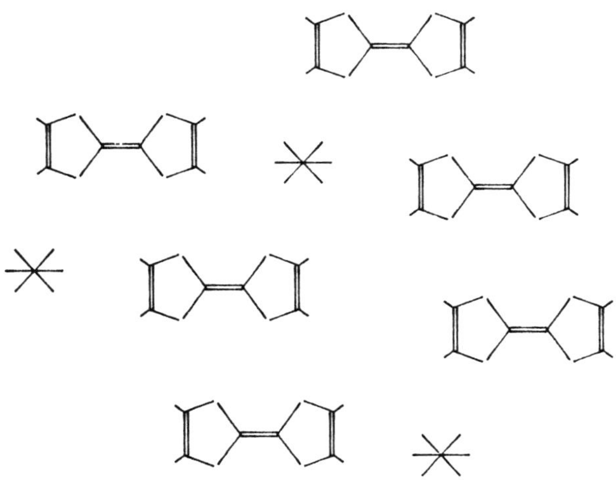

Fig. 2. Crystal structure of $(TMTSF)_2X$. View along the a-axis (schematic). Marks $*$ show the positions of anions.

t_a, along the a^*-axis is $\sim (1/4)$ eV while that along the second direction, t_b, is one order smaller and that the third, t_c, can usually be neglected.[3]

At ambient pressures, the conductivity along the a-axis of $(TMTSF)_2X$ is 400–800 (S/cm) at room temperature and increases with lowering temperature down to the metal to non-metal transition temperature T_c 10 K [4]. Just above T_c σ_a is as large as $10^2 - 10^6$ (σ_a (room temperature)). Increase of σ_a with ordering of anions is relevant in $(TMTSF)_2NO_3$ and also in quenched $(TMTSF)_2ClO_4$ on heating, as shown in Figure 3.

It was found by X-ray diffraction that lattice modulation is absent below the metal to non-metal transition; the transition at T_c was assigned as that of the SDW but not the CDW [11]. Later magnetic resonance studies have provided definite evidence for that mechanism [12].

The SDW wavevector has been determined from analysis of NMR signals

[3] Electronic structure of $(TMTSF)_2X$ has been fully discussed by Yamaji [10].

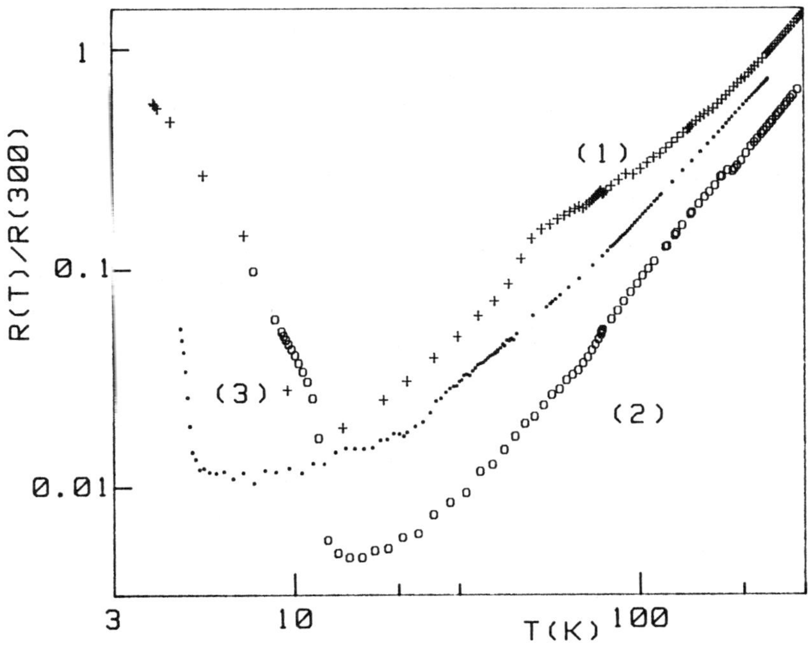

Fig. 3. Resistance of $(TMTSF)_2X$ normalized at 300 K: (1) X = NO_3, (2) PF_6 and (3) ClO_4 (quenched, measured on heating). Curves (1) and (3) are shifted vertically. Decrease of the resistivity below the anion ordering temperatures is found in (1) and (3). An example of 'resistance jump' is relevant in (2) at ~ 180 K.

of protons in the methyl groups of (TMTSF), by Takahashi *et al.* [13] and by Delrieu *et al.* [14]. Because of the small spin polarization amplitude and the small size of the crystals, neutron diffraction has not been tried so far. It is incommensurate with the lattice. The wavevector and magnetization amplitude are $0.5a^* + 0.24b^*$ and 0.08 μ_B/molecule in $(TMTSF)_2PF_6$ [13] and $0.5a^* + 0.12b^*$ and 0.12 μ_B/molecule in $(TMTSF)_2ClO_4$ [14][4] at 4 K.

Yamaji calculated the optimum nesting wavevector of the Fermi surface based on the tight-binding band model [10]. His result is qualitatively consistent with the SDW wavevector determined experimentally. It should be pointed out, however, that there is no reason why the a^*-component is definitely commensurate; as long as the second component is incommensurate, there is no gain of the commensurability energy. In fact, no sign of the commensurability pinning has been found in the conductivity works reviewed in the next section.

$(TMTSF)_2X$ are well known as the first organic superconductor; t_b in-

[4] The third component of the wavevector in $(TMTSF)_2PF_6$ has not been determined.

creases and T_c decreases under pressure and the superconducting phase appears above a critical pressure. In the phase diagram superconductivity is found just above the critical pressure. In $(TMTSF)_2ClO_4$, the low temperature phase is not the SDW but that of superconductivity, even at the ambient pressure, when the orientation of ClO_4 anions is well ordered in the lattice by careful cooling. New SDW phases have been found in high magnetic field and have been extensively studied. This 'field-induced' SDW is composed of various sub-phases; successive phase transitions are observed with varying magnetic field. A variety of interesting and novel characteristics have been found in these 'organic superconductors' and in the field-induced SDW, but they are beyond the scope of this paper.

3. Nonlinear Conductivity with Threshold and Narrow Band Noise

Positive experimental results for the SDW sliding were obtained for the first time by Tomic *et al.* [15]. They measured the electrical conductivity of $(TMTSF)_2NO_3$ crystals, which were cooled slowly enough to avoid any resistance jump. They found that the conductivity in low field is clearly ohmic, in contrast with the previous results on $(TMTSF)_2PF_6$, and that the conductivity increases sharply above the threshold electric field $E_T = 30 - 40$ mV/cm. They also found that E_T is higher and the onset of the nonlinearity is more gradual in less pure crystals, where $\sigma_a(T \gtrsim T_c)/\sigma_a$ (room temperature) was used as a measure of the purity, as in ordinary metals. Subsequently, a similar increase of the conductivity above threshold was observed in $(TMTSF)_2PF_6$, first by Kang *et al.* [16], and in $(TMTSF)_2ClO_4$ [17]. In the latter compound, the SDW is induced with $T_c = 6$ K, when the non-centrosylmmetric ClO_4 anions remain disordered by rapidly quenching the sample through the ordering temperature of 22 K [18]. Figure 4 shows the conductivity of $(TMTSF)_2ClO_4$ as a function of the applied field. While the low field σ_a is ohmic in $(TMTSF)_2ClO_4$ and $(TMTSF)_2NO_3$, that of most $(TMTSF)_2PF_6$ samples has been rather nonlinear, even after careful cooling. When its nonlinearity is not so strong as reported previously [7], however, E_T can be determined from the conductivity-field relation. In our experience, most crystals of $(TMTSF)_2PF_6$ are rather mechanically weak. Improvements of crystal growth conditions will be greatly appreciated.

From the above observation of the nonlinear conductivity with threshold and of the 'narrow band noise' presented below, it has been established that the SDW is capable of sliding in these materials. The threshold electric field for the nonlinear conduction corresponds to the critical field for the SDW to overcome the pinning by, e.g., impurities. The measured magnitude of E_T in pure crystals is of the order of 10 mV/cm, irrespective of the material. It might seem that this magnitude is too large for the pinned SDW, in which

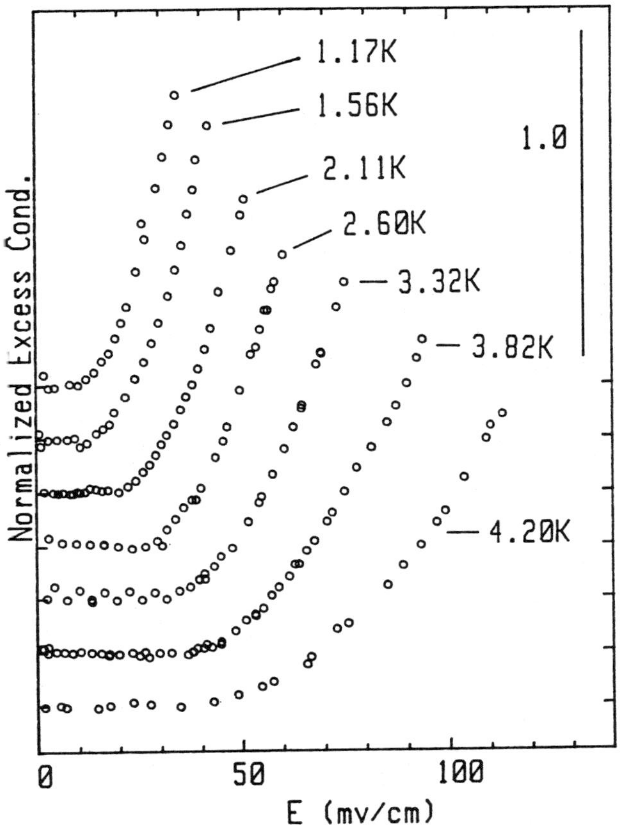

Fig. 4. Excess conductivity of quenched $(TMTSF)_2ClO_4$ normalized by the ohmic values. Each curve is shifted vertically. The vertical bar in the figure is to scale the normalized excess conductivity of 1.0.

the charge density is uniform when not distorted. Theoretical studies have concluded, however, that the E_T of the SDW pinned by impurities can be as large as that of the CDW.

The impurity pinning of the SDW has been studied theoretically by Takada [19], Tua and Ruvalds [20] and by Tutto and Zawadowski [21]. Recently Maki and Virosztek [22] have given explicit expressions for the threshold electric field E_T of the SDW pinned by impurities. The SDW deforms itself around charged impurity atoms and the induced net charge is coupled to the impurity. In a matrix with finite impurity concentration, the classification into weak and strong pinnning regimes, which has been developed for the CDW, is applied similarly. The threshold field for depinning is determined by the balance of externally applied electric force with the sticking force of the SDW to impurities. The former acts on the total condensed electrons and the latter is given by the above second-order coupling to non-magnetic impurities. The impurity potential is then a periodic function of the SDW phase and is proportional to $\Delta \tanh(\Delta/2k_BT)$ [20], where Δ is the gap parameter. Both the electric and sticking forces are considered to be dependent on temperature only through those of the gap parameter and of the density of condensed electrons, n_{SDW}. Though the detailed mechanism of pinning determines the functional form of the threshold as the function of the parametrized SDW energy gap, no significant temperature dependence of the threshold is expected at low temperature, e.g. at $T < T_c/2$, as long as the gap parameter behaves like that of a BCS superconductor.

Both Tomic *et al.* [15] and Kang *et al.* [16] claimed that E_T decreases with lowering temperature and is independent of temperature below $T_C/2$, in good agreement with the theoretical prediction by Maki and Virosztek [22]. On the other hand, a systematic study on $(TMTSF)_2ClO_4$, the SDW of which is induced by rapid quenching, made it clear that E_T is more strongly temperature-dependent as compared with the published results for $(TMTSF)_2NO_3$ and for $(TMTSF)_2PF_6$ and with the theoretical prediction by Maki and Virosztek. Because the conductivity increases more or less gradually, especially near E_T, care should be taken in the determination of the magnitude of E_T. Figure 5 shows the threshold field E_T as a function of temperature, determined from Figure 4. That figure also shows the theoretical curve, normalized at $T_c/2$, for weak impurity pinning in three dimensions calculated by Maki and Virosztek [21]. Although the magnitude of $E_T \sim 10$ mV/cm is of the same order of those observed in both $(TMTSF)_2NO_3$ and for $(TMTSF)_2PF_6$, the E_T of $(TMTSF)_2ClO_4$ is more sensitive to temperature irrespective of how it is determined; it decreases monotonously even below $T_c/2$ and its temperature derivative is larger near T_c.

From the naive picture of the SDW depinning, it is expected that the temperature dependence of E_T is common to these materials as long as the SDW wavevector is incommensurate; refinement of both experiment and theory is necessary. Recently, Huang and Maki [23] proposed that the temperature dependence of E_T is stronger than the previous calculation [22] when the Fermi surface does not nest perfectly. In a quasi-one dimensional system, including the present class of materials, the Fermi surface is a pair of planes, which are warped because of a small $t_\perp$ and are open to directions perpendicular to the one-dimensional axis. Yamaji proposed that, with a single nesting vector, the energy gap develops over the entire area of the Fermi surface and the SDW is stabilized as long as the 'imperfect nesting' parameter is smaller than a critical value. As a result, the system is insulating but the energies just below and above the gap are not constant but vary with $k_\perp$. Besides that, the single-particle energy gap is smaller than the order parameter, and it was shown that the temperature dependence

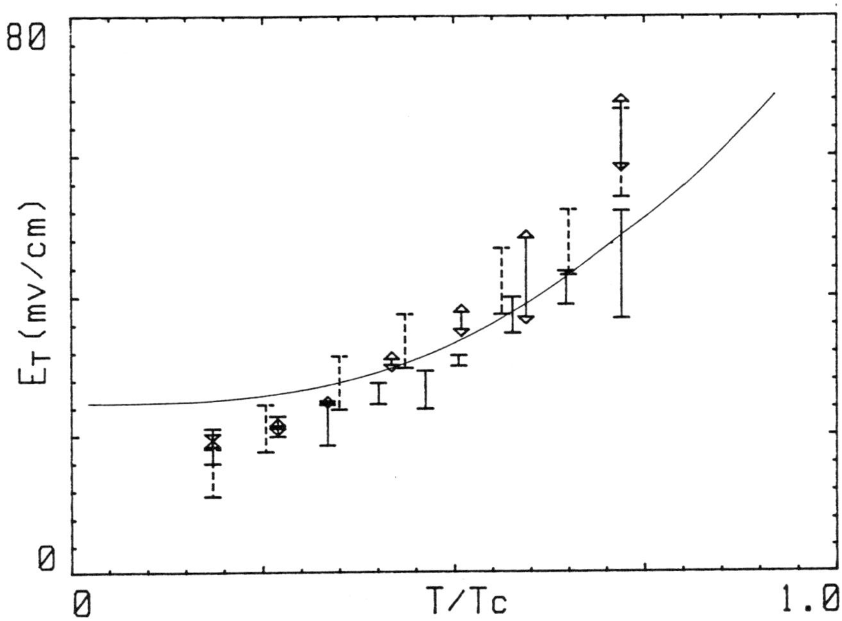

Fig. 5. Threshold electric field E_T of $(TMTSF)_2ClO_4$ normalized at $T_c/2$. The lower bound is defined as the field where deviation from the ohmic value is found in Figure 4, while the upper bound is determined by extrapolating the conductivity between 1.2 $\sigma(0)$ and 1.5 $\sigma(0)$ to $\sigma(0)$ base line. Theoretical result for three dimensional weak pinning by impurities is shown for comparison.

of the gap parameter is stronger than that of BCS [10]. Huang and Maki showed that the temperature dependence of E_T is stronger because of the gap parameter in cases of less 'perfect nesting'.

Although the above mechanism seems to reconcile the experimental results for $(TMTSF)_2ClO_4$, other possibilities must be pursued as well, e.g. if the optimum nesting vector varies with temperature and whether the assumption of $k_\perp$-independent gap is appropriate for quantitative analysis. The real Fermi surface is of lower symmetry and is nested less perfectly than is supposed in the model calculation. The mechanism of pinning of the SDW by impurities itself is still open to detailed studies. Other pinning mechanisms, e.g. by electric contacts, should be carefully examined, though no sign of commensurability pinning has been observed at present.

The magnitude of the non-ohmic conductivity and its field dependence should also provide important information. In the CDW, the conductivity has been measured well above E_T and its value in the high field limit has been successfully compared with the theoretical prediction that it is approximately equal to that in the normal state at the same temperature.

In $(TMTSF)_2X$, on the other hand, such a high field does not seem to be easily attained because of a remarkable effect of Joule heating. The excess conductivity near E_T is also open to future work; because of the current inhomogeneity within the sample no plausible conclusion has been obtained at present.

The inhomogeneous current density is a rather serious technical difficulty encountered in most samples. Though the conductivity at room temperature, measured before cooling, is convergent to the previously published value, it does not always mean that the SDW current density is uniform because the latter flow is more restricted along the crystallographic a-axis. If the path of the SDW flow is interrupted by a defect, its coherence is broken there and the SDW is strongly pinned. Another effect is that the field parallel to the SDW current path becomes inhomogeneous. Inhomogeneous current flow can easily be induced by current contacts, too; an effect similar to that produced by micro-cracks inside the sample should be observed if the current is injected only into a part of the sample cross-section. The same problem should have existed in the CDW system as well. But the samples of, e.g. trichalcogenides, such as $NbSe_3$ and TaS_3 are so large in length compared with the transverse dimensions that the homogeneity of current density is improved.

One of the prominent phenomena associated with the sliding is the voltage/current oscillation (or 'narrow band noise') under d.c. current/voltage bias in field higher than E_T. It has been observed in various CDW systems and has provided clear evidence for the CDW sliding. Though the origin is still not well elucidated, an intuitive explanation is given in terms of impurity pinning as follows. Since the charge density of the CDW is periodic in space, its potential energy is a periodic function of its translation or its phase in the field of distributed impurities. Because the CDW is coupled directly to the impurity, the force on it is modulated in time or in every translation by its wavelength when it moves with a constant velocity. As a result, the voltage oscillation superimposed on the d.c. signal is observed as the narrow band noise, with frequency proportional to the extra current by the condensate. In the SDW, as well, a similar phenomenon should be observed because of the periodic nature of its response to impurities. The SDW current density is given by $j_{\mathrm{SDW}} = n_{\mathrm{SDW}}\, ev$, where n_{SDW} and v are the density of condensed electrons to the SDW and its sliding velocity, respectively. Since the noise frequency F is equal to v/λ_{PIN}, where λ_{PIN} is the translational period of the pinning potential by impurities, the following well-known relation is found: $F/j_{\mathrm{SDW}} = 1/n_{\mathrm{SDW}}\, e\lambda_{\mathrm{PIN}}$. As long as the coupling of the SDW to non-magnetic impurities is of second order, it is expected that $\lambda_{\mathrm{PIN}} = (2\pi/q_a)/2$, one half of the SDW wavelength.

The narrow band noise was observed in quenched $(TMTSF)_2ClO_4$ [24] and recently in $(TMTSF)_2PF_6$ under d.c. current bias. Figure 6 shows the

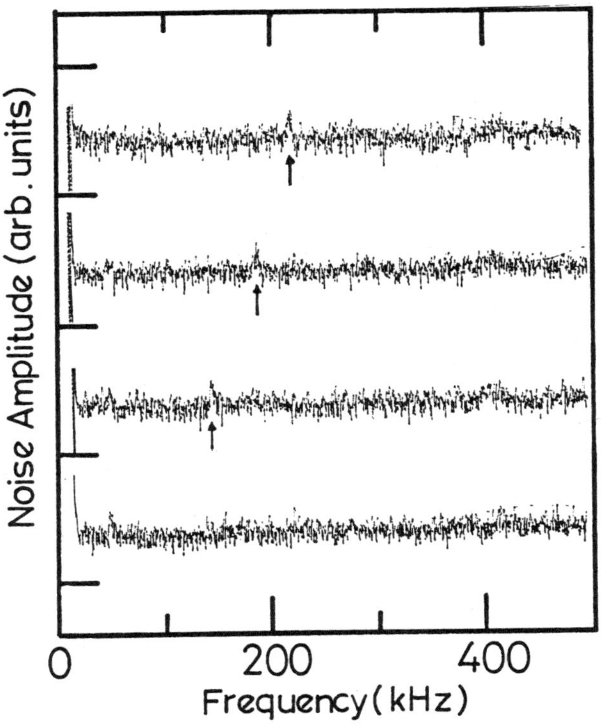

Fig. 6. Voltage noise in $(TMTSF)_2PF_6$ at 4 K. The peak frequency marked by arrows increases with d.c. current bias, while spurious peaks are found at fixed frequencies. Bias current is, from the bottom, 0, 41, 45 and 48 μA, respectively.

Fourier-analyzed voltage spectrum in the latter compound at 4 K. Above a critical current, sharp noise components are observed in the voltage response and their frequencies should increase linearly with the excess current, the latter being calculated using the conductivity below the threshold. For quenched $(TMTSF)_2ClO_4$, the electron density $n_{SDW} = 1.4 \times 10^{21}$ cm^{-1} at $T \ll T_c$ is given by the density of holes on the (TMTSF) columns. From this and the measured value of F/j_{SDW}, λ_{PIN} was given as 1.8 Å [24]. This is 1/8 of the SDW wavelength, or 1/4 of the pinning period λ_{PIN} expected for the second order impurity coupling. In some $(TMTSF)_2PF_6$ samples the relation between its frequency F and the excess current I_{SDW} was not linear; an anomalously small value of λ_{PIN} was given by the measured ratio F/I_{SDW} and the latter decreased with increasing I_{SDW}. Though qualitatively, the above observation provides definite evidence for the SDW sliding.

The smaller value of λ_{PIN} obtained experimentally is presumably due to an inhomogeneous flow of the SDW current; the SDW sliding does not occur through the entire sample and the SDW current density is underestimated. The nonlinear relation between F and I_{SDW} in $(TMTSF)_2PF_6$ samples is attributed to that the area, through which the SDW sliding occurs, increases with increasing I_{SDW} rather continuously. In such samples, the active area is actually less than 1/10 of the sample size; $(TMTSF)_2PF_6$ crystals used so far are less perfect than $(TMSTF)_2ClO_4$, though there is no reason to suppose that this is intrinsic. Because the SDW sliding in $(TMTSF)_2PF_6$ has been established, the nonlinear conductivity in it is not associated with a hot-electron effect.[5]

4. A.C. Conductivity and NMR

Quinlivan *et al.* [25] measured the surface impedance and obtained the a.c. conductivity $\sigma(\omega)$ of $(TMTSF)_2PF_6$ in the SDW state up to $\sim$ 150 MHz. They found a resonance peak of Re $\{\sigma(\omega)\}$ at low frequency of ~ 1 cm^{-1}, which is interpreted as the resonance frequency of the pinned SDW. The observed frequency range is consistent with the magnitude of E_T. They also obtained $\sigma(\omega)$ from Kramers–Kronig analysis by combining their data with optical observations. Another peak was found which corresponds to the single particle excitation across the SDW gap, which agrees with the gap obtained from the d.c. conductivity. They estimated the effective mass and relaxation time of the SDW rigid body. The most striking result is that the effective mass is $\sim 10^2\, m_e$, where m_e is the band mass. This magnitude is unexpectedly large. In the CDW systems the large effective mass, $\sim 10^3\, m_e$, reflects the fact that the lattice distortion is associated with the motion of condensed electron. On the other hand, there is no plausible mechanism of mass enhancement in the SDW which is derived from the electron–electron interaction.

Delrieu and Kinoshita [26] were successfull in observing a narrowing of ^{1}H NMR signals of quenched $(TMTSF)_2ClO_4$ under large d.c. current, which is further strong evidence for the SDW sliding. In the SDW state, the characteristic NMR lineshape is observed in the presence of a distributed internal field. They found that the signal becomes narrower with increasing d.c. current larger than a critical magnitude and that the lineshape is the same as that observed above T_c. The local field oscillates with the frequency v/λ_{SDW} when the SDW slides; narrowing occurs when the frequency is larger than the line width determined by the distributed local field. The remaining problem lies in the fact that the narrowing was observed below 2.5 K but not at 4 K. At higher temperature, the temperature of the sample itself under d.c. current becomes higher due to Joule heating, but in the measurement

[5] More exactly, the hot-electron effect is not the sole mechanism of the nonlinear conduction in $(TMTSF)_2PF_6$.

at 4 K the sample temperature was still lower than T_c because the lineshape was the same as that below T_c.

Acknowledgements

We acknowledge our students K. Ichimura, T. Shimizu, N. Hino, K. Fukuoka and H. Nagasawa for unpublished data presented in Sections 2 and 3, H. Anzai (Himeji Inst. Tech.), G. Saito (Kyoto U.), M. Tokumoto and N. Kinoshita (ETL) for supply of most crystals used in this work and of know-how concerning crystal growth techniques, and K. Hirao (Chemistry Dept., Hokkaido U.) for detailed instruction on organic chemistry.

References

1. A.W. Overhauser: 1962, *Phys. Rev.* **128**, 1437.
2. L. Corliss, J. Hastings and R. Weiss: 1959, *Phys. Rev. Lett.* **3**, 211; S.A. Werner, A. Arott and H. Kendrick: 1967, *Phys. Rev.* **155**, 528.
3. W.M. Lomer: 1962, *Proc. Roy. Soc.* **80**, 489.
4. K. Bechgaard, C.J. Jacobsen, K. Mortensen, H.J. Pedersen and N. Thorop: 1980, *Solid State Commun.* **33**, 1119.
5. (**a**: 1D) *Electronic Properties of Inorganic Quasi-One Dimensional Compounds*, Vols. 1 & 2, P. Monceau (ed.), Reidel, Dordrecht 1985; G. Gruner and A. Zettl: 1985, *Physics Reports* **119**, 117. (**b**: 2D) J.A. Wilson, F.J. DiSalvo and S. Mahajan: 1975, *Advances in Physics* **24**, 117.
6. M.W. Walsh, Jr., F. Wudl, G.A. Thomas, D. Nalewajek, J.J. Heeger, P.A. Lee and T. Poehler: 1980, *Phys. Rev. Lett.* **45**, 1874.
7. P.M. Chaikin, G. Gruner, E.M. Engler and R.L. Greene: 1980, *Phys. Rev. Lett.* **45**, 1874; P.M. Chaikin and M.-Y. Choi: 1982, *Mol. Cryst. Liq. Cryst.* **79**, 79.
8. E.M. Conwell and N.C. Banik: 1982, *Mol. Cryst. Liq. Cryst.* **79**, 95.
9. G. Gruner, A. Zawadowski and P.M. Chaikin: 1981, *Phys. Rev. Lett.* **46**, 511.
10. K. Yamaji, *Superconductivity in Magnetic and Exotic Materials*, T. Matsubara and A. Kotani (eds.), Springer Series in Solid State Sciences, Vol. 52, Springer, Heidelberg, 1984, p. 149.
11. J.P. Pouget, R. Moret, R. Comes, K. Bechgaard, J.M. Fabre and L. Giral: 1982, *Mol. Cryst. Liq. Cryst.* **79**, 129.
12. K. Mortensen, Y. Tomkiewicz, T.D. Schultz and E.M. Engler: 1981, *Phys. Rev. Lett.* **46**, 1234; J.B. Torrance, H.H. Pedersen and K. Bechgaard: 1982, *Phys. Rev. Lett.* **49**, 881.
13. T. Takahashi, Y. Maniwa, H. Kawamura and G. Saito: 1986, *J. Phys. Soc. Jpn.* **55**, 1364.
14. J.M. Delrieu, M. Rogers, Z. Toffano, A. Maradpour and K. Bechgaard: 1986, *J. Physique (Paris)* **47**, 839; J.M. Delrieu, M. Roger, C. Coulon, R. Laversanne and E. Dupart: 1988, *Synthetic Metals* **27**, B35.
15. S. Tomic, J.R. Cooper, D. Jerome and K. Bechgaard: 1989, *Phys. Rev. Lett.* **62**, 462.
16. W. Kang, S. Tomic, J.R. Cooper and D. Jerome: 1990, *Phys. Rev. B* **41**, 4862.
17. T. Sambongi, K. Nomura, T. Shimizu, K. Ichimura, K. Kinoshita, M. Tokumoto and H. Anzai: 1989, *Solid State Commun.* **72**, 817.

18. S. Tomic, D. Jerome, P. Mond and K. Bechgaard: 1983, *J. Physique (Paris)* **44**, C3–1083.
19. S. Takada: 1984, *J. Phys. Soc. Jpn.* **53**, 2193.
20. P.F. Tua and J. Ruvalds: 1985, *Phys. Rev. B* **32**, 4660.
21. I. Tutto and A. Zawadowski: 1988, *Phys. Rev. Lett.* **60**, 1442.
22. K. Maki and A. Virosztek: 1989, *Phys. Rev. B* **39**, 9640.
23. X. Huang and K. Maki: 1990, *Phys. Rev. B* **42**, 6498.
24. K. Nomura, T. Shimizu, K. Ichimura, T. Sambongi, M. Tokumoto, H. Anzai and N. Kinoshita: 1989, *Solid State Commun.* **72**, 1123.
25. D. Quinlivan, Y. Kim, K. Hoczer, G. Gruner and F. Wudl: 1990, *Phys. Rev. Lett.* **65**, 1816.
26. J.M. Delrieu and N. Kinoshita: 1990, *Comptes Rend.* **310**, 891.

DOPING AND DISORDER IN CONDUCTING POLYMERS

Y. WADA

Department of Physics, University of Tokyo, Bunkyo-ku, Tokyo 113, Japan

Abstract. We review experiments which have studied the structure of the doped polyacetylene with the help of X-ray diffractions. Particular emphasis is placed on the relation between polymer chains and dopant columns along the chain direction. It is concluded that the experiments have not ruled out the possibility that electrons in the polymer chains may be in more or less disordered circumstances.

1. Introduction

Linear *trans*-polyacetylene, $(CH)_x$, attracted much interest when it was successfully doped to show a dramatic increase in the electrical conductivity over twelve orders of magnitude from that of an insulator to a metal [1]. Naarmann and Theophilou reported success in preparing a sample with conductivity higher than 10^5 S/cm [2]. Unfortunately, nobody succeeded in reproducing such a metallic sample, following their method. Recently, Tsukamoto, Takahashi, and Kawasaki succeeded in obtaining samples with conductivity higher than 10^5 S/cm by iodine doping, using a modification of the Naarmann–Theophilou method [3]. The electrical resistivity, magnetoresistance, magnetic susceptibility and thermoelectric power have been measured [4]. It is important to refer to the well-known fact that the electrical conductivity rises steeply at a doping level near one percent. This is similar to the usual metal-insulator transition. However, no Pauli paramagnetism is observed at this level. The increase of the magnetism takes place at higher concentrations, near six percent. This was the first indication of solitons at the light doping level. Since the solitons do not satisfy the ordinary charge-spin relationship, a spinless soliton can be a charge carrier. A specific soliton-related transport mechanism was proposed by Kivelson [5], involving phonon-assisted electron hopping between soliton midgap states. This model was successful in giving a good agreement with experimental results on the temperature dependence of the d.c. conductivity as well as the a.c. conductivity [6]. The transition at the light doping level is gradual, without sudden changes. By contrast, the second transition, near six percent, looks like a first-order transition from a nonmagnetic to a fully metallic state [7]. Kivelson and Heeger proposed an interpretation of this transition in terms of a crossover from a lattice of charged solitons to a regular array of polaron-like distortions [8]. There was, however, an infrared absorption experiment which showed that the intensities of the three characteristic peaks, associated with solitons, increase more or less linearly with the impurity concentration up to eighteen percent [9]. This would indicate that the solitons

H. Aoki et al. (eds.), New Horizons in Low-Dimensional Electron Systems, 415–432.

are still playing a role at the high doping levels. Moreover, electron energy loss experiments showed that electronic levels spread well across the energy gap, whereas a polaron lattice would result in two narrow well-separated bands in the gap [10]. Mele and Rice had proposed another mechanism for the metal-insulator transition [11] which takes place in a one-dimensional chain as a result of two effects countering each other. The linear chain tends to be in the insulating state which is stabilized by Peierls dimerization. On the other hand, the random field of the ionized impurities suppresses the bond alternation to lower the energy. At a concentration of ten percent, the randomness takes over to give a non-vanishing density of states at the Fermi level. The bond alternation persists into the metallic states, giving rise to a gapless Peierls distortion. Three dimensional coupling is as important as the one dimensional disorder. It should be pointed out that they determined the form of the bond alternation order parameter intuitively and its amplitude by a variational calculation. Su determined the order parameter in a similar way [12]. Without taking the impurity potential into account, he solved the SSH Hamiltonian to find the merging of the gap states with the conduction band at six percent in the donor doping with the Fermi level at the top of the gap state. He found, however, that the electron wave functions are well localized at the Fermi energy. They extend over the system only when their energy is higher than 0.5Δ, 2Δ being the energy gap, relative to the Fermi energy. It should be mentioned that the discussion was not concerned with the question of whether the lattice configurations obtained are in their lowest energy states.

We have carried out a series of works to investigate the random doping effect [13–21]. Two types of dopants have been studied in a generalized version of the Su–Schrieffer–Heeger model. One is the site-type impurity which locally varies the site energy of π electrons. The other is the bond-type impurity which locally changes the electronic transfer integral. The system is dimerized when there is one electron per site. If the bond between the $2n$th site and the neighboring $(2n+1)$th is short, we have a long bond between the $(2n+1)$th site and the $(2n+2)$th site, and *vice versa*. This pattern extends over the chain. A site-type impurity at an even-numbered site gives a different effect on electrons from a site-type impurity at an odd-numbered site. Similarly, a bond-type impurity along a short bond gives a different effect from an impurity along a long bond. We have been interested in the formation of impurity bands in the Peierls gap, the change in the bond alternation order parameter, and a possible vanishing of the energy gap by the random doping. These properties were first investigated for a system with a uniform order parameter and half-filled electronic states with the help of the coherent-potential approximation (CPA) [13–18].

The bond-type impurity effects have been investigated in [13,14]. In [13], the impurity along a short bond is assumed to give an opposite effect on the electron transfer to the impurity along a long bond. When the former enhances the transfer, the latter deemphasizes it by the same amount, and *vice versa*. The enhancement pattern is in phase with the dimerization. These bond-type impurities are named *cis*-type. It has been found that there is no impurity band in the energy gap. In [14], each impurity has been supposed either to enhance or deemphasize the transfer with an equal probability. Half of the impurities are in phase with the dimerization and the other half are out of phase. Two impurity bands can be formed through the out-of-phase impurities. This type of disorder would take place when a bond is shortened or elongated by a local misalignment with neighboring impurity columns and polymer chains.

The site-type impurity problems have been studied in [15,16]. In [15], we have investigated the impurity distribution in which the impurities are at interstitial sites either always along the long bonds or always along the short bonds. Each impurity modifies the electronic site energy at the two ends of the bond by the same amount. This distribution simplifies the problem since it eliminates backward scattering of electrons. It would represent the case when the impurity releases an electron to another chain or accepts it from there. At low concentrations, an isolated impurity band is formed above the top of the valence band or below the bottom of the conduction band according to the sign of the impurity strength. At high concentrations and large impurity strengths, the order parameter and energy gap vanish. In [16], another impurity distribution is studied. Each impurity is assumed to be located at a lattice site. Half of them are at the even-numbered sites and the other half at the odd-numbered sites. They give rise to backward as well as forward scattering. There are no qualitative changes from the results of [15], which indicates the fact that the qualitative features of the site-type impurities are rather insensitive to the distributions. The situation becomes a little complex when the doping is compensated with the same amount of donors as acceptors being introduced. The former gives rise to attractive potentials to the electrons, while the latter gives repulsive potentials. In [14], they are assumed to occupy the interstitial sites as in [15]. At low doping levels, we have found two impurity bands, one of them being closer to the valence band and the other to the conduction band. This can be understood with the help of the results in [15]. Since there are impurities with different signs, one impurity band is formed by the impurities with one sign and the other band is formed by the other impurities with the other sign.

In general, the impurities would modify the electronic transfer integral and the site energy at the same time. It has been pointed out in [17] that, at low doping levels, the formation of impurity bands is suppressed when the strength of the bond component is larger than that of the site component. As the bond component stabilizes the dimerization, it becomes more difficult to disrupt the energy gap to gain the electronic energy. At a high doping level, however, the energy gap does vanish for strong site components [18].

Several problems remain to be examined. One of them is the question of how reliable the CPA results are, particularly when applied to low-dimensional systems. This has been explored with the help of numerical solutions of the Su–Schrieffer–Heeger model for the polyacetylene [19]. The equation for the electronic wavefunction is solved depending on the order parameter. The order parameter is, in turn, determined with the help of the wavefunctions. The whole problem is thus solved by an iteration. Since the dopants are assumed to be distributed randomly, an average over the distributions has to be performed for physical quantities which are obtained numerically for each distribution. We have selected one hundred samples with one electron per site. In each sample, the positions of the dopants are selected randomly. We have calculated lattice configurations, electron density distributions, and the wavefunctions near the energy gap for each type of doping. The averaging procedures are performed to obtain the order parameter and density of states. We have found remarkable agreements with the CPA results when the impurity concentrations are as low as several percent. There are distinctive features in the density of states of the valence or the conduction band. Wherever the CPA has predicted an isolated impurity band at low doping levels, there is a tail of the density of states in the energy gap. There is no such a tail where the CPA has not predicted it. Since the numerical solutions have been obtained at the concentrations of the order of several percent, the isolated impurity bands have not been observed, which is also consistent with the CPA result. It has thus become clear that the CPA calculations can describe the global features of the density of states even in the low-dimensional systems.

The second remaining problem is the fact that, according to the CPA, the energy gap vanishes at higher doping levels and stronger impurity strengths than the observed levels and strengths, respectively. These conclusions have been obtained since the CPA calculations have assumed a uniform order parameter for the half-filled systems. The metal–insulator transition takes place at a doping level of several percent. The density of the electrons deviates from the half-filled value by the corresponding amount. It should induce solitons and polarons which make the order parameter spatially variant. It would be rather hard for the CPA to study the system with a spatially varying order parameter. However, the numerical method can be applied to the problem. Harigaya and Terai have performed the calculations for the site-type acceptor impurities [20]. They have not assumed any form for the order parameter. This has been self-consistently determined together with the electronic wave functions. The lattice has thus been perfectly relaxed. This is the difference from Mele and Rice's variational relaxation. One hundred samples are selected with various random distributions of dopants. The order parameter of bond alternation is averaged over the samples to show that it remains constant at higher doping levels. The Mele–Rice order parameter indicated the gapless dimerization, but it was a monotonously decreasing function of the dopant concentration. The perfect lattice relaxation has shown that the Peierls distortion persists more strongly. Since the solitons, which are pinned by the impurities, have narrower widths than those of impurity-free solitons, the order parameter is larger in impurity-free regions than that in the impurity-free soliton lattice system. This leads to a stronger Peierls distortion. Each sample has an electronic energy spectrum with a nonvanishing energy gap. The sample average of the energy gap is tentatively defined by the difference between the minimum of the energies of the lowest unoccupied states among the samples and the maximum of those of the highest occupied states. It vanishes at a reasonable doping level when the impurity strength is of the order of the band width. The localization of the wavefunctions is also examined. With the help of the charge conjugation symmetry, which holds in Su's model [12], we can reinterpret his results for the acceptor doping. The Fermi level is at the top of the valence band. The wavefunctions below the Fermi level are localized if the energy difference is less than 0.5Δ from the Fermi level. The model of [20] does not have the symmetry because of the site-type impurity potential and it shows that the wavefunctions in the valence band are far more localized as far as the inverse participation number is concerned, while the wavefunctions extend over the system when their energy is higher than 1.9Δ above the Fermi level.

Harigaya and Terai have used the same method to study the doped system with impurities which have a Coulomb potential [21]. An anisotropic dielectric screening is assumed at long distances from the impurity. They have reproduced every conclusion they obtained with the site type impurity. Only the order parameter varies more gradually. The critical impurity strength, given with the site-type impurities, can be reasonably understood as representing an effect of several sites around the impurity. The results of these works suggest a possibility that the mechanism of the metal–insulator transition in the conducting polymers could be clarified with the help of the disorder effects.

There have been arguments that the transition has to be and has been understood without the disorder effects [22–24]. Such arguments proceed on the basis of the results of structure studies by X-rays, neutron scattering and electron scattering which are supposed to establish the fact that above doping levels of 0.4 percent the impurities are quite well ordered. In [22], Conwell, Mizes, and Jeyadev have studied 8.33 percent Na-doped polyacetylene. They have discussed 16.67 percent K-doped $(CH)_x$ in [23]. According to the X-ray measurements, the dopants form ordered columns in channels formed by polyacetylene chains. In [22], each sodium column is assumed to be surrounded by three polyacetylene chains. In each column, the Na^+ ions are distributed in a regular way. The spacing between the two neighboring ions is $4a$, a being the average distance between CH groups. In each polyacetylene chain, there are solitons whose spacing is $12a$. This configuration corresponds to the doping of $1/12 = 8.33$ percent. The form of the soliton lattice is taken intuitively and not relaxed. Taking account of the finite length of the chains of $\sim 100a$ and the Coulomb potential of the doping ions, they claim that a metallic behavior is obtained. In a sodium-doped system, the potential causes instability in electronic states located at the end regions of the polymer [22]. The doubly occupied end states are pushed close to the conduction band edge. The calculated energy levels thus give a level distance at the Fermi level as small as the level distance in the conduction or valence band. There remains, however, a gap of the order of the gap between the soliton band and the conduction band of the impurity-free soliton lattice system below the Fermi level. It would be questionable if this is energetically favorable, since the conventional Peierls theorem states that the Fermi level should be in a gap. In [23], the lattice structure for stage 1 is used. This is a tetragonal structure in which each potassium-ion column is surrounded by four planar, zig-zag polyacetylene chains. The chains make a checkerboard in a plane perpendicular to the chain direction and an ion column passes through the center of each check. There are two polymer chains per ion column. The spacing between the two neighboring ions is $3a$. In each polyacetylene chain the soliton spacing is $6a$. This gives a doping of $1/6 = 16.7$ percent. Two possibilities are considered for the arrangement of the solitons. One is named the staggered arrangement, in which the solitons in a chain are opposite every other ion. The solitons on the neighboring chains are opposite the remaining ions. The other one is called the symmetric arrangement, in which the polymer chains have the same soliton distribution. The solitons are halfway between ions. The authors claim that the chains are found to be metallic in all cases. In the staggered arrangement, the electronic level distance is fairly uniform at the Fermi level, but there is a small gap below the Fermi energy in the symmetric arrangement, as in [22]. Since the configuration of the lattice in the polymer chain is not relaxed and the total energy is not estimated, it remains to be seen if the lattice configurations are in their lowest energy states. In addition, it is not necessarily clear why the energy gap closes between the soliton band and the conduction band by the effect of the Coulomb potential. Therefore, it is not convincing enough to conclude that the average level spacing of ~ 0.14 eV around E_F is due to the finite-size effect and should vanish in the limit of a large size. Recently, Mizes and Conwell have carefully studied the effects of interchain couplings and couplings between the polymer chains and ion columns in the experimentally determined geometry [24]. The interchain couplings give a dispersion in the direction perpendicular to the one-dimensional bands. If a soliton band maximum achieves such a width with the help of the dispersion that it gets in touch with a broadened conduction band minimum, the gap would be closed. It has turned out, however, that, if the Coulomb potential of the ions were absent, the potassium-doping levels, required to give rise to the metallic phase, would be much greater than the experimental value. The authors claim that this conclusion is consistent with their thesis that the insulator–metal transition is due to Coulomb effects on finite chains.

An argument is presented against this thesis by Stafström [25]. He has essentially reiterated the calculations done in [22] for 8.33 percent sodium doping. Instead of the fixed dimerization order parameter, he has relaxed the lattice completely to find that the destabilization of the highest occupied states, caused by the end effects, is so small that it cannot give rise to the insulator–metal transition. He has then taken into account an intra-chain electron–electron interaction. This reduces the effect of the potentials by the dopants and other chains, as well as the end effects. It also reduces the energy gap, since the soliton width is increased and the interaction between adjacent solitons becomes stronger, which results in an increase in the soliton band width. Relaxation of the lattices on surrounding chains has also been taken into account in a self-consistent way. This makes the single-particle potential even smoother and extends the solitons to reduce the gap. In spite of these effects, the energy gap remains twice as large as the typical energy level spacing above the doping level of ten percent. Finally, he has shown that a disorder in the position of the dopant ions can reduce the gap, which is the average separation between the top of the soliton band and the bottom of the conduction band, to a value close to the level spacing. The maximum displacement of an ion, in the disorder, has been assumed to be a half of a carbon–carbon bond length for twenty different samples.

In the series of our works, we have assumed that the doping gives a random distribution of the dopants. It is the limit of strong disorder. On the other hand, Stafström has assumed a weak disorder. Both works would indicate the possibility that some kind of disorder is playing a role in the insulator–metal transition. It should, therefore, be worth studying how well the dopant arrangement has been established in a column by structure studies using X-ray diffraction and so on.

2. Review of Experiments

The basic structure of a single polymer chain of the undoped *trans*-polyacetylene $(CH)_x$ is a zigzag arrangement of the carbon atoms in a plane [26]. A C_2H_2 repeat distance is 2.48 Å, a carbon–carbon bond angle 125.3°, and a C–H bond distance 1.11 Å. The calculated carbon–carbon bond length is 1.396 Å for a chain without bond alternation. Actually the bond length is alternating. If the bond length between the nth and $(n+1)$th carbon atoms is long, the length between the $(n-1)$th and nth atoms is short, and *vice versa*. This pattern extends over the chain. When we introduce a coordinate for each CH group to describe their translations along the direction of the zigzag chain, as Su, Schrieffer, and Heeger used in their model, the average spacing between the coordinates of successive CH groups is 1.23 Å. The length reduction of the short bond is $2u_0 \cong 0.06$ Å. The long bond lengthens by the same amount. In the three-dimensional undoped *trans*-polyacetylene, the chains are arranged regularly. It has a monoclinic structure with $a = 4.24$ Å, $b = 7.32$ Å, $c = 2.46$ Å, and $\beta = 91.5°$. When projected onto a plane normal to the chain direction, a polymer chain passes through each corner of the unit cell which is almost rectangular with axes a and b, β being the angle between them. The planes of these chains make a setting angle of 55° with the a axis. The other set of chains passes through the centers of the cells with a setting angle 125°. As far as the dimerization pattern is not concerned, the zigzag patterns of chains are commensurate with each other. However, the dimerization is out of phase between the chains at the corners and the chains at the centers.

We confine our review to the donor doping with alkali-metal ions and the acceptor doping by iodine. Five alkali-metals have been used as dopants. They are Li^+, Na^+, K^+, Rb^+, and Cs^+, in order of the ion size.

2.1. Lithium doping

Leitner, Leising, Kahlert, Fink, and Fritsche used the high energy electron diffraction (HEED) method to find a doping induced peak about 0.95 $Å^{-1}$ in (010) equatorial reflection [27]. Although the doping level was not determined, the peak shifted to lower wavenumbers as the doping proceeded. This meant that the spacing between associated crystallographic planes increased. It was the unit cell parameter b. There were two additional prominent peaks in the equatorial scan. The positions of two lithium columns were determined tentatively in the projected plane normal to the chain direction. The arrangement of the polymer chains in the projected plane was assumed to be identical with the undoped polyacetylene, except for the increase in b. A meridian scan showed a (001) diffraction. A peak at ~ 5 $Å^{-1}$ in the undoped sample shifted to lower wavenumbers with the doping. This indicated that the lattice parameter c in the chain direction was elongated. There was another small peak at ~ 2 $Å^{-1}$. This meant that the average spacing between neighboring lithium ions is about 3.3 Å in a column. It is thus incommensurate with the lattice parameter. We note that the peak intensity is small and the peak is in a background with relatively high intensity in comparison with the (001) peak. The shift of the lattice parameter c was measured systematically over doping levels from 2 mole percent to 10 percent by Winokur, Moon, Heeger, Barker, and Bott, using the X-ray meridian scan of a (002) diffraction [28]. They did not observe any clear indication of alkali ordering within the columns.

A trigonal structure for the lithium complex was proposed by Murthy, Shacklette, and Baughman to reproduce their X-ray data [29]. They measured the samples with the doping level from 10 to 17 percent. The equatorial scan found five to six peaks. These profiles were reproduced by the trigonal structure with a distance between the neighboring columns $a' = 7.16$ Å. The unit cell dimension was $a = \sqrt{3}\ a' = 12.4$ Å. They took three C_2H_2 units along the chain axis in the unit cell. The chain axis dimension is $c = 3 \times 2.46$ Å $= 7.5$ Å. The trigonal structure can be derived from the lattice of the undoped polymer by a rotation of the polymer chains about their chain axis. Some of the chain layers remain essentially unchanged during the doping and only the chains between these invariant layers rotate by $\sim 60°$. We thus obtain an aggregate of equilateral triangles. An ion column passes through the center of the triangle. We denote stoichiometric phases by $(C_nH_n)_mLi$ where m means the number of polymer chains per ion column and n denotes the number of CH units along a given chain, which corresponds to the average spacing of lithium ions within the columns. If we denote the stoichiometry by $CHLi_y$, we have $y = (nm)^{-1}$. The stoichiometry is thus $(C_3H_3)_3Li$ which is $CHLi_{0.111}$. The trigonal structure was claimed to have a lower energy than the structure discussed in [27]. It should be pointed out, however, that the results of meridian scans were not reported in detail. The proposed commensurate structure between the polymer chains and ion columns was apparently a surmise. There was a statement in [29]; "it is not necessary for the lithium columns to be commensurate with the polyacetylene chains. In fact, we have shown in iodine-doped polyacetylene that the positions of the iodine atoms in neighboring columns are uncorrelated." Actually, the meridian scan in HEED showed that the chain structure is incommensurate with the column structure [27].

2.2. Sodium doping

Baughman, Murthy, Miller, Shacklette, and Metzger reported the results of early X-ray experiments which could determine only the structures on the projected plane perpendicular to the chain axis [30]. A tetragonal structure was assumed. Each sodium column was supposed to be surrounded by four polyacetylene chains. Projection of each chain makes one side of a square. The ion column passes through the center of the square. The length of the side was reported to be 5.98 Å for sodium. The lattice parameter along the chain direction was estimated with the help of the assumption of close packing of van der Waals atomic and ionic spheres.

Shacklette and Toth studied the doping processes by electrochemical methods [31]. The electrochemical doping was accomplished by forming a contact between polyacetylene films and an open mesh of either platinum or nickel, immersing the polymer in an electrolyte containing sodium, and applying an appropriate potential. The potential was changed stepwise. After each step, the current was relaxed, and then the circuit was opened and the potential of the polymer was recorded. The composition was determined from the number of Coulombs passed during the ion insertion. The open-circuit potential is related to the chemical potential of the polymer complex with respect to the reference electrode. As the doping level changes, the potential changes abruptly when a composition of an ordered phase has been developed. Additional dopants would change the chemical potential. At the doping level in a region of first-order transition, where two ordered phases exist, the potential does not change with the doping level. The electrochemical method showed that there are ordered phases given by $(C_4H_4)_3Na$, $(C_3H_3)_3Na$, and $(C_3H_3)_2Na$ in the ion insertion process and $(C_5H_5)_3Na$ in the extraction process. These values correspond to $y = 0.083$, 0.111, 0.167, and 0.067. The data supported the existence of a structure having a stoichiometry of three chains per ion column, in contrast to the conclusion in [30]. Changes in the number of chains per column give rise to the first order transitions. It would thus seem that the polyacetylene chains are commensurate with the ion columns. But it was pointed out in [31] that evidence was found for noninteger values of n being 3.33 and 3.67. A possibility was suggested that each column possesses a regular array of different commensurate spacings, which, however, has not been experimentally substantiated.

Another X-ray result was reported be Winokur, Moon, Heeger, Barker, Bott, and Shirakawa [32]. They performed the experiments on the sample *in situ* during electrochemical doping to find a sequence of reversible transitions as the doping level increases. At low levels, the sodium columns form a strongly modulated incommensurate lattice in the plane normal to the polymer chains. Below six percent, this incommensurate lattice coexists as a two-phase mixture with lightly (random) doped polyacetylene. At intermediate levels of about ten percent, the sodium lattice becomes commensurate, forming a $\sqrt{3}\times\sqrt{3}$ superlattice relative to the average triangular spacing of the rearranged $(CH)_x$ chains. At higher levels, the sodium superlattice transforms into a dense disordered state characterized as a highly modulated fluid. Mainly the equatorial ($hk0$) scans were carried out. At the intermediate levels, five peaks were observed. Three of them were attributed to (100), (110), and (200) reflections of a hexagonal polyacetylene structure. Another peak was due to a 'sodium' lattice (100) reflection with polyacetylene contributions. The last one was principally a modulation satellite resulting from interactions between the sodium and polyacetylene lattices. The $\sqrt{3}\times\sqrt{3}$ superlattice has the structure which Murthy *et al.* later applied to the lithium doping in [29]. At higher doping levels, the sodium (100) peak broadens rapidly and the last peak disappears. It led to the conclusion that the superlattice becomes a highly modulated fluid while the polyacetylene lattice remains intact. Only a preliminary result was reported for the meridian scans. In addition to the (002) of the polyacetylene chains, a broad peak was observed at 1.27 $Å^{-1}$. It was consistent with the formation of columns of sodium ions in which neighboring ions are separated by 4.9 Å with a coherence length of ~ 25 Å. This would suggest that there are some disorders in the ion arrangement in the column. The report in [28] did not observe any alkali ordering within the columns for sodium, either.

2.3. Potassium doping

In [30], the early X-ray data were analyzed, using the tetragonal structure applied to the sodium doping. The four polyacetylene chains, which surround a potassium column, are projected on a plane normal to the chain direction to make a square. The side has a length of 5.98 Å, the same value with the sodium doping.

In [31], the electrochemical method showed that there are ordered phases given by $(C_4H_4)_4K$, $(C_4H_4)_3K$, $(C_4H_4)_2K$, and $(C_3H_3)_2K$. During the ion-insertion process, all four phases were recognized. On the ion-extraction process, the phase $(C_4H_4)_3K$ was not observed. It should be noted that the phase $(C_4H_4)_4K$ is composed of the configuration with four polymer chains per ion column, as discussed in [30].

Another X-ray diffraction work was reported by Murthy, Shacklette, and Baughman [33]. The X-ray scans were reported for the samples with the dopant concentrations 6.2 percent, 11 percent, 12.5 percent, and 15 percent. It was shown that the scans for the 6.2 percent sample could be well reproduced by a mixture of the 'stage-2' complex and undoped *trans*-polyacetylene. The stage-2 complex has the symmetry in chain-axis projection of C_{mm} and contains four polyacetylene chains per potassium column. The unit cell in the plane normal to the chain direction is a rectangle formed by five potassium columns. Four of them pass through the corners and the remaining one through the center. The lengths of the sides are $a' = 8.0$ Å and $b' = 16.0$ Å. The setting angle is 45°. It is the angle between the plane of the polymer backbone and a plane normal to the a axis. The 'stage-2' terminology was introduced since there are two layers made of polyacetylene chains, which are parallel to the a axis, between the nearest two layers made of potassium columns. It should be noted that the scanning profile was presented after a subtraction of an additional background due to a diffuse forward scattering. This diffuse scattering was observed only in lightly doped samples and was presumably due to vacancies and disorders in the ion columns.

At the doping levels between 6.2 percent and 12.5 percent, a two phase structure of stage-1 and -2 structures was suggested to be most stable. At the level of 12.5 percent, a single phase is obtained with the stage-1 structure.

This is a tetragonal structure in which each potassium column is surrounded by four polyacetylene chains. Each polyacetylene chain is in contact with two ion columns. The spacing between the nearest two ion columns is 6 Å. This was named a stage-1 complex since there is one layer made of polyacetylene chains between the nearest two layers made of potassium columns. For a heavily doped and highly oriented film ($CHK_{0.125}$), the results of meridional scan were supplemented with those of an equatorial scan. All of the reflections in the equatorial scan were indexed on the basis of the stage-1 structure. In the meridional data, there were two peaks. One is due to the (001) reflection which corresponds to the diffraction spacing of 2.492 Å. The other is due to the (002) reflection at 1.246 Å [34]. The 2.492 Å spacing corresponds to the length of a C_2H_2 unit in the chain-axis direction. The data indicated the absence of a symmetry element which includes a translation operation of half the length of the C_2H_2 unit. It suggested the persistence of the chain dimerization. It should be noted that the meridional data did not reveal any peaks associated with the ion ordering in the chains and the background level was significantly higher than in the equatorial data. Although it is possible that such peaks would mainly be in the region of scattering so far forward that the observation was not carried out, the meridional data would be suggesting a fact that the disorder could be still working in the ion columns at this high doping level. In conjunction with the electrochemical data, the stage-2 complex was represented by $(C_nH_n)_4K$ with $n = 3$ and 4. The stage-1 complex was given by $(C_nH_n)_2K$ with $n = 3$ and 4.

Recently, Coustel, Bernier, and Fischer carried out a detailed study, using electrochemical methods together with conductivity measurements [35]. The calibration of the doping level might involve a 20 percent margin of error. As in [31], the open-circuit potential was measured as a function of the doping level y with which the doped polyacetylene was denoted by $(CHK_y)_x$. It exhibited the typical sequence of sloping regions and quasi-plateaus. The sloping regions signify the isostructural addition or subtraction of ionic charge, while the quasi-plateaus represent mixed-phase regions between neighboring ordered structures. The detailed measurements found four quasi-plateaus at $0.02 < y < 0.05$, $0.065 < y < 0.102$, $0.121 < y < 0.128$, and $0.145 < y < 0.152$ with open-circuit potentials of 1.4 V, 1.22 V, 1.05 V, and 0.62 V, respectively. The first two had been observed in [31]. With a help of the conductivity and ESR measurements, which gave strong evidence for a new effect near $y = 0.03$, it was concluded that there are at least six unique phases. One is the undoped phase. At $y = 0.03$, there would be a 'dilute stage-2' structure with $(C_8H_8)_4K$, which might be a transition state since it shows up only upon dedoping. A 'dense stage-2' structure exists over $0.05 < y < 0.065$ with $(C_4H_4)_4K$. There are three stage-1 phases. One is over $0.106 < y < 0.121$ with $(C_5H_5)_2K$. The second phase is over $0.128 < y < 0.148$ with $(C_4H_4)_2K$. The last one is over

$0.152 < y < 0.167$ with $(C_3H_3)_2K$. The last two stage-1 phases exhibit larger slopes than the first stage-1 phase with a lower doping level. This is not consistent with a prediction by a model which took the isostructural addition or subtraction of charge into account with a help of an effective capacitance [36]. Three possibilities were suggested to understand this problem. The first one is that the energy associated with commensurate lock along the chain direction is large enough to stabilize distinct 3D structures with identical 2D channel lattices. A second possibility is that the disposition of the (CH) chains about the channels is density dependent for a given two-dimensional superlattice. In fact, there are X-ray data for $y = 0.12$ which are not consistent with the stage-1 structures [37]. A third possibility is that one or more of the stage-1 phases is disordered along the chain axis, which might explain the decrease in the conductivity in the range $0.12 < y < 0.17$. These arguments have shown that, while there is no evidence for (a, b) plane disorder or incommensurability at any doping levels, the ion distribution in the ion columns might still have some disorder at various doping levels.

2.4. Rubidium doping

The early X-ray data were analyzed in [30], using the tetragonal structure. The side of the square on the plane normal to the chain direction was 6.19 Å.

A recent X-ray work was reported for $y \sim 0.06$ in [33]. It found six crystalline reflections which could not be indexed with the help of the tetragonal structure. The data were interpreted on a basis of a monoclinic version of the stage-2 structure for the potassium-doped polyacetylene. It has the symmetry in chain-axis projection of $P2$ and contains four polyacetylene chains per rubidium column. The spacing between the nearest two ion columns along the b direction was 9.6 Å. An ion column in the neighboring layer makes an isosceles triangle together with the above two columns on the plane normal to the chain direction. The length of the equilateral sides is 8.6 Å. No information was available for the ion distribution in each column.

2.5. Cesium doping

The analyses in [30] assigned a length of 6.43 Å to the side of the square on the plane normal to the chain, in the tetragonal structure.

A recent X-ray work was reported for $y = 0.15$ in [38]. The data were analyzed on the basis of three dimensionally ordered channels. Fifteen peaks representing 27 reflections were described by a tetragonal pseudocell with $\bar{a} = \bar{b} = 9.093$ Å, $\bar{c} = 7.950$Å containing 24(CH) and 4Cs units. The cesium sublattice has a body-centered tetragonal structure. The intrachannel Cs–Cs separation is $\bar{c}/2 = 3.975$ Å and adjacent Cs channels are staggered by $\bar{c}/4$. The cesium columns are thus incommensurate with the polyacetylene chains, if the c-axis projected C–C distance remains at its undoped *trans*-polyacetylene value of 1.23 Å. In two-dimensional systems, an incommensurability this large usually produces 'modulation peaks' whose wavevectors reflect the competing periodicities, etc., if there is coupling between the two sublattices. Since such peaks were not observed, it was concluded that the c-axis correlations within a column are entirely controlled by Cs^+–Cs^+ interactions. As the cesium columns were strongly correlated in the (a, b) plane, the composite structure was described as two interpenetrating three-dimensional lattices which 'float' with respect to each other along the c-axis. Since the closest commensurate phase would have three (CH) units per Cs along the c-axis, the data were finally analyzed on a body-centered tetragonal pseudocell in which each cesium channel is surrounded by four (CH) chains. This produces a commensurate structure with four Cs and 24 (CH) per cell, since the cell must contain an even number of (CH) units. The data were fairly well reproduced by this structure for $(hk0)$ reflections. The observed intensities of $(h02)$ were reduced to a half or a quarter of the calculated values. These would suggest that the Debye–Waller factor for c-axis motion is significantly larger than the (a, b) plane component. The amplitude of ion thermal motion within a column is larger than the transverse mode.

These discussions are suggesting a fact that a polyacetylene chain in a doped sample would be better described in circumstances which are somewhat disordered than in regular distributions of static dopants.

2.6. Iodine doping

Iodine has been used as a typical acceptor for the doping of *trans*-polyacetylene. The early X-ray data were analyzed in [30]. The detailed discussions were given in [39]. Experiments were performed for partially oriented films with several different sets of doping conditions. Doping levels were, however, not precisely determined. Equatorial lines were commonly observed at spacings of 3.6 – 4.0 Å and 7.7 – 8.2 Å. A third equatorial line was observed at 13.8 – 14.3 Å for samples that had been annealed at room temperature. A meridional line was commonly observed at 3.1 Å. This reflection was not symmetrical about the maximum, but rose sharply in intensity on the low angle side of the maximum and tailed off gradually towards higher angles. Such an asymmetry is consistent with the existence of numerous faults corresponding to displacements with vector component in the chain-axis direction. The line showed that an average iodine–iodine distance is 3.1 Å along the meridional direction, if the faults are randomly distributed. A second meridional line was observed at 4.65 – 4.7 Å for samples doped at 0.30 Torr iodine. Using the C_2H_2 length of 2.48 Å in the undoped *trans*-polyacetylene, the number of iodines per C_2H_2 unit was $2.48/3.1 = 0.8$. It is important to realize that the dopants can have two species I_3^- and I_5^-, as demonstrated by Raman, ultraviolet, and other experiments. The equatorial line with the spacing 7.7 – 8.2 Å had been attributed to a first-stage complex of the doped polyacetylene which has a layer structure and the layers containing iodine columns are separated by a layer containing only polyacetylene chains. The line with the spacing 3.6 – 4.0 Å were also due to the first-stage complex and undoped polymers. The 13.8 – 14.3 Å line indicated a separation of iodine-rich layers by three close-packed planes of undoped polyacetylene chains. The structure was named the fully intercalated third-stage compound, when the dopant-rich layers are occupied by only iodine columns and the cell has the same parameter as the undoped polyacetylene parallel to the intercalation plane and normal to the chain axis. The composition was then given by $(CHI_{0.13})_x$. Likewise, a composition $(CHI_{0.056})_x$ was obtained for the third-stage complex with an alternate occupation of column sites in the dopant plane by iodine columns and polymer chains. This was named the partially intercalated third-stage compound. X-ray diffraction intensities were calculated on the basis of these structures with the I_3^- species. Comparison with observed intensities was difficult due to sample-to-sample variation in line intensities and spacings.

Since Riekel, Hasslin, Menke, and Roth were not able to observe the 14 Å spacing in their neutron-diffraction experiments [40], Murthy, Miller, and Baughman repeated the experiments, using aqueous KI/I_2 doping, to confirm their earlier data and propose new structures [41]. Highly oriented films were doped. Meridional scans found at least nine reflections in addition to the earlier 4.7 and 3.1 Å spacings. These reflections were superposed on broad amorphous scatterings, centered at 3.0 and 1.5 Å, due to disordered iodine atoms. It should again be noted that scanning profiles were obtained after a subtraction of the additional large diffuse background to obtain a flat base line for these iodine-doped samples. This additional background is of the form which would arise from randomly distributed scatterers of size $\sim$ 3 Å. This was tentatively attributed to vacancies in iodine columns, as it was in the weak potassium doping [33]. Equatorial scans found the earlier 15, 7.7, and 3.8 Å spacings. The 7.7 Å spacing could be obtained from the first-stage complex. There are two possibilities. One is the fully intercalated complex in which the dopant layer consists only of iodine atoms. It gives a composition of $(CHI_{0.39})_x$. The other is the partially intercalated first-stage complex with $(CHI_{0.13})_x$. Equatorial scans for the sample with $y = 0.14$ were well reproduced by this complex. In [39], the 15 Å spacing had been interpreted by a third-stage complex. But the calculated intensity of the 15 Å reflection turned out to be much higher than the observed value. To overcome this difficulty, a new model was proposed for the structures at intermediate dopant levels. Layers completely filled with iodine columns were denoted as F layers. The layers partially filled with iodine columns were denoted as P layers. The layers of polyacetylene chains without dopants were U layers. The previously proposed fully intercalated third-stage complex would be written $(UUUP)_n$. The new stacking sequence was $(UPUF)_n$ which has the stoichiometry of $(CHI_{0.24})_n$. Equatorial diffraction intensities for the sample with $y = 0.28$ were well reproduced by this model. It was indicated that the neutron experiment [40] could not find the 15 Å spacing, since the $(UPUF)_n$ phase could be obtained only for $y \sim 0.24$. As discussed in [39], the 3.1 Å reflection along the meridian corresponded to an average iodine–iodine distance in the iodine columns. The asymmetric shape of the reflection suggested that the positions of the iodine atoms in neighboring columns are uncorrelated. In order to explain the 4.7 Å spacing and other nine reflections in the meridional scans, a model was proposed. The iodine column was an array $(-I_3^- -I_5^- -I_3^-)_n$. Using published distances between iodines in I_3^- and I_5^- species, the distance between the I_3^- and I_5^- species and that between two $(I_3^- -I_5^- -I_3^-)$ species were adjusted so as to reproduce the observed data. The agreement between the observed and calculated intensities was remarkably good.

3. Conclusions

We have reviewed the experimental works which have studied the structures of doped *trans*-polyacetylene. Special emphasis has been placed on the question of how well the dopant arrangement has been established in the dopant columns. It is remarkable to find that there is no evidence for (a, b) plane disorder or incommensurability at any y for all the dopants. Information on the ion arrangement in the columns is far less clear, except for the iodine doping with $y \sim 0.1$. Even in the iodine doping, the positions of the iodine atoms in neighboring iodine columns would be uncorrelated. Commensurability has often been assumed between the ion–ion distance in the column and the $(CH)_2$ repeat dimension in the polymer chain. But this has not been substantiated. No information has been available on the correlation between the ion arrangement and possible soliton lattices on the polymer chains. It remains to be seen if future experiments will be performed to give reliable results of *in situ* $(00l)$ meridional scans with highly oriented samples which would conclude that electrons in the polymer chains are not in disordered circumstances. The present experimental status would rather indicate an opposite conclusion that their circumstances would be better approximated by more or less disordered states.

There is certainly no evidence about a possible connection of disorder and the insulator-metal transition. But, as discussed in Section 1, it is rather hard to diminish the Peierls gap to zero without the disorder effect. Since there are theoretical indications that the effect might be working in the transition, it should be carefully taken into account as one of possible mechanisms of the transition insofar as there is no experimental evidence that the transition is not triggered by disorder.

Acknowledgements

The author would like to acknowledge the steady friendship with Professor Hiroshi Kamimura which has lasted for four decades and will last in the future. He thanks K. Harigaya and A. Terai for useful discussions.

References

1. H. Shirakawa, E.J. Louis, A.G. MacDiarmid, C.K. Chiang and A.J. Heeger: 1977, *Chem. Commun.*, 578; C.K. Chiang, C.R. Fincher, Jr., Y.W. Park, A.J. Heeger, H. Shirakawa, E.J. Louis, S.C. Gau, and A.G. MacDiarmid: 1977, *Phys. Rev. Lett.* **39**, 1098.
2. N. Theophilou and H. Naarmann, *Conducting Polymers*, L. Alcacer (ed.), D. Reidel (1987); H. Naarmann and N. Theophilou: 1987, *Synth. Metals* **22**, 1.
3. J. Tsukamoto, A. Takahashi and K. Kawasaki: 1990, *Japan. J. Appl. Phys.* **29**, 125.
4. Y. Nogami, H. Kaneko, H. Ito, T. Ishiguro, T. Sasaki, N. Toyota, A. Takahashi and J. Tsukamoto: 1991, *Phys. Rev. B* **43**, 11829; Y. Nogami, H. Kaneko, T. Ishiguro, A. Takahashi, J. Tsukamoto and N. Hosoito, preprint.
5. S. Kivelson: 1981, *Phys. Rev. Lett.* **46**, 1344.
6. A.J. Epstein, H. Rommelmann, M. Abkowitz and H. W. Gibson: 1981, *Phys. Rev. Lett.* **47**, 1549.
7. J. Chen, T.-C. Chung, F. Moraes and A. J. Heeger: 1985, *Solid State Commun.* **53**, 757.
8. S. Kivelson and A.J. Heeger: 1985, *Phys. Rev. Lett.* **55**, 308.
9. X.Q. Yang, D.B. Tanner, M.J. Rice, H.W. Gibson, A. Feldblum and A.J. Epstein: 1987, *Solid State Commun.* **61**, 335; see also Y.H. Kim and A.J. Heeger: 1989, *Phys. Rev. B* **40**, 8393.
10. J. Fink, N. Nucker, B. Scheerer, A. vomFelde, and G. Leising, *Electronic Properties of Conjugated Polymers*, H. Kuzmany, M. Mehring and S. Roth (eds.), Springer-Verlag, 1987.
11. E.J. Mele and M.J. Rice: 1981, *Phys. Rev. B* **23**, 5397.
12. W.P. Su: 1983, *Solid State Commun.* **47**, 947.
13. K. Harigaya, Y. Wada and K. Fesser: 1990, *Phys. Rev. B* **42**, 1268.
14. K. Harigaya: 1990, *J. Phys. Soc. Japan* **59**, 1348.
15. K. Harigaya, Y. Wada and K. Fesser: 1990, *Phys. Rev. B* **42**, 1276.
16. K. Harigaya, Y. Wada and K. Fesser: 1990, *Phys. Rev. B* **42**, 11303.
17. K. Harigaya, Y. Wada and K. Fesser: 1989, *Phys. Rev. Lett.* **63**, 2401; 1990, *ibid.* **65**, 806(R).
18. K. Harigaya, Y. Wada and K. Fesser, *Strongly Coupled Plasma Physics*, S. Ichimaru (ed.), Elsevier, 1990, p. 255.
19. K. Harigaya, A. Terai, Y. Wada and K. Fesser: 1991, *Phys. Rev. B* **43**, 4141.
20. K. Harigaya and A. Terai: 1991, *Solid State Commun.* **78**, 335.
21. K. Harigaya and A.Terai, to be published in *Phys. Rev. B*.
22. E.M. Conwell, H.A. Mizes and S. Jeyadev: 1989, *Phys. Rev. B* **40**, 1630.
23. E.M. Conwell, H.A. Mizes and S. Jeyadev: 1990, *Phys. Rev. B* **41**, 5067.

24. H.A. Mizes and E.M. Conwell: 1991, *Phys. Rev. B* **43**, 9053.
25. S. Stafström: 1991, *Phys. Rev. B* **43**, 9158.
26. A. Karpfen and J. Petkov: 1979, *Theor. Chim. Acta* (Berlin) **53**, 65; C.R. Fincher, Jr., C.-E. Chen, A.J. Heeger, A.G. MacDiarmid and J.B. Hastings: 1982, *Phys. Rev. Lett.* **48**, 100.
27. O. Leitner, G. Leising, H. Kahlert, J. Fink and H. Fritzsche, *Electronic Properties of Conjugated Polymers III*, H. Kuzmany, M. Mehring and S. Roth (eds.), Springer-Verlag, 1989. HEED results were given for potassium and cesium doped samples by O. Leitner, H. Kahlert, G. Leising, J. Fink and H. Fritzsche: 1989, *Synth. Met.* **28**, D225.
28. M.J. Winokur, Y.B. Moon, A.J. Heeger, J. Barker and D.C. Bott: 1988, *Solid State Commun.* **68**, 1055.
29. N.S. Murthy, L.W. Shacklette and R.H. Baughman: 1989, *Phys. Rev. B* **40**, 12550.
30. R.H. Baughman, N.S. Murthy, G.G. Miller, L.W. Shacklette and R.M. Metzger: 1983, *J. de Phys.* **44**, C3-53.
31. L.W. Shacklette and J.E. Toth: 1985, *Phys. Rev. B* **32**, 5892.
32. M. Winokur, Y.B. Moon, A.J. Heeger, J. Barker, D.C. Bott and H. Shirakawa: 1987, *Phys. Rev. Lett.* **58** 2329; 1987, *ibid.* **59**, 608.
33. N.S. Murthy, L.W. Shacklette and R.H. Baughman: 1990, *Phys. Rev. B* **41**, 3708. A different stage-2 structure had been proposed by D. Billaud, J. Ghanbaja and C. Goulon: 1987, *Synth. Met.* **17**, 497. See also D. Djurado, J.E. Fischer, P.A. Heiney, J. Ma, N. Coustel and P. Bernier: 1989, *Synth. Met.* **34**, 683.
34. N.S. Murthy, L.W. Shacklette and R.H. Baughman: 1987, *J. Chem. Phys.* **87**, 2346.
35. N. Coustel, P. Bernier and J.E. Fischer: 1991, *Phys. Rev. B* **43**, 3147.
36. A. Metrot: 1983, *Synth. Met.* **7**, 177; A. Metrot and M. Tihli: 1985, *ibid.* **12**, 517; 1988, **23**, 19.
37. F. Saldi, J. Ghanbaja, D. Begin, M. Lelaurain and D. Billaud: 1989, *C.R. Acad. Sci.* **309**, 671; J.E. Fischer, P.A. Heiney, J. Ma, *Synth. Met.*, to be published.
38. J. Ma, D. Djurado, J.E. Fischer, N. Coustel and P. Bernier: 1990, *Phys. Rev. B* **41**, 2971.
39. R.H. Baughman, N.S. Murthy, G.G. Miller and L.W. Schacklette: 1983, *J. Chem. Phys.* **79**, 1065.
40. C. Riekel, H.W. Hasslin, K. Menke and S. Roth: 1984/85, *Synth. Met.* **10**. 31.
41. N.S. Murthy, G.G. Miller and R.H. Baughman: 1988, *J. Chem. Phys.* **89**, 2523.

ELECTRONIC STRUCTURES OF ICOSAHEDRAL CLUSTERS AND CRYSTALS

TETSUO HATAKEYAMA
Department of Physics, Chiba University, Yayoi, Chiba 260, Japan

Abstract. The electronic structures of icosahedral boron clusters and icosahedral aluminum clusters are investigated using the Hartree–Fock method to clarify the origin of the icosahedral crystal structure of boron and AlMn quasicrystals. The calculated results for icosahedral boron clusters show that 'multi-center bonds' are formed in the icosahedral boron cluster if we assume that the atomic distance between boron atoms is the same that in the crystal structure of boron. Further, from the calculated results for icosahedral aluminum clusters, we can conclude that the abnormally short atomic distance between aluminum atoms of icosahedral units in AlMn quasicrystals and related alloys is due to the formation of multi-center bonds.

We also calculated the electronic structures of α-rhombohedral boron, which has an icosahedron as a unit cell, to investigate the effect of the surrounding atoms on electronic structures of the icosahedral cluster. The calculated results show that two types of inter-icosahedral bonds and the multi-center intra-icosahedral bond appear in the crystalline α-rhombohedral boron.

1. Introduction

In this article we present the results on the electronic structures of boron and aluminum icosahedra and icosahedral crystals calculated from the first principles and show how an icosahedral structure is stabilized.

The first aim in this article is to clarify the bonding mechanism of elemental boron. Boron is a semiconductor that is made up purely of a Group III element in the periodic table. Its crystal structure consists of a three-dimensional skeleton of boron atoms built from icosahedral units [1]. In an icosahedron which is contructed by twelve boron atoms, each atom has five nearest neighbors. In this respect the bonding character is different from the ordinary covalent bond. On the other hand, the packing ratio of an icosahedral structure without a center atom is low, so that this structure is not advantageous for forming ordinary metallic bonds. The qualitative picture for bonding of icosahedral boron has already been discussed by Longuet-Higgings *et al.* in 1955 [2]. They have shown that the thirteen molecular orbitals contribute to the formation of intra-icosahedral bonding by the empirical tight-binding method. In this article we examine the bonding mechanism in an icosahedron from the first principles.

The second aim is to calculate the electronic structure of an icoshedral aluminum cluster in order to elucidate the origin of making aluminum icoshadra which is found in the AlMn alloy. In 1984 Schechtman *et al.* found that quenched AlMn Alloys show a new kind of ordered structures called

H. Aoki et al. (eds.), New Horizons in Low-Dimensional Electron Systems, 433–444.

'quasicrystals' [3]. Much attention has been focused on this new material. We note that the structure of the quasicrystal may be considered as the natural structural analogues of elemental boron which consist of the periodic configuration of icosahedral boron clusters. In fact, some quasicrystals composed of aluminum and transition metals are known as hierarchical packing of icosahedral units, called Mackay icosahedra. The inner shell of a Mackay icosahedron is a small icosahedron which is constructed from twelve Al atoms. Since the number of valence electrons in boron and aluminum atoms is three, one can expect something in common as regards the bonding mechanism of icosahedral structure. Thus we think that the difference between electronic structures of boron and aluminum icosahedra are the key for the mechanism of formation of quasicrystals.

Further, to investigate the effect of surrounding atoms on the electronic structures of the icosahedral cluster, we have calculated the electronic structures of α-rhombohedral boron and aluminum with the α-rhombohedral structure according to the local density functional formalism. As far as we know this is the first report of the electronic structure calculation of α-rhombohedral boron from first principles.

2. Electronic Structures of Icosahedral Boron Clusters

In Figure 1 we show the energy levels of an icosahedral boron cluster calculated by the Hartree–Fock method [4] as a function of the B–B bond length. The obtained energy levels are labeled by the irreducible representations of icosahedral group. The electrons occupy the energy levels up to the $2T_{1u}$ state completely and the $1T_{2u}$ state partially when the B–B bond length is 1.73 Å. For other cases the electrons occupy the energy levels up to the $1T_{2u}$ state completely and the $2T_{1u}$ state partially. From this figure we can conclude that the $1T_{2u}$ state has an antibonding character because the $1T_{2u}$ levels increase as the B–B bond length decreases. Further, to investigate the effect of the $1T_{2u}$ state on the stability of the icosahedron, we have calculated the total energy of a B_{12}^{2+} cluster for the following two configurations:

(i) The $1T_{2u}$ level is occupied and the $2T_{1u}$ level is unoccupied, and

(ii) the $1T_{2u}$ level is unoccupied and the $2T_{1u}$ level is occupied.

We find that when the cluster takes on electron configuration (ii), the peaks in the charge density appear on the faces of the icosahedron. The bond with this feature has a pair of electrons shared by more than three atoms, and is called a 'multi-center bond'. It may be considered as covalent. When electron configuration (i) is taken, the peaks in the charge density on the faces of the icosahedron are smoothed out. This is because the wavefunction of the $1T_{2u}$ level has large amplitudes, mainly on the vertices of the

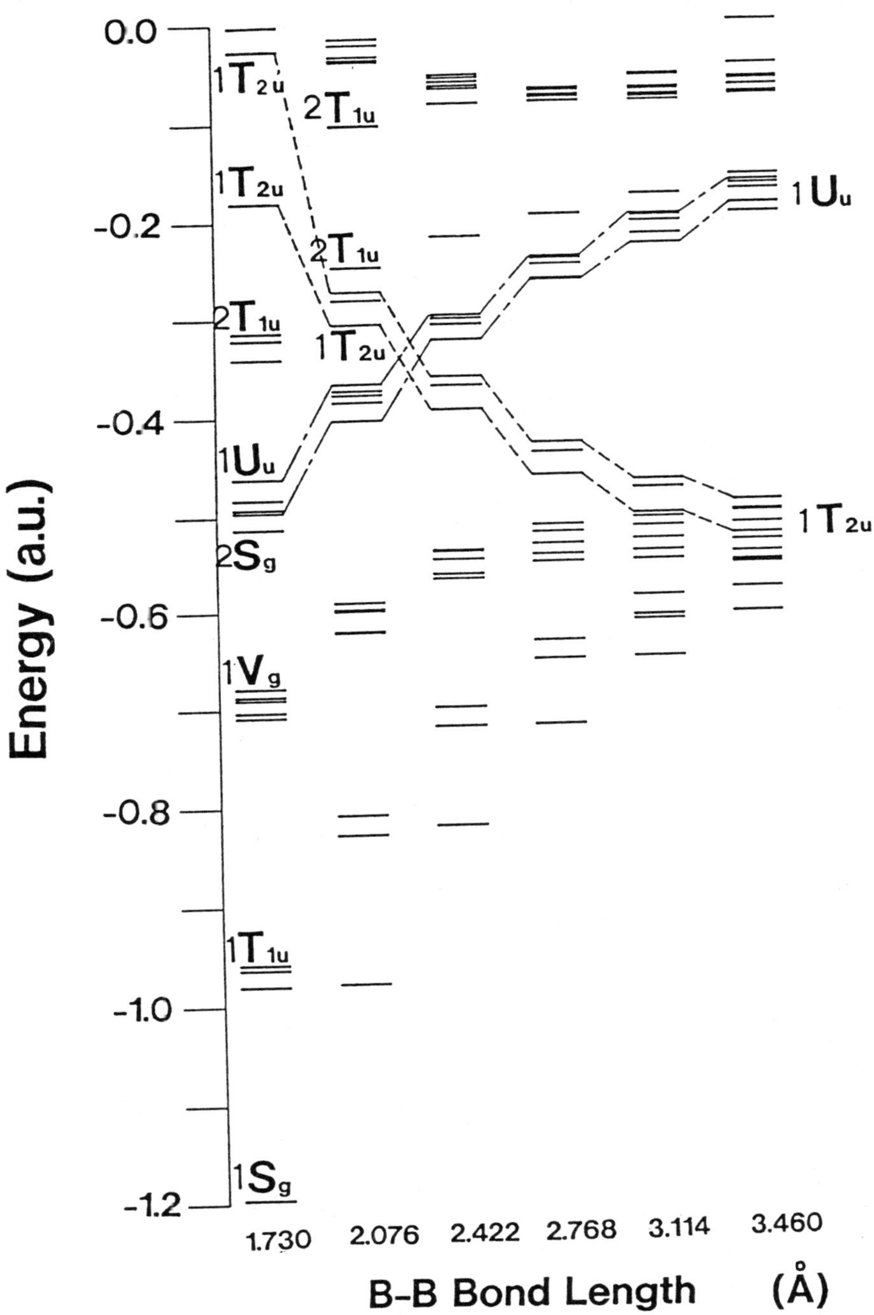

Fig. 1. Energy levels of an icosahedral boron cluster as a function of the B–B bond length.

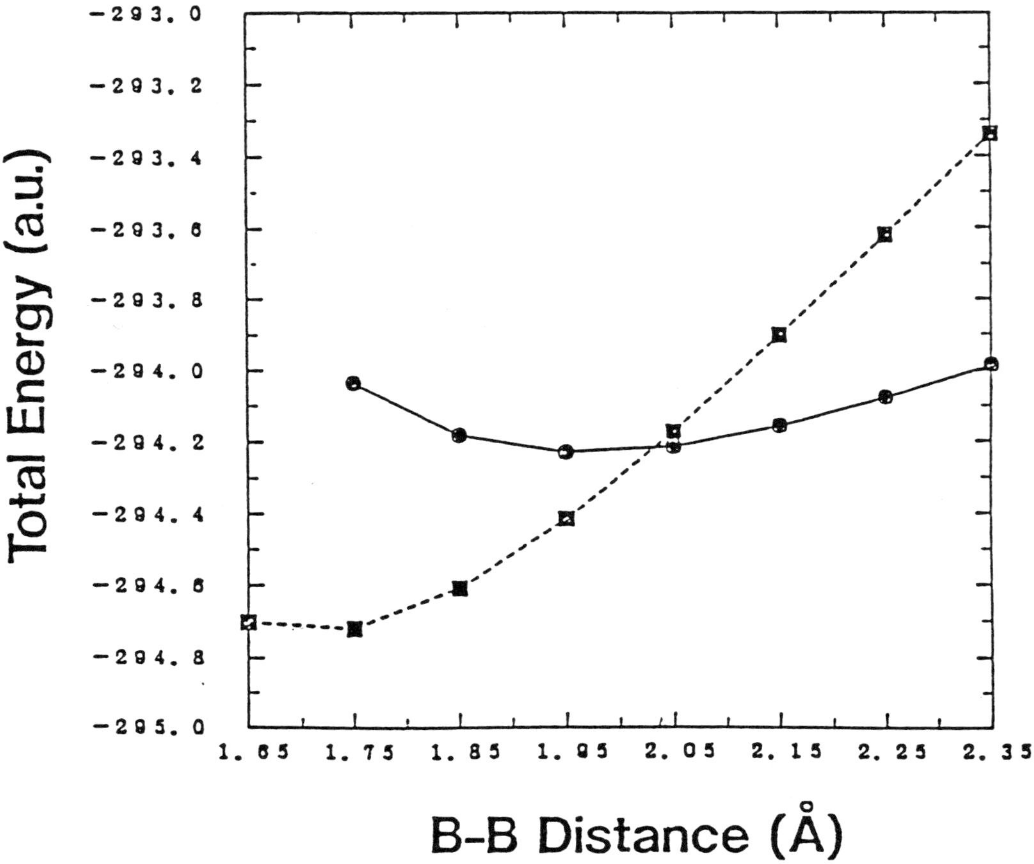

Fig. 2. The total energy of a B_{12}^{2+} cluster. The solid line represents the total energy as a function of the B–B bond length for electron configuration (i). The broken line represents that for electron configuration (ii).

icosahedron. This configuration may be considered as metallic in character. In Figure 2 we show the total energy of the B_{12}^{2+} cluster as a function of the B–B bond length. At large B–B separation, configuration (i) has a lower energy. There is a transition to configuration (ii) with decreasing the B–B bond length. From this we conclude that the B_{12}^{2+} cluster changes from a metallic state to a semiconducting state as the bond length is decreased. This transition takes place at 2.0 Å.

3. Electronic Structures of Icosahedral Aluminum Clusters

In Figure 3 we show how the energy levels of an icosahedral Al_{12}^{2+} cluster change with the Al-Al bond length. In Figure 4 we show the total energy

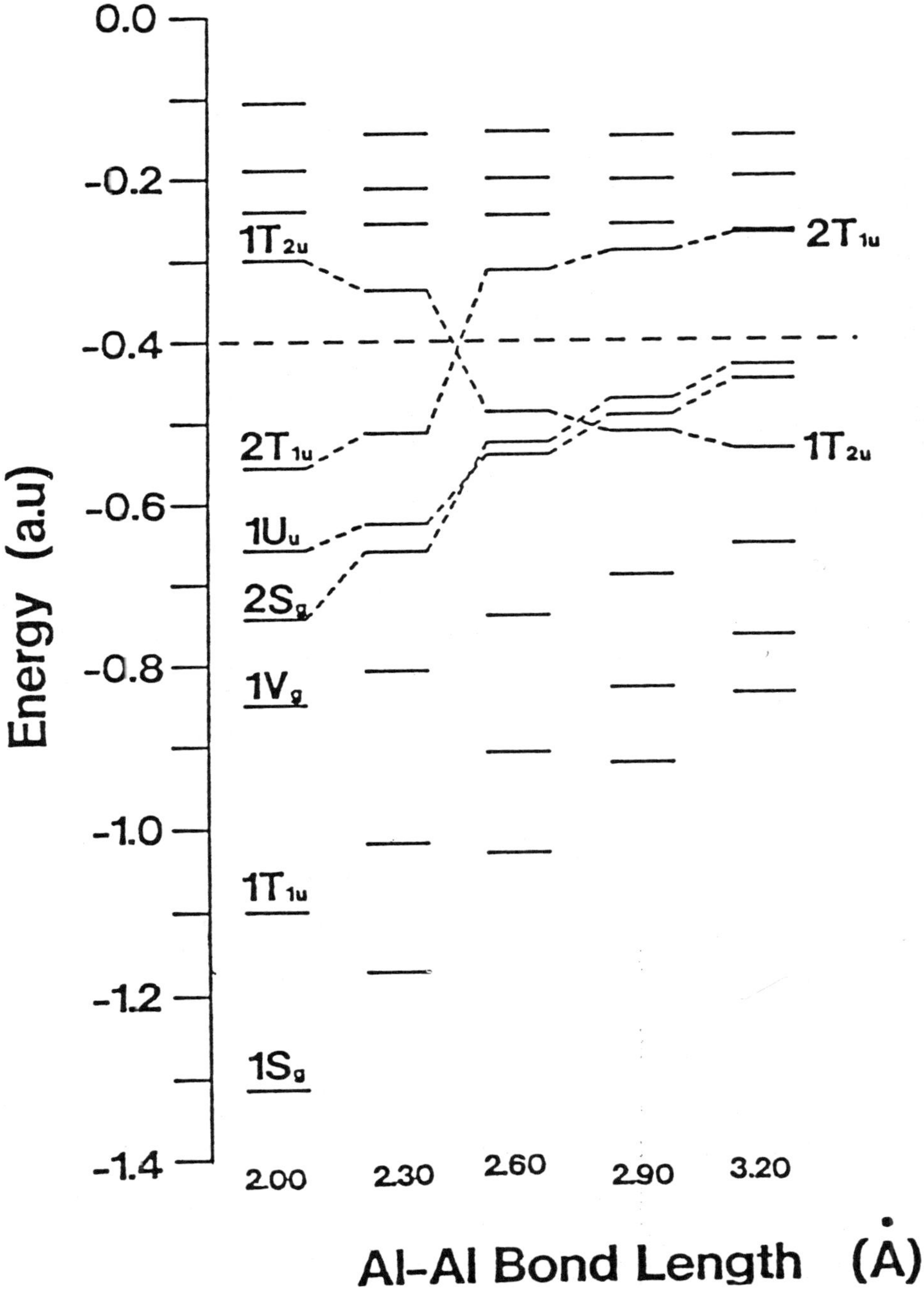

Fig. 3. Energy levels of an icosahedral Al_{12}^{2+} cluster as a function of the Al–Al bond length.

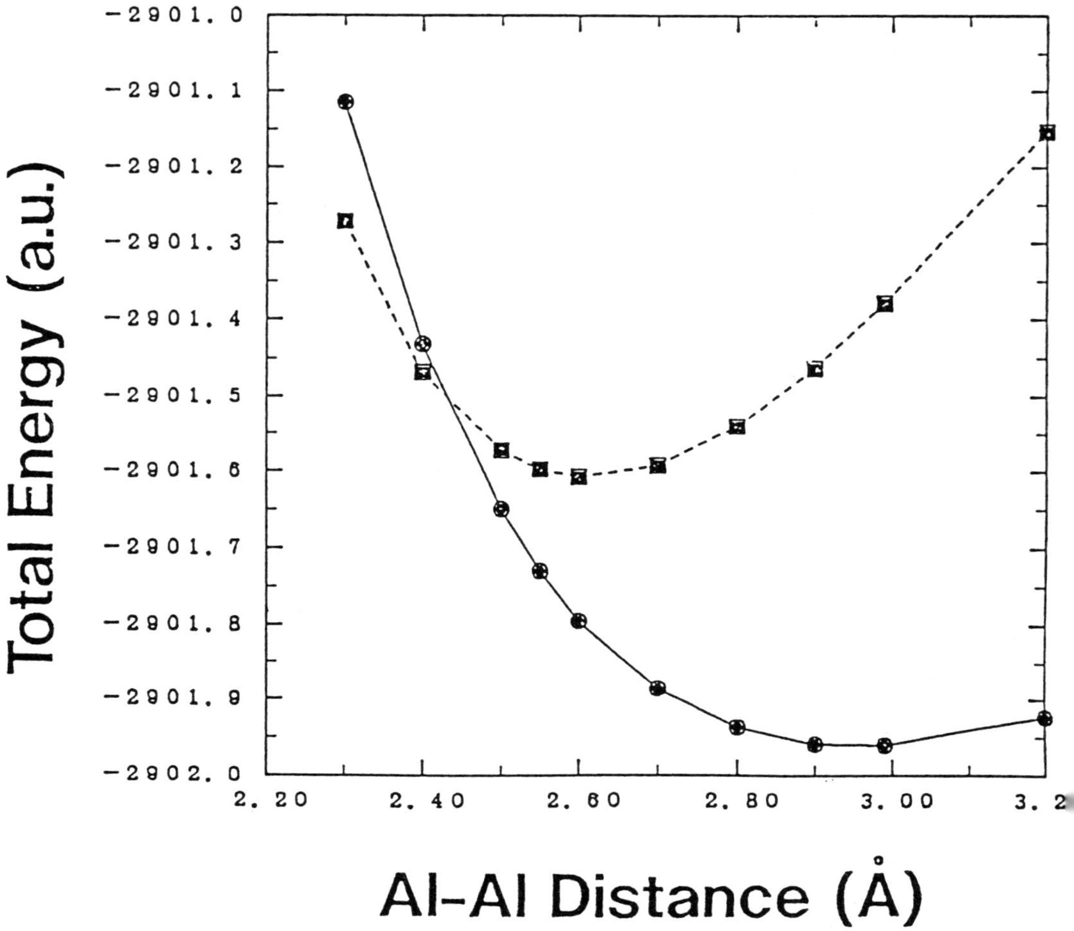

Fig. 4. The total energy of an Al_{12}^{2+} cluster. The solid line represents the energy for electron configuration (i). The broken line represents that for electron configuration (ii).

of the Al_{12}^{2+} cluster for the two electron configurations as a function of the Al–Al bond length. In these calculations the same trends in the behavior are obtained. From Figure 4 we can conclude that an icosahedral aluminum cluster is more inclined to be metallic, compared with the icosahedral boron cluster. Figure 4 also shows that, when the Al–Al bond length is less than 2.50 Å, 'multi-center bonds' are formed in the icosahedral aluminum cluster. It should be noted that, in the aluminum icosahedron in AlMn alloy, the distance between the Al atoms is much shorter than in the fcc pure aluminum structure, namely 2.55 Å [5] instead of 2.86 Å.

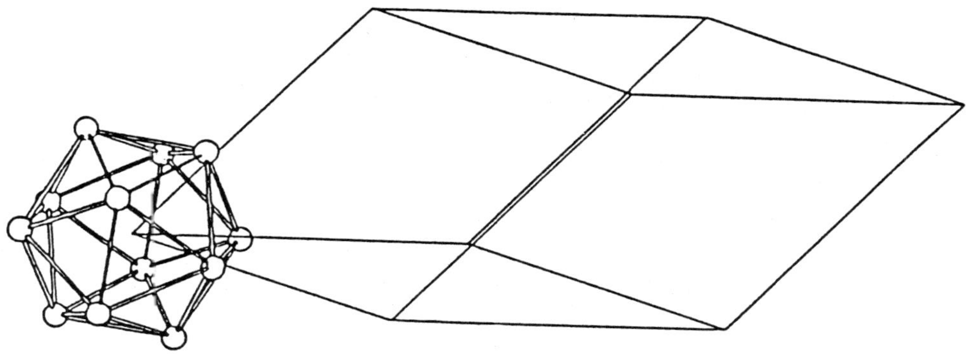

Fig. 5. The unit cell of α-rhombohedral boron.

4. Electronic Structures of α-Rhombohedral Boron

In this section we present the calculated results for the electronic structures of α-rhombohedral boron by the LDA (local density approximation) self-consistent pseudopotential method [6]. These results are the first non-empirical band structure calculations reported. The calculation is performed for the unit cell shown in Figure 5. We note that the skeleton of icosahedra in this structure may be considered as a slightly deformed face centered cubic closed packing of icosahedra; icosahedra form a triangular lattice in the [111] layer which is orthogonal to the longest diagonal of the rhombohedral cell, as shown in Figure 6.

In Figure 7 the energy band structure of α-rhombohedral boron is shown, and Figure 8 shows its Brillouin zone. In this energy band structure the bands up to 18th band are fully occupied and the 19th band is empty, showing that α-rhombohedral boron is a semiconductor with an indirect gap of 1.4 eV. Unfortunately the band gap of α-rhombohedral boron has not been observed definitely because of difficulty in preparing a pure crystal [1].

In Figure 9 the contour map of the calculated total charge density is shown. In this figures the amplitudes of the charge density are plotted on the xy-, yz- and zx-planes. The ellipsoidal figure shows the stereo contour map of the wavefunction projected on a spherical surface. The radius of the sphere is taken as the distance between the center and each boron atom, that is the radius of the icosahedron. x-, y- and z-axes are shown at the top of this figure with an icosahedron. The z-axis coincides with the [111] direction of rhombohedral cell. The x-axis coincides with one of the two-fold axes and the yz-plane is the mirror plane.

As seen in Figure 9, the charges are localized on the face of the icosahe-

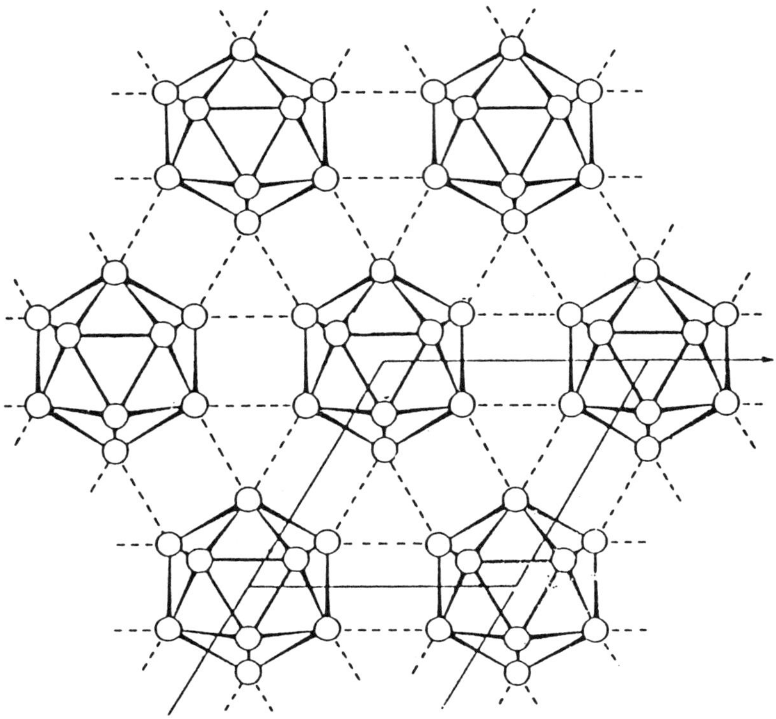

Fig. 6. Crystal structure of α-rhombohedral boron in the [111] layer.

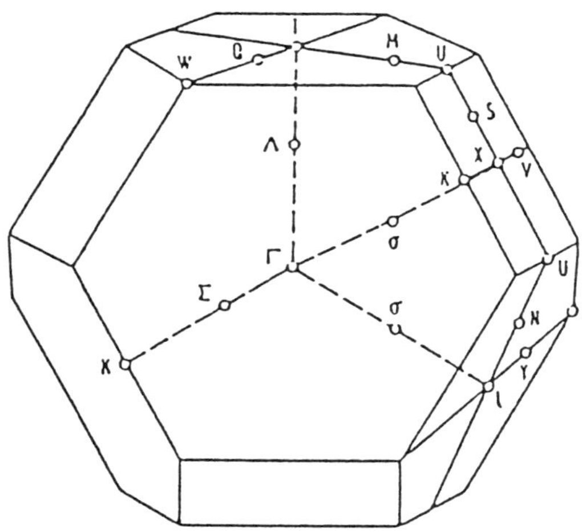

Fig. 7. The calculated band structure of α-rhombohedral boron.

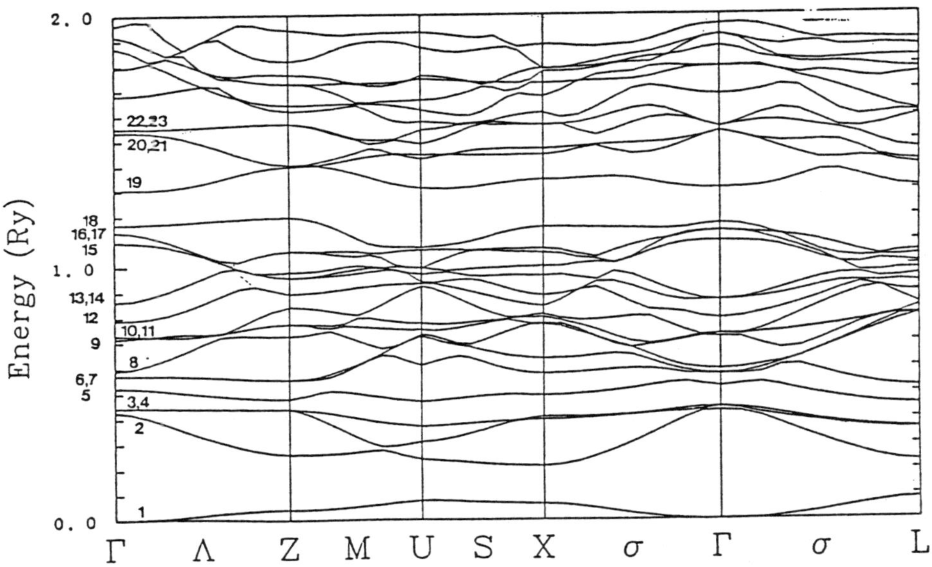

Fig. 8. The Brillouin zone of α-rhombohedral boron.

dron and between icosahedra along the rhombohedral axes. The existence of localized charge density on the faces of the icosahedron suggests the formation of multi-center bonds in the icosahedron, even in a crystal. The charge density distribution of α-rhombohedral boron has been measured by the X-ray method and the experimental results have shown the existence of the localized charge density on triangular faces of an icosahedron [7]. Thus the present theoretical result is consistent with experiments.

Next we pay attention to the charges localized on the inter-icosahedral region along the rhombohedral axis which is shown in Figure 9 [yz-plane]. The shape of this localized charge is similar to that of the covalent bonds in silicon with a diamond structure. As shown in Figure 5, one of the rhombohedral axes almost coincides with one of the three pseudo-five-fold axes of an icosahedron, and six atoms in an icosahedron are situated along the rhombohedral axes. The localized charge in the inter-icosahedral region is formed by the bonding state between p orbitals of these boron atoms. We note that the distance between the nearest boron atoms on the two neighboring icosahedra along rhombohedral axis is 1.70 Å while the B–B bond length inside the icosahedron is 1.75 – 1.8 Å. This fact suggests that the bond strength for intra-icosahedron and inter-icosahedron are of the same order of magnitude.

Further we consider the charge density between icosahedra on the same [111] layer. As seen in Figure 6, three neighboring icosahedra within the

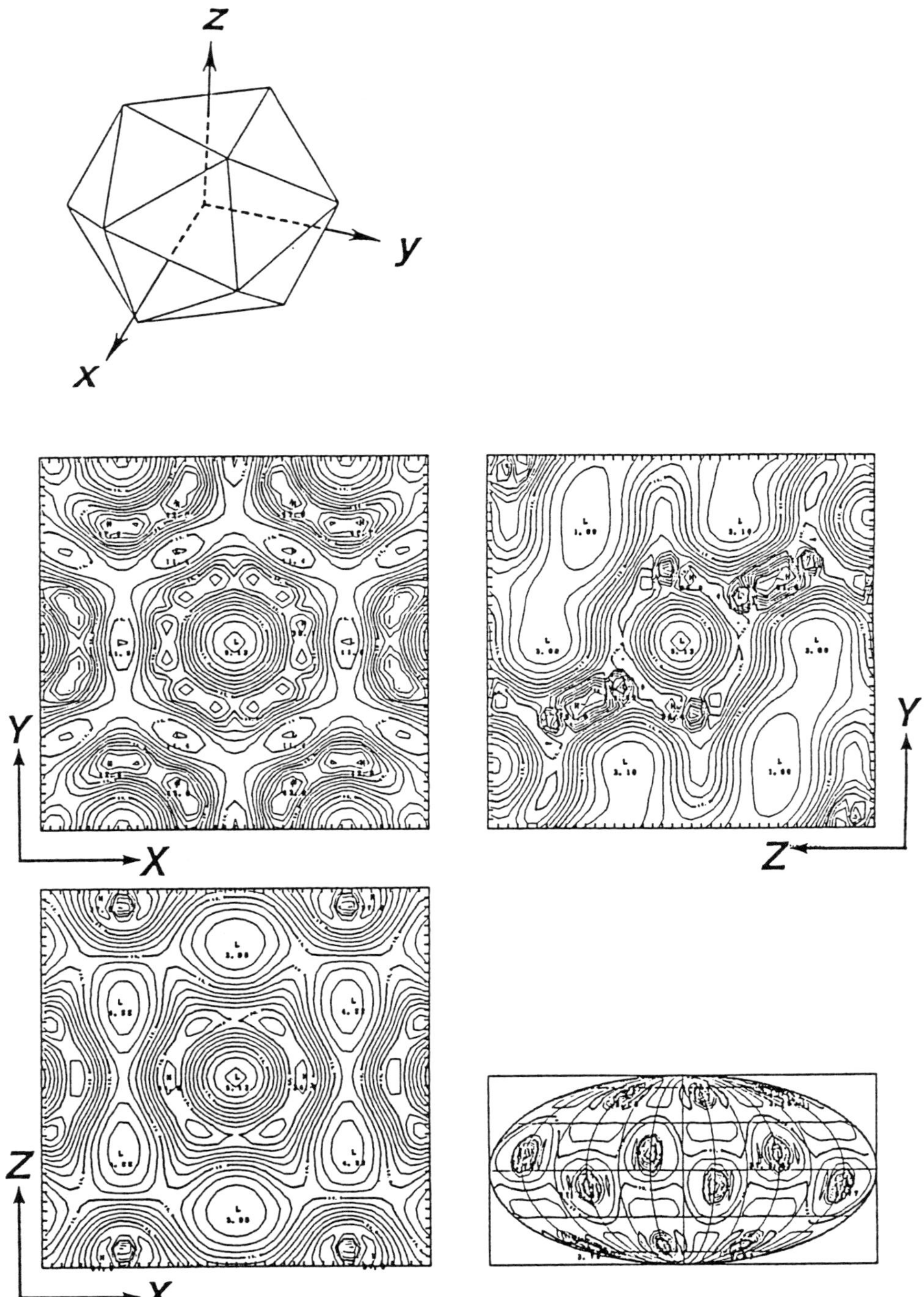

Fig. 9. The calculated total charge density of α-rhombohedral boron.

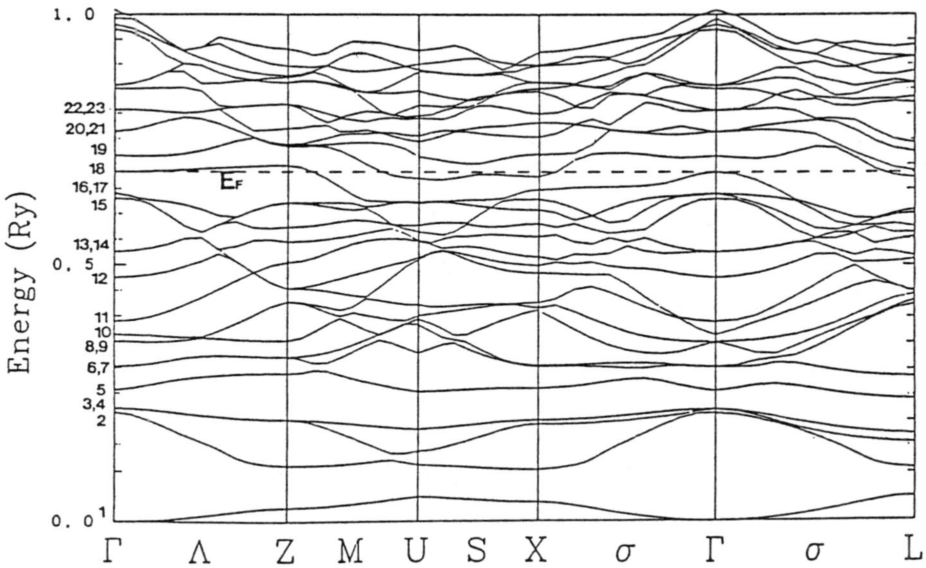

Fig. 10. The calculated band structure of α-rhombohedral aluminum.

[111] layer make a triangle the vertices of which are the boron atoms in each icosahedron. We note that distance between boron atoms in this triangle is 2.02 Å. We see from Figure 9 [xy-plane] that the charge density in this region is higher than those in the other regions between icosahedra within [111] layer. But the shape of charge density within the [111] layer is different from that between the icosahedra on the different [111] layers.

Summarizing the above calculated results on the charge density we can conclude that two types of inter-icosahedral bonds and the multi-center intra-icosahedral bond appear in the crystalline α-rhombohedral boron.

In order to clarify the difference between aluminum and boron we have calculated the band structures of aluminum with the α-rhombohedral structure. The calculated results of α-rhombohedral aluminum for the optimized crystal structure which are shown in Figure 10 indicate that α-rhombohedral aluminum is a metal. The 18th band, corresponding to the $1U_u$ state in an icosahedral cluster, is partially filled, so that the multi-center bonds are destroyed. Thus this feature is different from that of α-rhombohedral boron.

5. Conclusions

Motivated by the experimental observations of quasicrystals, we have performed a first principles calculation on the electronic structures of icosahedral boron and aluminum. We find that both systems change from a metal to a semiconductor with decreasing bond lengths. In the semiconductor limit, the electronic character is determined by the emergence of multi-center bonds.

Acknowledgements

We would like to thank Professor H. Kamimura, Professor N. Shima, Dr. T. Nakayama and Dr. K. Shiraishi. This work was partially supported by a Grant-in-Aid from the Ministry of Education, Science and Culture of Japan. The numerical calculations were done on the HITAC S-820 at the computer center of the University of Tokyo. We have used the library program GSCF3 for the Hartree–Fock calculation.

References

1. H. Werheit: 1988, *Prog. Crystal Growth and Charact.* **16**, 179 and references therein.
2. H.C. Longuet-Higgins and M. de V. Roberts: 1955, *Proc. R. Soc. A* **230**, 110.
3. D. Schechtman, I. Blech, D. Gratias and J.W. Cahn: 1984, *Phys. Rev. Lett.* **53**, 1951.
4. Program GSCF3, Library Program, The Computer Centre, the University of Tokyo (1985).
5. M. Audier and P. Guyot: 1986, *Phil. Mag. B* **53**, L43.
6. K. Shiraishi, Doctoral Thesis, University of Tokyo, (1987).
7. B. Morosin, A.W. Mullendore, D. Emin and G.A. Slack: 1985, *AIP Conf. Proc.* **140**, 70.

NEGATIVE DONOR CENTERS IN HIGH MAGNETIC FIELDS AND QUANTUM WELLS

STEVEN G. LOUIE and TAO PANG

Department of Physics, University of California, Materials Sciences Division, Lawrence Berkeley Laboratory, Berkeley, California 94720, U.S.A.

Abstract. The properties of negative donor centers D^- in quantum wells and strong magnetic fields are analyzed by an effective mass model. The model is solved exactly by a diffusion quantum Monte Carlo method from the three-dimensional limit to the two-dimensional limit. Available experimental results of D^- centers in bulk GaAs and GaAs/$Ga_{1-x}Al_x$ as quantum wells are compared with the theoretical results, and excellent agreement is found between theory and magneto-optical experiments. The strong increase in the binding energy of D^- centers due to the reduction of the dimensionality is quantitatively predicted. Enhancement of electron correlation effect in the two-dimensional systems is discussed.

1. Introduction

One of the most important and interesting areas in semiconductor physics is the study of impurities to which Professor Hiroshi Kamimura is a pioneer and leading contributor [1]. Recently the electronic structure and properties of donor centers in quantum wells and strong magnetic fields have become a subject of intensive study because many novel effects in these systems, such as the quantum Hall effects, metal-insulator transitions and electron localizations, are intimately related to phenomena of high magnetic fields and impurity states [2]. In this article, we discuss results of a calculation of the properties of negative donor D^- centers in multiple quantum well structures and strong magnetic fields.

A D^- center in semiconductor systems is formed by a neutral donor center (D^0), such as a Si atom in GaAs, trapping an extra bound electron [3]. For homogeneous semiconductors, such as GaAs, InP or InSb, states associated with the D^- centers can be thought as the states of two electrons of effective mass m^* under a central Coulomb potential screened by the dielectric constant [1]. This effective mass model has been highly successful in discussing neutral shallow donors. For example, results from variational calculations (such as the peak position of the calculated far-infrared absorption lines for D^0 centers in GaAs/GaAlAs quantum wells) are in good agreement with experimental observations [4]. However, owing to the complexity of electron–electron correlation effects in D^- centers, it has been shown that the variational methods that have been highly successful for D^0 centers do not give very accurate results for D^- centers [3].

Comparison [3] of experimental data on D^- states in bulk GaAs with previous variational calculations [5] showed that the experimental binding

H. Aoki et al. (eds.), New Horizons in Low-Dimensional Electron Systems, 445–454.

energies are systematically larger than the computed values, and the discrepancy increases with increasing applied magnetic field strength. Recently, we have studied D^- centers in bulk GaAs with the effective mass model using a diffusion quantum Monte Carlo (DQMC) method [6]. The exact ground-state and binding energies are evaluated from the simulations. It is found that the new theoretical values for the binding energies are in significantly better agreement with the experimental data. This has led to the conclusion that D^- centers in a bulk homogeneous semiconductor like GaAs are well described by the effective mass model. The discrepancy between experimental data and previous variational results is purely due to the inadequacy in previous trial wavefunctions in including electron-electron correlation effects.

Most recently, D^- states have been identified experimentally in 100 Å and 200 Å wide GaAs/$Ga_{1-x}Al_xAs$ quantum wells with $x = 0.25$ and $x = 0.3$ [7,8]. Magneto-optical measurements were carried out with an applied field perpendicular to the well structures. The results show that D^- centers are observable at fields larger than 4 T, and there is a strong enhancement in the binding energy of D^- centers in quantum wells. Our DQMC study of D^- centers in quantum wells has also given a very good agreement between the effective mass theory and experiments, even though other effects such as differences in effective masses and dielectric constants in the barrier and well regions and the effects of non-parabolic bands in narrow well cases have showed increased influence [6].

In the work discussed here, the effective mass model is applied to study D^- and D^0 centers in quantum wells from the three-dimensional (3D) limit to the two-dimensional (2D) limit using the DQMC approach. The DQMC approach in principle provides the exact ground-state properties of D^- centers in the effective mass model, apart from statistical noises. In Section 2, the effective mass model for D^- centers in quantum wells is given; in Section 3, the basic aspects of the DQMC method is discussed; in Section 4, the numerical results for the ground-state energy and binding energy of D^- centers in quantum wells from the 3D limit to the 2D limit are presented; and in Section 5, some implications of the numerical results are discussed.

2. Effective Mass Model for D^- Centers

For a D^- center located at the center of a single quantum well and with an applied field along the z-direction perpendicular to the well, the Hamiltonian in the effective mass model is given by [6]:

$$H = H(1) + H(2) + V(r_{12}) + \frac{\gamma}{2\eta} L_z , \qquad (1)$$

with

$$H(i) = -\frac{1}{2\eta}\,\nabla_i^2 + \frac{\gamma^2}{8\eta}\,(x_i^2 + y_i^2) + U(r_i) + V_Q(z_i)\,, \tag{2}$$

where η is the ratio of the effective mass m^* in a specific region to that of bulk GaAs; $\gamma = \hbar\omega_c/2Ry^*$ is the magnetic field in the effective atomic units of GaAs with ω_c the cyclotron frequency and Ry^* the effective Rydberg; $V(r)$ and $U(r)$ are the interaction between two electrons and the interaction of an electron with the central impurity charge respectively, including all image charges due to the difference in dielectric constants in the well, ε_1 and in the barrier, ε_2 [4,9]; L_z is the angular momentum along the field direction; and $V_Q(z)$ is the quantum well potential:

$$V_Q(z) = \begin{cases} V_0\,, & |z| > d/2 \\ 0\,, & |z| \le d/2 \end{cases} \tag{3}$$

with V_0 the potential barrier height and d the well width. Only the $L_z = 0$ case is considered here. Unless specified, the effective atomic units of GaAs with $m_0^* = 0.067m_e$ and $\varepsilon_0 = 12.53$ will be used.

The binding energy of D^- centers can be written as

$$E_b = E_{D^0} + E_{Q\gamma} - E_{D^-}\,, \tag{4}$$

where E_{D^0} is the ground-state energy of D^0, $E_{Q\gamma}$ is the ground-state energy for a free electron in the quantum $V_Q(z)$ and magnetic field γ, and E_{D^-} is the ground-state energy of Equation (1).

3. Diffusion Quantum Monte Carlo Method

In order to obtain the exact ground-state energy of Equation (1) and (2), a quantum Monte Carlo method is employed to simulate the Schrödinger equation as a diffusion equation of a time dependent probability in configuration space [10].

In the DQMC approach, for a two-electron system, one takes a variational wavefunction $\Phi(\mathbf{R})$ with $\mathbf{R} = (\mathbf{r}_1, \mathbf{r}_2)$ as a trial wavefunction for the ground state and constructs a time dependent probability density, $F(\mathbf{R}, t)$:

$$F(\mathbf{R}, t) = \Psi(\mathbf{R}, t)\;\Phi(\mathbf{R})\,, \tag{5}$$

with

$$\Psi(\mathbf{R}, t) = \exp\left[\int_0^t E_n(t')\,\mathrm{d}t' - Ht\right]\Phi(\mathbf{R})\,, \tag{6}$$

where $E_n(t)$ is an as-yet-undetermined c-number which can be considered as a normalization constant for the moment. As long as $\Phi(\mathbf{R})$ is not orthogonal to the exact ground state of H, $\Psi(\mathbf{R}, t)$ will approach the ground state as t goes to infinity. Defining a time dependent expectation value $E(t)$:

$$E(t) = \frac{< \Phi(\mathbf{R})|H|\Psi(\mathbf{R},t) >}{< \Phi(\mathbf{R})|\Psi(\mathbf{R},t) >} = \frac{\int d\mathbf{R} F(\mathbf{R},t)\ \varepsilon(\mathbf{R})}{\int d\mathbf{R} F(\mathbf{R},t)} , \tag{7}$$

where $\varepsilon(\mathbf{R}) = \Phi^{-1}(\mathbf{R})\ H\Phi(\mathbf{R})$ is the local energy in configuration space, then $E(t)$ becomes the exact ground-state energy of H as t goes to infinity [10].

For the D^- center, since the ground state is a singlet of two electrons, the wavefunction has no nodal structure, that is, $F(\mathbf{R},t)$ is always positively defined. Therefore, Equation (7) is equivalent to an average over a statistical weight $F(\mathbf{R},t)$:

$$E(t) = \frac{1}{M} \sum_i^M \varepsilon(\mathbf{R}_i) , \tag{8}$$

with $\mathbf{R}_i$ generated according to $F(\mathbf{R},t)$. In principle, $E(t)$ in Equation (8) can be sampled by the Monte Carlo method exactly in the limit of zero time step intervals. However, in practice, a zero time step interval is impossible. So the choice of the total simulation time and the time step interval is governed by the accuracy required for the results. Generally, the statistical error of the Monte Carlo data is given by the total simulation time. The time step interval is chosen so that the error introduced will be within the statistical error of the data. The choice of the trial state $\Phi(\mathbf{R})$ will influence the speed of convergence of the results. Simulations with better trial states converge faster. The trial state used in this work has the form [6,9]:

$$\Phi(\mathbf{r}_1,\mathbf{r}_2) = f(r_{12})\ [g_1(r_1)\ g_2(r_2) + g_1(r_2)\ g_2(r_1)] \prod_{i=1}^{2} \phi(z_i)\ \kappa(\rho_i)\ w(r_i) , \tag{9}$$

with $f(r) = \exp[ar/(1+br)]$, $g_i(r) = \exp[-c_i r^2/(d_i + e_i r)]$, $\phi(z) = \cos(kz)$ in the well and $\phi(z) = \cos(kd/2)\ \exp[\lambda(d/2 - |z|)]$ outside the well, $\kappa(\rho) = \exp(-\mu\rho^2)$, and $w(r) = \exp[\alpha r/(1+\beta r)]$. All the parameters a, b, c_i, d_i, e_i, k, λ, μ, α and β are optimized by a variational Monte Carlo simulation first before Φ is used as a guiding wavefunction in the DQMC runs. Details on the variational parameters and the algorithm used in this work will be given elsewhere [9].

4. Simulation Results

4.1. D^- CENTERS IN THREE DIMENSIONS

Applying the method described above to the D^- centers, the ground-state energy can be obtained with very high accuracy. In Figure 1, the DQMC

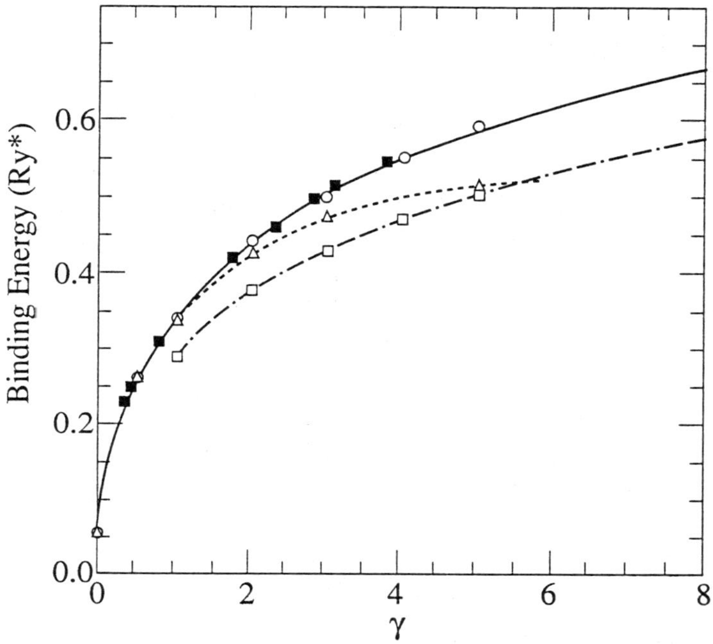

Fig. 1. Binding energies of D^- centers in GaAs from present work (○) compared with the experimental data of Reference [3] (■) and previous calculations of Refrence [1] (□) and Reference [5] (△). The error bars of present results are within the circles.

results [6] together with results from two previous calculations using variational methods [1,5] are compared to the experimental results for the case of Si doped GaAs [3]. It is clear that previous theoretical calculations have underestimated the binding energy and are increasingly less accurate as the field strength increases. The excellent agreement of the DQMC results with the experimental data indicates that the effective mass model describes the D^- centers in GaAs very well. This result provides a definitive answer to the question of the causes for the discrepancy between the experimental data of D^- in GaAs [3] and previous theoretical results [1,5]. It was argued that central cell and perhaps other crystal field effects might be significant in this case, as they are in other systems such as D^- centers in Si [11]. Figure 1 shows that the discrepancy discussed in Reference [3] is purely due to the inadequacy of the trial wavefunctions adopted in the previous calculations.

We have also calculated the electron density distribution for the D^- center at different magnetic field strength [9]. The field dependence of the root mean square radius in the field direction, r_z, and that in the perpendicular direction, r_{xy}, has been analyzed using the simulation data [6]. The compe-

TABLE I
DQMC simulation binding energy of D^- and D^0 centers in quantum wells from the 3D limit to the 2D limit for the specific case of $m_1^* = m_2^*$ and $\varepsilon_1 = \varepsilon_2$ (see text). The energies are in effective Rydberg Ry* of GaAs; 1 Ry* = 5.806 meV. The Monte Carlo error bars are estimated at 2 units of the last digit in the data.

	$\gamma = \frac{\hbar\omega_c}{2\ \text{Ry}^*}$	3D	200 Å	100 Å	2D
D^-					
	0	0.056	0.23	0.29	0.511
	1	0.34	0.65	0.77	1.17
	3	0.50	0.94	1.13	1.82
D^0					
	0	1.00	1.74	2.09	4.00
	1	1.66[a]	2.52	2.92	4.91
	3	2.33[a]	3.36	3.89	6.35

[a] Variational results from Reference [17].

tition between the repulsive electron-electron interaction and the attractive central potential leads to a configuration that the two electrons in the system stay in two different orbitals [12], that is, g_1 and g_2 in Equation (9) with different characteristic lengths, $r_{01} = e_1/c_1$ and $r_{02} = e_2/c_2$. At low fields, r_{01} is around 1 and r_{02} is around 2 in units of the effective Bohr radius. Under a strong magnetic field, the orbitals are pushed inwards in the xy-planes *as well as* in the z-direction [6]. From variational studies, one finds that the second orbital gradually merges with the first orbital in increasingly high magnetic field [1,9], that is, $r_{02} \to r_{01}$.

4.2. D^- CENTERS IN TWO DIMENSIONS

Before we analyze the simulation results for the quantum well cases, let us examine the results of another limiting case: the two-dimensional case. The ground-state energy of a 2D D^0 center is -4 Ry* in the zero field. In Table I, the binding energies of the last electron for D^0 and D^- centers in the 2D limit with various magnetic field strengths are provided in comparison with the 3D case. As can easily be seen, the binding energy in the reduced dimension cases is significantly enhanced. The binding energy of a 2D D^- center at zero field is almost an order of magnitude higher than the corresponding 3D case. In zero magnetic field, the binding energy obtained from DQMC simulation for a 2D D^- center is 0.511 ± 0.002 Ry*, which is about 6% higher than a previous variational result of 0.48 Ry* [13]. The qualitative behavior of a two-orbital structure in the electron charge density distribution of 2D D^- centers in low field is found to be similar to the case in 3D.

TABLE II
Binding energy of D^- and D^0 centers in GaAS/$Ga_{0.75}Al_{0.25}$As quantum well. (a) $m_1 = m_2$, $\varepsilon_1 = \varepsilon_2$; (b) $m_1 \neq m_2$, $\varepsilon_1 \neq \varepsilon_2$; (c) Same as (b) but with nonparabolicity of GaAs conduction band included (see text). Experimental data are from Reference [7]. Energy is in effective Rydberg of GaAs; Monte Carlo error bars are estimated at 0.02 Ry*.

	Magnetic field	Binding Energy (Ry*)			
		Theory			Expt.
	$\gamma = \hbar\omega_c/2$ Ry*	(a)	(b)	(c)	
D^-					
	0	0.29	0.33	0.35	–
	1	0.77	0.77	0.83	0.94
	3	1.13	1.11	1.22	1.28
D^0					
	0	2.09	2.16	2.23	–
	1	2.92	3.00	3.13	–
	3	3.89	3.94	4.18	–

4.3. D^- CENTERS IN QUANTUM WELLS

As is known from previous analyses of the electronic structure of quantum well systems, a simple two-dimensional model in general does not represent the quantum well case very well [14]. More careful considerations of the quantum well potential and other effects have to be taken into account.

From the comparison of our DQMC results for 3D D^- centers with experimental results in Figure 1, it is clear that the effective mass model describes D^- centers in GaAs quite well, provided that electron correlation is treated accurately. But since we do not know how much deviation would be caused by complications introduced in GaAs/$Ga_{1-x}Al_x$As quantum wells, one cannot make an *a priori* assessment regarding the effective mass model for D^- centers in quantum wells.

Recent experiment [7,8] has given evidence that the binding energies of D^- centers located around the center of a 100 Å GaAs/$Ga_{0.75}Al_{0.25}$As quantum well is strongly enhanced although signatures for D^- states were only observed for $B > 4$ T. In order to compare with these experimental data [7,8], the experimental parameters for the quantum wells studied in References [7] and [8] are used to obtain the results summarized in Table II. The calculations were carried out in a single quantum well with well width 100 Å and Al concentration $x = 0.25$. The barrier height is chosen as $0.65\Delta E_g$ with $\Delta E_g = 1.247x$ eV as the band gap mismatch between GaAs and $Ga_{1-x}Al_x$As [4]. The effective mass and the dielectric constant in the well are chosen as those of bulk GaAs. Two typical situations have been studied: (1) the case with the effective mass m_2^* and dielectric constant ε_2 in the barrier taken to be the same as those in the well; (2) the case with $m_2^* = (0.067 + 0.083x)\, m_e$ and $\varepsilon_2 = 12.53 - 2.73x$ [4]. The binding energies of D^- in these two cases are given in columns a and b in Table II. The corresponding results for D^0 are also given. It is necessary to include all the images of the central impurity charge and those of the electrons in the Hamiltonian in Equation (1) when $\varepsilon_1 \neq \varepsilon_2$ [4,9]. When $m_1^* \neq m_2^*$, the boundary conditions $\Phi_1 = \Phi_2$ and $m_2^* \partial\Phi_1/\partial z = m_1^* \partial\Phi_2/\partial z$ at the interfaces are imposed on the guiding wavefunction along the z-direction. The effect of the energy dependence of the effective mass [16] to the binding energy is also estimated by rescaling the results to the energy dependent effective mass, and the modified results are given in column c in Table II. For GaAs the energy dependent effective mass is taken as $m^*(E) = (0.0665 + 0.0436E + 0.236E^2 - 0.147E^3)\, m_e$, where E is the single particle energy above the band edge in eV [16]. For the ground state of D^-, there is no well-defined single particle energy. So half of the total energy is assigned to each electron in order to include this energy dependent effect.

At zero magnetic field, the theory shows a dramatic increase, by a factor of seven, for the binding energy of D^- in the 100 Å quantum well as compared to that in bulk GaAs. The calculated electron density distributions in the quantum well structure show that the electrons are further confined by the well potential [9]. Combined with the confinement due to the magnetic field, it leads to the strong enhancement of the binding energy of D^- in a quantum well. The theoretical results for the binding energy agree well with the experimental results at high fields where there are data. The calculated values are however about 5–10% lower than the experimental estimates [7,8]. The effective well width in the sample may be smaller than the layer thickness due to a non-square like potential, which might partially account for the difference.

A representative study [6] was also carried out for a D^- center in a 200 Å GaAs/$Ga_{0.75}Al_{0.25}As$ well at $B = 9$ T, or $\gamma = 1.34$. The calculated binding energy from our DQMC is 4.5±0.1 meV which is excellent agreement with the estimated value [8] of 4.45 ± 0.25 meV from the experimental data [15] of a 200 Å well at the same magnetic field, even though the effect of the different effective masses and dielectric constants in the well and barrier and the effects of nonparabolic band structure are not taken into account. This excellent agreement encouraged us to pursue more theoretical studies for various magnetic field strength and well width. Several calculations for 100 Å and 200 Å wells are done and the results are given in Table I. To illustrate the transition from 3D case to 2D case, the binding energies for D^- at various well widths and magnetic fields are plotted in Figure 2.

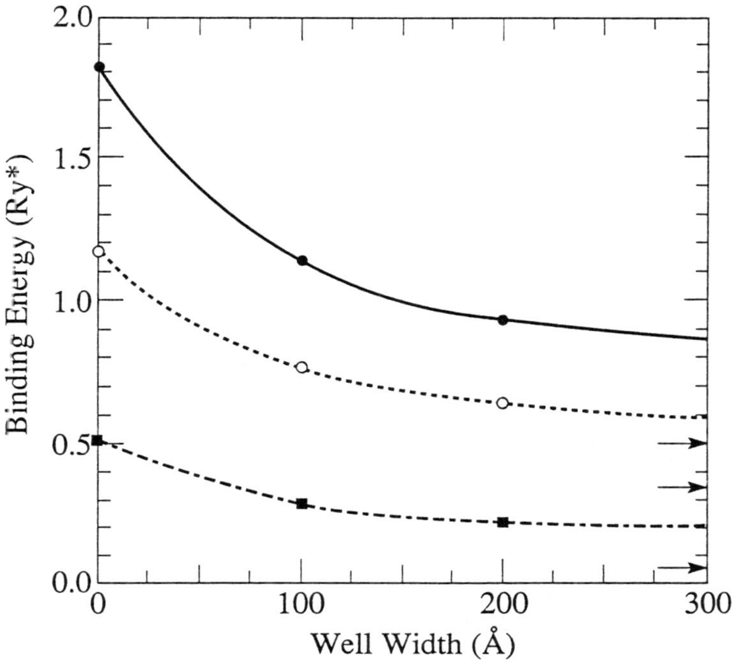

Fig. 2. Binding energy of D^- centers in GaAs/$Ga_{0.75}Al_{0.25}$As quantum wells from 2D to 3D limit for the specific case of $m_1^* = m_2^*$ and $\varepsilon_1 = \varepsilon_2$ at $\gamma = 0$ (■), $\gamma = 1$ (○) and $\gamma = 3$ (●). The results for the 3D case are indicated by arrows. The curves are guides for the eyes.

5. Conclusions

In this paper, we have reviewed some of the results for D^- centers in quantum wells obtained using the DQMC approach. It is found that D^- states in bulk semiconductors such as Si doped GaAs are well described by the effective mass model. Comparing the present DQMC results with previous calculations shows that electron–electron correlation effects are important in these systems and simple trial wavefunctions in general would not give a complete treatment of these correlation effects. On the other hand, the DQMC method applied here in principle has treated electron correlations exactly. For D^- centers in quantum wells, the binding energy is in agreement with experimental data from multi-quantum well samples to within 10% after taking into account several important effects. The present theoretical results reenforce the identification made in References [7] and [8] that the observed special feature in the magneto-optical spectra indeed arises from the ionization of D^- centers in the quantum well. However, the actual details of the formation of the D^- centers are still a complex matter. For a

100 Å well width and barrier width superlattice of GaAs/$Ga_{0.75}Al_{0.25}As$, the binding energy of the shallow donor (D^0) in the barrier center is found to be in fact larger than the binding energy of the D^- centers in the well. The question of the mechanism for the source of the surplus electrons needed for the formation of the D^- centers observed in References [7] and [8] remains to be answered.

Acknowledgements

Fruitful discussions with Drs. S. Huant and G. Martinez are gratefully acknowledged. This work was supported by NSF Grant No. DMR88-18404 and by the director, Office of Energy Research, Office of Basic Energy Sciences, Materials Science Division of DOE under Contract No. DE-AC03-76SF00098. SGL acknowledges support from the J.S. Guggenheim Foundation and TP acknolwedges support from the Miller Institute for Basic Research in Science. Cray time was provided by NSF at the San Diego Supercomputer Center.

References

1. A. Natori and H. Kamimura: 1978, *J. Phys. Soc. Jpn.* **44**, 1216; see also, H. Kamimura and H. Aoki, *The Physics of Interacting Electrons in Disordered Systems*, Chapter 2, Clarendon, Oxford, 1989.
2. See, for example, articles in *High Magnetic Fields in Semiconductor Physics II*, G. Landwehr (ed.), Springer-Verlag, Berlin, 1989.
3. For a recent review, see. S.P. Najda, C.J. Armistead, C. Trager, and R.A. Strading: 1989, *Semicond. Sci. Technol.* **4**, 439.
4. S. Fraizzoli, F. Bassani and R. Buczko: 1990, *Phys. Rev. B* **41**, 5096.
5. D.M. Larsen: 1979, *Phys. Rev. Lett.* **42**, 742; 1979, *Phys. Rev. B* **12**, 5217.
6. T. Pang and S.G. Louie: 1990, *Phys. Rev. Lett.* **65**, 1635.
7. S. Huant and S.P. Najda and B. Etienne: 1990, *Phys. Rev. Lett.* **65**, 1486.
8. S. Huant, S.P. Najda, W. Knap, G. Martinez, B. Etienne, C.J. Langerak, J. Singleton, R.A. Thomeer, G. Hai, F.M. Peeters and J.T. Devreese, in *Proceedings of the 20th International Conference on the Physics of Semiconductors*, Thessaloniki, Greece, August 6–10, 1990, p. 1369.
9. T. Pang and S.G. Louie (unpublished).
10. P.J. Reynolds, D.M. Ceperley, B.J. Alder and W.A. Lester. Jr.: 1982, *J. Chem. Phys.* **77**, 5593.
11. For a review, see A.K. Ramdas and S. Rodriguez: 1981, *Rep. Prog. Phys.* **44**, 1297.
12. S. Chandrasekhar: 1944, *J. Astrophys.* **100**, 176.
13. D.E. Phelps and K.K. Bajaj: 1983, *Phys. Rev. B* **27**, 4883.
14. T. Ando, A.B. Fowler and F. Stern: 1982, *Rev. Mod. Phys.* **54**, 437.
15. E. Glaser, B.V. Shanabrook, R.L. Hawkins, W. Beard, J.-M. Mercy, B.D. McCombe and D. Musser: 1987, *Phys. Rev. B* **36**, 8185.
16. R.M. Kolbas, Ph.D. Thesis, University of Illinois at Urbana-Champaign (1979); U. Ekenberg: 1987, *Phys. Rev. B* **36**, 6152.
17. D. Cabib, E. Fabbi and G. Fiorio: 1972, *Il Nuovo Cimento* **10B**, 185.

POINT DEFECTS: THE ZERO-DIMENSION THREE-DIMENSIONAL OBJECTS

SOKRATES T. PANTELIDES
IBM Research Division, Thomas J. Watson Research Center, Yorktown Heights, New York 10598, U.S.A.

Silence, silence, contemptible Creature.
You call yourself the All in All, but you
are the Nothing.
[An inhabitant of Flatland addressing the
Point in Pointland, from Flatland, by
E.A. Abbott, 1882.]

Abstract. Point defects are 'zero-dimensional' objects, i.e. they have a finite extent in all three dimensions. They of course have an internal structure that can have full three dimensional character or reduced dimensionality. This paper contains reflections on the field's evolution over the last twenty years and on current and future trends.

1. Introduction

This book is about low-dimensional electron systems. Most natural materials are, however, intrinsically three-dimensional. There are exceptions, namely materials in which the atoms are arranged in two-dimensional sheets with only weak forces holding the sheets together. Graphite, intercalated graphite, and the recently discovered superconducting oxides are such materials. Other natural two-dimensional systems are crystal surfaces. There are also natural one-dimensional electron systems – materials made up of loosely coupled strands of atoms. Molecules of all kinds are natural zero-dimensional systems in the sense that they have a finite extent in all three dimensions.

The world of low-dimensional systems has been enriched in recent years by artificial structures. The interface between silicon and silicon dioxide traps what is effectively a two-dimensional electron gas. The advent of layer-by-layer growth of thin films and superlattices has led to a wealth of two-dimensional artificial structures. Then, advances in techniques to pattern structures have led to one-dimensional quantum wires and zero-dimensional quantum dots. These are the exciting new horizons in artificial low-dimensional systems. Artificially created clusters of atoms are also an exciting new world.

Three-dimensional solids, however, hide another whole world of reduced dimensionality: the world of imperfections or defects. There are two-dimensional defects, namely stacking faults and grain boundaries. There are one-dimensional defects, namely dislocations. Finally, there is the world of zero-dimensional defects, the so-called point defects.

H. Aoki et al. (eds.), New Horizons in Low-Dimensional Electron Systems, 455–464.

The zero-dimensional world is not, of course, without dimensions! Molecules, clusters, quantum dots, and point defects are not point objects any more than atoms or nuclei or protons are. Quantum dots and point defects are merely finite in all three dimensions. A finite, usually small number of atoms is involved and the corresponding electronic wavefunctions are localized.

This article focuses on point defects in semiconductors. It is not meant to be a technical review. The literature abounds with such articles. Instead, in the true spirit of a Festschrift, this article is more a celebration of point defects. It is perhaps one person's reflection on the field after twenty years of personal involvement. It is a look back and a look ahead.

2. 'Large' and 'Small' Point Defects

How large are point defects? It depends on what you count! Take, for example, a single phosphorus atom that takes the place of a silicon atom in a silicon crystal. If one looks at atoms, this defect is as point-like as they come: only one atom is disturbed, with the neighboring silicon atoms essentially staying at their perfect-crystal positions. The electronic system, however, is perturbed over a very large radius. Phosphorus has an extra proton which produces a Coulomb field. The electrons surrounding it are polarized and produce a screening field. Phosphorus of course also has an extra electron. This electron can be bound in a localized state with a hydrogenic wavefunction that extends over many shells of neighboring atoms. There is in fact a whole Rydberg-like series of such electronic states whose energy levels lie very close to the conduction-band edge. They are known as shallow levels. The wavefunctions consist of hydrogenic 'envelope' functions multiplied by the Bloch function of the band edge. The excited hydrogenic states have, as in the free hydrogen atom, larger and larger radii, approaching infinity. And we are still talking about 'point' defects!

Another class of point defects are characterized by a strong potential that is localized over a few atoms, unlike the long-range Coulombic potential of shallow impurities. A missing atom (vacancy), for example, or an extra atom at an interstitial site (interstitial) introduce such potentials. From an atomic-structure point of view, these defects are actually somewhat more extended than shallow impurities, because the neighboring atoms tend to relax away from the perfect-crystal positions. These defects typically also have electrons bound in localized states, but their wavefunctions tend to be far more localized than those of shallow impurities. In turn, the corresponding energy levels tend to lie deep in the band gap. These 'deep' levels can actually lie anywhere and can coincidentally lie close to a band edge. They are not, however, associated with that band edge the way a true shallow level is.

Deep centers in general also have shallow Rydberg states. These states exist because, when one of the deeply bound electrons or holes is removed, the center left behind is charged, giving rise to a long-range Coulombic tail that always binds hydrogenic excited states. Thus, even the deep point defects have something about them that is infinite!

3. Shallow Impurities

The role of shallow impurities as dopants was recognized back in the 1940s: they act like hydrogen atoms (electrons bound to fixed positive charges or holes bound to fixed negative charges) embedded in a dielectric so that the effective Rydberg is small. At room temperature, virtually all such impurities are ionized with the free electrons or holes enhancing the conductivity. These impurities were studied extensively in the 1950s and 1960s. A variety of optical experiments measured the Rydberg spectra of the common shallow impurities in the common semiconductors. Effective-mass theory was derived rigorously and shown to describe accurately all but the 'ground' state of the hydrogenic series. In this theory, the localized state is derived from the Bloch function of the nearest band extrema and the impurity potential is approximated by a screened Coulomb potential $U(r) = -e^2/\varepsilon r$. The discrepancies in the ground state, labeled central-cell corrections or chemical shifts, were recognized to arise primarily from two effects: deviations from the screened Coulomb potential, which does not describe accurately the actual impurity potential in the region of the central cell; contributions from interactions among wavefunctions derived from different equivalent band extrema, when such exist (e.g. donors in Si, Ge, GaP), known as multivalley or valley-orbit interactions.

Despite claims made by many authors, including my own claims in the early 1970s, inclusion of the central-cell effects and multivalley interactions has not yet been done accurately by any calculation. In the late 1970s, a number of papers focused on improving the way multivalley potential interactions are included, but did not agree on the form of multivalley kinetic-energy terms. One set of authors used multivalley equations that do not contain multivalley kinetic-energy terms, whereas other authors used a particular form for multivalley kinetic-energy terms without addressing the question of their derivation, validity or even explicit form [1]. The debate seems to have persisted (see, e.g. Reference [2]). A rigorous derivation of full multivalley equations was actually published in 1979 [3], but these equations have not yet been used by anyone for actual calculations. The issue goes beyond just the form of the equations. Advocates of no multivalley kinetic-energy terms find a critical shallow-deep instability that gets triggered by a slight change in the material parameters. The other camp argues that the instability is an artifact of neglecting multivalley kinetic-energy terms. Since the DX centers in AlGaAs alloys exhibit a shallow-deep instability, one would like to know its true origin.

The claims about a shallow-deep instability that is inherent in the effective-mass equations are also suspect because of the use of a screened Coulomb potential together with Bloch functions derived from a pseudopotential band structure calculation. The impurity potential must be consistent with the band structure and must, therefore, have a pseudopotential form (difference between the pseudopotentials of the impurity and host atoms) [3]. The Coulombic form in the central cell is an arbitrary form and any resulting instability caused by it may have nothing to do with real impurities. In conclusion, the whole issue of shallow-deep instability needs to be addressed with rigorously derived effective-mass equations [3] with an appropriate pseudopotential. Similarly, central-cell corrections for acceptors need to be calculated by using internally consistent impurity and crystal pseudopotentials and Bloch functions.

In more recent years, a great deal of attention has been paid to shallow impurities in superlattices and quantum wells. The additional confinement gives rise to a variety of interesting phenomena. There is one aspect of theory, however, that I believe is amiss. The concept of a position-dependent effective mass $m^*(r)$ is often introduced. In all such cases, the corresponding effective-mass equations are not derived. Typically, some other paper is referenced where such an equation appeared. To my knowledge, no one has given a rigorous derivation of an effective-mass equation containing a position-dependent effective mass [4]. I have looked into this question carefully many times, both in the context of bulk impurities and impurities in alloyed superlattices. I am convinced that, though the concept is plausible, it is actually *not possible* to derive effective-mass equations with a position-dependent effective mass. I am still looking for a paper that will prove this assertion wrong.

In experiments, the current frontier is primarily the effect of confinement on shallow impurities in quantum wells, delta doping (a sheet of shallow impurities), and other intriguing phenomena arising from artificial structures. There is a nice collection of papers on this subject in the Proceedings of the most recent International Conference on the Physics of Semiconductors [5].

4. Deep Impurities and Defects

Deep impurities [6] were recognized back in the 1950s and 1960s to act primarily as traps for electrons or holes or as recombination centers for electron–hole pairs. Their concentrations were typically smaller than those of shallow impurities and existing optical techniques did not work very well. The only efficient technique was electron paramagnetic resonance, which yielded information on the atomic structure of the underlying defect but not its energy levels. The technique was used with great success for lattice defects, primarily vacancies and vacancy-impurity complexes, that could be induced at will by irradiation with energetic electrons. The technique was also used to study the spin states of transition-metal impurities, but could not give information about energy levels, capture cross sections and the like.

In the early 1970s, a number of related techniques were developed based on *pn* junctions. In the region of the junction, the Fermi level passes right through the middle of the energy gap so that one can easily occupy or empty deep levels by simple biases and/or temperature scans. The resulting changes could be detected by monitoring either the effect on the current or the junction capacitance. One could also use light to change the charge state and then again monitor the junction capacitance. These techniques proved to be extremely powerful and yielded a great deal of information about capture cross sections, energy levels, even reactions involving many deep impurities and defects. Most notable was the discovery that some defect reactions are enhanced by electron–hole recombination. In more recent years, optical techniques have also been refined so that they are capable of getting detailed spectra with sharp structures, including the Rydberg series that are always present in principle when the center left behind is charged.

Theoretical understanding of deep impurities was relatively primitive throughout the 1960s and most of the 1970s. The structure of the vacancy and related defects was understood qualitatively in terms of the dangling bonds on the neighbors and simple symmetry arguments. Elementary understanding was also possible for the *d* states of transition-metal impurities. Actual calculations, however, were scarce and based on rather drastic approximations whose validity was not scrutinized.

Late in the 1970s, techniques were developed that made possible calculations of the electronic structure of deep centers at the same level of sophistication that was possible for bulk perfect crystals and crystal surfaces. For a number of years, the focus was on energy levels and wavefunctions. The basic structure of deep states was elucidated in great detail. Then followed calculations of total energies so that actual atomic arrangements could be determined through energy minimization. Formation and migration energies also became accessible. Most recently, with the advent of fully quantum-mechanical molecular dynamics calculations, it has been possible to actually monitor the detailed migration of an impurity atom or self-interstitial step by step.

The 1980s saw an explosion of interesting research on deep centers in semiconductors. On one hand, theory was making enormous strides, as several groups used state-of-the-art techniques to probe the electronic and atomic structure of all manner of point defects and impurities. Experimentalists, on the other hand, armed with an array of mature and tested techniques, were ready to tackle real problems. A few tough ones presented themselves and drove much of the excitement. Two of them in particular, the so-called EL2 defect in GaAs and the DX centers in AlGaAs, sprung on the scene in the late 1970s, at the same time that theoretical activity was picking up steam. They remained at center stage throughout the 1980s and still continue to attract significant attention.

DX centers form in AlGaAs alloys and exhibit a split personality, appearing either as shallow or deep depending on subtle variations in hydrostatic pressure, alloy composition, or even local alloy composition. It was originally suggested that the center is a complex of a shallow donor and an unknown deep defect X (hence the name DX), but more recently the X part has been discarded. It is now believed that isolated substitutional donors can actually undergo large atomic relaxations into a configuration with a deep energy level.

EL2, a native defect in GaAs, exhibits a deep level that is responsible for making semiinsulating material. It was discovered as one of several electron traps (its name simply stands for electron trap #2). It also exhibits a split personality of a different kind. Under low-temperature illumination, it was found to undergo a transformation into another state without any levels in the energy gap. Originally, the defect was suspected to be oxygen, but that was ruled out when it was found that a sample could have a lot more EL2 than oxygen impurities. It is now generally believed to be an arsenic antisite (arsenic at a normally gallium site) which can transform into another configuration by large lattice relaxation.

The literature on DX and EL2 is enormous [5,7]. Every conceivable technique has been used to probe them. First-principles calculations have shown that indeed the simple defects (isolated antisite in the case of EL2 and isolated substitutional donors in the case of DX centers) can undergo a transformation to a distorted configuration. A number of observed properties are nicely understood, but vexing questions still remain. For example, for DX centers, theory says that the distorted state of a substitutional donor is stable only if it contains two electrons, i.e. the center has zero net spin. Yet, experiments have found a static paramagnetic susceptibility associated with the deep state. In the case of EL2, the picture of an isolated antisite that simply distorts is not consistent with ENDOR data and miscellaneous other observations. Suggestions that another defect e.g. a vacancy or an interstitial are associated with the antisite defect, at least in the second configuration, persist.

In addition to EL2 and the DX centers, there have been a few more star attractions in recent years. One of them is hydrogen. This ubiquitous impurity was believed to be electrically inactive and harmless in crystalline silicon. It was known to 'tie off' dangling bonds in amorphous silicon by forming Si–H bonds. It was believed to behave in a similar fashion in defective regions of crystalline silicon where dangling bonds might exist and also at crystalline surfaces. As such, it was regarded as a passivator, a healing agent that could only help in the active region of silicon wafers. In ultrapure germanium, there was evidence that hydrogen could attach itself to other impurities and either passivate them or activate them (e.g., turn electrically inactive isovalent impurities into donors), but the implications of those results for hydrogen in silicon were not considered in depth.

In 1983, there were reports that hydrogen could enter the active region of CMOS devices and deactivate the boron-doped *p*-type silicon. That discovery unleashed the scientific defects community into a frenzy. Experimental and theoretical results came in at a fast pace and soon mapped out the details of hydrogen motion and reactions with a variety of impurities. A number of papers in References [5] and [7] and a recent volume dedicated to hydrogen [8] capture the excitement, which, by and large, has by now abated. A number of questions remain, as of course is always the case in any field.

Another class of defects that sought star status in recent years were the so-called thermal donors in silicon. They are a whole family of defects that were discovered to form during the metallization process after the active devices are fabricated on silicon wafers. There are strong indications that they are clusters of three or four oxygen atoms and exhibit shallow-donor behavior. Electron spin resonance experiments by different groups have yielded somewhat inconsistent information about the structure of these defects and theorists have yet to come up with any arrangement of oxygen atoms that has a shallow donor state. Interest in these centers has faded substantially because of the lack of progress.

All the defects that have been star attractions share one thing in common. They were first brought to light, so to speak, by technology. That was true of EL2, the DX centers, hydrogen, the thermal donors, and a number of others. It does not at all mean, however, that these were the most vexing defects for technology. It simply means that those were the defects that captured the scientists' imagination, or, better yet, those were the defects for which a number of complementary experimental and theoretical techniques could be used. Technology, of course, finds ways to deal with or get around problem defects.

One truly major defect-related problem for technology is diffusion. Dopant impurities need to be put in the right place (often by diffusion) and then need to stay put or move predictably during other processing steps. The number of processing steps is constantly increasing, more than a hundred by the latest count, so that trial and error is not adequate. Device and circuit designers rely extensively on computer simulations. The writers of these computer programs need to know what the microscopic mechanisms of diffusion actually are. In particular they need to know what intrinsic defects (vacancies, interstitials, etc.) mediate dopant diffusion. They need to know the diffusivities of the intrinsic defects, the reactions that occur between the intrinsic defects and dopant atoms, the diffusivities of any resulting complexes (e.g. vacancy–dopant pairs), and so on. Most experiments to probe these issues have in fact been carried out by engineers who need this information. In the end, of course, the needs of technology are met by a set of assumptions that are tested against actual data and are adjusted and massaged until they do the required task. As devices get smaller and more complex, there is, naturally, need for improved assumptions, better understanding of the underlying mechanisms.

Diffusion of dopants in silicon is a field that has been riven by controversy, largely because too many authors drew 'definitive' conclusions from incomplete data. As Bertrand Russell once said, "the most savage controversies are those about matters as to which there is no good evidence either way." Indeed, that has been the case. The most savage controversy was whether vacancies or interstitials mediate dopant diffusion. In the 1970s, there were two camps with uncompromising stands. I recall at one Gordon conference there was a sign-up sheet for people to register their preference! A scrutiny of the literature of that era sadly reveals that the same experimental data were sometimes quoted as evidence for contradictory assumptions by advocates of the two camps. As the 1980s rolled in, there were already strong experimental indications that both defects play a role. First-principles calculations in the mid-1980s and beyond have slowly established this fact. Theory also suggested that a defectless mechanism, direct exchange between neighboring atoms, may also play a role.

Though theory has been chipping away at the diffusion problem, experimentalists in the scientific defects community all but ignored the problem. Virtually all data come from the engineering world. The reason, I believe, is that the common defects techniques have not been very useful. Defects experimentalists are used to studying stable defects that just sit there, or undergo transformation under controlled conditions. It is true that vacancies and vacancy–pairs have been artificially created by electron bombardment and studied by the electron spin resonance technique at very low temperatures where these defects stay put, but not much of that relates to diffusion. In order to understand diffusion one needs to measure concentrations and diffusivities of various defects. The standard optical and electrical measurements that have proved powerful probes of the atomic and electronic structure and properties of defects are not very helpful for concentrations and diffusivities.

Similar difficulties exist with sorting out the mechanisms of gettering unwanted impurities from the active region of a silicon wafer. This is also a technologically important problem, but progress in understanding the underlying mechanisms has been slow. It is essentially a problem of diffusivities and reactions which can be probed only in limited ways by existing techniques.

5. The New Frontiers

We already discussed the open issues and newest frontiers in shallow impurities. In the arena of deep centers, there will always be defects whose atomic structure and peculiar properties are not understood. There certainly are open questions about the old timers among defects, namely EL2, DX, etc. Other, perhaps less popular defects, abound. One might ask, however, if there are any generic major issues that drive the agenda of the defects community. A number of such generic issues did drive the community in the past: understanding the structure and properties of *sp*-bonded impurities; understanding the structure and properties of transition-metal impurities; understanding the structure and properties of vacancies and self-interstitials; sorting out the properties of metastable defects such as EL2, DX, pairs between transition-metal impurities and dopant atoms, etc. Theory is now mature and has provided a detailed description of the basic structure and properties of representative defects. What then lies ahead? Perhaps metastability still hides many faces and will help keep the torch alive. Perhaps technology will unveil new strange defects. In my opinion, the field of point defects is now a mature one, with a great deal under the belt. There are mopping-up operations for more detail, for resolving left-over puzzles in otherwise well understood behavior. There are old unresolved problems, such as the mechanisms of self-compensation in II–VI compounds which typically can be doped either n-type or p-type, but not both. There are perhaps new vistas yet to be uncovered involving point defects in quantum wells or on surfaces. Only time will tell.

Acknowledgements

This work was supported in part by the U.S. Office of Naval Research under Contract No. N00014-84-C-0396.

References

1. See, for example, M. Altarelli: 1981, *Phys. Rev. Lett.* **46**, 205 and references therein. See also Y.C. Chiang, T.C. McGill and D.L. Smith: 1981, *Phys. Rev. B* **23**, 4169.
2. See, e.g., L. Resca: 1987, *Phys. Rev. B* **36**, 4452; N. Chand, T. Henderson, J. Klem, W.T. Masselink, R. Fischer, Y.C. Chiang and H. Morkoc: 1987, *Phys. Rev. B* **36**, 445.
3. S.T. Pantelides: 1979, *Solid State Commun.* **30**, 65.
4. The nearest thing to a derivation I have seen is in a paper by T. Gora and F. Williams in *Handbook on Semiconductors*, W. Paul (ed.), North Holland, Amsterdam, 1982, Vol. 1, p. 639. A number of crucial steps are omitted. The final equation does not follow from those given.
5. *The Physics of Semiconductors*, E.M. Anastassakis and J.D. Joannopoulos (eds.), World Scientific, Singapore, 1990, three volumes.

6. For a detailed review of the evolution of the field of deep centers, including a lengthy list of references to the original papers, see S.T. Pantelides, in *Deep Centers in Semiconductors*, S.T. Pantelides (ed.), Gordon and Breach, New York, 1986, p. 1. See also References [5] and [7].
7. *Impurities, Defects and Diffusion in Semiconductors: Bulk and Layered Structures*, D.J. Wolford, J. Bernholc and E.E. Haller (eds.), Materials Research Society, Pittsburgh, 1990.
8. *Hydrogen in Semiconductors*, N.M. Johnson and J.I. Pankove (eds.), Academic Press, New York, 1991.

SUBJECT INDEX

Physics and Chemistry of Materials with Low-Dimensional Structures

Previously published under the Series Title:
PHYSICS AND CHEMISTRY OF MATERIALS WITH LAYERED STRUCTURES

1. R.M.A. Lieth (ed.): *Preparation and Crystal Growth of Materials with Layered Structures*. 1977 ISBN 90-277-0638-7
2. F. Lévy (ed.): *Crystallography and Crystal Chemistry of Materials with Layered Structures*. 1976 ISBN 90-277-0586-0
3. T. J. Wieting and M. Schlüter (eds.): *Electrons and Phonons in Layered Crystal Structures*. 1979 ISBN 90-277-0897-5
4. P.A. Lee (ed.): *Optical and Electrical Properties*. 1976 ISBN 90-277-0676-X
5. F. Hulliger: *Structural Chemistry of Layer-Type Phases*. Ed. by F. Lévy. 1976 ISBN 90-277-0714-6
6. F. Lévy (ed.): *Intercalated Layered Materials*. 1979 ISBN 90-277-0967-X

Published under:
PHYSICS AND CHEMISTRY OF MATERIALS WITH LOW-DIMENSIONAL STRUCTURES
SERIES A: LAYERED STRUCTURES

7. V. Grasso (ed.): *Electronic Structure and Electronic Transitions in Layered Materials*. 1986 ISBN 90-277-2102-5
8. K. Motizuki (ed.): *Structural Phase Transitions in Layered Transition Metal Compounds*. 1986 ISBN 90-277-2171-8

PHYSICS AND CHEMISTRY OF MATERIALS WITH LOW-DIMENSIONAL STRUCTURES
SERIES B: QUASI-ONE-DIMENSIONAL STRUCTURES

B1. P. Monceau (ed.): *Electronic Properties of Inorganic Quasi-One-Dimensional Compounds*. Part I: Theoretical. 1985 ISBN 90-277-1789-3
B2. P. Monceau (ed.): *Electronic Properties of Inorganic Quasi-One-Dimensional Compounds*. Part II: Experimental. 1985 ISBN 90-277-1800-8
B3. H. Kamimura (ed.): *Theoretical Aspects of Band Structures and Electronic Properties of Pseudo-One-Dimensional Solids*. 1985 ISBN 90-277-1927-6
B4. J. Rouxel (ed.): *Crystal Chemistry and Properties of Materials with Quasi-One-Dimensional Structures*. A Chemical and Physical Synthetic Approach. 1986 ISBN 90-277-2057-6

Discontinued.

PHYSICS AND CHEMISTRY OF MATERIALS WITH LOW-DIMENSIONAL STRUCTURES
SERIES C: MOLECULAR STRUCTURES

C1. I. Zschokke (ed.): *Optical Spectroscopy of Glasses*. 1986 ISBN 90-277-2231-5
C2. J. Fünfschilling (ed.): *Relaxation Processes in Molecular Excited States*. 1989 ISBN 07923-0001-7

Discontinued.

Physics and Chemistry of Materials with Low-Dimensional Structures

9. L.J. de Jongh (ed.): *Magnetic Properties of Layered Transition Metal Compounds*. 1990
ISBN 0-7923-0238-9
10. E. Doni, R. Girlanda, G. Pastori Parravicini and A. Quattropani (eds.): *Progress in Electron Properties of Solids*. Festschrift in Honour of Franco Bassani. 1989
ISBN 0-7923-0337-7
11. C. Schlenker (ed.): *Low-Dimensional Electronic Properties of Molybdenum Bronzes and Oxides*. 1989
ISBN 0-7923-0085-8
12. R. H. Friend (ed.): *Conducting Polymers*. 1991 (forthcoming)
13. H. Aoki, M. Tsukada, M. Schlüter and F. Lévy (eds.): *New Horizons in Low-Dimensional Electron Systems*. A Festschrift in Honour of Professor H. Kamimura. 1992
ISBN 0-7923-1302-X

KLUWER ACADEMIC PUBLISHERS – DORDRECHT / BOSTON / LONDON